MODELLING, SIMULATION AND CONTROL OF TWO-WHEELED VEHICLES

Automotive Series

Series Editor: Thomas Kurfess

Modelling, Simulation and Control of Two-Wheeled Vehicles	Tanelli, Corno and Savaresi	February 2014
Modeling and Control of Engines and Drivelines	Eriksson and Nielsen	February 2014
Advanced Composite Materials for Automotive Applications: Structural Integrity and Crashworthiness	Elmarakbi	December 2013
Guide to Load Analysis for Durability in Vehicle Engineering	Johannesson and Speckert	November 2013

MODELLING, SIMULATION AND CONTROL OF TWO-WHEELED VEHICLES

Mara Tanelli, Matteo Corno and Sergio M. Savaresi
Informazione e Bioingegneria, Politecnico di Milano, Italy

This edition first published 2014

©2014 John Wiley & Sons Ltd

Registered office

John Wiley & Sons Ltd, The Atrium, Southern Gate, Chichester, West Sussex, PO19 8SQ, United Kingdom

For details of our global editorial offices, for customer services and for information about how to apply for permission to reuse the copyright material in this book please see our website at www.wiley.com.

The right of the author to be identified as the author of this work has been asserted in accordance with the Copyright, Designs and Patents Act 1988.

All rights reserved. No part of this publication may be reproduced, stored in a retrieval system, or transmitted, in any form or by any means, electronic, mechanical, photocopying, recording or otherwise, except as permitted by the UK Copyright, Designs and Patents Act 1988, without the prior permission of the publisher.

Wiley also publishes its books in a variety of electronic formats. Some content that appears in print may not be available in electronic books.

Designations used by companies to distinguish their products are often claimed as trademarks. All brand names and product names used in this book are trade names, service marks, trademarks or registered trademarks of their respective owners. The publisher is not associated with any product or vendor mentioned in this book.

Limit of Liability/Disclaimer of Warranty: While the publisher and author have used their best efforts in preparing this book, they make no representations or warranties with respect to the accuracy or completeness of the contents of this book and specifically disclaim any implied warranties of merchantability or fitness for a particular purpose. It is sold on the understanding that the publisher is not engaged in rendering professional services and neither the publisher nor the author shall be liable for damages arising herefrom. If professional advice or other expert assistance is required, the services of a competent professional should be sought.

Library of Congress Cataloging-in-Publication Data

Tanelli, Mara.
  Modelling, simulation and control of two-wheeled vehicles / Mara Tanelli, Sergio Savaresi and Matteo Corno.
     pages cm
  Includes bibliographical references and index.
  ISBN 978-1-119-95018-9 (cloth)
  1. Motorcycles – Dynamics. I. Savaresi, Sergio M. II. Corno, Mauro, 1970- III. Title.
  TL243.T36 2014
  629.2'31 – dc23
                                                                              2013036260

A catalogue record for this book is available from the British Library.

ISBN 9781119950189

Typeset in 10/12pt TimesLTStd by Laserwords Private Limited, Chennai, India
Printed and bound in Malaysia by Vivar Printing Sdn Bhd

Contents

About the Editors   xi

List of Contributors   xiii

Series Preface   xv

Introduction   xvii

Part One   TWO-WHEELED VEHICLES MODELLING AND SIMULATION

1   Motorcycle Dynamics   3
Vittore Cossalter, Roberto Lot, and Matteo Massaro

1.1   Kinematics   3
       1.1.1   Basics of Motorcycle Kinematics   3
       1.1.2   Handlebar Steering Angle and Kinematic Steering Angle   5
1.2   Tyres   6
       1.2.1   Contact Forces and Torques   7
       1.2.2   Steady-State Behaviour   9
       1.2.3   Dynamic Behaviour   11
1.3   Suspensions   13
       1.3.1   Suspension Forces   13
       1.3.2   Suspensions Layout   14
       1.3.3   Equivalent Stiffness and Damping   16
1.4   In-Plane Dynamics   18
       1.4.1   Pitch, Bounce and Hops Modes   18
       1.4.2   Powertrain   22
       1.4.3   Engine-to-Slip Dynamics   24
       1.4.4   Chatter   27
1.5   Out-of-Plane Dynamics   29
       1.5.1   Roll Equilibrium   29
       1.5.2   Motorcycle Countersteering   29
       1.5.3   Weave, Wobble and Capsize   33
1.6   In-Plane and Out-of-Plane Coupled Dynamics   40
       References   41

2	**Dynamic Modelling of Riderless Motorcycles for Agile Manoeuvres**	**43**
	Yizhai Zhang, Jingang Yi, and Dezhen Song	
2.1	Introduction	43
2.2	Related Work	44
2.3	Motorcycle Dynamics	45
	2.3.1 Geometry and Kinematics Relationships	45
	2.3.2 Motorcycle Dynamics	48
2.4	Tyre Dynamics Models	51
	2.4.1 Tyre Kinematics Relationships	51
	2.4.2 Modelling of Frictional Forces	52
	2.4.3 Combined Tyre and Motorcycle Dynamics Models	54
2.5	Conclusions	55
	Nomenclature	55
	Appendix A: Calculation of M_s	56
	Appendix B: Calculation of Acceleration $\dot{v}_G$	57
	Acknowledgements	57
	References	57
3	**Identification and Analysis of Motorcycle Engine-to-Slip Dynamics**	**59**
	Matteo Corno and Sergio M. Savaresi	
3.1	Introduction	59
3.2	Experimental Setup	60
3.3	Identification of Engine-to-Slip Dynamics	61
	3.3.1 Relative Slip	72
	3.3.2 Throttle Dynamics	73
3.4	Engine-to-Slip Dynamics Analysis	73
	3.4.1 Throttle and Spark Advance Control	74
	3.4.2 Motorcycle Benchmarking	75
3.5	Road Surface Sensitivity	78
3.6	Velocity Sensitivity	79
3.7	Conclusions	80
	References	80
4	**Virtual Rider Design: Optimal Manoeuvre Definition and Tracking**	**83**
	Alessandro Saccon, John Hauser, and Alessandro Beghi	
4.1	Introduction	83
4.2	Principles of Minimum Time Trajectory Computation	86
	4.2.1 Tyre Modelling	87
	4.2.2 Engine and Drivetrain Modelling	88
	4.2.3 Brake Modelling	89
	4.2.4 Wheelie and Stoppie	90
4.3	Computing the Optimal Velocity Profile for a Point-Mass Motorcycle	90
	4.3.1 Computing the Optimal Velocity Profile for a Realistic Motorcycle	96
	4.3.2 Application to a Realistic Motorcycle Model	100

4.4	The Virtual Rider	102
	4.4.1 The Sliding Plane Motorcycle Model	102
4.5	Dynamic Inversion: from Flatland to State-Input Trajectories	103
	4.5.1 Quasi-Static Motorcycle Trajectory	104
	4.5.2 Approximate Inversion by Trajectory Optimization	106
4.6	Closed-Loop Control: Executing the Planned Trajectory	107
	4.6.1 Manoeuvre Regulation	107
	4.6.2 Shaping the Closed-Loop Response	112
	4.6.3 Interfacing the Maneuver Regulation Controller with the Multibody Motorycle Model	113
4.7	Conclusions	115
4.8	Acknowledgements	116
	References	116

5 The Optimal Manoeuvre — 119
Francesco Biral, Enrico Bertolazzi, and Mauro Da Lio

5.1	The Optimal Manoeuvre Concept: Manoeuvrability and Handling	121
	5.1.1 Optimal Manoeuvre Mathematically Formalised	123
	5.1.2 The Optimal Manoeuvre Explained with Linearized Motorcycle Models	124
5.2	Optimal Manoeuvre as a Solution of an Optimal Control Problem	133
	5.2.1 The Pontryagin Minimum Principle	136
	5.2.2 General Formulation of Unconstrained Optimal Control	137
	5.2.3 Exact Solution of a Linearized Motorcycle Model	139
	5.2.4 Numerical Solution and Approximate Pontryagin	142
5.3	Applications of Optimal Manoeuvre to Motorcycle Dynamics	145
	5.3.1 Modelling Riders' Skills and Preferences with the Optimal Manoeuvre	146
	5.3.2 Minimum Lap Time Manoeuvres	148
5.4	Conclusions	152
	References	152

6 Active Biomechanical Rider Model for Motorcycle Simulation — 155
Valentin Keppler

6.1	Human Biomechanics and Motor Control	156
	6.1.1 Biomechanics	157
	6.1.2 Motor Control	159
6.2	The Model	161
	6.2.1 The Human Body Model	161
	6.2.2 The Motorcycle Model	165
	6.2.3 Steering the Motorcycle	166
6.3	Simulations and Results	167
	6.3.1 Rider's Vibration Response	168
	6.3.2 Lane Change Manoeuvre	170
	6.3.3 Path Following Performance	170

	6.3.4 Influence of Physical Fitness	176
	6.3.5 Analysing Weave Mode	176
	6.3.6 Provoking Wobble Mode	178
	6.3.7 Road Excitation and Ride Comfort	178
6.4	Conclusions	179
	References	180

7 A Virtual-Reality Framework for the Hardware-in-the-Loop Motorcycle Simulation 183
Roberto Lot and Vittore Cossalter

7.1	Introduction	183
7.2	Architecture of the Motorcycle Simulator	184
	7.2.1 Motorcycle Mock-up and Sensors	184
	7.2.2 Realtime Multibody Model	185
	7.2.3 Simulator Cues	186
	7.2.4 Virtual Scenario	188
7.3	Tuning and Validation	188
	7.3.1 Objective Validation	190
	7.3.2 Subjective Validation	190
7.4	Application Examples	191
	7.4.1 Hardware- and Human-in-the-Loop Testing of Advanced Rider Assistance Systems	192
	7.4.2 Training and Road Education	194
	References	194

Part Two TWO-WHEELED VEHICLES CONTROL AND ESTIMATION PROBLEMS

8 Traction Control Systems Design: A Systematic Approach 199
Matteo Corno and Giulio Panzani

8.1	Introduction	199
8.2	Wheel Slip Dynamics	202
8.3	Traction Control System Design	206
	8.3.1 Supervisor	206
	8.3.2 Slip Reference Generation	208
	8.3.3 Control Law Design	209
	8.3.4 Transition Recognition	212
8.4	Fine tuning and Experimental Validation	212
8.5	Conclusions	218
	References	219

9 Motorcycle Dynamic Modes and Passive Steering Compensation 221
Simos A. Evangelou and Maria Tomas-Rodriguez

9.1	Introduction	221
9.2	Motorcycle Main Oscillatory Modes and Dynamic Behaviour	222

9.3	Motorcycle Standard Model	224
9.4	Characteristics of the Standard Machine Oscillatory Modes and the Influence of Steering Damping	226
9.5	Compensator Frequency Response Design	228
9.6	Suppression of Burst Oscillations	233
	9.6.1 Simulated Bursting	233
	9.6.2 Acceleration Analysis	235
	9.6.3 Compensator Design and Performance	237
9.7	Conclusions	240
	References	240

10 Semi-Active Steering Damper Control for Two-Wheeled Vehicles — 243
Pierpaolo De Filippi, Mara Tanelli, and Matteo Corno

10.1	Introduction and Motivation	243
10.2	Steering Dynamics Analysis	245
	10.2.1 Model Parameters Estimation	248
	10.2.2 Comparison between Vertical and Steering Dynamics	251
10.3	Control Strategies for Semi-Active Steering Dampers	252
	10.3.1 Rotational Sky-Hook and Ground-Hook	253
	10.3.2 Closed-Loop Performance Analysis	255
10.4	Validation on Challenging Manoeuvres	257
	10.4.1 Performance Evaluation Method	257
	10.4.2 Validation of the Control Algorithms	258
10.5	Experimental Results	266
10.6	Conclusions	267
	References	268

11 Semi-Active Suspension Control in Two-Wheeled Vehicles: a Case Study — 271
Diego Delvecchio and Cristiano Spelta

11.1	Introduction and Problem Statement	271
11.2	The Semi-Active Actuator	272
11.3	The Quarter-Car Model: a Description of a Semi-Active Suspension System	275
11.4	Evaluation Methods for Semi-Active Suspension Systems	277
11.5	Semi-Active Control Strategies	279
	11.5.1 Sky-hook Control	279
	11.5.2 Mix-1-Sensor Control	280
	11.5.3 The Ground-Hook Control	280
11.6	Experimental Set-up	281
11.7	Experimental Evaluation	281
11.8	Conclusions	289
	References	289

12 Autonomous Control of Riderless Motorcycles — 293
Yizhai Zhang, Jingang Yi, and Dezhen Song

12.1	Introduction	293

12.2	Trajectory Tracking Control Systems Design	294
	12.2.1 External/Internal Convertible Dynamical Systems	294
	12.2.2 Trajectory Tracking Control	297
	12.2.3 Simulation Results	301
12.3	Path-Following Control System Design	305
	12.3.1 Modelling of Tyre–Road Friction Forces	306
	12.3.2 Path-Following Manoeuvring Design	306
	12.3.3 Simulation Results	308
12.4	Conclusion	315
	Acknowledgements	317
	Appendix A: Calculation of the Lie Derivatives	317
	References	318

13 Estimation Problems in Two-Wheeled Vehicles — 319
Ivo Boniolo, Giulio Panzani, Diego Delvecchio, Matteo Corno, Mara Tanelli, Cristiano Spelta, and Sergio M. Savaresi

13.1	Introduction	319
13.2	Roll Angle Estimation	320
	13.2.1 Vehicle Attitude and Reference Frames	322
	13.2.2 Experimental Set-up	324
	13.2.3 Accelerometer-Based Roll Angle Estimation	325
	13.2.4 Use of the frequency separation principle	328
13.3	Vehicle Speed Estimation	329
	13.3.1 Speed Estimation During Traction Manoeuvres	331
	13.3.2 Experimental Setup	331
	13.3.3 Vehicle Speed Estimation via Kalman Filtering and Frequency Split	331
	13.3.4 Experimental Validation	335
13.4	Suspension Stroke Estimation	337
	13.4.1 Problem Statement and Estimation Law	337
	13.4.2 Experimental Results	339
13.5	Conclusions	342
	References	342

Index — 345

About the Editors

Mara Tanelli

Mara Tanelli was born in Lodi, Italy, in 1978. She is an Assistant Professor of Automatic Control at the Dipartimento di Elettronica, Informazione e Bioingegneria of the Politecnico di Milano, Italy, where she obtained the Laurea degree in Computer Engineering in 2003 and the PhD in Information Engineering in 2007. She also holds an MSc in Computer Science from the University of Illinois at Chicago.

Her main research interests focus on control systems design for vehicles, energy management of electric vehicles, control for energy-aware IT systems and sliding mode control.

She is co-author of more than 100 peer-reviewed scientific publications and seven patents in the above research areas. She is also co-author of the monograph *Active braking control systems design for vehicles*, published in 2010 by Springer. She is a Senior Member of the IEEE and a member of the Conference Editorial Board of the IEEE Control Systems Society.

In the past few years, she has gained considerable experience in industrial projects carried out in collaboration with leading manufacturers of four- and two-wheeled vehicles, that involved – besides her scientific research activities – prototyping, implementation and experimental testing.

Matteo Corno

Matteo Corno was born in Italy in 1980. He received his MSc degree in Computer and Electrical Engineering (University of Illinois) and his PhD cum laude degree with a thesis on active stability control of two-wheeled vehicles (Politecnico di Milano) in 2005 and 2009, respectively.

He is currently an Assistant Professor at the Dipartimento di Elettronica, Informazione e Bioingegneria, Politecnico di Milano, Italy.

In 2011, his paper "On Optimal Motorcycle Braking" was awarded the best-paper prize for *Control Engineering Practice*, published in the period 2008–2010.

In 2012 and 2013, he co-founded two highly innovative start-ups: E-Novia and Zehus.

His current research interests include dynamics and control of vehicles; lithium-ion battery modelling; estimation and control; and modelling and control of human-powered electric vehicles. He has held research positions at Thales Alenia Space, University of Illinois, Harley Davidson, University of Minnesota, Johannes Kepler University in Linz, and TU Delft. He is author or co-author of more than 50 peer-reviewed scientific publications and of six patents. Most of his publications are the result of strong industrial collaboration with leading companies in the automotive and motorcycle industries.

Sergio M. Savaresi

Born in Manerbio, Italy, in 1968, Sergio Savaresi holds an MSc in Electrical Engineering and a PhD in Systems and Control Engineering, both from the Politecnico di Milano, and an MSc in Applied Mathematics from Università Cattolica, Brescia. After receiving his PhD, he became a consultant for McKinsey & Co., Milan Office. He has been a Full Professor in Automatic Control since 2006, and has been visiting scholar at Lund University, Sweden; University of Twente, the Netherlands; Canberra National University, Australia; Minnesota University at Minneapolis, USA and Johannes Kepler University, Linz, Austria.

He is an Associate Editor of several international journals and he has been on the international program committees of many international conferences.

His main research interests are in the areas of vehicles control, automotive systems, data analysis and modelling, nonlinear control and industrial control applications. He is co-author of the monographs *Active Braking Control Systems Design for Vehicles*, and *Semi-Active Suspension Control For Vehicles*. He is also author or co-author of more than 300 peer-reviewed scientific publications and of 28 patents.

He is the Chair of the Systems and Control Section of Politecnico di Milano, and Head of the MOVE research team (http://move.dei.polimi.it). He is a Lecturer on a Masters Course in "Automation and Control in Vehicles", and has been Principal Investigator in more than 100 research cooperation projects between Politecnico di Milano and private companies, mostly in the fields of automotive and motorcycle dynamics and control.

List of Contributors

Alessandro Beghi, Department of Information Engineering, University of Padova, Italy

Enrico Bertolazzi, Department of Industrial Engineering, University of Trento, Italy

Francesco Biral, Department of Industrial Engineering, University of Trento, Italy

Ivo Boniolo, Dipartimento di Ingegneria, Università degli Studi di Bergamo, Italy

Vittore Cossalter, University of Padova, Italy

Mauro Da Lio, Department of Industrial Engineering, University of Trento, Italy

Pierpaolo De Filippi, Dipartimento di Elettronica, Informazione e Bioingegneria, Politecnico di Milano, Italy

Diego Delvecchio, Dipartimento di Elettronica, Informazione e Bioingegneria, Politecnico di Milano, Italy

Simos A. Evangelou, Department of Electrical and Electronic Engineering, and Imperial College, London, UK

John Hauser, Department of Electrical, Computer and Energy Engineering, University of Colorado Boulder, USA

Valentin Keppler, Biomotion Solutions GbR and Department of Sports Science, University of Tübingen, Germany

Roberto Lot, University of Padova, Italy

Matteo Massaro, University of Padova, Italy

Giulio Panzani, Dipartimento di Elettronica, Informazione e Bioingegneria, Politecnico di Milano, Italy

Maria Tomas-Rodriguez, School of Engineering and Mathematical Sciences, City University, London, UK

Alessandro Saccon, Department of Mechanical Engineering, Eindhoven University of Technology, The Netherlands

Dezhen Song, Texas A&M University, USA

Cristiano Spelta, Dipartimento di Ingegneria, Università degli Studi di Bergamo, Italy

Jingang Yi, Rutgers University, USA

Yizhai Zhang, Rutgers University, USA

Series Preface

The motorcycle is the most prevalent form of mechanized transportation on the planet. In its human-powered form, the bicycle, it is one of the first pragmatic and useful vehicles that most people encounter. The dynamics of two-wheeled vehicles have been studied for many years, and provide the foundation for most vehicle dynamic analyses. Not only are these dynamics fundamental to the transportation sector, but they are quite elegant in nature, linking various aspects of kinematics, dynamics and physics. In fact, the dynamics of the motorcycle and bicycle are inherently linked to their functionality; one cannot easily balance these vehicles unless they are in motion!

Modelling, Simulation and Control of Two-Wheeled Vehicles is a comprehensive text of the dynamics, modelling and control of motorcycles. It provides a broad and in-depth perspective of all the necessary information required to fully understand, design and utilize the motorcycle. Topics covered in this text range from basic two-wheeled dynamics that are used as the foundation for most vehicle dynamic analyses to advanced control and estimation theory applied to fully developed complex systems models. This text is part of the *Automotive Series* whose primary goal is to publish practical and topical books for researchers and practitioners in industry, and for postgraduate or advanced undergraduates in automotive engineering. The series addresses new and emerging technologies in automotive engineering, supporting the development of the next generation transportation systems. The series covers a wide range of topics, including design, modelling and manufacturing, and it provides a source of relevant information that will be of interest and benefit to people working in the field of automotive engineering.

Modelling, Simulation and Control of Two-Wheeled Vehicles presents a number of different design and analysis considerations related to motorcycle transportation systems including integration dynamics, agile manoeuvring systems integration, rider biomechanical models, passive and active steering control and autonomous control of riderless motorcycles. The theory and supporting applications are second to none, as are the authors of this wonderful book. The text provides a strong foundational basis for motorcycle design and development, and is a welcome addition to the Automotive Series.

Thomas Kurfess
August 2013

Introduction

Mara Tanelli, Matteo Corno, and Sergio M. Savaresi
Dipartimento di Elettronica, Informazione e Bioingegneria, Politecnico di Milano, Italy

Plenty of books have been written on modeling, simulation and control on four-wheeled vehicles (cars, in short). As such, one would be tempted to ask: is a "two-wheeled-specific" book really needed or missing? The editors and authors of this book strongly believe that the answer is *yes*.

A thorough technical motivation for this answer will be implicitly given throughout the book. A simpler, somehow naive, but effective answer, however, is: we *drive* a car, but we *ride* a two-wheeled vehicle: this crucial difference highlights that they just cannot be "similar" vehicles. In the field of vehicle modeling and automatic control, the majority of scientists and practitioners have been working on automotive (car)-related modeling and control problems. As such, one may think that moving from four to two-wheeled vehicles just requires a small re-casting (re-modeling, re-design of the controllers, re-tuning, etc.) effort. This sort of prejudice typically vanishes when dealing with real problems, on real two-wheeled vehicles. Most of the authors of this book have been through this enlightening process. Discovering that two-wheeled vehicles are not just a "subset of cars" is both challenging and fascinating.

This book helps the reader discover all the peculiar features of modeling and control of this very special class of vehicles.

The potential interest for a book specifically dealing with two-wheeled vehicles is amplified by the current and future mobility trends: traffic congestion in urban and metropolitan areas and the need to reduce energy consumption and pollutant emission are pushing towards a strong downsizing of vehicles used for urban mobility. The number of E-Bikes, scooters, motorcycles, narrow-track vehicles (tilting or non-tilting) is expected to grow exponentially in the next decades, especially around large metropolitan areas. Along this trend, two-wheeled vehicles can play a key role: they have the appealing features of being light and having a very small energy footprint. Thus, there are good chances that the two-wheeled vehicles market will soon compete (in volume, and, possibly, in technology) with the today larger and more advanced automotive market.

Such an expansion, then, will see an increasing interest in finding innovative and original solutions for solving many challenging problems that deal with the dynamic analysis and control of such vehicles, and this book can be one of the first comprehensive answer to such needs.

In this respect, this book lies in the class of edited-books, namely books that are a collection of chapters written by different authors. The price to pay is a limited homogeneity of notation and presentation style, but their main advantage is that a single book embeds the perspective of an (almost) entire scientific community, rather than that of a single research group.

This book has been carefully conceived in order to provide at the same time a broad perspective and a rigorous structure. Its contents have been clearly divided into two parts: in the first part, the modeling and simulation issues are considered, while in the second one the problem of controlling (mostly by feedback electronic control systems) the vehicle is analyzed. In many cases, there are pairs of chapters written by the same authors: one in the first, one in the second part, stressing the fact that modeling and control are just two sides of the same coin. The *valuable* coin is the dynamic behavior of a two-wheeled vehicle: weird, exhilarating, challenging. We want to understand it. We want to control it.

Organization of the book

The two parts of the book are organized as follows.
Part one:

- Chapter 1 (by Vittore Cossalter, Roberto Lot and Matteo Massaro – University of Padova) is a comprehensive and introductory chapter that describes all the aspects of the kinematic and dynamic behavior of a two-wheeled vehicle.
- Chapter 2 (by Yizhai Zhang and Jingang Yi – Rutgers University – and by Dezhen Song – Texas A&M University) further develops the modeling topic, with a special focus on a reduced-order model suited for modeling fast-dynamics maneuvers.
- Chapter 3 (by Matteo Corno and Sergio M. Savaresi – Politecnico di Milano) explores the field of black-box control-oriented modeling, by presenting a case study of direct identification from experiments of the engine-to-slip dynamics, ancillary to traction-control design. The design-of-experiment in this context represents a major issue and is described in detail.
- Chapter 4 (by Alessandro Saccon – TU Eindhoven – John Hauser – University of Colorado Boulder – and Alessandro Beghi – University of Padova) and Chapter 5 (by Francesco Biral, Enrico Bertolazzi and Mauro Da Lio – University of Trento) present, with different approaches, the problem of simulating the motorcycle dynamics in a time-optimal maneuver. This problem is a combination of dynamics modeling, optimization and optimal control issues. This topic is highly relevant not only for the purpose of automatic (electronic) feedback control, but mostly for better understanding the sensitivity of the performance of a motorcycle, with respect to different parameter configurations.
- Chapter 6 (by Valentin Keppler – University of Tubingen) deeply explores the issue of rider modeling and simulation. This issue is a key element of two-wheeled vehicles simulation, since the rider is so deeply linked with the vehicle dynamics that the two elements can hardly be simulated separately. In this chapter, the rider simulation is dealt with a sophisticated bio-mechanics approach.
- Chapter 7 (by Vittore Cossalter and Roberto Lot – University of Padova) ends part one by presenting a research work that can be considered in-between simulation and control: the development of a virtual-reality system for the hardware-in-the-loop simulation of vehicle

dynamics. The system described in this chapter might have multi-faceted applications: it can be used as a design and testing tool for advanced electronic control, as a rider-training system and ... even as a sophisticated and fun-to-use *toy*).

Part two:

- Chapter 8 (by Matteo Corno and Giulio Panzani – Politecnico di Milano) presents a complete (from model-based design to experimental validation) design procedure for a traction control system of a high-performance motorcycle. Traction control is the most used electronic control system in high-end motorcycles today, and has a key role both on the safety and the performance of a sport motorcycle. This Chapter is somehow the continuation of Chapter 3.
- Chapter 9 (by Simos A. Evangelou – Imperial College – and Maria Tomas-Rodriguez – City University of London) focuses on the key issue of steer dynamics, with an approach that aims to improve the dynamic behavior by passive mechanical elements.
- Chapter 10 (by Pierpaolo De Filippi, Mara Tanelli and Matteo Corno – Politecnico di Milano) addresses again the problem of steer dynamics, proposing a solution that employs closed-loop electronic control systems and relies on a semi-active damping technology.
- Chapter 11 (by Diego del Vecchio – Politecnico di Milano, and Cristiano Spelta – University of Bergamo) focuses on vertical and pitch dynamics, by presenting a complete case-study of semi-active suspension control design. Semi-active suspensions have been first presented in motorcycle applications at the end of 2012, and they constitute – today – one of the fastest-growing electronic-control technology in motorcycles).
- Chapter 12 (by Yizhai Zhang and Jingang Yi – Rutgers University – and by Dezhen Song – Texas A&M University) is the natural continuation of Chapter 2, and the problem of designing an electronic-rider for the autonomous control of a 2-wheeled vehicle is analyzed.
- Chapter 13 (Ivo Boniolo – University of Bergamo, Giulio Panzani, Diego del Vecchio, Matteo Corno and Mara Tanelli – Politecnico di Milano, Cristiano Spelta – University of Bergamo – and Sergio M. Savaresi – Politecnico di Milano) is a sort of appendix chapter, where three important problems of variable estimation (or software sensing) are considered: roll-angle estimation, vehicle-speed estimation, and suspension stroke estimation. Variable estimation from indirect measurements is, today, a key element for the optimization of sensors layout, both for reducing the cost and for improving the safety of the control systems.

This overview of chapters shows that the book provides a broad perspective on all the main modeling, simulation and control issues of modern two-wheeled vehicles. Moreover, the style and content of the chapters (with a good balancing between theory and experimental results) make this book potentially useful for both practitioners and researchers. From a technological and industrial point of view, the content of the book is up-to-date: it contains the latest technologies both in terms of electronic control systems (traction control, suspension control, steer-damping control), vehicle-dynamics optimization, rider-modeling and virtual-reality hardware-in-the-loop frameworks.

A last comment about the authorship: Italy is largely represented and this reflects the fact that the motorcycle (and bicycle) Italian industry has been, and still is, one of the most vital and technologically advanced worldwide, with a vast number of large, medium and small bike and motorbike companies and prestigious brands. UK, USA and Germany are also represented, consistently with the location of the main motorcycle industries. The most evident missing contribution is from Japan, that has expressed, in the last 30 years, an enormous industrial power and potential, but, somehow, this potential has not been equally represented in the academic research activities (which, in this field, do exist but are quite fragmented).

A final comment for the reader: the books has been conceived for being readable both end-to-end or by cherry-picking some chapters. Each chapter is almost completely self-consistent, with the (partial) exception of the twin-chapters (one in part one, one in part two) written by the same authors.

Part One

Two-Wheeled Vehicles Modelling and Simulation

1

Motorcycle Dynamics

Vittore Cossalter, Roberto Lot, and Matteo Massaro
University of Padova, Italy

This chapter aims at giving a basic insight into the two-wheeled vehicle dynamics to be applied to vehicle modelling and control. The most relevant kinematic properties are discussed in Section 1.1, the peculiarities of motorcycle tyres are reported in Section 1.2, the most popular suspension schemes are presented in Section 1.3, while Sections 1.4 and 1.5 are devoted to the analysis of the vehicle in-plane and out-of-plane vibration modes. Finally, Section 1.6 highlights the coupling between in-plane and out-of-plane dynamics.

1.1 Kinematics

From the kinematic point of view, every mechanical system consists of a number of rigid bodies connected to each other by a number of joints. Each body has six degrees of freedom (DOF) since its position and orientation in the space are fully defined by six parameters, such as the three coordinates of a point (x, y, z) and three angles (yaw, roll, pitch). When a joint is included, the number of DOFs reduces according to the type of joint: the revolute joint (e.g., the one defining the motorcycle steering axis) inhibits five DOFs, the prismatic joint (e.g., the one defining the telescopic fork sliding axis) inhibits five DOFs, the wheel–road contact joint inhibits three DOFs when pure rolling is assumed (only three rotations about the contact point are allowed while no sliding is permitted), or one DOF when longitudinal and lateral slippage is allowed (the only constraint being in the vertical direction, where the compenetration between the wheel and the road is avoided).

1.1.1 Basics of Motorcycle Kinematics

Two-wheeled vehicles can be considered spatial mechanisms composed of six bodies:

- the rear wheel;
- the swingarm;

- the chassis (including saddle, tank, drivetrain, etc.);
- the handlebar (including rear view mirrors, headlamp, the upper part of the front suspension, etc.);
- the front usprung mass (i.e., the lower part of the front suspension, front brake calliper, etc.);
- the front wheel.

These bodies are connected each other and with the road surface by seven joints:

- a contact joint between the rear wheel and the road surface;
- a revolute joint between the rear wheel and the swingarm, to give the rear wheel spin axis;
- a revolute joint between the swingarm and the chassis, to give the swingarm pivot on the chassis;
- a revolute joint between the chassis and the handlebar, to give the steering axis;
- a prismatic joint between the handlebar and the front unsprung, to give the sliding axis of the telescopic fork;
- a revolute joint between the front unsprung and the front wheel, to give the front wheel spin axis;
- a contact joint between the front wheel and the road plane.

Therefore, the two-wheeled vehicle has nine DOFs, given the 20 DOFs inhibited by the four revolute joints, five DOFs inhibited by the prismatic joint and the two DOFs inhibited by the two contact joints (tyre slippage allowed), subtracted from the 36 DOFs related to the six rigid bodies. It is also common to include the rear and front tyre deformation due to the tyre compliance, and consequently the number of DOFs rises to 11.

Among the many different sets of 11 parameters that can be selected to define the vehicle configuration, it is common (e.g. Cossalter et al. 2011b, 2011c) to use the ones depicted in Figure 1.1: position and orientation of the chassis, steering angle, front suspension travel, swingarm rotation and wheel spin rotations.

Finally, it is worth mentioning that these DOFs are related to the gross motion of the vehicle, while additional DOFs are necessary whenever some kind of vehicle structural flexibility is considered, e.g. Cossalter et al. (2007b).

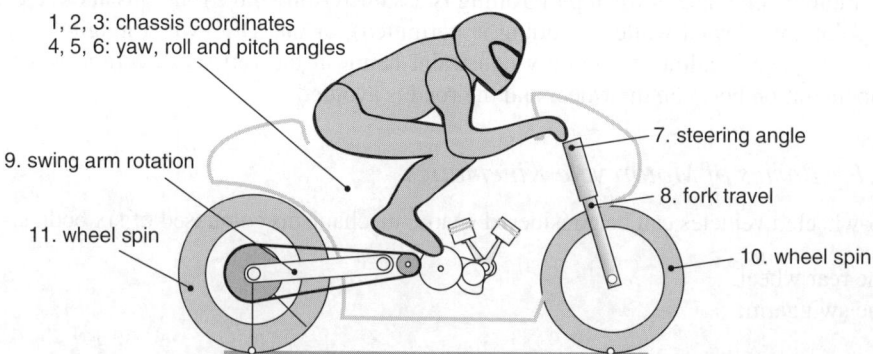

Figure 1.1 Degrees of freedom of a two-wheeled vehicle

Some geometric parameters such as the wheelbase p, normal trail a_n and caster angle ε, are very important when it comes to the vehicle stability, manoeuvrability and handling. In more detail, the wheelbase is the distance between the contact points on the road and usually ranges between 1.2 and 1.6 m, the normal trail is the distance between the front contact point and the steering axis (usually 80–120 mm) and the caster angle is the angle between the vertical axis and the steering axis (usually 19–35°).

In general, an increase in the wheelbase, assuming that the other parameters remain constant, leads to an unfavourable increase in the flexional and torsional deformability of the frame (this may reduce vehicle manoeuvrability), an unfavourable increase in the minimum curvature radius, a favourable decrease in the load transfer during accelerating and braking (this makes wheelie and stoppie more difficult) and a favourable increase in the directional stability of the motorcycle.

The trail and the caster angle are especially important inasmuch as they define the geometric characteristics of the steering head. The definition of the properties of manoeuvrability and directional stability of two-wheeled vehicles depend on these two parameters, among others. Small values of trail and caster characterize sport vehicles, while higher values are typical of touring and cruiser vehicles. The trail and caster are related to each other by the following relationship:

$$a_n = R_f \sin \varepsilon - d, \qquad (1.1)$$

where R_f is the front tyre radius and d is the fork offset; see Figure 1.2.

Finally, it is worth noting that all these parameters are usually given for the nominal (standstill) trim configuration, while they change as the vehicle speed, longitudinal and lateral accelerations change.

1.1.2 Handlebar Steering Angle and Kinematic Steering Angle

While the driver operates the handlebar steering angle, the vehicle cornering behaviour is determined by the projection on the road surface of the angle between the rear and front

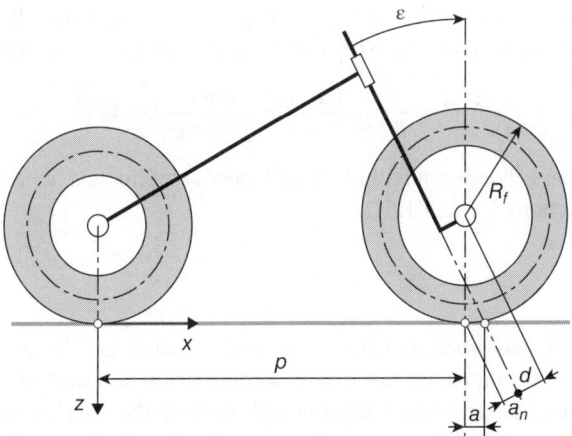

Figure 1.2 Wheelbase, caster angle and trail

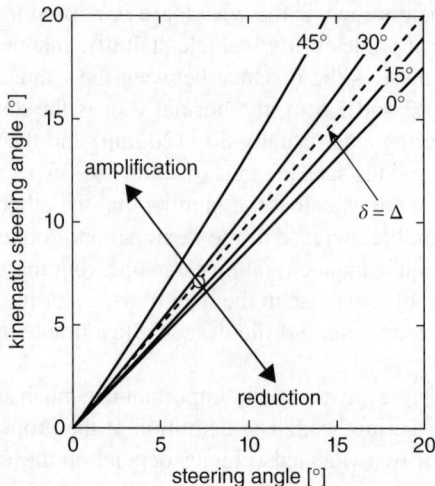

Figure 1.3 Kinematic steering angle Δ as a function of the handlebar steering angle δ for different values of the roll angle ϕ

wheel planes, the so-called kinematic steering angle. In two-wheeled vehicles, the relationship between the handlebar and kinematic steering angles varies appreciably with the roll angle. In particular, the steering mechanism is attenuated (i.e. the kinematic angle is lower than the handlebar angle) up to a certain value of the roll angle (close to the value of the caster angle), then it is amplified (i.e. the kinematic angle is higher than the handlebar angle); see Figure 1.3 for example.

The following simplified expression can be used to estimate the kinematic steering angle Δ from the handlebar steering angle δ, the caster angle ε and the roll angle ϕ:

$$\Delta = \arctan\left(\frac{\cos \varepsilon}{\cos \varphi} \tan \delta\right) \tag{1.2}$$

The local curvature of the vehicle trajectory C (or the turning radius R_c) can be estimated from the kinematic angle Δ and the wheelbase p using the following expression:

$$C = \frac{1}{R_c} \cong \frac{\tan \Delta}{p} = \frac{\cos \varepsilon}{p \cos \varphi} \tan \delta \tag{1.3}$$

Note that Equation 1.3 does not include the effect of tyre slippage, whose contribution will be described in Sections 1.2 and 1.5.2.

1.2 Tyres

The performance of two-wheeled vehicles is largely influenced by the characteristics of their tyres. Indeed, the control of the vehicle's equilibrium and motion occurs through the generation of longitudinal and lateral forces resulting from the rider's actions on the steering mechanism, throttle and braking system. The peculiarity of motorcycle tyres is that they work with camber angles up to 50° and even more, while car tyres rarely reach 10°.

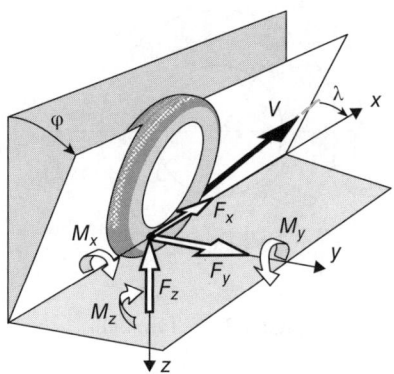

Figure 1.4 Tyre forces and torques

1.2.1 Contact Forces and Torques

From a macroscopic viewpoint, the interaction of the tyre with the road can be represented by a system composed of three forces and three torques, as in Figure 1.4:

- a longitudinal force F_x (positive if driving and negative if braking);
- a lateral force F_y;
- a force F_z normal to the road surface;
- an overturning moment M_x;
- a rolling resistance moment M_y;
- a yawing moment M_z.

Experimental observations show that the force and torque generation is mainly related to the following input quantities:

- tyre longitudinal slip κ;
- tyre lateral slip λ;
- tyre camber angle ϕ;
- tyre radial deflection ζ_R;
- tyre spin rate ω.

Therefore we can write:

$$\begin{aligned} F_x &= F_x(\kappa, \lambda, \phi, \zeta_R, \omega) \\ F_y &= F_y(\kappa, \lambda, \phi, \zeta_R, \omega) \\ M_x &= M_x(\kappa, \lambda, \phi, \zeta_R, \omega) \\ M_y &= M_y(\kappa, \lambda, \phi, \zeta_R, \omega) \\ M_z &= M_z(\kappa, \lambda, \phi, \zeta_R, \omega) \end{aligned} \quad (1.4)$$

with the longitudinal force F_x mainly related to longitudinal slip κ, lateral force F_y mainly related to the lateral slip λ and the camber angle ϕ, overturning moment M_x mainly related to the camber angle ϕ, rolling resistance mainly related to the wheel spin rate ω and yawing moment mainly related to the lateral slip λ and camber angle ϕ.

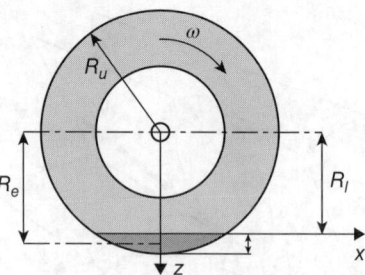

Figure 1.5 Tyre radii

The longitudinal slip (positive when driving and negative when braking) is defined as:

$$\kappa = \frac{\omega R_e - V_x}{V_x}, \quad (1.5)$$

where V_x is the tyre longitudinal velocity, ω is the tyre spin rate and R_e is the tyre effective rolling radius. In particular, the effective rolling radius R_e can be computed from the freely rolling tyre as

$$R_e = V_x/\omega. \quad (1.6)$$

Note that the effective rolling radius does not coincide with either the tyre loaded radius R_l or the the tyre unloaded radius R_u; see Figure 1.5. This should not be surprising since the tyre is not a rigid body. Experimental observations show that $R_l < R_e < R_u$. However, a common assumption is $R_l = R_e$.

Sometimes a slightly different formulation of longitudinal slip is adopted:

$$\kappa' = \frac{\omega R_e - V_x}{\omega R_e} \quad (1.7)$$

It can easily be shown that

$$\kappa' = \kappa/(1+\kappa) \quad (1.8)$$

and the relative difference between the two is:

$$\varepsilon = (\kappa' - \kappa)/\kappa = -\kappa' \quad (1.9)$$

which is typically lower than 5% in normal conditions (i.e. no skidding).

The lateral slip is defined as:

$$\lambda = -\arctan \frac{V_y}{V_x} \quad (1.10)$$

where V_y is the lateral velocity of the tyre and V_x is the longitudinal velocity. The sign is chosen to give positive force for positive slip.

Sometimes another input quantity is considered, the turn slip ϕ_t:

$$\phi_t = -\frac{1}{R_c} = -\frac{\dot{\psi}}{V_x} \quad (1.11)$$

where R_c is the curvature of the tyre contact point path and $\dot{\psi}$ is the yaw rate. This quantity is important only at very low speed and therefore is not considered in the following sections.

1.2.2 Steady-State Behaviour

A widely used model for computing the steady-state tyre forces and moment is based on the so-called Magic Formula (Pacejka 2006). The general form is:

$$y(x) = D \sin[C \arctan\{Bx - E(Bx - \arctan Bx)\}] \quad (1.12)$$

where $y(x)$ passes through the origin $x = y = 0$, reaches a maximum and subsequently tends to a horizontal asymptote; see Figure 1.6. For given values of the coefficients B, C, D, E, the curve shows an anti-symmetric shape with respect to the origin. To allow the curve to have an offset with respect to the origin (e.g. because of ply-steer and conicity of the tyre), two shifts S_H and S_V can be introduced:

$$\begin{aligned} Y &= y(x) + S_V \\ x &= X + S_H \end{aligned} \quad (1.13)$$

Coefficient $D > 0$ represents the peak value of the curve, while the product BCD corresponds to the slope of the curve at the origin (e.g. the lateral slip stiffness when the lateral force is reported in the vertical y axis and the lateral slip is reported in the horizontal x axis). The shape factor $C > 0$ determines the shape of the resulting curve. The factor B is used to determine the slope at the origin and is called the stiffness factor. The factor $E \leq 1$ is introduced to control the curvature at the peak and at the same time the horizontal position of the peak. The various factors depend on the tyre normal load F_z (or tyre radial deflection).

In particular, the slope of the lateral force BCD_y is especially sensitive to load variation, and is usually modelled as follows (Figure 1.7):

$$BCD_y = p_1 \sin(2 \arctan(F_z/p_2)) \quad (1.14)$$

The sideslip stiffness attains a maximum p_1 at a normal load $F_z = p_2$.

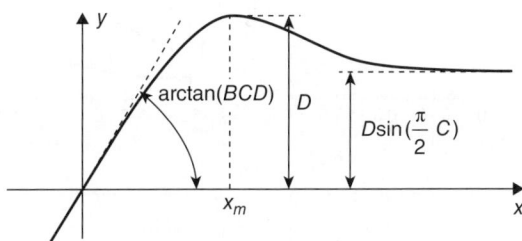

Figure 1.6 Main parameters of the tyre Magic Formula

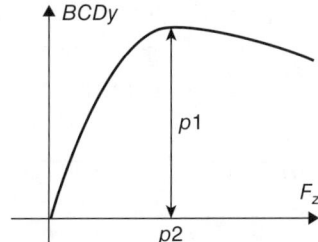

Figure 1.7 Tyre cornering stiffness as a function of normal load

Another widely used tyre formula is the Burckhardt model (Kiencke and Nielsen 2001):

$$y(x) = \vartheta_1(1 - e^{-x\vartheta_2}) - x\vartheta_3 \qquad (1.15)$$

Again, the curve typically passes through the origin $x = y = 0$, reaches a maximum and subsequently decreases. An offset can be added, following the same approach used above.

Typical tyre curves are depicted in Figure 1.8.

A fundamental concept when dealing with tyre behaviour is the coupling between longitudinal and lateral forces on the contact patch. In practice, the tyre gives the maximum longitudinal (lateral) force when in pure longitudinal (lateral) slip condition. Indeed, the theoretical analysis on physical models (Pacejka 2006) shows that the tyre longitudinal and lateral force generation depends on the following theoretical slip quantities:

$$\sigma_x = \frac{\kappa}{1+\kappa} \qquad \sigma_y = \frac{\tan \lambda}{1+\kappa} \qquad (1.16)$$

rather than on the practical slip quantities κ and λ, and that there exists a total slip:

$$\sigma = \sqrt{\sigma_x^2 + \sigma_y^2} \qquad (1.17)$$

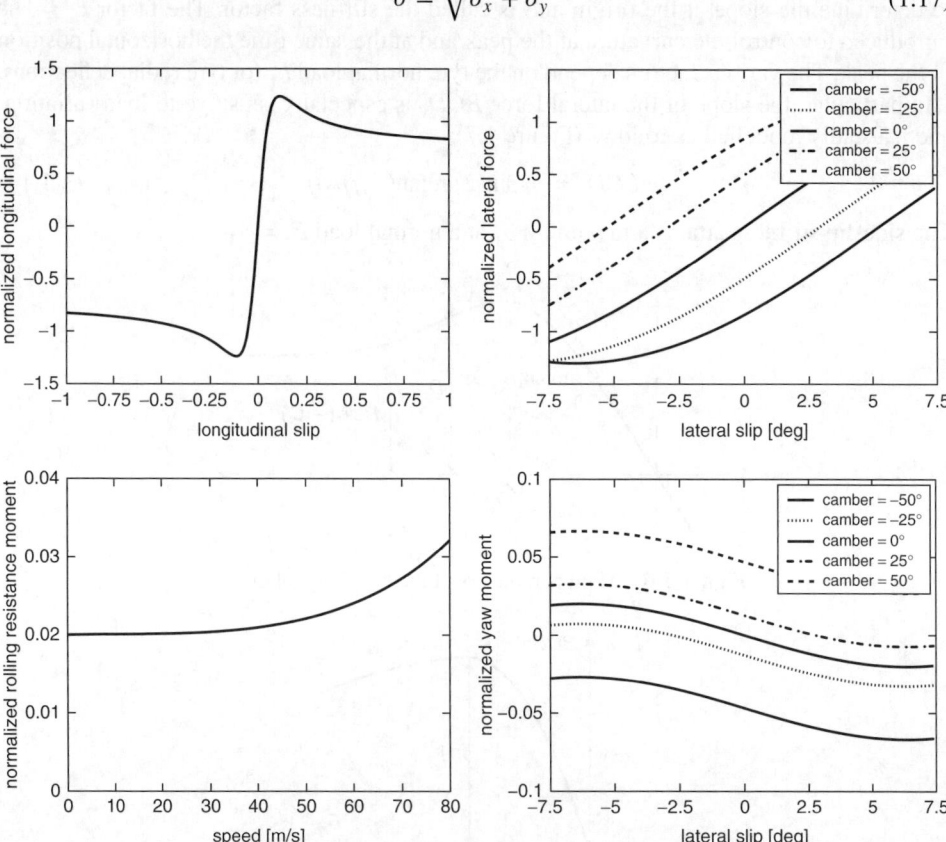

Figure 1.8 Typical tyre curves

which defines the maximum friction force available from the tyre. The corresponding total force can be split between the longitudinal and lateral directions, according to the slip σ_x and σ_y. Also, the effect of camber can be included into the sideslip as follows:

$$\lambda^* = \lambda + \frac{k_\phi}{k_\lambda}\phi \tag{1.18}$$

and the formulas for forces read:

$$F_x = \frac{\sigma_x}{\sigma}F_{x0}(\sigma) \quad F_y = \frac{\sigma_y}{\sigma}F_{y0}(\sigma) \tag{1.19}$$

where F_{x0} and F_{y0} are the longitudinal and lateral forces in pure slip condition.

There is also a newer empirical approach to modelling force coupling (Pacejka 2006). To describe the effect of combined slip on the lateral force and longitudinal force characteristics, the following hill-shaped function G is employed:

$$G = D\cos(C\arctan(Bx)) \tag{1.20}$$

where x is either the longitudinal slip κ or the lateral slip λ (or $\tan\lambda$). The coefficient D is the peak value, C determines the height of the hill's base and B influences the sharpness of the hill, which is the main factor responsible for the shape of the function. The formulas in combined slip conditions read

$$F_x = G_x F_{x0}(\kappa) \quad F_y = G_y F_{y0}(\lambda) \tag{1.21}$$

1.2.3 Dynamic Behaviour

The relationships between the tyre inputs (slips, camber, load/deflection and spin) and the tyre outputs (forces and torques) described in the previous section hold in steady-state conditions. However, the tyre forces do not arise instantaneously: to appear the tyre needs to travel a certain distance, which depends on the tyre characteristics. The physical reason is the tyre flexibility, and the related behaviour can be explained as follows.

We consider a tyre whose contact point has longitudinal velocity $V_x + \dot{\zeta}_x$ and lateral velocity $V_y + \dot{\zeta}_y$, where V_x and V_y are the velocities of the contact point when neglecting tyre deformation, while $\dot{\zeta}_x$ and $\dot{\zeta}_y$ are the deflection velocities; see Figure 1.9.

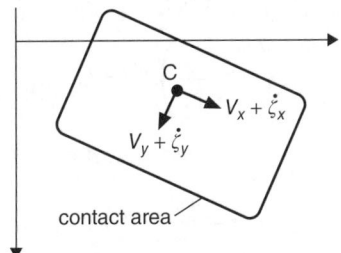

Figure 1.9 Tyre contact area with deflections

The observed longitudinal slip (e.g. with sensors on the rim) is

$$\kappa = \frac{\omega R_e - V_x}{V_x} \qquad (1.22)$$

with ω the rim spin rate and R_e the effective rolling radius, while the actual (or instantaneous) longitudinal slip experienced by the contact point is

$$\kappa_i = \frac{\omega R_e - \dot{\zeta}_x - V_x}{V_x} \qquad (1.23)$$

and therefore

$$\kappa_i = \kappa - \frac{\dot{\zeta}_x}{V_x} \qquad (1.24)$$

Similarly, the observed lateral slip is

$$\lambda = -\arctan \frac{V_y}{V_x}, \qquad (1.25)$$

while the actual (or instantaneous) lateral slip is

$$\lambda_i = -\arctan \frac{V_y + \dot{\zeta}_y}{V_x}. \qquad (1.26)$$

Under small angle assumption it is

$$\lambda_i = \lambda - \frac{\dot{\zeta}_y}{u} \qquad (1.27)$$

At the tyre–road contact point, the slip-induced longitudinal and lateral forces balance the deflection-induced forces. Under small slips assumption the following relationships hold:

$$k_\kappa \kappa_i F_z = k_{\zeta_x} \zeta_x \quad k_\lambda \lambda_i F_z = k_{\zeta_y} \zeta_y \qquad (1.28)$$

where k_κ and k_λ are the lateral slip stiffness and longitudinal slip stiffness respectively, k_{ζ_x} and k_{ζ_y} are the lateral and longitudinal structural stiffness and F_z is the tyre normal load. When introducing Equations 1.24 and 1.27 into 1.28 one obtains:

$$k_\kappa \left(\kappa - \frac{\dot{\zeta}_x}{V_x} \right) F_z = k_{\zeta_x} \zeta_x \qquad (1.29)$$

$$k_\lambda \left(\lambda - \frac{\dot{\zeta}_y}{V_x} \right) F_z = k_{\zeta_y} \zeta_y \qquad (1.30)$$

which yields, after the elimination of carcass deflections ζ_x and ζ_y

$$F_{x0} = \frac{\sigma_\kappa}{u} \dot{F}_x + F_x \qquad (1.31)$$

$$F_{y0} = \frac{\sigma_\alpha}{u} \dot{F}_y + F_y \qquad (1.32)$$

where F_x and F_y are the actual tyre forces (i.e. computed with the instantaneous slip κ_i and λ_i), while F_{x0} and F_{y0} are the tyre forces computed with the practical slips κ and λ and:

$$\sigma_\kappa = \frac{k_\alpha F_z}{k_{\zeta_x}} \quad \sigma_\alpha = \frac{k_\lambda F_z}{k_{\zeta_y}}. \tag{1.33}$$

In practice, there is a deformation-induced lag between F_x, F_y and F_{x0}, F_{y0}. The resulting first-order differential equations are called *relaxation equations*, with $\sigma_{\kappa,\alpha}$ the relaxation lengths. Equation 1.33 shows that the longitudinal (lateral) relaxation length increases with the longitudinal (lateral) slip stiffness and with the normal load, while it reduces with the longitudinal (lateral) structural stiffness. The relaxation length represents the space that the wheel has to cover in order for the force to be 63% of the steady-state force. Typical values of relaxation length are in the range 0.10–0.4 m, the higher values corresponding to higher tyre normal load and higher speeds.

The equations above describe the effect of flexibilities on the longitudinal force due to longitudinal slip, and lateral force due to lateral slip. Actually, in two-wheeled vehicles there is a significant component of the lateral force related to the camber angle ϕ. Therefore, under small ϕ assumption, Equation 1.28 becomes

$$(k_\lambda \lambda_i + k_\phi \phi) F_z = k_\zeta \zeta_y \tag{1.34}$$

where k_ϕ is the camber stiffness, and after substitution Equation 1.32 gives

$$F_{y0} + \frac{\sigma_\alpha}{V_x} F_z \dot{\phi} = \frac{\sigma_\alpha}{V_x} \dot{F}_y + F_y. \tag{1.35}$$

Finally, it is worth mentioning the gyroscopic couple that arises as a result of the time rate of change of the tyre camber distortion, the wheel spin rate and the belt inertia. This effect is visible for certain types of tyre at high speeds, and leads to an increase of the observed relaxation length σ_α (De Vries and Pacejka 1998).

1.3 Suspensions

Suspensions serve several purposes such as contributing to the vehicle's road-holding/handling, keeping the rider comfortable and reasonably well isolated from road noise. These goals are generally at odds. In addition, the suspensions affect the vehicle's trim while accelerating, braking, turning and so on. The proper choice of front and rear suspension characteristics depends on many parameters: the weight of the rider and the vehicle, the position of the centre of gravity, the characteristics of stiffness and vertical damping of the tyres, the geometry of the motorcycle, the conditions of use, the road surface, the braking performance, the engine power and the driving technique, among others.

1.3.1 Suspension Forces

The total force F exerted by the spring–damper group is the sum of the following different actions:

$$F = F_e + F_d + F_f + F_p \tag{1.36}$$

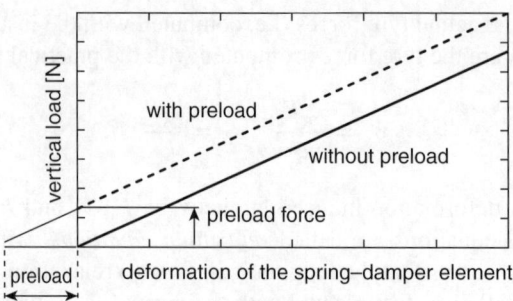

Figure 1.10 Elastic force as a function of preload

where F_e is the elastic force exerted by the coil spring and/or air spring (or different elastic components), F_d is the damping force exerted by the shock absorber, F_f is the friction force and F_p is the end-stroke pad force.

Preloading is commonly used to adjust the initial position of the suspension with the weight of the vehicle and rider acting on it. It consists of a precompression of the spring: as a consequence, if the spring is stressed with forces that are lower than or equal to the preload, it is not compressed. In practice, this adjustment shifts the curve of the elastic force as a function of travel F_e; see Figure 1.10.

1.3.2 Suspensions Layout

Several suspension layouts have been used over the years and the following sections present a brief overview.

1.3.2.1 Front Suspension Types

The most widespread front suspension is the telescopic fork (Figure 1.11a). It is made up of two telescopic sliders which run along the interior of two fork tubes and form a prismatic

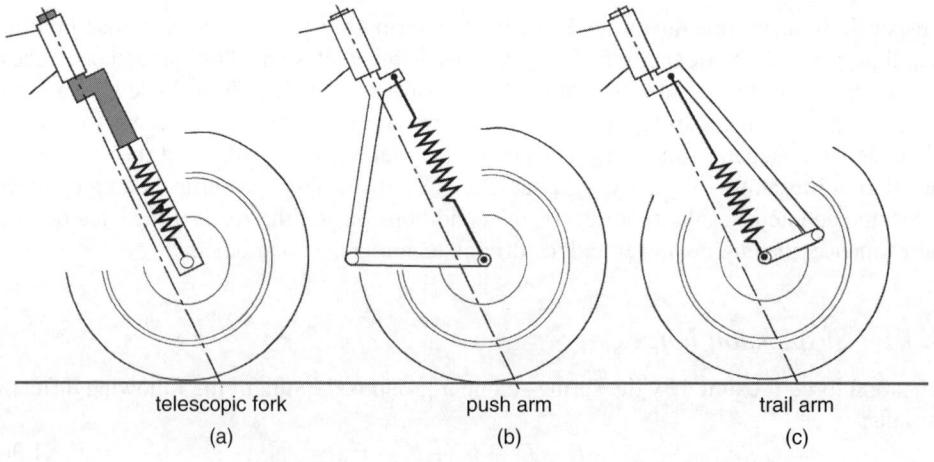

Figure 1.11 Example of front suspensions

joint between the unsprung mass of the front wheel and the sprung mass of the chassis. The telescopic fork is characterized by limited inertia around the axis of the steering head.

Two limitations of the telescopic fork are the impossibility of attaining progressive force displacement and the rather high values of the unsprung mass that is an integral part of the wheel. To overcome the typical defects of the telescopic fork, alternative solutions have been proposed: push arm (Figure 1.11b), trail arm (Figure 1.11c) and four-bar linkage (like the BMW Duolever).

The front arm suspension and four-bar linkage suspensions can be designed so as to present total or partial anti-dive behaviour in braking conditions. Further, the absence of a prismatic joint eliminates the typical dry friction problems of telescopic forks.

1.3.2.2 Rear Suspension Types

The classic rear suspension is composed of a swingarm (a rocker made up of two oscillating arms) with two spring–damper elements, one on each side (Figure 1.12a). The main advantages are the simplicity of construction and the modest reactive forces transmitted to the chassis. Among its disadvantages are a poorly progressive force–displacement characteristic and the possibility that the two spring–damper units generate different forces and therefore torsional stress on the swingarm.

An alternative is represented by the cantilever mono-shock system, characterized by only one spring–damper unit. However, this suspension does not enable a progressive force–displacement characteristic and the positioning of the spring–shock absorber unit close to the engine can cause problems with the absorber's heat dissipation.

The introduction of a four-bar linkage in the rear suspension makes it easier to obtain the desired stiffness curves. Different attachment points of the spring–damper elements can be chosen: for example, in the Kawasaki Uni-Trak the suspension element is between the rocker and the chassis (Figure 1.12b), in the Suzuki Full-Floater it is between the rocker and the swingarm and in the Honda Pro-Link the element is between the connecting rod and the swingarm. Modest unsprung masses are obtained, as well as large wheel amplitude, but langer reactive forces are exchanged between the various parts of the four-bar linkage.

The four-bar linkage (Figure 1.12c) is also the basis of a suspension used especially on the final shaft transmission with universal joints (e.g. the BMW Paralever). The wheel is attached to the connecting rod of the four-bar linkage. The suspension acts as if it were composed of a very long fork fastened to the chassis at the centre of rotation (the point of intersection of the axes of the two rockers). An additional small four-bar linkage can be

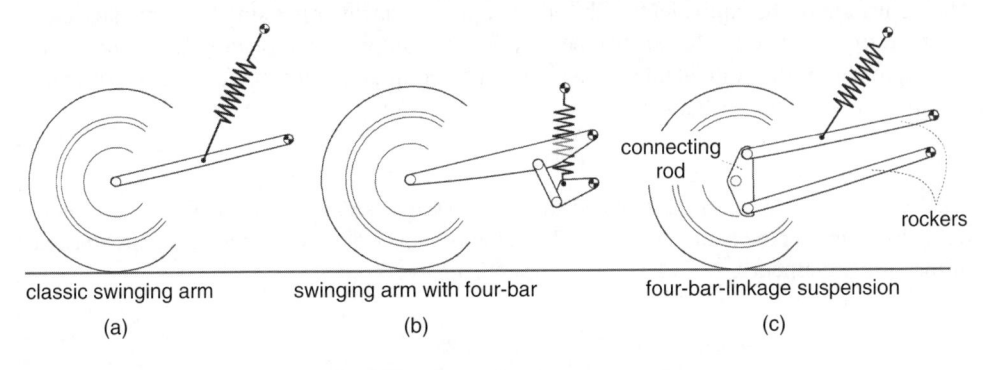

Figure 1.12 Example of rear suspensions

added to provide a suitable attachment point for the spring–shock element and thus a proper suspension behaviour.

1.3.3 Equivalent Stiffness and Damping

From a dynamics point of view, the vehicle can be considered as a main sprung body (chassis and rider) connected to two unsprung bodies (wheels) with two elastic systems (front and rear suspension). Also, rather than the characteristics of the spring–damper units, it is important to consider the characteristics of the suspensions in terms of wheel vertical displacement as a function of the vertical force applied. Therefore it is useful to reduce the real suspensions to equivalent, simpler, suspensions represented by two vertical spring–damper elements that connect the unsprung masses to the sprung mass. The parameters defining the equivalent suspension are: reduced stiffness, reduced damping, dependence of the reduced stiffness on the vertical displacement (progressive/regressive suspension), maximum travel and preloading.

To derive the equivalent (or reduced) stiffness, we consider the expression of the variation of the spring force F_e as a function of the travel:

$$F_e = F_0 + k_L(L - L_0) \tag{1.37}$$

where F_0 is the spring force at the initial suspension travel L_0, k_L is the stiffness at L_0 and L is the travel after variation. The power balance between the actual spring force F_e and its equivalent vertical force F_z at the wheel centre is

$$F_z \dot{z} = F_e \dot{L} = F_e \frac{\partial L}{\partial z} \dot{z} \tag{1.38}$$

Therefore

$$F_z = F_e \frac{\partial L}{\partial z} = F_e \tau, \tag{1.39}$$

where τ is the velocity ratio, between the suspension travel velocity and the wheel vertical velocity. The equivalent stiffness k_z is

$$k_z = \frac{\partial F_z}{\partial z} = k_L \tau^2 + F_e \frac{\partial \tau}{\partial z}. \tag{1.40}$$

When assuming a constant velocity ratio the expression simplifies to

$$k_z = k_L \tau^2. \tag{1.41}$$

The derivation of the equivalent (reduced) damping is carried out using the same approach, and therefore Equation 1.40 can also be used for the damping by replacing k_z, k_L with c_z, c_L.

The preload of the equivalent suspension can be computed with the following expression (again from force power balance):

$$F_z|_{L=0} = F_e|_{L=0} \tau \tag{1.42}$$

Finally, the dependence of the reduced stiffness on the vertical displacement can be affected either by changing the characteristic of the velocity ratio τ or by changing the characteristics of the spring element k_L.

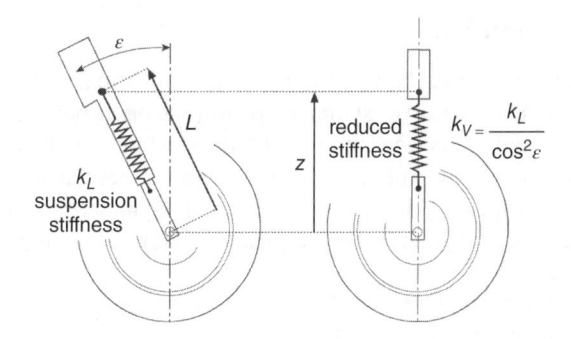

Figure 1.13 Reduced stiffness for the telescopic fork

1.3.3.1 Front

The equivalent stiffness and damping can be easily computed for the widely spread telescopic fork (Figure 1.13). The velocity ratio is derived by considering the geometric relationship between the fork travel and the wheel vertical displacement:

$$\tau = \frac{\partial L}{\partial z} = \frac{1}{\cos(\varepsilon)} \qquad (1.43)$$

where ε is the caster angle. Under the assumption of constant stiffness k_L and damping c_L coefficients, and constant velocity ratio τ, the reduced values are

$$k_z = \frac{k_L}{\cos^2\varepsilon} \qquad c_z = \frac{c_L}{\cos^2\varepsilon} \qquad (1.44)$$

In more complex linkages, the velocity ratio is computed numerically from a kinematic analysis of the mechanism. When it is not constant, or the spring/damper coefficients are not constant, the full Equation 1.40 should be used.

Usually the suspension stiffness k_L is in the range 13–25 kN/m, and the equivalent stiffness k_z in the range 15–37 kN/m, while the damping coefficient c_L is in the range 500–2000 Ns/m, and the equivalent c_z in the range 550–2200 Ns/m, with the velocity ratio τ in the range 1.05–1.25 and the caster angle in the range 19–35°.

1.3.3.2 Rear

The velocity ratio of a rear suspension featuring a linkage depends on many parameters and sometimes cannot be expressed analytically. In any case, the ratio can be easily computed numerically from a kinematic analysis of the mechanism. Typical values of τ for a swingarm with a four-bar linkage are in the range 0.3–0.6. Usually the suspension stiffness k_L is in the range 100–150 kN/m, and the equivalent stiffness k_z in the range 10–55 kN/m, while the damping coefficient c_L is in the range 5–15 kNs/m and the equivalent c_z in the range 450–5400 Ns/m.

1.4 In-Plane Dynamics

Vehicle dynamics can be divided between in-plane and out-of-plane dynamics. The former involve the motion of the vehicle in its symmetry plane (e.g. pitch, bounce, suspension travel) and mostly affect the riding comfort and road-holding, while the latter involve the lateral motion of the vehicle (e.g. yaw, roll, steer) and strongly affect the stability and safety. In straight running, two-wheeled vehicles are substantially symmetric and in-plane and out-of-plane motions are decoupled (therefore they can be examined separately), whereas while cornering strong interactions occur.

In this section, different in-plane models of increasing complexity are presented to highlight the main vehicle dynamics involved.

In practice, these dynamics are excited by road undulations and/or by the inertial forces generated while accelerating/braking. Suppose that the vehicle travels with constant velocity V_x on a road with equidistant irregularities (e.g. the bays of a viaduct), Figure 1.14. The time Δt required to cover the distance L between two irregularities (length of the bay) is equal to

$$\Delta t = \frac{L}{V_x} \tag{1.45}$$

and represents the period of the external excitation. A resonance condition occurs whenever the excitation frequency is equal to the natural frequency of one of the in-plane vibration modes of the vehicle. As an example, with $L = 12$ m and $V_x = 24$ m/s, it is $\Delta t = 0.5$ s, so the excitation has a frequency of 2 Hz. In general, several frequency components are present at the same time, depending on the road characteristics.

1.4.1 Pitch, Bounce and Hops Modes

Among the 11 DOFs necessary to fully define the vehicle trim (see Section 1.1), only seven are involved in the in-plane dynamics: longitudinal and vertical motion of the chassis, pitch of the chassis, suspension travel, wheels rotation. If we assume that the vehicle is traveling at constant speed (which is a common assumption when dealing with comfort analysis), the DOFs further reduce to four: vertical motion of the chassis, pitch angle and suspension travel. This DOFs are related to four physical vibration modes: *bounce*, *pitch*, wheel *front hop* and *rear hop*. Bounce is mainly related to the vertical motion of the chassis, pitch is mainly related to the pitch of the chassis and hops are mainly related to the wheels' vertical

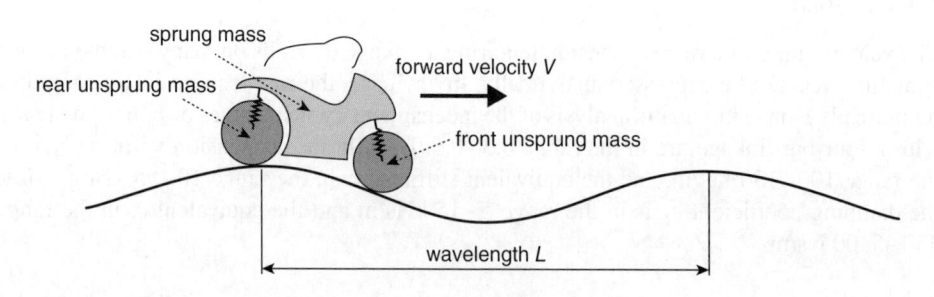

Figure 1.14 Road undulation

Motorcycle Dynamics

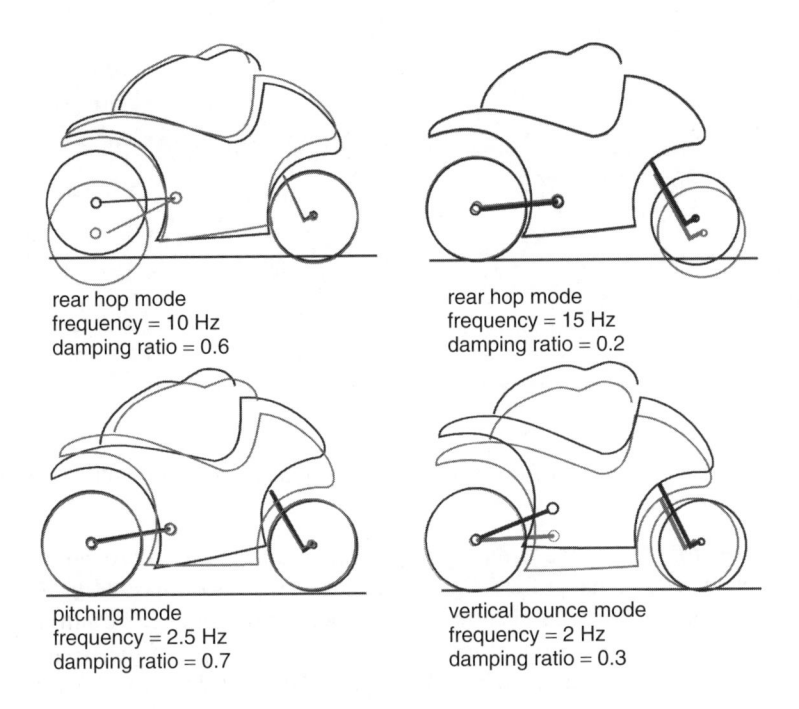

rear hop mode
frequency = 10 Hz
damping ratio = 0.6

rear hop mode
frequency = 15 Hz
damping ratio = 0.2

pitching mode
frequency = 2.5 Hz
damping ratio = 0.7

vertical bounce mode
frequency = 2 Hz
damping ratio = 0.3

Figure 1.15 In-plane vibration modes

motion. In practice, every mode involves some contribution from all four DOFs. Figure 1.15 depicts an example of in-plane vibration modes, with typical values of natural frequency and damping ratio.

In order to model in-plane dynamics, several simple models are commonly used in addition to complex multibody models. The most popular are reported in the following sections.

1.4.1.1 Half-Vehicle Models

The half-vehicle model is a simple yet widespread model used to analyse the suspension–tyre dynamics. The name is due to the fact that only one tyre and one suspension are considered. Two versions are used: one DOF (or SDF) and two DOF.

In the simplest version (Figure 1.16) the model features a mass suspended by a spring–damper element. The mass may represent either the sprung mass (whose share is computed from the whole vehicle mass by considering the tyre loads distribution) or the unsprung mass (wheel rim, brake calliper, etc.). In the former case the spring–damper element represents the suspension, while the tyre compliance is neglected (Figure 1.16a), in the latter case the spring–damper element represents the tyre compliance while the suspension dynamics are neglected (Figure 1.16b). The undamped natural frequency f_0, damped frequency f and damping ratio ζ are

$$f_0 = \frac{1}{2\pi}\sqrt{\frac{k_z}{m}} \quad f = f_0\sqrt{1-\zeta^2} \quad \zeta = \frac{c_z}{2m(2\pi f_0)} \tag{1.46}$$

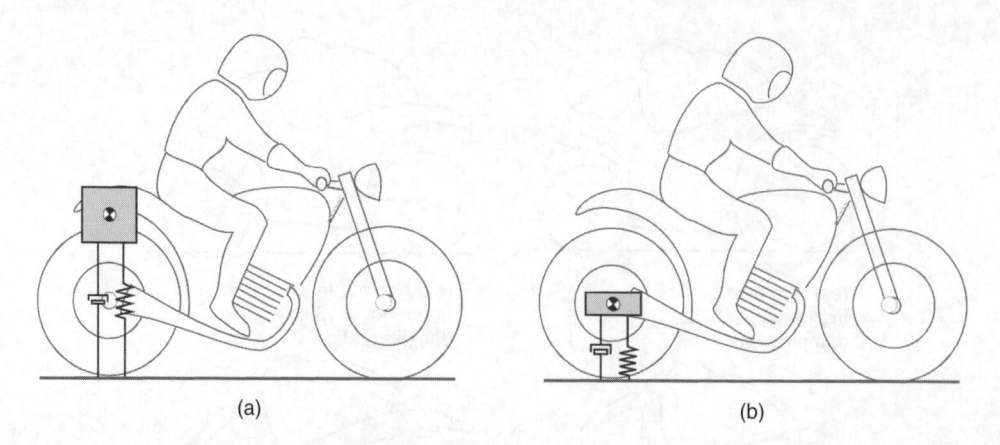

Figure 1.16 Half-vehicle model with (a) one DOF for sprung mass and (b) for unsprung mass

where k_z is the stiffness, c_z the damping coefficient and m the mass. As an example, we consider a vehicle with a mass of 200 kg, including two wheels of mass 15 kg each, a rider with a mass of 80 kg and the whole centre of mass exactly in the middle. We aim at estimating the natural frequency of the front wheel–suspension system, given the front suspension reduced vertical stiffness (see Section 1.3) $k_z = 15$ kN/m and the tyre radial stiffness of $k_T = 180$ kN/m. If we want the model to capture the suspension mode, we use $m = 140$ kg and $k_z = 15$ kN/m, obtaining a natural frequency $f_0 = 1.64$ Hz. Another option is to use a combination of the suspension spring and tyre spring:

$$k_{z'} = \frac{k_z k_T}{k_z + k_T} \tag{1.47}$$

In this case, the stiffness reduces to $k_{z'} = 14$ kN/m and the natural frequency to $f_0 = 1.59$ Hz. Otherwise, if we want the model to capture the wheel hop mode, we use $m = 15$ kg and $k_z = 180$ kN/m, obtaining a natural frequency of 17 Hz.

In the version with two DOFs (Figure 1.17), the model features two masses, representing the sprung mass m_s and unsprung mass m_u, and two spring–damper elements, representing

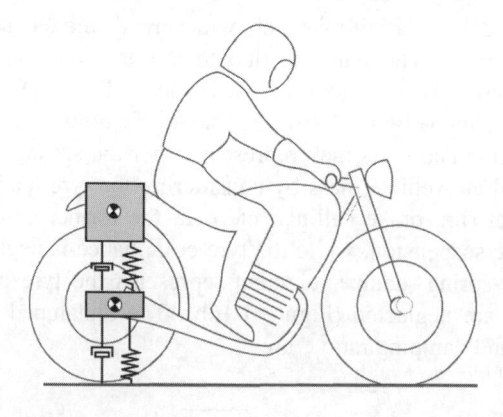

Figure 1.17 Half-vehicle model with two DOFs

the suspension characteristic and the tyre radial compliance. Therefore two modes influencing each other are captured. The expressions for the natural undamped frequencies are:

$$f_{01,02}^2 = \frac{1}{4\pi^2} \frac{-a_1 \pm \sqrt{a_1^2 - 4a_2 a_0}}{2a_2} \qquad (1.48)$$

with:

$$a_2 = m_s m_u \qquad a_1 = k_z(m_s + m_u) + k_T m_u \qquad a_0 = k_z k_T. \qquad (1.49)$$

Using the vehicle parameters defined above, gives $f_{01} = 1.58$ Hz (suspension mode) and $f_{02} = 18.15$ Hz (wheel hop mode).

1.4.1.2 Full-Vehicle Models

The full-vehicle model has four DOFs and is depicted in Figure 1.18. Since there are no analytic and compact expressions for the system natural frequencies, the equations of motion are reported as:

$$\mathbf{M}\{\ddot{\mathbf{x}}\} + \mathbf{C}\{\dot{\mathbf{x}}\} + \mathbf{K}\{\mathbf{x}\} = 0 \qquad (1.50)$$

with:

$$\mathbf{x} = \begin{Bmatrix} z \\ \mu \\ z_F \\ z_R \end{Bmatrix} \qquad \mathbf{M} = \begin{bmatrix} m & 0 & 0 & 0 \\ 0 & I_{yG} & 0 & 0 \\ 0 & 0 & m_f & 0 \\ 0 & 0 & 0 & m_r \end{bmatrix} \qquad (1.51)$$

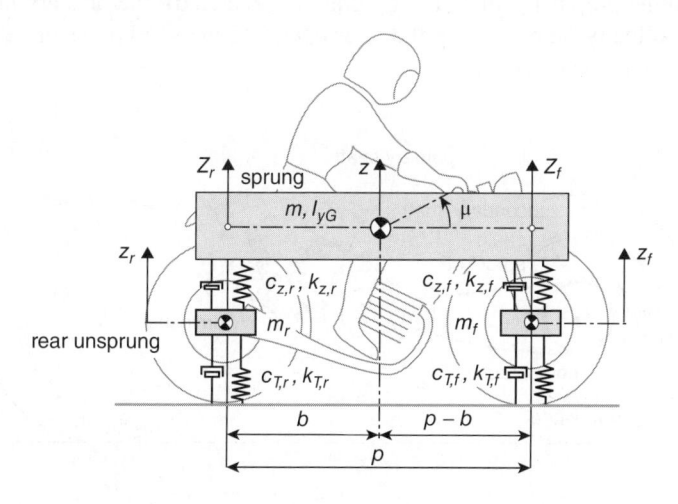

Figure 1.18 Full-vehicle model with four DOFs

$$\mathbf{C} = \begin{bmatrix} c_{z,f} + c_{z,r} & c_{z,f}(p-b) - c_{z,r}b & -c_{z,f} & -c_{z,r} \\ c_{z,f}(p-b) - c_{z,r}b & (p-b)^2 c_{z,f} + c_{z,r}b^2 & -c_{z,f}(p-b) & c_{z,r}b \\ -c_{z,f} & (-p+b)c_{z,f} & c_{z,f} + c_{T,f} & 0 \\ -c_{z,r} & c_{z,r}b & 0 & c_{z,r} + c_{T,r} \end{bmatrix}$$

$$\mathbf{K} = \begin{bmatrix} k_{z,f} + k_{z,r} & -k_{zV,r}b + k_{z,f}(p-b) & -k_{z,f} & -k_{z,r} \\ -k_{z,r}b + k_{z,f}(p-b) & (p-b)^2 k_{z,f} + k_{z,r}b^2 & -k_{z,f}(p-b) & k_{z,r}b \\ -k_{z,f} & (-p+b)k_{z,f} & k_{z,f} + k_{T,f} & 0 \\ -k_{z,r} & k_{z,r}b & 0 & k_{z,r} + k_{T,r} \end{bmatrix}$$

where m_f and m_r are the front and rear unsprung masses respectively, m is the sprung mass, $k_{z,f}$ and $k_{z,r}$ are the front and rear suspension reduced stiffness, $c_{z,f}$ and $c_{z,r}$ are the front and rear suspension reduced damping, $k_{T,f}$ and $k_{T,r}$ are the front and rear tyre radial stiffness and $c_{T,f}$ and $c_{T,r}$ are the front and rear tyre radial damping.

Note that the undamped radian frequencies ω can be numerically derived from

$$|\mathbf{K} - \omega^2 \mathbf{M}| = \mathbf{0} \tag{1.52}$$

while for the full modal analysis the system must be reduced to a standard first-order formulation before computing the eigenvalues.

1.4.2 Powertrain

Powertrain dynamics involve the fluctuation of the vehicle's longitudinal velocity and wheel rotations, the three DOFs not considered in the models presented in the previous section. Figure 1.19 depicts a common motorcycle powertrain layout, which includes the crankshaft (where the engine propulsive or braking torque is generated), the primary and secondary shafts (whose velocity ratio is set by the rider operating the gearbox lever), the chain final transmission and the rear wheel.

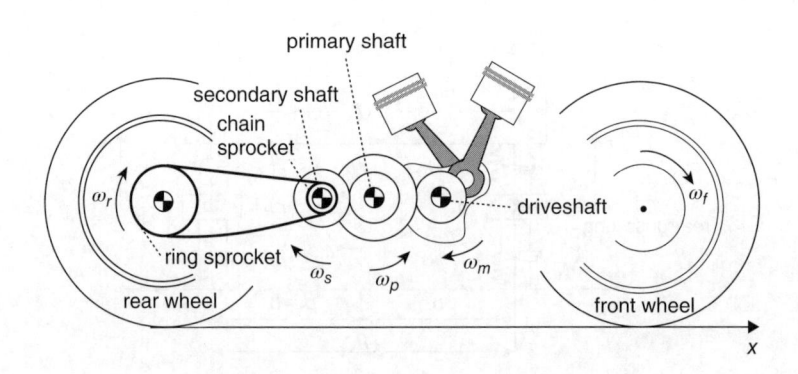

Figure 1.19 Motorcycle powertrain

Motorcycle Dynamics

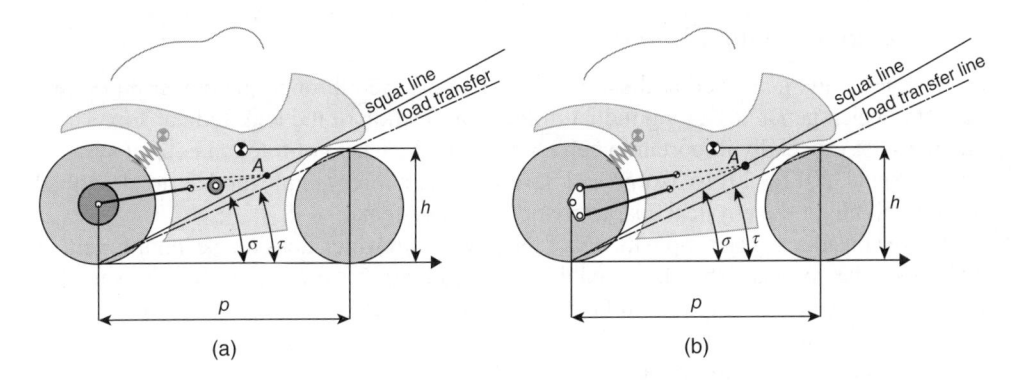

Figure 1.20 Squat and load transfer lines

It is worth noting that the geometry of the final transmission strongly affects the vehicle trim while varying the propulsive force. In particular, it can be shown (Cossalter 2006) that the variation of the trim of the vehicle rear end (with respect to the standstill configuration) mainly depends on a parameter called the squat ratio R:

$$R = \frac{\tan \tau}{\tan \sigma} \qquad (1.53)$$

where τ is the angle of the load transfer line and σ is the angle of the squat line; see Figure 1.20, where the lines are depicted for the swingarm with chain final transmission (Figure 1.20a) and for the four-bar linkage with shaft final transmission (Figure 1.20b).

In more detail, the transfer load line represents the direction of the transfer tyre force acting on the rear contact point. The load transfer is originated by the vehicle acceleration and/or aerodynamic forces. Therefore two transfer lines can be computed: in the case of significant acceleration, the transfer line is the load line, whereas for mild acceleration, the transfer line is the aerodynamic line:

$$\tan \tau = \frac{h_g}{p} \quad \text{or} \quad \tan \tau = \frac{h_a}{p} \qquad (1.54)$$

where h_g is the height of the vehicle centre of mass, h_a is the height of the aerodynamic centre and p is the wheelbase.

As regards the squat line, it passes through the rear tyre contact point and the point A, which is the intersection of the swingarm axis with the chain axis, in the case of a final transmission with chain and swingarm (Figure 1.20a), the swingarm pivot on the chassis in the case of a final transmission with shaft and swingarm, and the intersection of the two rockers in the case of a final transmission with shaft and four-bar linkage (Figure 1.20b).

When $R = 1$ (often a design target) there is no variation of the trim of the vehicle rear end while changing the tyre thrust force F_x. In practice, there may be a small variation, due to the theoretical assumptions. When increasing the longitudinal force with $R < 1$ the rear suspension extends, while in the case of $R > 1$ the rear suspension compresses.

1.4.3 Engine-to-Slip Dynamics

The engine torque generated at the crankshaft is transferred through the powertrain to the rear tyre, which generates a longitudinal force as a function of the longitudinal slip. These dynamics are especially important when it comes to the design of traction control systems (Massaro et al. 2011a, 2011b, Corno and Savaresi 2010): three simple models are described below to highlight the physical characteristics of the system.

First we consider a very simple model of the powertrain which does not account for either the tyre or the sprocket absorber flexibilities; see Figure 1.21(a). Note that the sprocket absorber is a device usually placed between the rear wheel sprocket and the rim, with the aim of damping the torsional vibration of the transmission system. This simple model is presented because it is widespread.

The engine torque T is applied to the crankshaft, it is transmitted through the gearbox (according to the selected gear ratio) to the output shaft (whose spin rate is β in Figure 1.21a), it passes through the chain to the rear wheel rim (whose angular velocity is ω) and then to the contact point. The external forces acting on the model are the tyre longitudinal force F_x and the tyre normal load F_z, but rolling resistance is neglected for simplicity. The equation of motion reads:

$$(I_w + I_t)\dot{\omega} = \tau T - R\, F_x \tag{1.55}$$

where I_w is the rear wheel spin inertia, I_t is the transmission inertia reduced to the rear wheel, ω is the spin rate of the wheel rim, τ is the whole transmission ratio, T the engine torque at the crankshaft, R the longitudinal force arm (assumed equal to the rolling radius) and F_x the longitudinal force.

In more detail, the transmission inertia reduced to the rear wheel is computed from the engine spin inertia I_e (plus clutch, starter, etc.), the gearbox primary shaft spin inertia I_p and the gearbox output shaft spin inertia I_o, given the primary ratio τ_p (between the crankshaft and the primary shaft of the gearbox), the gear ratio τ_g (between the primary and the output shaft of the gearbox thus depending on the selected gear) and the final ratio τ_f (between the output shaft and the rear wheel):

$$I_t = ((I_e \tau_p^2 + I_p)\tau_g^2 + I_o)\tau_f^2 \tag{1.56}$$

Moreover the product of the primary ratio, the gearbox ratio and the final ratio is defined as the whole transmission ratio:

$$\tau = \tau_p \tau_g \tau_f \tag{1.57}$$

and represents the ratio between the engine spin rate and the rear wheel spin rate.

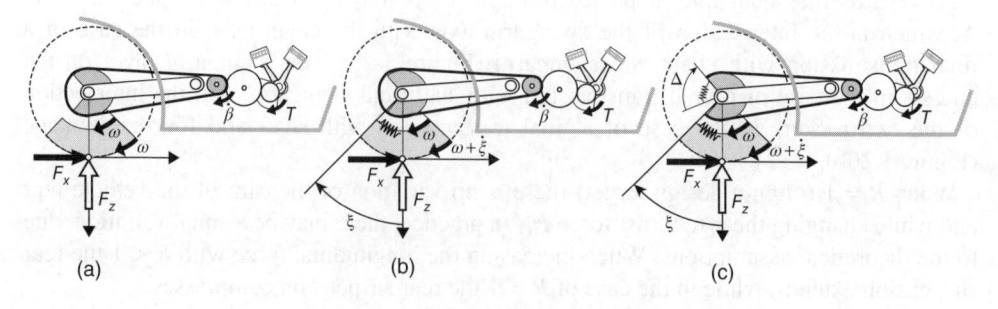

Figure 1.21 Half-vehicle model

For computation of the longitudinal road–tyre force, the full non linear formula is linearized about the steady-state condition (i.e. steady-state longitudinal slip κ_{ss} and steady-state longitudinal force $F_{x,ss}$), thus giving the following relationship between the actual force $F_{x,\kappa}$ and the slip κ:

$$F_{x,\kappa} = F_{x,ss} + k_\kappa(\kappa - \kappa_{ss})F_z \quad (1.58)$$

where k_κ is the slope of the longitudinal slip curve at the linearization point (also called the normalized longitudinal slip stiffness at the linearization point) and F_z is the tyre normal load. The longitudinal slip is defined as:

$$\kappa = \frac{\omega R - V_x}{V_x} \quad (1.59)$$

The model equation can be written in state space formulation:

$$\begin{cases} \dot{\mathbf{x}} = \mathbf{A}\mathbf{x} + \mathbf{B}T \\ \kappa = \mathbf{C}\mathbf{x} \end{cases} \quad (1.60)$$

and the transfer function between engine torque and longitudinal slip can be expressed as:

$$H(s) = \frac{K}{T}(s) = \mathbf{C}(s\mathbf{I} - \mathbf{A})^{-1}\mathbf{B} \quad (1.61)$$

where $\mathbf{I}$ is the identity matrix, s the Laplace variable and:

$$\mathbf{A} = \left[-\frac{R^2 k_\kappa F_z}{(I_w + I_t)V_x}\right] \quad \mathbf{B} = \left[\frac{\tau}{I_w + I_t}\right] \quad \mathbf{C} = \left[\frac{R}{V_x}\right] \quad \mathbf{x} = [\omega]. \quad (1.62)$$

Therefore the system has one pole at:

$$p = -\frac{R^2 k_\kappa F_z}{(I_w + I_t)V_x}. \quad (1.63)$$

Since all the parameters of Equation 1.63 are always positive but k_κ, the plant stability (i.e., the sign of the pole) is bound to the sign of the normalized longitudinal slip stiffness k_κ. In particular, the system is unstable when the slip stiffness k_κ is negative, and this usually happens only for high values of slip (skidding condition), after the peak of the force–slip curve, which usually occurs for slip values in the range 0.1–0.2. Finally, it is worth highlighting that the plant dynamic is very fast at low speeds, $p(V \to 0) = \infty$, and very slow at high speeds, $p(V \to \infty) = 0$.

As a second step, the tyre circumferential compliance is also considered; see Figure 1.21(b). With respect to the previous model, now the rim angular velocity differs from the tyre circumferential angular velocity, because of the deflection ξ. As a consequence, the longitudinal slip expression changes to

$$\kappa_i = \frac{(\omega + \dot{\xi})R - V_x}{V_x}. \quad (1.64)$$

In practice, when it comes to road tests the tyre circumferential deflection is not considered and the slip is computed according to Equation 1.59. For this reason, it is common to refer to Equation 1.59 as practical slip, since this is the slip which is measured in practice, and to Equation 1.64 as instantaneous slip (Lot 2004), since this is the actual slip at the contact point, which is the physical reason of the tyre longitudinal force. In other words, the instantaneous slip generates the longitudinal force, while the practical slip is what is observed. Therefore, the instantaneous slip κ_i is used to compute the road–tyre force while the practical slip κ is observed when computing the engine-to-slip transfer function Equation 1.61 and when comparing numerical results with road tests. The physical effect of the flexibility is to generate a phase lag between the practical slip and the actual force. Indeed, the actual force is in phase with the instantaneous slip.

A spring–damper element is used to take into account the flexibility of the tyre, and in particular the following expression relates the tyre circumferential deflection ξ and deflection rate $\dot{\xi}$ to the tyre longitudinal force F_x:

$$F_{x,el} = -(k_\xi \xi + c_\xi \dot{\xi})R \tag{1.65}$$

At the contact point, there is force equilibrium between the force due to the elastic deflection $F_{x,el}$ and the force due to the slippage $F_{x,\kappa}$:

$$F_{x,el} = F_{x,\kappa} \tag{1.66}$$

The system now has two equations (1.56 and 1.66) and two state variables (ω and ξ). The following state space matrices are found:

$$\mathbf{A} = \begin{bmatrix} -\dfrac{R^2 k_\kappa N c_\xi}{(I_w + I_t)(k_\kappa N + c_\xi V)} & \dfrac{R^2 k_\kappa N k_\xi}{(I_w + I_t)(k_\kappa N + c_\xi V)} \\ \dfrac{k_\kappa N}{k_\kappa N + c_\xi V} & -\dfrac{V k_\xi}{k_\kappa N + c_\xi V} \end{bmatrix}$$

$$\mathbf{B} = \begin{bmatrix} \dfrac{\tau}{I_w + I_t} \\ 0 \end{bmatrix} \quad \mathbf{C} = \begin{bmatrix} \dfrac{R}{V} & 0 \end{bmatrix} \quad \mathbf{x} = \begin{bmatrix} \omega \\ \xi \end{bmatrix} \tag{1.67}$$

When inspecting the engine-to-slip transfer function (Equation 1.61), it turns out that at null longitudinal speed the system is vibrating with undamped natural frequency $f_{1,2}$ and damping ratio $\zeta_{1,2}$:

$$f_{1,2} = \frac{1}{2\pi} R \sqrt{\frac{k_\xi}{I_w + I_t}} \quad \zeta_{1,2} = \frac{c_\xi R}{2\sqrt{k_\xi(I_w + I_t)}} \tag{1.68}$$

As the speed increases, the frequency reduces and the damping increases up to a critical velocity (usually in the range 30–60 m/s):

$$V_{cr} = \frac{(c_\xi R + 2\sqrt{k_\xi(I_w + I_t)})F_z k_\kappa R}{k_\xi(I_w + I_t)} \tag{1.69}$$

Above this the system is no longer vibrating (the two poles turn from complex conjugate pairs to real poles).

It is worth noting that there is an alternative approach to account for this tyre force lag. Instead of considering the tyre flexibility, it is possible to add a first-order differential equation (*relaxation equation*, see Section 1.2.3) which replaces Equation 1.66 :

$$\frac{\sigma_\kappa}{V_x}\dot{F}_{x,\kappa} + F_{x,\kappa} = F^0_{x,\kappa} \tag{1.70}$$

where σ_κ is the relaxation length, V_x is the longitudinal velocity, $F_{x,\kappa}$ the actual longitudinal force, $F^0_{x,\kappa}$ the longitudinal force computed with the practical slip of Equation 1.59. The two approaches give similar results when

$$\sigma_\kappa = \frac{k_\kappa F_z}{k_\xi} \tag{1.71}$$

As a third step, a flexible sprocket absorber is introduced between the rear wheel chain sprocket and the rear wheel rim, in addition to the compliant tyre; see Figure 1.21(c). The following expression is used to compute the absorber torque T_a as a function of its deflection Δ:

$$T_a = k_a \Delta + c_a \underline{\Delta} \tag{1.72}$$

where k_a is the absorber stiffness, c_a the damping coefficient and $\underline{\Delta}$ the absorber deflection rate. The state space matrices now read:

$$\mathbf{A} = \begin{bmatrix} -\dfrac{R^2 k_\kappa N c_\xi}{I_w(k_\kappa N + c_\xi V)} & \dfrac{R^2 k_\kappa N k_\xi}{I_w(k_\kappa N + c_\xi V)} & \dfrac{k_a}{I_w} & \dfrac{c_a}{I_w} \\ -\dfrac{k_\kappa N}{k_\kappa N + c_\xi V} & -\dfrac{V k_\xi}{k_\kappa N + c_\xi V} & 0 & 0 \\ 0 & 0 & 0 & 1 \\ \dfrac{R^2 k_\kappa N c_\xi}{I_w(k_\kappa N + c_\xi V)} & -\dfrac{R^2 k_\kappa N k_\xi}{I_w(k_\kappa N + c_\xi V)} & -k_a\left(\dfrac{1}{I_w}+\dfrac{1}{I_t}\right) & -c_a\left(\dfrac{1}{I_w}+\dfrac{1}{I_t}\right) \end{bmatrix}$$

$$\mathbf{B} = \begin{bmatrix} 0 \\ 0 \\ 0 \\ \dfrac{\tau}{I_t} \end{bmatrix} \quad \mathbf{C} = \begin{bmatrix} \dfrac{R}{V} & 0 & 0 & 0 \end{bmatrix} \quad \mathbf{x} = \begin{bmatrix} \omega & \xi & \Delta & \underline{\Delta} \end{bmatrix}^T \tag{1.73}$$

No compact expressions for poles are available, but they can be easily computed numerically from Equation 1.73. The four complex poles of the system are associated with two torsional vibrating modes, which may be either identified in the tyre circumferential and sprocket absorber deflection, or in the wheel and transmission spin.

Finally, it should be noted that when the engine-to-slip dynamics are of interest for frequencies above 30 Hz, the tyre belt dynamics should also be added to the model (thus increasing the number of state variables above four).

1.4.4 Chatter

The chatter of motorcycles is a vibration phenomenon which appears during braking and consists of a vibration of the rear and front unsprung masses with frequency in the range

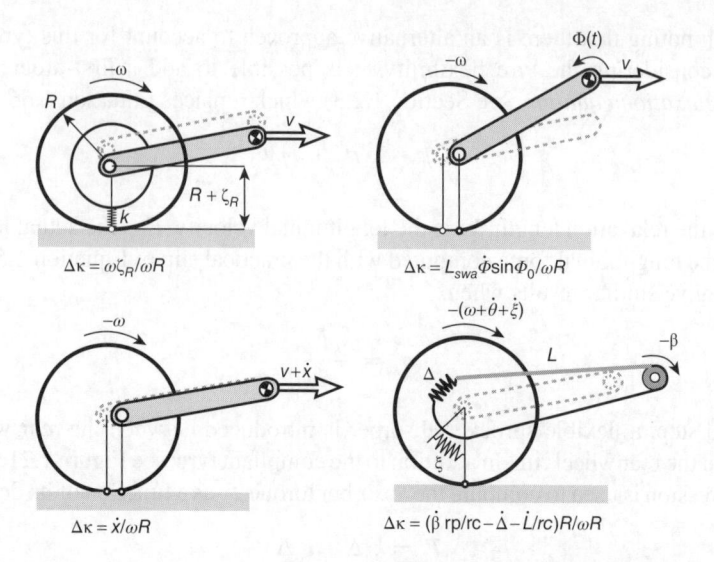

Figure 1.22 Mechanisms generating longitudinal fluctuating slips

17–22 Hz, depending on vehicle characteristics. This vibration could be very strong and acceleration of the unsprung masses can reach 5–10 g (Cossalter et al. 2012, 2008).

A physical explanation of the phenomenon is as follows. The braking manoeuvre may be seen as a composition of two motions: the non-vibrating braking gross motion (i.e. the sequence of equilibrium positions at chosen deceleration and different speeds), and the vibrating motion (i.e. the vehicle oscillation around these equilibrium positions). During braking, the kinetic energy of the gross motion is lost, but if there is an unstable mode part of this energy which is transferred to the in-plane dynamics, so the chatter appears. In practice, when riders start braking there is a transient and the vehicle start vibrating around the equilibrium position. These oscillations lead to a fluctuation of the rear longitudinal slip, i.e. to a variation of the longitudinal force which may drive energy into the system, depending on the phase lag between the rear tyre longitudinal force and fluctuations of the contact point position. The longitudinal slip fluctuations may be grouped into four main origins; see Figure 1.22: a) the radial deflection of the rear tyre ζ_R, b) the fluctuation of the swingarm rotation Φ, c) the fluctuation of the longitudinal speed $\dot{x}$ and d) the fluctuation of the sprocket absorber deflection Δ.

The tyre torsion flexibility and the sprocket absorber flexibility are essential factors to capture the chatter instability, and the chatter vibration depends on the braking style (only front brake, front brake plus rear brake, with or without engine braking, etc.). In more detail, the powertrain flexibility is related to a vibration mode (*transmission* mode) which involves the powertrain inertia and may become unstable under certain motion condition (e.g., while braking with certain deceleration at certain speeds). When inspecting the shape of this vibration mode, components related to the tyre vertical load fluctuation are found, which explain the chattering behaviour.

It is worth stressing that this instability can appear on a perfectly flat road and with perfectly balanced wheels. However, it is expected that both road unevenness and wheel imbalance can further excite the vibration. In particular, since the wheel spin frequency f_w is:

$$f_w = \frac{V_x}{2\pi R} \tag{1.74}$$

with V_x is the vehicle speed and R the rolling radius, the typical speed range where the road and/or the wheels may further excite the *transmission* mode is 32–42 m/s (115–150 km/h).

1.5 Out-of-Plane Dynamics

Lateral dynamics involve the lateral displacement of the vehicle, yaw, roll and steer angles. In straight running, these dynamics are decoupled from in-plane dynamics, whereas while cornering strong interactions occur. The out-of-plane behaviour has been extensively investigated over the years because it is related to the vehicle stability and safety (Cossalter 2006, 2011b).

In this section, first the two-wheeled steering static response is presented, then the most important vibration modes are discussed.

1.5.1 Roll Equilibrium

The equilibrium roll angle ϕ can be estimated from the curve radius R_c and the vehicle speed V_x as

$$\phi = \arctan \frac{V_x^2}{gR_c}, \tag{1.75}$$

where g is the gravitational acceleration, when considering that the resultant of the centrifugal force and the weight force passes through the line joining the two road–tyre contact points, under the assumption of think-disk tyre, null steering angle and negligible tyre/engine gyroscopic effects. When including the effect of the tyre cross-section, Equation 1.75 changes to (Cossalter 2006):

$$\phi = \arctan \frac{V^2}{gR_c} + \Delta\phi$$
$$\Delta\phi = \arcsin \frac{t \cdot \sin\left(\arctan \frac{V^2}{gR_c}\right)}{h-t} \tag{1.76}$$

which highlights that the actual roll angle increases as the tyre cross-section increases (wide tyre) and the centre of gravity lowers, with t the tyre cross-section and h the height of the centre of gravity; see Figure 1.23.

1.5.2 Motorcycle Countersteering

The lateral dynamics of single-track vehicles is mainly controlled by rider's steering action, even though the rider's body movements can give additional contribution. There is a significant difference when comparing the rider's steering action on a two-wheeled vehicle with that of a driver in a four-wheeled vehicle. In a four-wheeled vehicle the driver turns the steering wheel right (clockwise) to enter a right turn by applying a clockwise steering

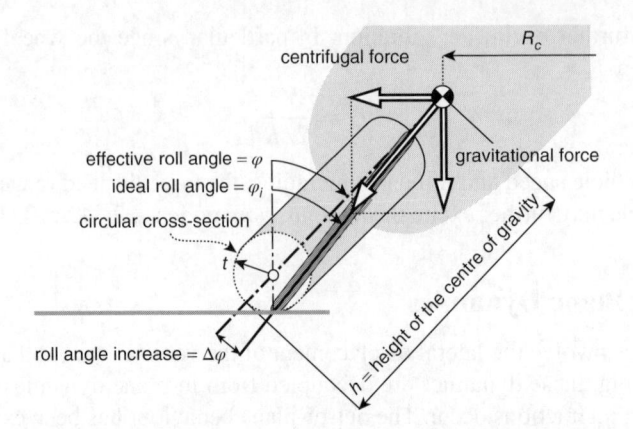

Figure 1.23 Roll equilibrium

torque, and keeps a clockwise steering torque and steering angle all through the turn; however, on a two-wheeled vehicle a completely different approach is used. In order to enter a right turn, the rider has to steer the handlebar angle left (counter-steering manoeuvre) by applying a counter-clockwise steering torque to the handlebar. As a consequence both a centrifugal force on the vehicle and a front tyre lateral force are generated, thus the vehicle leans right into the turn. At this moment, the rider can steer the handlebar into the turn. The final steering torque may be either clockwise or counter-clockwise, depending the vehicle characteristics. The preferred behaviour is to have a counter-clockwise steering torque and a clockwise steering angle when riding a clock wise turn, and motorcycle engineers used to adjust some vehicle parameters to obtain the desired steering torque behaviour.

It is worth noting that the counter-steering manoeuvre can be avoided by using body movement to enter the turn (Bertolazzi et al. 2007). However, steering torque is rather more efficient than body movement torque. Skilled riders take advantage of both mechanisms, while most of the everyday riders ignore the counter-steering approach.

The steering torque applied by the rider to the handlebar is the reaction to the many force contributions (Cossalter et al. 2010). When considering the equilibrium of the front frame in steady-state condition, the rider's steering torque reacts to the forces and torques depicted in Figure 1.24: the tyre longitudinal force, lateral force, vertical load, tyre rolling torque, yawing torque, weight and centrifugal force of the front assembly, and the gyroscopic torque. These contributions can be divided into aligning components and misaligning components. When the aligning components prevail, the rider's handlebar torque is inward to the turn, that is the rider pulls the handlebar with the hand inside the turn. But when the misaligning contributions prevail, the rider's handlebar torque is outward to the turn, that is the rider pushes the handlebar with the hand inside the turn. The tyre lateral force, tyre rolling torque, front frame centrifugal force and wheel gyroscopic effect give self-aligning torques around the steering axis, whereas the tyre longitudinal (braking) force, tyre normal force, tyre yaw torque and front frame weight force give misaligning torques. It is worth noting that in the unusual case where the front frame centre of gravity is behind the steering axis the centrifugal contribution becomes misaligning and the weight contribution aligning. Note that when braking with

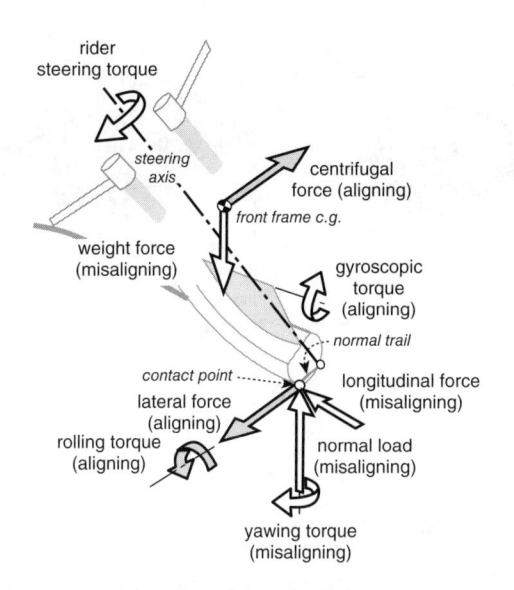

Figure 1.24 Forces and torques acting on the front frame

the front tyre, a significant misaligning component arises, which adds to that related to the longitudinal friction.

The main contributions to the whole steering torque are those related to the tyre normal load and to the tyre lateral force, which almost balance each other. As a consequence of the tyre carcass hysteresis, a rolling torque opposes the wheel rotation and a friction force arises. The longitudinal friction force on the front tyre leads to a small misaligning contribution to the steering torque, whose presence is due to the fact that the tyre carcass is not a thin disk, and therefore the longitudinal force has a non-null component when the vehicle is leaned because of the migration of the contact point on the tyre carcass. The tyre rolling torque also gives an (aligning) contribution to the steering torque related to the fact that when the vehicle is leaned, the rolling torque projection on the steering axis is no longer null. Another important contribution is that related to the tyre yaw torque. This contribution to the steering torque is always misaligning and significant. The gyroscopic contribution (always aligning) to the steering torque arises because the front wheel is both rolling about its spin axis and yawing as a consequence of the cornering manoeuvre. Finally, minor components are those related to the front frame weight (misaligning), and to the front frame centrifugal effect (aligning).

1.5.2.1 Understeering and Oversteering

The steering behaviour of the vehicle depends on many parameters, and in particular on the tyre properties. Indeed, when considering tyres slippage, the effective steering angle becomes (Figure 1.25):

$$\Delta^* = \Delta + \lambda_R + \lambda_F \tag{1.77}$$

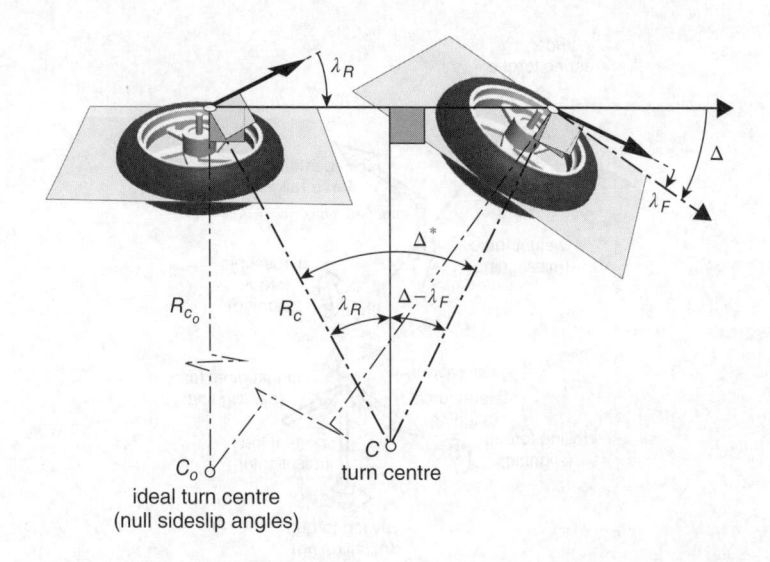

Figure 1.25 Kinematic steering angle Δ and effective steering angle Δ^*

with Δ from Equation 1.2, λ_R the rear tyre sideslip angle and λ_F the front tyre sideslip angle. In practice, the effective steering angle Δ^* is equal to the kinematic steering angle Δ only when both tyres have the same sideslip, that is $\lambda_R = \lambda_F$ (or when there is no sideslip, which is a reasonable assumption at very low speeds). Otherwise, it is smaller or larger, giving an understeering or oversteering behaviour respectively.

The steering behaviour can be expressed by means of the steering ratio ξ:

$$\xi = \frac{R_{c0}}{R_c} = \frac{\Delta^*}{\Delta} \approx 1 + \frac{\lambda_r - \lambda_f}{\Delta} \qquad (1.78)$$

where R_{c0} is the kinematic radius of curvature and R_c is the actual radius of curvature.

The vehicle's steering behaviour can be defined as follows:

- neutral: $\xi = 1 \iff \lambda_R = \lambda_F$;
- oversteering: $\xi > 1 \iff \lambda_R > \lambda_F$;
- understeering: $\xi < 1 \iff \lambda_R < \lambda_F$.

In case of oversteering vehicles, there may be a critical speed V_{cr} where $\Delta = 0$ and therefore $\xi \to \infty$. Above this speed the vehicle is ridden in counter-steering, i.e. the rider has to keep the handlebar turned outside the turn (as in speedway races).

Unlike in four-wheeled vehicles, considerable camber angles (up to 50–60°) are present while cornering with two-wheeled vehicles. Therefore the steering ratio is also significantly affected by the characteristics of the lateral force as a function of camber (in addition to the characteristics of the force as a function of lateral slip). This can be effectively highlighted when expressing the steering ratio for a simple vehicle model under the assumption of small roll angles and linear tyre behaviour:

$$F_{y,R} = (k_{\lambda,R}\lambda_R + k_{\phi,R}\phi_F)F_{z,R} \quad F_{y,F} = (k_{\lambda,F}\lambda_F + k_{\phi,F}\phi_F)F_{z,F} \qquad (1.79)$$

where $F_{y,R}$ and $F_{y,F}$ are the rear and front tyre lateral forces respectively, $k_{\lambda,R}$ and $k_{\lambda,F}$ are the rear and front sideslip stiffnesses, $k_{\phi,R}$ and $k_{\phi,F}$ are the rear and front tyre camber stiffnesses, λ_R and λ_F are the rear and front sideslip angles ϕ_R and ϕ_F are the rear and front tyre camber angles and $F_{z,R}$ and $F_{z,F}$ are the rear and front tyre normal loads. Solving Equation 1.79 for the sideslip angles gives:

$$\lambda_R = \frac{1}{k_{\lambda_R}}\left(\frac{F_{y,R}}{F_{z,R}} - k_{\varphi_R}\varphi_R\right) \quad \lambda_F = \frac{1}{k_{\lambda_F}}\left(\frac{F_{y,F}}{F_{z,F}} - k_{\varphi_F}\varphi_F\right) \quad (1.80)$$

In steady turning, under the assumption of small camber angles, we have

$$\frac{F_{y,R}}{F_{z,R}} \approx \phi_R \approx \phi_F \approx \frac{F_{y,F}}{F_{z,F}} \quad (1.81)$$

which can be used in Equation 1.80 to give

$$\lambda_R \approx \frac{1-k_{\varphi_R}}{k_{\lambda_R}}\varphi \quad \lambda_F \approx \frac{1-k_{\varphi_F}}{k_{\lambda_F}}\varphi \quad (1.82)$$

When using Equation 1.82 in Equation 1.78, the steering ratio becomes

$$\xi = \frac{1}{1 - \left(\frac{1-k_{\varphi_R}}{k_{\lambda_R}} - \frac{1-k_{\varphi_F}}{k_{\lambda_F}}\right)\frac{V_x^2}{gR}} \quad (1.83)$$

which shows that it is possible to mitigate the oversteering behaviour by increasing the rear tyre camber stiffness and reducing the front tyre camber stiffness, as well as increasing the rear tyre sideslip stiffness and reducing the front tyre sideslip stiffness.

Finally, note that other authors define the steering behaviour differently, using the variation of the steering ratio (as a function of the vehicle speed V_x while travelling on a curve with constant radius R_c) rather than its absolute value:

- neutral: $\left(\frac{\partial \xi}{\partial V}\right)_{R_c} = 1$;
- oversteering: $\left(\frac{\partial \xi}{\partial V}\right)_{R_c} > 1$;
- understeering: $\left(\frac{\partial \xi}{\partial V}\right)_{R_c} < 1$.

With this latter definition, there may be oversteering also with $\lambda_R < \lambda_F$, where the previous definition defines understeering. In practice, we define oversteering as a condition where, on a steady turn, the rear tyre sideslip increases more than the front sideslip when increasing the vehicle speed.

1.5.3 Weave, Wobble and Capsize

In straight motion out-of-plane dynamics involves four of the eleven DOFs necessary to fully define the vehicle trim (see Section 1.1): the lateral displacement of the vehicle, the yaw, roll and steer angles. These DOFs combine to give three well-known vibration modes: *capsize*, *weave* and *wobble*.

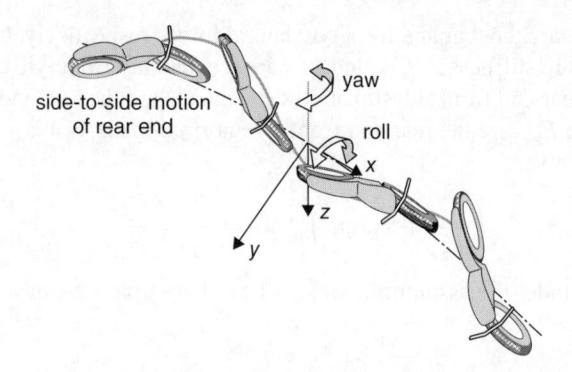

Figure 1.26 Weave mode

Capsize is a non-vibrating mode which mainly consists of a roll motion combined with a lateral displacement (plus some less important steering and yaw movements). Depending on the vehicle characteristics it can be stable at low speeds, then slightly unstable above a critical speed, or slightly unstable over the whole speed range. In any case, this mode is easily controlled by rider.

Weave is an oscillation of the entire motorcycle (Figure 1.26), with frequency rising with speed from 0 up to 3–4 Hz depending on the vehicle characteristics. It is usually unstable for speeds up to 6–7 m/s, then it is stable and tends to be poorly damped at high speeds (>100 km/h or 28 m/s). High speed instability may occur both hands-off or hands-on the handlebars, with the former having the potential to be more stable than the latter under the same set of running conditions (Massaro et al. 2012). More precisely, at zero speed the weave consists of two non-vibrating modes: *body capsize* and *steering capsize*. These two modes coalesce to generate the vibrating weave at a speed in the range 0–1 m/s.

Body capsize is a capsize of the whole vehicle an inverted pendulum-like roll instability. Its time constant τ_{bc} can be estimated with a simple inverted pendulum model:

$$\tau_{bc} = \sqrt{\frac{I_x + Mh^2}{Mgh}} \tag{1.84}$$

where I_x is the whole vehicle moment of inertia around the centre of mass, M is the whole vehicle mass, h is the height of the vehicle centre of mass and g is the gravitational acceleration.

Steering capsize is a steering instability, due to the misaligning effects of both the front frame mass and the front tyre normal load. Its time constant can be estimated with a simple front frame model:

$$\tau_{sc} = \sqrt{\frac{(M_f g b_f + F_{z,f} a_n) \sin \varepsilon}{I_f}} \tag{1.85}$$

where M_f is the front frame mass, b_f is the distance of the front frame centre of mass from the steering axis, $F_{z,f}$ is the front tyre normal load, a_n is the normal trail and ε is the caster angle.

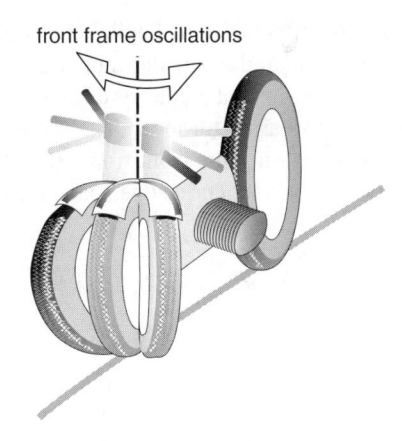

Figure 1.27 Wobble mode

Finally, *wobble* is a vibrational mode dominated by the oscillation of the front steering assembly around the steering axis (Figure 1.27) in the range 6–10 Hz. In practice, wobble mode is usually visible with hands off the handlebars (Cossalter et al. 2007b) and, depending on front assembly structural compliance and wheel imbalance, may become perceptible to the rider at low to medium speeds.

Summarizing, the typical stability behaviour of a motorcycle at constant speed in straight line motion is the following. At standstill, the vehicle is unstable because of body capsize and steering capsize. Indeed, riders put their feet on the road to prevent the vehicle from falling over. For speeds up to 6–7 m/s the motorcycle is unstable because of weave mode and therefore a rider action is needed to stabilize the vehicle. Above 6–7 m/s the vehicle enters a self-stabilizing zone, so no rider action is necessary to stabilize the system (the two-wheeled vehicle also keeps upright with the rider's hands off the handlebars). At increasing speeds, the capsize may become slightly unstable (with hands off the handlebar), but this instability is easily stabilized by rider's passive action on the handlebars. If the vehicle has been properly designed, there should not be any weave or wobble instability in normal riding conditions. However, high speed and cargo loading promote poorly damped weave while wobble may be triggered by wheel imbalance and certain road surfaces. Moreover, these two vibration modes are dangerous because riders may have difficulty in controlling them. In addition, most of the parameters improving the stability of one, worsen the other. This is why their behaviour must be carefully considered when designing the vehicle.

1.5.3.1 Rigid Bodies Model

In this section the results of a basic model made of rigid bodies are presented to discuss the lateral vibration modes of the vehicle (*www.multibody.net*). Since the aim is to study the stability in straight motion at constant speed, the presence of the suspensions can be neglected and only four DOFs are required: lateral displacement y, yaw angle ψ, roll angle ϕ and steering angle δ. Moreover, the tyre properties can be linearized, since only small lateral slips and the roll angle will be employed by a vehicle vibrating around the upright trim. Tyre pure rolling can be assumed in the longitudinal direction, since there is no interest in the longitudinal dynamics. The effect of tyre flexibility (and the related force lag, see Section 1.2.3) is accounted for by including in the model the front and rear tyre lateral deflection ζ_R and ζ_F. Note that the non-instantaneous tyre behaviour is essential for capturing wobble. The only rider input is the steering torque τ. The model has four rigid bodies: the rear frame (including rider and swingarm), the front frame (handlebar and suspension) and wheels. Its equations of motion can be written in state space form:

$$\dot{\mathbf{x}} = \mathbf{Ax} + \mathbf{Bu}$$
$$\mathbf{x} = \{y, \psi, \underline{\psi}, \phi, \underline{\phi}, \delta, \underline{\delta}, \zeta_R, \zeta_F\}^T \quad (1.86)$$
$$\mathbf{u} = \{\tau\}$$

where $\underline{y}, \underline{\psi}, \underline{\phi}, \underline{\delta}$ represent the derivatives of the corresponding variables.

The stability is analysed by computing the eigenvalues of $\mathbf{A}$ in Equation 1.86 at different speeds, see the dotted lines of Figure 1.28. The hands-off vehicle stability reads as follows. Up to a speed of 0.6 m/s the vehicle is unstable because of both body capsize (real positive eigenvalue with time constant 0.30 s at 0.1 m/s) and steering capsize (real positive eigenvalue with time constant 0.17 s at 0.1 m/s). At 0.6 m/s the two unstable modes coalesce to give the unstable weave (complex conjugate pair eigenvalues), whose frequency rises from 0 at 0.6 m/s to 0.4 Hz at 7 m/s. From here on the weave is stable and its frequency rises with speed (3 Hz at 40 m/s). Note that at high speeds the mode moves towards the instability area. Wobble has a frequency in the range 7–8 Hz and is unstable for speeds higher than 22 m/s. Capsize is stable in the whole speed range. There is also another mode, usually called rear wobble or weave 2: it is vibrating for speeds up to 25 m/s, then it splits into two non-vibrating modes. This mode is not very interesting because it is very stable over the whole speed range, and therefore is not shown in Figure 1.28.

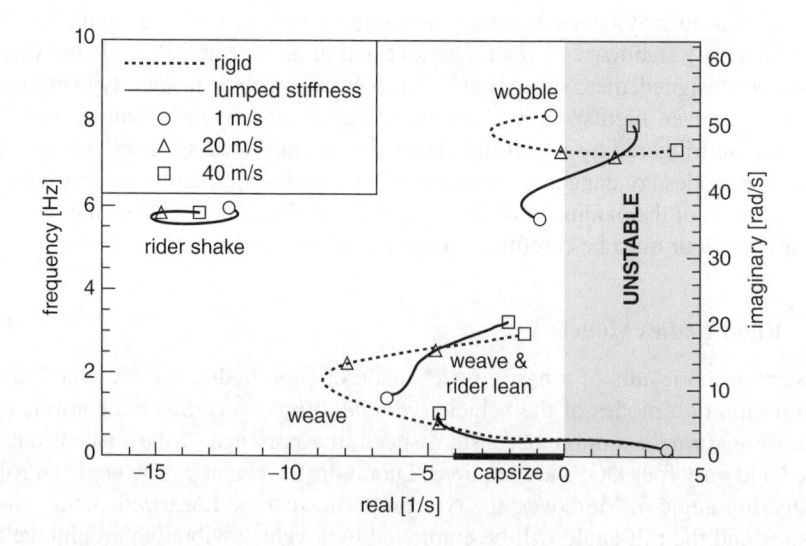

Figure 1.28 Example of motorcycle root-locus as a function of speed

1.5.3.2 Effect of the Frame Compliance

The frame compliance and rider passive mobility can significantly affect the stability of two-wheeled vehicles (Cossalter et al. 2007b) includes an experimental validation). In particular, the most important structural flexibility is that of the front assembly with respect to the rear assembly, while the rider passive roll vibration is the most important rider passive motion affecting stability.

The model used in the previous section is extended to include the main vehicle compliance (the equations and state matrices are reported at *www.multibody.net*). In particular, several DOFs are added to account for: swingarm torsion and bending flexibility (α_R and β_R), front frame bending and torsion flexibility (α_F and β_F) and rider lateral and roll passive motion on the saddle (y_p and ϕ_p). The motion along these new DOFs is restrained by means of spring–damper elements tuned to replicate the real vehicle compliance properties. The additional DOFs are related to the same number of complex conjugate pairs representing the structural modes of vibration. Even more important, these DOFs enter the weave and wobble which affects their stability.

In particular, the front frame bending strongly affects the *wobble* stability (Figure 1.29). The rigid vehicle model predicts a stable wobble at low speeds (up to 22 m/s in the vehicle considered here) and unstable wobble at high speeds, but when including the front flexibility the model predicts an instable vibration mode at low speeds and a stable mode at high speeds. In other words, the front frame flexibility reverses the behaviour of wobble stability as a function of speed. The physical reason of the stabilization effect is the additional gyroscopic effect generated by the deflection rate combining with the wheel spin inertia and spin rate (Cossalter et al. 2007b).

The effect of swingarm bending and torsion flexibility on weave stability is depicted in Figures 1.30 and 1.31 respectively. In the vehicle analysed, the bending has a negligible to positive effect on high speed weave stability (again the stabilizing effect is related to the additional gyroscopic effect induced by bending deflection), while the torsion flexibility worsens the high speed weave stability.

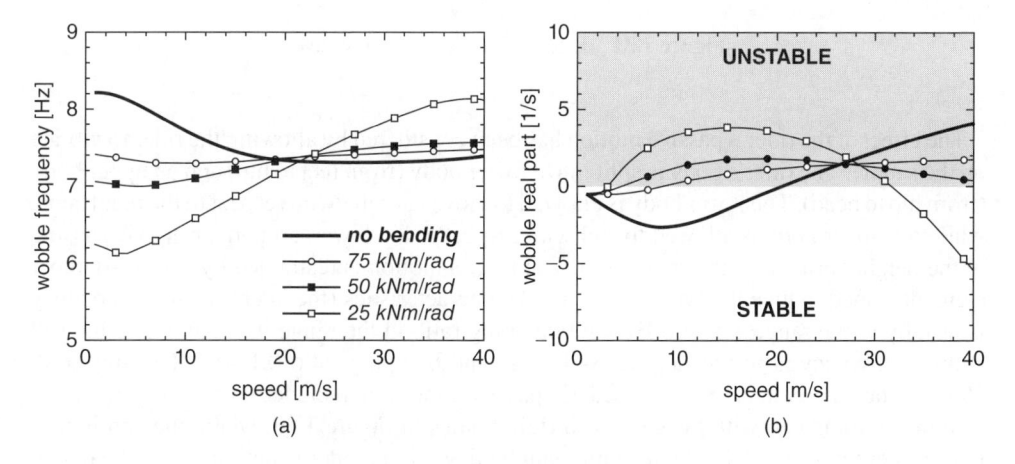

Figure 1.29 Effect of front bending on wobble

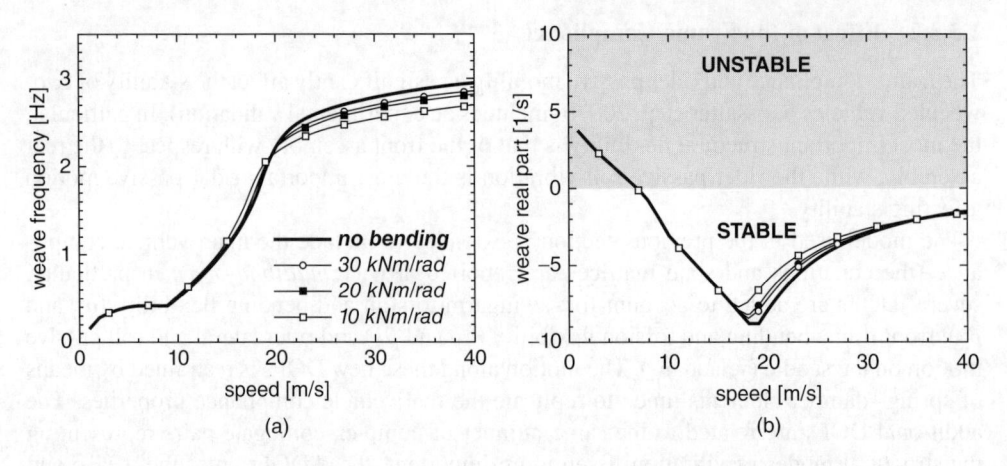

Figure 1.30 Effect of rear bending on weave

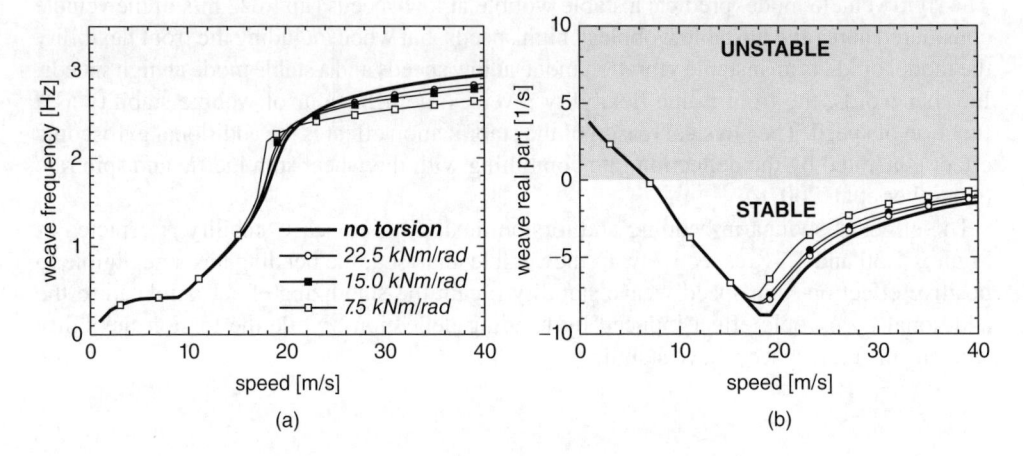

Figure 1.31 Effect of rear torsion on weave

The effect of the rider's passive motion has been considered by allowing the rider to vibrate on the saddle. The rider's body is split into a lower body (from feet to hip) and an upper body (from hip to head). The lower body is allowed to move laterally with respect to the rear frame, while the upper body is allowed to roll with respect to the lower, around a roll axis passing in the neighbourhood of the rider's hip. The rider's motion is restrained by spring–damper elements tuned to give the typical rider modal characteristics (the lateral natural frequency is usually in the range 3.5–4.0 Hz with damping ratio in the range 0.3–0.6, while the roll natural frequency is in the range 0.8–1.5 Hz and damping ratio 0.1–0.3 (Nishimi et al. 1985, Katayama et al. 1987). The rider's passive motion is included in the model, and its stability is compared with the case with rigid bodies in Figure 1.28, where the continuous lines represent the model with the compliant bodies and the rider's motion, while the dotted lines represent the model with rigid bodies. The rider's lateral motion is related to a new vibration mode named *rider shake* while the roll motion deeply merges with the weave, whose curve is now split. In practice, the rider's motion stabilizes the wobble at low speed and significantly stabilizes the high speed weave, since the effect of roll on weave is the more important contribution.

When accelerating, the weave mode tends to increase its frequency while that of the wobble reduces, and a coupling between them may also take place. When braking, the wobble mode increases its frequency and usually reduces its damping while the weave reduces its frequency and damping. It is worth stressing that all the considerations reported in this section refers to the free vehicle stability, that is, they remain valid when the rider has hands off the handlebar or gently grasping the handlebar (Massaro et al. 2012).

1.5.3.3 Effect of the Rider Impedance

In practice, the rider has their hands on the handlebar while riding the vehicle. This may change the two-wheeled vehicle stability because of the loop created between the rider's body, the rear frame (where the rider sits) and the front frame (where the rider's hands are). The effect of such rider passive steering impedance on vehicle stability has only recently been investigated (Sharp and Limebeer 2004, Cossalter et al. 2011a, Massaro and Cole 2012, Massaro et al. 2012, with only the last including a comparison with experimental road tests).

To model this effect, the rider's upper body is allowed to yaw with respect to their lower body and the handlebar. The motion is restrained by spring–damper elements properly tuned to give the experimentally measured rider's modal properties, Figure 1.32.

The effect of rider impedance is to stabilize the wobble mode and to destabilize the high speed weave mode. The effect is similar to that of a steering damper. However, while an ideal steering damper generates a steering torque τ proportional to and in phase with the

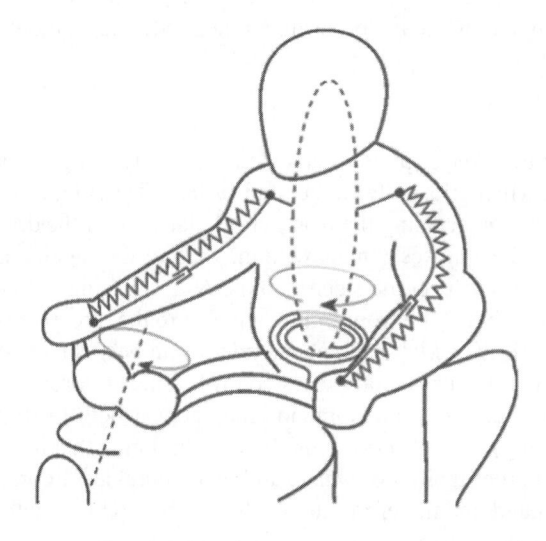

Figure 1.32 Rider impedance model

steering angle rate $\dot{\delta}$:

$$\tau = c_d \dot{\delta} \qquad (1.87)$$

the steering torque generated by the rider changes with frequency, both in magnitude and in phase. Moreover, while the steering damper generates equal and opposite torques on the front assembly and on the rear assembly, the rider does it only at low and high frequency, because of the presence of the inertia.

1.6 In-Plane and Out-of-Plane Coupled Dynamics

In the previous two sections in-plane and out-of-plane dynamics have been treated separately, using simplified models in order to highlight the most important vehicle behaviour. However there are conditions (e.g. entering a curve while braking, Massaro 2011, Massaro and Lot 2010) where it is essential to capture the coupling between in-plane and out-of-plane dynamics using full vehicle models like *FastBike* (Cossalter et al. 2011b; Cossalter et al. 2011c). Below three typical examples of coupled dynamics are reported.

Cornering Stability

When cornering, the in-plane and out-of-plane dynamics couple (Cossalter et al. 2004) and all the vehicle vibration modes include both in-plane and out-of-plane DOFs. The interactions may be predicted from straight line motion stability, when in-plane and out-of-plane modes have eigenvalues close to each other in the root locus. A typical coupling is between *bounce/pitch* and *weave*: in this condition also the set-up of the suspension can be used to control the cornering oscillations. Similarly, *wobble* may combine with wheel *front hop* mode, to give a steering oscillation combined with the suspension oscillation.

It is also important to highlight that, while in straight line motion the in-plane modes are almost speed independent, while in cornering all the modes depend on the speed of travel.

Kick Back

Road undulations or transverse joints, such as on motorway bridges, may unload the front wheel of the vehicle, which lifts up from the road surface. The rider usually reacts automatically with a steering action bringing the front wheel plane out of the driving direction of the bike. When the front wheel makes contact with the road surface again, the front frame is not in force equilibrium with respect the steering axis. As a consequence an impulsive force is generated, which 'kicks back' the front frame opposite to the direction of the steering angle (Lot and Massaro 2007). The kick back phenomenon can be so heavy that the rider cannot control the handlebar and consequently loses control of the motorcycle. The steering angle can reach very high values. This phenomenon may appear both in straight-running (especially when accelerating) or while cornering. It is worth noting that it is not related to a new vibration mode, but to the stability of weave and wobble modes. In practice, the less stable mode is the most excited and therefore the resulting vibration may either be in the wobble or in the weave frequency range.

High Side

This phenomenon is due to the interaction between the rear tyre lateral force and the longitudinal force. It can happen during a braking manoeuvre while entering a curve or during a thrusting manoeuvre while exiting from a curve (Cossalter et al. 2007a).

For example, to exit the curve the rider starts to thrust the rear wheel, therefore the longitudinal driving force increases and adds to the existing lateral force. If the total tyre friction force reaches the limit value, the rear wheel loses grip and therefore the rear of the motorcycle moves outwards from the turn. The rider reacts by reducing the throttle opening, so the thrust force reduces suddenly, and the rear tyre regains its grip. The existing large sideslip (originated by the earlier grip loss), generates a lateral force impulse that violently pushes the motorcycle upwards and may even throw the rider out of the saddle. In any case, weave mode is strongly excited by the impulse action.

References

Bertolazzi E, Biral F, Da Lio M and Cossalter V 2007 The influence of riders upper body motions on motorcycle minimum time maneuvering *Multibody Dynamics 2007, ECCOMAS Thematic Conference*.

Corno M and Savaresi S 2010 Experimental identification of engine-to-slip dynamics for traction control applications in a sport motorbike. *European Journal of Control* **16**(1), 88–108.

Cossalter V 2006 *Motorcycle Dynamics*. Lulu.

Cossalter V, Aguggiaro A, Debus D, Bellati A and Ambrogi A 2007a Real cases motorcycle and rider race data investigation: fall behavior analysis *20th International Technical Conference on Enhanced Safety of Vehicles: Innovations for Safety Opportunities and Challenges*.

Cossalter V, Doria A, Lot R and Massaro M 2011a The effect of rider's passive steering impedance on motorcycle stability: Identification and analysis. *Meccanica* **46**(2), 279–292.

Cossalter V, Lot R and Maggio F 2004 The modal analysis of a motorcycle in straight running and on a curve. *Meccanica* **39**(1), 1–16.

Cossalter V, Lot R and Massaro M 2007b The influence of frame compliance and rider mobility on the scooter stability. *Vehicle System Dynamics* **45**(4), 313–326.

Cossalter V, Lot R and Massaro M 2008 The chatter of racing motorcycles. *Vehicle System Dynamics* **46**(4), 339–353.

Cossalter V, Lot R and Massaro M 2011b An advanced multibody code for handling and stability analysis of motorcycles. *Meccanica* **46**(5), 943–958.

Cossalter V, Lot R and Massaro M 2012 The significance of powertrain characteristics on the chatter of racing motorcycles.

Cossalter V, Lot R, Massaro M and Peretto M 2010 Motorcycle steering torque decomposition, vol. 2, pp. 1257–1262.

Cossalter V, Lot R, Massaro M and Sartori R 2011c Development and validation of an advanced motorcycle riding simulator. *Proceedings of the Institution of Mechanical Engineers, Part D: Journal of Automobile Engineering* **225**(6), 705–720.

De Vries E and Pacejka H 1998 Motorcycle tyre measurements and models. *Vehicle System Dynamics* **29** (Suppl.), 280–298.

Katayama K, Aoki A, Nishimi T and Okayama T 1987 Measurement of structural properties of riders. *SAE paper 871229*.

Kiencke U and Nielsen L 2001 *Automotive Control Systems*. Springer-Verlag.

Lot R 2004 A motorcycle tire model for dynamic simulations: Theoretical and experimental aspects. *Meccanica* **39**(3), 207–220.

Lot R and Massaro M 2007 The kick-back of motorcycles: Experimental and numerical analsysis *Multibody Dynamics 2007, Eccomas Thematic Conference*.

Massaro M 2011 A nonlinear virtual rider for motorcycles. *Vehicle System Dynamics* **49**(9), 1477–1496.

Massaro M and Cole DJ 2012 Neuromuscular-steering dynamics: motorcycle riders vs car drivers.

Massaro M and Lot R 2010 A virtual rider for two-wheeled vehicles, pp. 5586–5591.

Massaro M, Lot R and Cossalter V 2011a On engine-to-slip modelling for motorcycle traction control design. *Proceedings of the Institution of Mechanical Engineers, Part D: Journal of Automobile Engineering* **225**(1), 15–27.

Massaro M, Lot, R. Cossalter V, Brendelson J and Sadaukas J 2012 Numerical and experimental investigation on the significance of rider on motorcycle weave. *Vehicle System Dynamics* **50**(S1), 215–227.

Massaro M, Sartori R and Lot R 2011b Numerical investigation of engine-to-slip dynamics for motorcycle traction control applications. *Vehicle System Dynamics* **49**(3), 419–432.

Nishimi T, Aoki A and Katayama K 1985 Analysis of straight running stability of motorcycles. *SAE paper 856124*.

Pacejka HB 2006 *Tyre and Vehicle Dynamics* 2nd edn. Butterworth-Heinemann, Oxford.

Sharp R and Limebeer D 2004 On steering wobble oscillations of motorcycles. *Proceedings of the Institution of Mechanical Engineers, Part C: Journal of Mechanical Engineering Science* **218**(12), 1449–1456.

2

Dynamic Modelling of Riderless Motorcycles for Agile Manoeuvres

Yizhai Zhang[a], Jingang Yi[a], and Dezhen Song[b]
[a]*Rutgers University, USA*
[b]*Texas A&M University, USA*

2.1 Introduction

Single-track vehicles, such as motorcycles and bicycles, have high manoeuvrability and strong off-road capabilities. In environments such as deserts, forests and mountains, the mobility of single-track vehicles significantly outperforms that of double-track vehicles. The recent demonstration of the Blue Team's autonomous motorcycle (Figure 2.1a) in the 2005 DARPA Grand Challenge autonomous ground vehicles competition has shown an example of the high-agility of the single-track platform (Levandowski et al. 2006).

Although the extensive study of the motorcycle dynamics has revealed some knowledge about motorcycle platforms under steady motion, modelling and control of motorcycles for agile manoeuvres, such as those by professional racing riders, still remains a challenging task due to the motorcycle's intrinsically unstable platform and complex tyre–road interaction. Professional motorcycle riders can push the safety limits of the tyre–road interaction, and maintain the vehicles at high performance while still preserving safety. As a first step towards understanding the high-performance capabilities of the human drivers and designing autonomous agile manoeuvres, the objective of this chapter and Chapter 12 is to develop a new modelling and control scheme for an autonomous motorcycle. Compared to with existing studies of motorcycle dynamics and control, the main contribution of this study is the new modelling and control system design with integrated motorcycle dynamics with tyre–road interaction. First, we relax the common zero lateral velocity

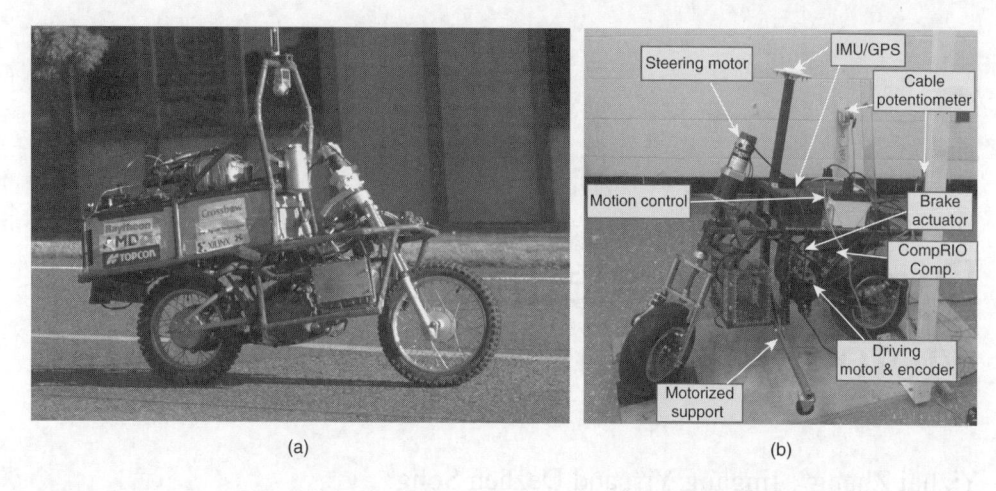

Figure 2.1 (a) The Blue Team autonomous motorcycle. (b) A Rutgers autonomous pocket bike

non-holonomic constraint for the wheel contact points of the motorcycle system. This nonholonomic constraint is not realistic for high-fidelity vehicle modelling (Limebeer and Sharp 2006). The existence of non-zero lateral velocity is particularly useful for capturing motorcycle dynamics in agile manoeuvres. Second, we explicitly consider the tyre–road interaction for designing control algorithms because of the importance of the tyre–road interaction on motorcycle dynamics. To our knowledge, there is no study that explicitly considers this kind of tyre dynamics in motorcycle control system design. The presented work in this chapter is an extension of the work in (Yi et al. 2009). Based on the new dynamics, in the next chapter, we extend the control system design in (Getz 1995 and Yi et al. 2006) for trajectory tracking and path-following manoeuvres.

The remainder of the chapter is organized as follows. We review related work in Section 2.2. In Section 2.3, we discuss dynamic modelling of a riderless motorcycle. In Section 2.4, we present a motorcycle tyre dynamics model and then integrate the tyre dynamics with the motorcycle dynamics. Finally, we conclude the chapter in Section 2.5.

2.2 Related Work

Mathematically modelling of a bicycle or a motorcycle has been an active research area for many years. Although some modelling differences have been discussed in Limebeer and (Sharp 2006), from control system design aspects, we consider bicycles and motorcycles as similar platforms, and hence do not explicitly distinguish between them. There is a large body of work that studies motorcycle stability and dynamics, and readers can refer to two recent review papers, one from a historical development viewpoint (Limebeer and Sharp 2006) and the other from a control-oriented perspective (Åström et al. 2005).

The motorcycle/bicycle models are obtained from two approaches, the inverted pendulum modelling approach and the multibody dynamic modelling approach (Åström et al. 2005). For example, a simple second-order dynamic model is presented in Lowell and McKell 1982) to study the balance stability of a bicycle. Several researchers have studied motorcycle dynamics using multibody dynamics (Cossalter 2002; Cossalter and Lot 2002; Kessler 2004;

Sharp 2001). The model developed in (Cossalter and Lot 2002) is very comprehensive and contains various vehicle components. The model has been implemented in a simulation package called *FastBike* for the purposes of real-time simulations. Multibody dynamics models are not suitable for the control system design due to their complexity while the inverted-pendulum models overly simplify the problem and do not capture all of the dynamics and geometric characteristics.

In (Sharp 1971) and (Getz 1995), mathematical models of a motorcycle are discussed using (constrained) Lagrange's equations. In (Biral et al. 2003), experimental study of motorcycle handling is compared with the mathematical dynamics model of a motorcycle with rider. Stability and steering characteristics of a motorcycle are typically discussed using a linearization approach with consideration of a constant velocity (Åström et al. 2005; Cossalter et al. 2004; Fajans 2000; Jones 1970; Limebeer and Sharp 2006; Meijaard et al. 2007; Sharp 2001). A *non-minimum phase* property (unstable poles and zeros in motorcycle dynamics) in these analyses explains the counter-steering phenomenon and other steering stability observations. In (Jones 1970), it is also demonstrated experimentally the in significance of the gyroscopic effect of the front wheel.

The concept of an autonomous bicycle without a rider has been proposed by several researchers (Beznos et al. 1998; Getz 1995; Levandowski et al., 2006; Lee and Ham 2002; Tanaka and Murakami 2004; Yi et al., 2006). In this chapter, we extend the modelling and control design in (Getz 1995 and Yi et al. 2006). For the modelling part, we take a constrained Lagrangian approach to capture the nonlinear dynamics of a motorcycle. Besides the consideration of control-oriented modelling approach that captures fundamental properties of the motorcycle platform with a manageable complexity, several new features have been adopted and developed. First, we relax the zero lateral velocity of the wheel contact points, and these modelling relaxations allow wheel sliding in the models, which provides more realistic vehicle modelling (Limebeer and Sharp 2006). Second, we explicitly consider the tyre–road interaction for designing control algorithms because of the importance of the tyre–road interaction in motorcycle dynamics (Lot 2004). The study of (Hauser and Saccon 2006) is probably the closest work to ours. They employ a nonholonomic motorcycle dynamic model and focus on the performance and manoeuvrability analysis of motorcycles using the tyre–road interaction characteristics from passenger vehicles.

2.3 Motorcycle Dynamics

Figure 2.1b shows the Rutgers autonomous motorcycle prototype. The motorcycle is rear-wheel driving. Steering and velocity control are considered as control inputs for the riderless autonomous motorcycle. Weight shifting is not considered as one actuation mechanism that human riders do because the Blue Team motorcycle had previously demonstrated an effective manoeuvrability only through vehicle steering and velocity control (Levandowski et al. 2006).

2.3.1 Geometry and Kinematics Relationships

The riderless motorcycle is considered as a two-part platform: a rear frame and a steering mechanism. Figure 2.2 shows a modelling schematic of the vehicle. We consider the following modelling assumptions: (1) the wheel/ground is a point contact, and thickness and geometry of the motorcycle tyre are neglected; (2) the motorcycle body frame is considered as a point mass; and (3) the motorcycle moves on a flat plane and vertical motion is neglected, namely, no suspension motion.

We denote C_1 and C_2 as the front and rear wheel contact points with the ground, respectively. As illustrated in Figure 2.2a, three coordinate systems are used: the navigation frame $\mathcal{N}$ (X, Y, Z axis fixed on the ground), the wheelbase moving frame (x, y, z axis fixed along line C_1C_2), and the rear body frame $\mathcal{B}$ (x_B, y_B, z_B axis fixed on the rear frame). For the frame $\mathcal{B}$, we use (3-1-2) Euler angles and represent the motion by yaw angle ψ and roll angle φ. We denote unit vector sets for the three coordinate systems as ($\boldsymbol{I}, \boldsymbol{J}, \boldsymbol{K}$), ($\boldsymbol{i}, \boldsymbol{j}, \boldsymbol{k}$) and ($\boldsymbol{i}_B, \boldsymbol{j}_B, \boldsymbol{k}_B$), respectively. It is straightforward to obtain that

$$\begin{bmatrix} \boldsymbol{i}_B \\ \boldsymbol{j}_B \\ \boldsymbol{k}_B \end{bmatrix} = \begin{bmatrix} 1 & 0 \\ 0 & \boldsymbol{R}(\varphi) \end{bmatrix} \begin{bmatrix} \boldsymbol{i} \\ \boldsymbol{j} \\ \boldsymbol{k} \end{bmatrix} = \begin{bmatrix} 1 & 0 \\ 0 & \boldsymbol{R}(\varphi) \end{bmatrix} \begin{bmatrix} \boldsymbol{R}(\psi) & 0 \\ 0 & 1 \end{bmatrix} \begin{bmatrix} \boldsymbol{I} \\ \boldsymbol{J} \\ \boldsymbol{K} \end{bmatrix} \quad (2.1)$$

$$= \begin{bmatrix} c_\psi & s_\psi & 0 \\ -c_\varphi s_\psi & c_\varphi c_\psi & s_\varphi \\ s_\varphi s_\psi & -s_\varphi c_\psi & c_\varphi \end{bmatrix} \begin{bmatrix} \boldsymbol{I} \\ \boldsymbol{J} \\ \boldsymbol{K} \end{bmatrix},$$

where the rotation matrix

$$\boldsymbol{R}(x) = \begin{bmatrix} c_x & s_x \\ -s_x & c_x \end{bmatrix}$$

and $c_x := \cos x$, $s_x := \sin x$ for angle x.

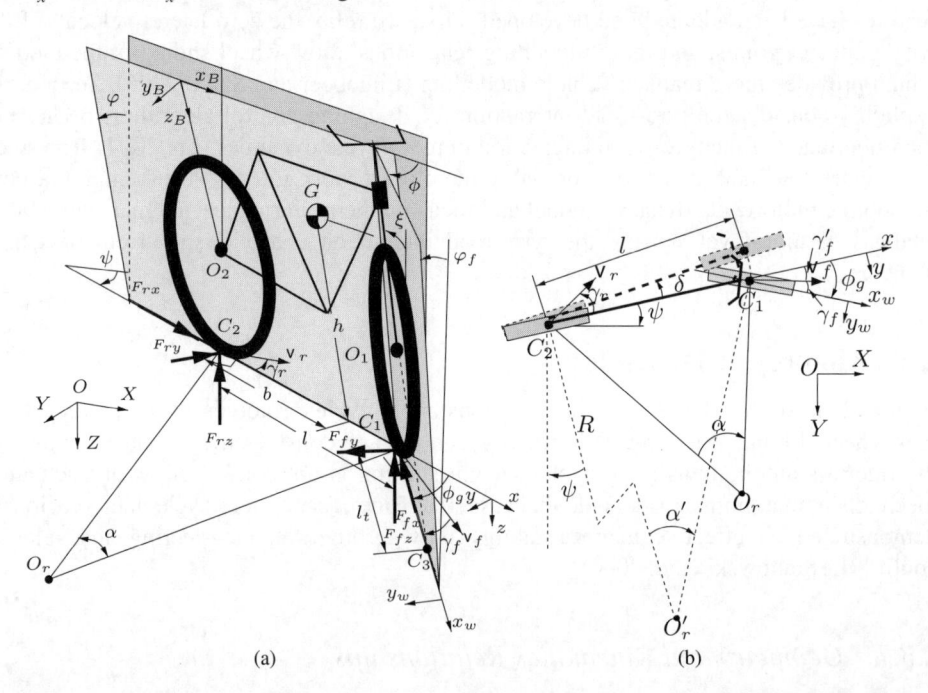

Figure 2.2 A schematic of the riderless motorcycle/bicycle. (a) Kinematic and dynamic modelling schematic. (b) Top view of the motorcycle/bicycle kinematic steering mechanism

We consider the trajectory of point C_2, denoted by its coordinates (X, Y) in $\mathcal{N}$, as the motorcycle position. The orientation of the coordinate systems and the positive directions for angles and velocities follow the conversion of the SAE standard (Meijaard et al. 2007).

We consider the instantaneous rotation centre of the motorcycle motion on the horizontal plane. Let O_r denote the instantaneous rotation centre and O_r' denote the neutral instantaneous rotation centre, which is the intersection point of the perpendicular lines of the front and rear wheel planes; see Figure 2.2. Under the neutral turning condition (Cossalter 2002), the slip angles of the front and rear wheels are the same, that is, $\lambda_f = \lambda_r$, and then the rotation centre angles for O_r and O_r' are equal to the kinematic steering angle ϕ_g, namely, $\alpha = \alpha' = \phi_g$. Let R denote the instantaneous radius of the trajectory of point C_2 under neutral turning conditions. We define σ as the kinematic steering variable as

$$\sigma := \tan \phi_g = \frac{l}{R}. \tag{2.2}$$

From the geometry of the front wheel steering mechanism (Cossalter, 2002), we find the following relationship:

$$\tan \phi_g c_\varphi = \tan \phi c_\xi. \tag{2.3}$$

If we assume a small roll and steering angles, then from (2.3), we obtain an approximation

$$\dot{\sigma} c_\varphi = \dot{\phi} c_\xi. \tag{2.4}$$

The motion of the motorcycle on the XY plane can be captured by the generalized coordinates $(X, Y, \psi, \varphi, \sigma)$. Note that the use of variable σ is to capture the steering impact on the motorcycle dynamics. The non-holonomic constraint of the rear wheel and the motion trajectory geometry imply the yaw kinematics equality

$$v_{rx} = R\dot{\psi} = \frac{l}{\sigma}\dot{\psi}. \tag{2.5}$$

From a differential geometry viewpoint,* we can partition the generalized velocities of the motorcycle as base velocities $\dot{r} = [\dot{\varphi}, v_{rx}, v_{ry}, \dot{\sigma}]^T$ and fibre velocities $\dot{s} = \dot{\psi}$. We then write the constraints in (2.5) simply as

$$\dot{s} + A(r, s)\dot{r} = 0, \tag{2.6}$$

where $A(r, s) = \begin{bmatrix} 0 & -\frac{\sigma}{l} & 0 & 0 \end{bmatrix}$.

Due to the steering mechanism and caster angle, the height of the mass centre of gravity of the motorcycle is changing under steering. As shown in Figure 2.2b, the height change Δh_G of the centre of gravity G due to the steering action can be calculated as (Yi et al. 2006)

$$\Delta h_G = \delta b s_\varphi \approx \frac{b l_t \sigma c_\xi}{l} s_\varphi, \tag{2.7}$$

where we use a small angle approximation $\sigma \approx \phi_g$ from the relationship (2.2).

---

* We here take the description of the base-fibre structure of non-holonomic dynamical systems with symmetry in (Bloch 2003).

Remark 1 *The height change Δh_G of the gravity centre G due to steering given in (2.7) is an approximation. A more accurate modelling of Δh_G with experimental validation is given in (Zhang et al. 2011). The model of Δh_G given in (Zhang et al. 2011) considers the effect of the tyre size without using a small angle approximation and the resultant relationship between Δh_G and s_φ is not linear as shown in (2.7). However, we still use the simplified model (2.7) to design the trajectory tracking and path-following controllers in Chapter 12 and the results can be readily extended to the realistic steering model in (Zhang et al. 2011).*

Remark 2 *In (Getz 1995 and Hauser and Saccon 2006), the steering axis is assumed to be vertical. This assumption simplifies the motorcycle dynamics and neglects a significant geometric stabilization mechanism, which is the 'motorcycle trail' (denoted as l_t in Figure 2.2a and discussed in (Cossalter 2002; Fajans 2000; Jones 1970; Lowell and McKell 1982). The resulting model of the motorcycle dynamics cannot capture the influence of the steering angle ϕ on the roll dynamics when $v_{rx} = 0$. Namely, we cannot use steering to stabilize the motorcycle. Such an observation is also pointed out in Åström et al. 2005).*

Given the roll angle φ and the steering angle ϕ, the camber angle of the front wheel is approximated as

$$\varphi_f = \varphi + \phi s_\xi. \tag{2.8}$$

We consider the relationship between velocities of point C_2 and the front wheel centre O_1. We write the position vector $\boldsymbol{r}_{O_1} = \boldsymbol{r}_{C_2} + \boldsymbol{\rho}_{C_2 O_1}$, where $\boldsymbol{r}_{C_2}$ is the position vector of point C_2 and $\boldsymbol{\rho}_{C_2 O_1} = l \boldsymbol{i}_B - r \boldsymbol{k}_B = l \boldsymbol{i} + r s_\varphi \boldsymbol{j} - r c_\varphi \boldsymbol{k}$ is the relative position vector of G. The angular velocity of the rear frame is represented as $\boldsymbol{\omega} = \dot{\varphi} \boldsymbol{i} + \dot{\psi} \boldsymbol{k}$. Thus, we obtain

$$\boldsymbol{v}_{O_1} = \dot{\boldsymbol{r}}_{C_2} + \boldsymbol{\omega} \times \boldsymbol{\rho}_{C_2 O_1} = (v_{rx} - r\dot{\psi} s_\varphi)\boldsymbol{i} + (v_{ry} + l\dot{\psi} + r\dot{\varphi} c_\varphi)\boldsymbol{j} + r\dot{\varphi} s_\varphi \boldsymbol{k}. \tag{2.9}$$

2.3.2 Motorcycle Dynamics

We use the constrained Lagrangian method in (Bloch 2003) to obtain the dynamic equation of the motion of the riderless motorcycle. We consider the motorcycle as two parts: a rear frame with mass m and a steering mechanism with the mass moment of inertia J_s. The Lagrangian L of the motorcycle is calculated as

$$L = \frac{1}{2} J_s \dot{\phi}^2 + \frac{1}{2} m \boldsymbol{v}_G \cdot \boldsymbol{v}_G - mg(hc_\varphi - \Delta h_G). \tag{2.10}$$

To calculate the mass centre velocity, we take a similar approach as in (2.9) and obtain

$$\boldsymbol{v}_G = (v_{rx} - h\dot{\psi} s_\varphi)\boldsymbol{i} + (v_{ry} + b\dot{\psi} + h\dot{\varphi} c_\varphi)\boldsymbol{j} + h\dot{\varphi} s_\varphi \boldsymbol{k}.$$

Plugging the above equations and (2.4)–(2.7) into (2.10), we obtain

$$L = \frac{J_s}{2c_\xi^2} \dot{\sigma}^2 + \frac{1}{2} m[(v_{rx} - h\dot{\psi} s_\varphi)^2 + (v_{ry} + b\dot{\psi} + h\dot{\varphi} c_\varphi)^2 + h^2 \dot{\varphi}^2 s_\varphi^2]$$

$$- mg\left(hc_\varphi - \frac{bl_t c_\xi}{l} \sigma s_\varphi\right). \tag{2.11}$$

Incorporating the constraints (2.6), we obtain the constrained Lagrangian L_c as[†]

$$L_c = \frac{J_s}{2c_\xi^2}c_\varphi^2\dot{\sigma}^2 + \frac{1}{2}m\left\{\left[\left(1-\frac{h}{l}\sigma s_\varphi\right)^2 + \frac{b^2}{l^2}\sigma^2\right]v_{rx}^2 + v_{ry}^2 + \frac{2b}{l}\sigma v_{rx}v_{ry} + \right.$$
$$\left. \frac{2bh}{l}c_\varphi\sigma\dot{\varphi}v_{rx} + 2hc_\varphi\dot{\varphi}v_{ry} + h^2\dot{\varphi}^2\right\} - mg\left(hc_\varphi - \frac{bl_tc_\xi}{l}\sigma s_\varphi\right). \qquad (2.12)$$

The moment M_s on the rotating axis is obtained as

$$M_s = \frac{l_t}{\sqrt{1+(l_t/r)^2}}(F_{fy}c_{\varphi_f} - F_{fz}s_{\varphi_f}). \qquad (2.13)$$

The detailed calculation of (2.13) is given in Appendix A.

The equations of motion using the constrained Lagrangian are obtained as (Bloch, 2003)[‡]

$$\frac{d}{dt}\frac{\partial L_c}{\partial \dot{r}^i} - \frac{\partial L_c}{\partial r^i} + A_i^k\frac{\partial L_c}{\partial s^k} = -\frac{\partial L}{\partial \dot{s}^l}C_{ij}^l\dot{r}^j + \tau^i, \; i,j=1,\ldots,4, \qquad (2.14)$$

where τ^i are the external forces/torques, A_i^k is the element of connection $A(r,s)$ at the kth row and ith column and C_{ij}^l denote the components of the curvature of $A(r,s)$ as

$$C_{ij}^l = \frac{\partial A_i^l}{\partial r^j} - \frac{\partial A_j^l}{\partial r^i} + A_i^k\frac{\partial A_j^l}{\partial s^k} - A_j^k\frac{\partial A_i^l}{\partial s^k}. \qquad (2.15)$$

From state variable σ, from (2.14), we obtain the steering dynamics as

$$\frac{d}{dt}\left(\frac{J_s}{c_\xi^2}c_\varphi^2\dot{\sigma}\right) - \frac{mgl_tbc_\xi}{l}s_\varphi = \tau_s + M_s. \qquad (2.16)$$

Considering a position feedback control of the steering angle directly, we can reduce the dynamic equation (2.16) by a kinematic steering system as

$$\dot{\sigma} = \omega_\sigma, \qquad (2.17)$$

where the input ω_σ is considered as the virtual steering velocity and given by dynamic extension

$$\dot{\omega}_\sigma = \frac{c_\xi^2}{J_sc_\varphi^2}(\tau_s + M_s) - 2\tan\varphi\dot{\varphi}\dot{\sigma} + \frac{mgl_tbc_\xi^3}{lJ_s}s_\varphi.$$

Similarly, we obtain the roll dynamics equation

$$\frac{bh\sigma}{l}c_\varphi\dot{v}_{rx} + hc_\varphi\dot{v}_{ry} + h^2\ddot{\varphi} + \left(1 - \frac{h\sigma}{l}s_\varphi\right)\frac{h\sigma c_\varphi}{l}v_{rx}^2$$
$$-g\left(hs_\varphi + \frac{l_tbc_\xi}{l}\sigma c_\varphi\right) = -\frac{bh}{l}c_\varphi v_{rx}\omega_\sigma, \qquad (2.18)$$

[†] Readers can refer to (Bloch 2003) for the definition of the constrained Lagrangian L_c and also Chapter 5 of (Bloch 2003) for the Lagrange–d'Alembert principle for non-holonomic constrained dynamical systems.

[‡] Here the summation convention is used where, for example, if s is of dimension m, then $A_i^k\frac{\partial A_j^l}{\partial s^k} \equiv \Sigma_{k=1}^m A_i^k\frac{\partial A_j^l}{\partial s^k}$.

longitudinal dynamics equation

$$\left[\left(1-\frac{h\sigma}{l}s_\varphi\right)^2+\frac{b^2\sigma^2}{l^2}\right]\dot{v}_{rx}+\frac{b\sigma}{l}\dot{v}_{ry}+\frac{bh\sigma}{l}c_\varphi\dot{\varphi}-2\left(1-\frac{h\sigma}{l}s_\varphi\right)\frac{h\sigma}{l}c_\varphi\dot{\varphi}v_{rx}$$

$$-\frac{bh\sigma}{l}s_\varphi\dot{\varphi}^2=-\left[-2\left(1-\frac{h\sigma}{l}s_\varphi\right)\frac{h}{l}s_\varphi v_{rx}+\frac{2b^2\sigma}{l^2}v_{rx}+\frac{b}{l}v_{ry}+\frac{bh}{l}c_\varphi\dot{\varphi}\right]\omega_\sigma$$

$$+\frac{1}{m}F_{rx}-\frac{1}{m\sqrt{1+\sigma^2}}(F_{fx}+\sigma F_{fy})-\frac{1}{m}C_d v_{rx}^2 \qquad (2.19)$$

and lateral dynamics equation

$$\frac{b\sigma}{l}\dot{v}_{rx}+\dot{v}_{ry}+hc_\varphi\ddot{\varphi}-hs_\varphi\dot{\varphi}^2=-\frac{bv_{rx}}{l}\omega_\sigma-\frac{1}{m}F_{ry}$$

$$+\frac{1}{m\sqrt{1+\sigma^2}}(F_{fy}-\sigma F_{fx}). \qquad (2.20)$$

In (2.19), C_d is the aerodynamic drag coefficient.

Let $\dot{q} := [\dot{\varphi}\ v_{rx}\ v_{ry}]^T$ denote the generalized velocity of the motorcycle and we rewrite the dynamic equations (2.18)–(2.20) in a compact matrix form as

$$\mathbf{M}\dot{q} = \mathbf{K}_m + \mathbf{B}_m \begin{bmatrix} \omega_\sigma \\ F_{fx} \\ F_{fy} \\ F_{rx} \\ F_{ry} \end{bmatrix}, \qquad (2.21)$$

where matrices

$$\mathbf{M} = \begin{bmatrix} M_{11} & M_{12} \\ M_{21} & M_{22} \end{bmatrix} = \begin{bmatrix} h^2 & \frac{bh\sigma}{l}c_\varphi & hc_\varphi \\ \frac{bh\sigma}{l}c_\varphi & \left(1-\frac{h\sigma}{l}s_\varphi\right)^2+\frac{b^2\sigma^2}{l^2} & \frac{b\sigma}{l} \\ hc_\varphi & \frac{b\sigma}{l} & 1 \end{bmatrix}, \qquad (2.22)$$

$$\mathbf{K}_m = \begin{bmatrix} -\left(1-\frac{h\sigma}{l}s_\varphi\right)\frac{h\sigma c_\varphi}{l}v_{rx}^2+g\left(hs_\varphi+\frac{l_t b c_\varepsilon}{l}\sigma c_\varphi\right) \\ 2\left(1-\frac{h\sigma}{l}s_\varphi\right)\frac{h\sigma}{l}c_\varphi\dot{\varphi}v_{rx}+\frac{bh\sigma}{l}s_\varphi\dot{\varphi}^2-\frac{1}{m}C_d v_{rx}^2 \\ hs_\varphi\dot{\varphi}^2 \end{bmatrix},$$

and

$$\mathbf{B}_m = \begin{bmatrix} -\frac{bh}{l}c_\varphi v_{rx} & 0 & 0 & 0 & 0 \\ B_\omega & -\frac{1}{m\sqrt{1+\sigma^2}} & -\frac{\sigma}{m\sqrt{1+\sigma^2}} & \frac{1}{m} & 0 \\ -\frac{bv_{rx}}{l} & -\frac{\sigma}{m\sqrt{1+\sigma^2}} & \frac{1}{m\sqrt{1+\sigma^2}} & 0 & -\frac{1}{m} \end{bmatrix}.$$

In the above matrix $\mathbf{B}_m$,

$$B_\omega = 2\left[\left(1 - \frac{h\sigma}{l}s_\varphi\right)\frac{h}{l}s_\varphi - \frac{b^2\sigma}{l^2}\right]v_{rx} - \frac{b}{l}v_{ry} - \frac{bh}{l}c_\varphi\dot\varphi.$$

It is clear that the control inputs in (2.17) and (2.21) are the virtual steering velocity ω_σ and the wheel traction/braking forces $\boldsymbol{F}_f$ and $\boldsymbol{F}_r$.

2.4 Tyre Dynamics Models

In this section, we discuss how to capture the motorcycle tyre–road interaction. We particularly present a friction force modelling scheme for motorcycle dynamics (2.21).

2.4.1 Tyre Kinematics Relationships

Figure 2.3 illustrates the kinematics of the tyre–road contact. Let $\boldsymbol{v}_c = v_{cx}\boldsymbol{i} + v_{cy}\boldsymbol{j} + v_{cz}\boldsymbol{k}$ and $\boldsymbol{v}_o = v_{ox}\boldsymbol{i} + v_{oy}\boldsymbol{j} + v_{oz}\boldsymbol{k}$ denote the velocities of the contact point and the wheel centre in frame $\mathcal{B}$, respectively. We define the longitudinal slip ratio λ_s and the lateral sideslip ratio λ_γ, respectively, as

$$\lambda_s := \frac{v_{cx} - r\omega_w}{v_{cx}}, \quad \lambda_\gamma := \tan\gamma = -\frac{v_{cy}}{v_{cx}}, \qquad (2.23)$$

where ω_w is the wheel angular velocity and γ is the sideslip angle.

For the front wheel, the camber angle is defined by (2.8), and the velocity relationship between C_1 and the wheel centre O_1 in $\mathcal{B}$ is then

$$v_{fx} = v_{fox} + r\dot\psi s_\varphi, \quad v_{fcy} = v_{foy} - r\dot\varphi_f c_\varphi, \quad v_{fz} = v_{foz} - r\dot\varphi_f s_\varphi. \qquad (2.24)$$

Using the relationships (2.9) and (2.8), we simplify the above velocity calculation and obtain

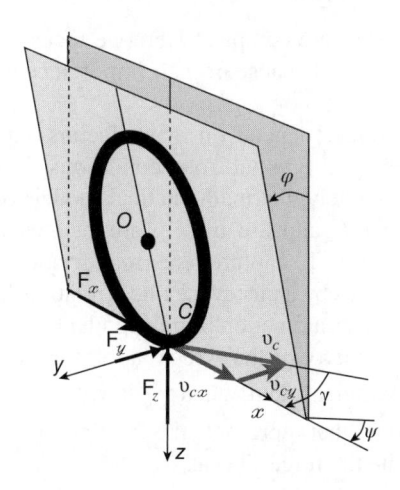

$$v_{fx} = v_{rx}, \quad v_{fy} = v_{ry} - r\dot\phi s_\xi c_\varphi + l\dot\psi. \qquad (2.25)$$

Figure 2.3 Schematic of the tire kinematics

From the sideslip ratio (2.23) of the rear wheel, we have

$$\lambda_{r\gamma} = \tan\gamma_r = -\frac{v_{ry}}{v_{rx}} = -\frac{v_{fy}}{v_{fx}} - \frac{r\dot\phi s_\xi c_\varphi - l\dot\psi}{v_{rx}} = \tan\gamma'_f - \frac{r\tan\xi c_\varphi^2}{v_{rx}}\omega_\sigma + \sigma, \qquad (2.26)$$

where $\gamma'_f := \phi_g - \gamma_f$ and $\tan\gamma'_f = -\frac{v_{fy}}{v_{fx}}$; see Figure 2.2. We also use relationships (2.4) and (2.5) in the last step above. Moreover, from (2.2) and the geometry and kinematics of the front wheel (Figure 2.2), we have

$$\sigma = \tan\phi_g = \tan(\gamma'_f + \gamma_f) \approx \tan\gamma'_f + \tan\gamma_f = \lambda_{r\gamma} + \frac{r\tan\xi c_\varphi^2}{v_{rx}}\omega_\sigma - \sigma + \lambda_{f\gamma}.$$

Therefore, we obtain the relationship between the front and rear wheel sideslip ratios as

$$\lambda_{f\gamma} = 2\sigma - \frac{r\tan\xi c_\varphi^2}{v_{rx}}\omega_\sigma - \lambda_{r\gamma}. \qquad (2.27)$$

Similarly, we obtain the slip ratio calculation of the front wheel as follows. The longitudinal velocity of point C_1 is calculated as

$$v_{fx_w} = v_{fx}c_{\phi_g} + v_{fy}s_{\phi_g} \approx v_{rx}c_{\phi_g} + (v_{ry} + \sigma v_{rx})s_{\phi_g}$$
$$= \frac{1}{\sqrt{1+\sigma^2}}[(1+\sigma^2)v_{rx} + \sigma v_{ry}].$$

Then, by the definition (2.23), we obtain the front wheel longitudinal slip ratio

$$\lambda_{fs} = 1 - \frac{r\omega_f}{v_{fx_w}} = 1 - \frac{r\sqrt{1+\sigma^2}}{(1+\sigma^2)v_{rx} + \sigma v_{ry}}\omega_f. \qquad (2.28)$$

2.4.2 Modelling of Frictional Forces

Tyre–road friction force models are complex. Here we focus on modelling of the longitudinal force F_x and lateral force F_y because of their importance in motorcycle dynamics and motion control.

The tyre–road frictional forces depend on many factors, such as slip and slip angles, vehicle velocity, normal load and tyre and road conditions. It is widely accepted that the pseudo-static relationships, namely, the mathematical models of the longitudinal force F_x and slip λ, and the lateral force F_y and slip angle γ, are the most important characteristics to capture the tyre–road interaction. To capture tyre–road friction characteristics, we propose to approximate the friction forces by a piecewise linear relationship shown in Figure 2.4. Let $F(x)$ denote the frictional force as a function of independent variable x. The piecewise linear function $F(x)$ captures the property of the tyre–road forces: when $0 \leq x \leq x_m$, $F(x) = kx$, where k is the stiffness coefficient, and when $x_m < x \leq x_{\max}$, $F = \frac{(1-\alpha_x)F_m}{x_m - x_{\max}}(x - x_m) + F_m$, where $0 \leq \alpha_x \leq 1$ is a constant that represents the fraction of the force at $x_{\max}$ of the maximum force F_m. We thus write the force $F(x)$ as

$$F(x) = k(a_1 + a_2 x), \qquad (2.29)$$

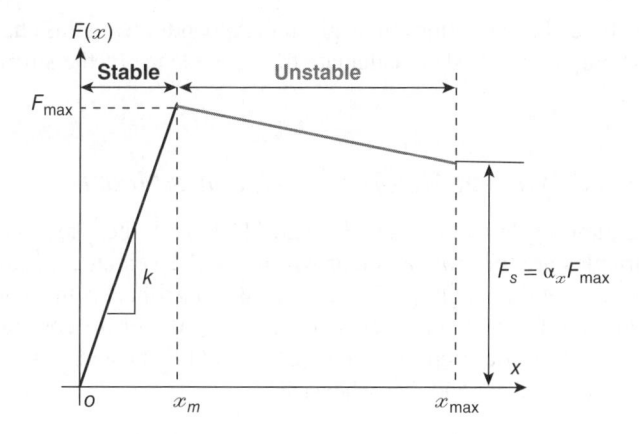

Figure 2.4 Linear approximation of the tyre–road frictional force $F(x)$

where

$$a_1 = \begin{cases} 0 & 0 \leq x \leq x_m \\ \frac{(x_{max}-x_m)x_m}{x_{max}-x_m} & x_m < x \leq x_{max} \end{cases}, \quad a_2 = \begin{cases} 1 & 0 \leq x \leq x_m \\ \frac{-(1-\alpha_x)x_m}{x_{max}-x_m} & x_m < x \leq x_{max} \end{cases}.$$

With the force model (2.29), we write the longitudinal force as

$$F_x(\lambda_s) = k_\lambda [a_{1\lambda} + a_{2\lambda}\text{sign}(\lambda_s)\lambda_s], \tag{2.30}$$

where the function $\text{sign}(x) = 1$ for $x \geq 0$ and -1 otherwise is used to capture both positive (braking) and negative (traction) forces for $F_x(\lambda_s)$. For the lateral force, due to the large camber angle of the motorcycle tyres, we have

$$F_y(\lambda_{eq}) = k_\gamma [a_{1\gamma} + a_{2\gamma}\text{sign}(\lambda_{eq})\lambda_{eq}], \tag{2.31}$$

where the equivalent sideslip ratio is

$$\lambda_{eq} = \tan\gamma_{eq} = \tan\left(\gamma + \frac{k_\varphi}{k_\gamma}\varphi\right) \approx \lambda_\gamma + \frac{k_\varphi}{k_\gamma}\tan\varphi.$$

The values of the longitudinal, cornering and cambering coefficients, k_λ, k_γ, k_φ, depend on the normal load F_z. Due to the acceleration and deceleration, the normal load F_z often changes during motion. For front and rear wheels, the normal loads F_{fz} and F_{rz} are obtained respectively as

$$F_{fz} = \frac{b}{l}mg - \frac{h}{l}m\dot{v}_{Gx}, \quad F_{rz} = \frac{l-b}{l}mg + \frac{h}{l}m\dot{v}_{Gx}, \tag{2.32}$$

where $\dot{v}_{Gx}$ is the longitudinal acceleration of the motorcycle at the mass centre G. The relationship between $\dot{v}_{Gx}$ and the acceleration of point C_2 is obtained as

$$\dot{v}_{Gx} = \dot{v}_{rx} - v_{ry}\dot{\psi} - h\dot{\psi}s_\varphi - b\dot{\psi}^2 - 2h\dot{\psi}\dot{\varphi}c_\varphi.$$

The calculation of the above relationship is given in Appendix B. In this chapter, we use the tyre models in (Sharp et al. 2004) to calculate the dependence of the stiffness coefficients on the normal load.

2.4.3 Combined Tyre and Motorcycle Dynamics Models

We combine the motorcycle dynamics (2.17) and (2.21) with the tyre dynamics. The controlled input variables are the front and rear wheel angular velocities, namely, ω_f and ω_r, respectively, and the steering angle ϕ. Note that the driving wheel is the rear wheel, and we can apply only braking for the front wheel, namely, $F_{fx} \geq 0$. For the control system design, we consider the pseudo-static friction models (2.30) and (2.31), and therefore we write the longitudinal forces at the front and rear wheels as

$$F_{fx} = F_{1f} + F_{2f}\lambda_{fs}, \quad F_{rx} = F_{1r} + F_{2r}\lambda_{rs} \tag{2.33}$$

and lateral forces

$$F_{fy} = F_{3f} + F_{4f}\left(\lambda_{f\gamma} + \frac{k_{f\varphi}}{k_{f\gamma}}\tan\varphi_f\right), \quad F_{ry} = F_{3r} + F_{4r}\left(\lambda_{r\gamma} + \frac{k_{r\varphi}}{k_{r\gamma}}\tan\varphi\right), \tag{2.34}$$

where $F_{1i} = k_{i\lambda}a_{1i\lambda}$, $F_{2i} = k_{i\lambda}a_{2i\lambda}\mathrm{sign}(\lambda_{is})$, $F_{1i} = k_{i\lambda}a_{1i\lambda}$, $F_{2i} = k_{i\lambda}a_{2i\lambda}\mathrm{sign}(\lambda_{is})$, $i = f, r$, and $a_{ji\lambda}, a_{ji\gamma}, j = 1, 2$, are the longitudinal and lateral force model parameters defined in (2.29), respectively.

Plugging (2.33) and (2.34) into (2.21) and using the relationship (2.27), we obtain

$$\mathbf{M}(q,\sigma)\dot{q} = \mathbf{K}(\dot{q},q,\sigma) + \mathbf{B}u, \tag{2.35}$$

where input $u := [\omega_\sigma \; u_\lambda^T]^T$, $u_\lambda = [\lambda_{fs} \; \lambda_{rs}]^T$, matrix

$$\mathbf{K} = \begin{bmatrix} K_1 \\ \hdashline K_2 \end{bmatrix} = \begin{bmatrix} (K_m)_1 \\ (K_m)_2 - \dfrac{F_{1f}}{m\sqrt{1+\sigma^2}} - \dfrac{\sigma}{m\sqrt{1+\sigma^2}}F_{34} + \dfrac{F_{1r}}{m} \\ (K_m)_3 - \dfrac{\sigma F_{1f}}{m\sqrt{1+\sigma^2}} + \dfrac{1}{m\sqrt{1+\sigma^2}}F_{34} - \dfrac{F_{ry}}{m} \end{bmatrix}, \tag{2.36}$$

$(K_m)_i$ is the ith row of matrix $\mathbf{K}$, $F_{34} = F_{3f} + F_{4f}\left(\lambda_{f\gamma} + \dfrac{k_{f\varphi}}{k_{f\gamma}}(2\sigma - \lambda_{r\gamma})\right)$, and

$$\mathbf{B} = \begin{bmatrix} B_{11} & \vdots & B_{12} \\ \hdashline B_{21} & \vdots & B_{22} \end{bmatrix} = \begin{bmatrix} -\dfrac{bh}{l}c_\varphi v_{rx} & \vdots & 0 & 0 \\ B_\omega + \dfrac{r\sigma F_{4f}\tan\xi\, c_\varphi^2\, k_{f\varphi}}{mv_{rx}k_{f\gamma}\sqrt{1+\sigma^2}} & \vdots & -\dfrac{F_{2f}}{m\sqrt{1+\sigma^2}} & \dfrac{F_{2r}}{m} \\ -\dfrac{bv_{rx}}{l} - \dfrac{rF_{4f}\tan\xi\, c_\varphi^2\, k_{f\varphi}}{mv_{rx}k_{f\gamma}\sqrt{1+\sigma^2}} & \vdots & \dfrac{\sigma F_{2f}}{m\sqrt{1+\sigma^2}} & 0 \end{bmatrix}. \tag{2.37}$$

Remark 3 *We assume that the motorcycle is rear wheel driven and thus the front wheel cannot produce any traction force. We consider the following distribution rule for braking and traction strategy between two wheels. The rear tyre can produce both traction and braking forces, while the front tyre can only produce braking force. If a braking force is needed, the front tyre would first be used to brake and produce the amount of braking force needed. If the needed braking force cannot be fully generated by the front tyre after the slip ratio reaches λ_{sm}, the rear tyre will then brake to produce the necessary extra braking force.*

In Chapter 12, we will develop a trajectory tracking and balancing control for dynamics (2.35).

2.5 Conclusions

In this chapter, we presented a new nonlinear dynamic model for autonomous motorcycles for agile manoeuvres. The proposed model is obtained through a constrained Lagrange modelling approach. Comparing with the existing riderless motorcycle models, the new features of the proposed motorcycle dynamics model are twofold. First we relaxed the assumption of zero lateral velocity constraints at tyre contact points and thus the model can be used for agile manoeuvres when wheels run with both large longitudinal slips and lateral sideslips. Second, we considered the motorcycle tyre models and extended the previously developed motorcycle dynamics. The control inputs for the proposed motorcycle dynamics are the front wheel steering angle and the angular velocities for the front and rear wheels. The trajectory tracking and path following control systems design is based on the new dynamic model and presented in Chapter 12.

Nomenclature

X, Y, Z	A ground-fixed coordinate system.
x, y, z	A wheel base line moving coordinate system.
x_w, y_w, z_w	A front wheel plane coordinate system.
x_B, y_B, z_B	A rear frame body coordinate system.
C_1, C_2	Front and rear wheel contact points on the ground.
F_{fx}, F_{fy}, F_{fz}	Front wheel contact forces in the x, y, z directions.
F_{rx}, F_{ry}, F_{rz}	Rear wheel contact forces in the x, y, z directions.
v_f, v_r	Velocity vectors of the front and rear wheel contact points, respectively.
v_{fx}, v_{fy}	Front wheel contact point C_1 velocities along the x and y directions, respectively.
v_{rx}, v_{ry}	Rear wheel contact point C_2 velocities along the x and y directions, respectively.
v_{fx_w}, v_{fy_w}	Front wheel contact point C_1 velocities along the x_w and y_w directions, respectively.
v_X, v_Y	Rear wheel contact point C_2 velocities along the X and Y directions, respectively.
ω_f, ω_r	Wheel angular velocities of the front and rear wheels, respectively.
v_G	Velocity vector of the motorcycle frame (with rear wheel set).

γ_f, γ_r	Slip angles of the front and rear wheels, respectively.
λ_f, λ_r	Longitudinal slip values of the front and rear wheels, respectively.
φ, ψ	Rear frame roll and yaw angles, respectively.
φ_f	Front steering wheel plane camber angle.
ϕ	Motorcycle steering angle.
ϕ_g	Motorcycle kinematic steering angle (projected steering angle on the ground plane).
σ	Front kinematic steering angle variable.
m	Total mass of the motorcycle rear frame and wheel.
J_s	Mass moment of rotation of the steering fork (with the front wheel set) about its rotation axis.
l	Motorcycle wheelbase, i.e., distance between C_1 and C_2.
l_t	Front steering wheel trail.
h	Height of the motorcycle centre of mass.
r	Front and rear wheel radius.
δ	Rear frame rotation angle from its vertical position.
ξ	Front steering axis caster angle.
R	Radius of the trajectory of point C_2 under neutral steering turns.
C_d	Aerodynamics drag coefficient.
$k_\lambda, k_\gamma, k_\varphi$	Longitudinal, lateral and camber stiffness coefficients of motorcycle tyres, respectively.
$L(L_c)$	(Constrained) Lagrangian of the motorcycle systems.

Appendix A: Calculation of M_s

We consider the front wheel centre O_1 and the projected steering axis point C_3 on the ground surface. Since the frictional moment is independent of the coordinate system, we set up a local coordinate system $x_f y_f z_f$ by rotating the coordinate system xyz around the z axis with an angle ϕ_g (origin at contact point C_1). Let $(\boldsymbol{i}_f, \boldsymbol{i}_f, \boldsymbol{i}_f)$ denote the unit vectors along the x_f, y_f, z_f directions, respectively.

In the new coordinate system, we obtain the coordinates of O_1 and C_3 as $(0, rs_{\varphi_f}, -rc_{\varphi_f})$ and $(l_t, 0, 0)$, respectively. We write the front wheel friction force vector $\boldsymbol{F}_f$ as

$$\boldsymbol{F}_f = -F_{fx}\boldsymbol{i}_f - F_{fy}\boldsymbol{j}_f - F_{fz}\boldsymbol{k}_f$$

and the vector $\boldsymbol{r}_{C_3 C_1} = -l_t \boldsymbol{i}_f$. The directional vector $\boldsymbol{n}_{O_1 C_3}$ of the steering axis O_1, C_3 is then

$$\boldsymbol{n}_{O_1 C_3} = \frac{l_t \boldsymbol{i}_f - rs_{\varphi_f}\boldsymbol{j}_f + rc_{\varphi_f}\boldsymbol{k}_f}{\sqrt{l_t^2 + r^2}}.$$

Therefore, the friction moment M_s about the steering axis is calculated as

$$M_s = (\boldsymbol{r}_{C_3 C_1} \times \boldsymbol{F}_f) \cdot \boldsymbol{n}_{O_1 C_3} = \frac{l_t}{\sqrt{1 + (l_t/r)^2}} (F_{fy} c_{\varphi_f} - F_{fz} s_{\varphi_f}).$$

Appendix B: Calculation of Acceleration $\dot{v}_G$

Taking the time derivative of the mass centre velocity v_G and considering the moving frame xyz's angular velocity $\boldsymbol{\omega} = \dot{\varphi}\boldsymbol{i} + \dot{\psi}\boldsymbol{k}$, we obtain

$$\dot{v}_G = \frac{\delta v_G}{\delta t} + \boldsymbol{\omega} \times v_G = (\dot{v}_{rx} - h\dot{\psi}s_\varphi - h\dot{\psi}\dot{\varphi}c_\varphi)\boldsymbol{i} + (\dot{v}_{ry} + b\ddot{\psi} + h\ddot{\varphi}c_\varphi$$
$$- h\dot{\varphi}^2 s_\varphi)\boldsymbol{j} + (h\ddot{\varphi}s_\varphi + h\dot{\varphi}^2 c_\varphi)\boldsymbol{k} + (\dot{\varphi}\boldsymbol{i} + \dot{\psi}\boldsymbol{k}) \times v_G$$
$$= (\dot{v}_{rx} - v_{ry}\dot{\psi} - h\ddot{\psi}s_\varphi - b\dot{\psi}^2 - 2h\dot{\psi}\dot{\varphi}c_\varphi)\boldsymbol{i} + (\dot{v}_{ry} + v_{rx}\dot{\psi} + b\ddot{\psi} + h\ddot{\varphi}c_\varphi - h\dot{\psi}^2 s_\varphi$$
$$- 2h\dot{\varphi}^2 s_\varphi)\boldsymbol{j} + (v_{ry}\dot{\varphi} + h\ddot{\varphi}s_\varphi + b\dot{\psi}\dot{\varphi} + 2h\dot{\varphi}^2 c_\varphi)\boldsymbol{k},$$

where $\frac{\delta v_G}{\delta t}$ denotes the derivative of v_G by treating the xyz coordinate as a fixed frame.

Acknowledgements

The authors thank Dr N Getz at Inversion Inc. for his helpful suggestions and support. The authors are grateful to Prof. S Jayasuriya of Drexel University, Dr EH Tseng and Dr J Lu at Ford Research and Innovation Center for their helpful discussions and suggestions.

References

Åström K, Klein R and Lennartsson A 2005 Bicycle dynamics and control. *IEEE Control Syst. Mag.* **25**(4), 26–47.
Beznos A, Formal'sky A, Gurfinkel E, Jicharev D, Lensky A, Savitsky K and Tchesalin L 1998 Control of autonomous motion of two-wheel bicycle with gyroscopic stabilisation *Proc. IEEE Int. Conf. Robot. Autom.*, pp. 2670–2675, Leuven, Belgium.
Biral F, Bortoluzzi D, Cossalter V and Da Lio M 2003 Experimental study of motorcycle transfer functions for evaluating handling. *Veh. Syst. Dyn.* **39**(1), 1–25.
Bloch A 2003 *Nonholonomic Mechanics and Control*. Springer, New York, NY.
Cossalter V 2002 *Motorcycle Dynamics*. Race Dynamics, Greendale, WI.
Cossalter V and Lot R 2002 A motorcycle multi-body model for real time simulations based on the natural coordinates approach. *Veh. Syst. Dyn.* **37**(6), 423–447.
Cossalter V, Lot R and Maggio F 2004 The modal analysis of a motorcycle in straight running and on a curve. *Meccanica* **39**, 1–16.
Fajans J 2000 Steering in bicycles and motorcycles. *Amer. J. Phys.* **68**(7), 654–659.
Getz N 1995 *Dynamic inversion of nonlinear maps with applications to nonlinear control and robotics* PhD thesis Dept Electr. Eng. and Comp. Sci., Univ. Calif. Berkeley, CA.
Hauser J and Saccon A 2006 Motorcycle modelling for high-performance manoeuvering. *IEEE Control Syst. Mag.* **26**(5), 89–105.
Jones D 1970 The stability of the bicycle. *Phys. Today* **23**(4), 34–40.
Kessler P 2004 *Motorcycle navigation with two sensors* Master's thesis Dept Mech. Eng., Univ. California Berkeley, CA.
Levandowski A, Schultz A, Smart C, Krasnov A, Chau H, Majusiak B, Wang F, Song D, Yi J, Lee H and Parish A 2006 Ghostrider: Autonomous motorcycle *Proc. IEEE Int. Conf. Robot. Autom. (Video)*, Orlando, FL.
Limebeer D and Sharp R 2006 Bicycles, motorcycles, and models. *IEEE Control Syst. Mag.* **26**(5), 34–61.
Lot R 2004 A motorcycle tires model for dynamic simulations: Theoretical and experimental aspects. *Meccanica* **39**, 207–220.
Lowell J and McKell H 1982 The stability of bicycles. *Amer. J. Phys.* **50**(12), 1106–1112.

Meijaard J, Papadopoulos J, Ruina A and Schwab A 2007 Linearized dynamics equations for the balance and steer of a bicycle: A benchmark and review. *Proc. Royal Soc. A* **463**, 1955–1982.

Lee S and Ham W 2002 Self-stabilzing strategy in tracking control of unmanned electric bicycle with mass balance *Proc. IEEE/RSJ Int. Conf. Intell. Robot. Syst.*, pp. 2200–2205, Lausanne, Switzerland.

Sharp R 1971 The stability and control of motorcycles. *J. Mech. Eng. Sci.* **13**(5), 316–329.

Sharp R 2001 Stability, control and steering responses of motorcycles. *Veh. Syst. Dyn.* **35**(4-5), 291–318.

Sharp RS, Evangelou S and Limebeer DJN 2004 Advances in the modelling of motorcycle dynamics. *Multibody Syst. Dyn.* **12**, 251–283.

Tanaka Y and Murakami T 2004 Self sustaining bicycle robot with steering controller *Proc. 2004 IEEE Adv. Motion Contr. Conf.*, pp. 193–197, Kawasaki, Japan.

Tanaka Y and Murakami T 2009 A study on straight-line tracking and posture control in electric bicycle. *IEEE Trans. Ind. Electron.* **56**(1), 159–168.

Yi J, Song D, Levandowski A and Jayasuriya S 2006 Trajectory tracking and balance stabilization control of autonomous motorcycles *Proc. IEEE Int. Conf. Robot. Autom.*, pp. 2583–2589, Orlando, FL.

Yi J, Zhang Y and Song D 2009 Autonomous motorcycles for agile maneuvers: Part I: Dynamic modeling *Proc. IEEE Conf. Decision Control*, pp. 4613–4618, Shanghai, China.

Zhang Y, J. Li, Yi J and Song D 2011 Balance control and analysis of stationary riderless motorcycles *Proc. IEEE Int. Conf. Robot. Autom.*, pp. 3018–3023, Shanghai, China.

3

Identification and Analysis of Motorcycle Engine-to-Slip Dynamics

Matteo Corno and Sergio M. Savaresi
Dipartimento di Elettronica, Informazione e Bioingegneria, Politecnico di Milano, Italy

This chapter presents a method for identifying the engine-to-slip dynamics of motorcycles. The proposed identification protocol returns dynamic models useful for the design of traction control systems for powered two-wheelers. After the description of the experimental setup, the method is presented. The results obtained for a racing motorcycle are used to illustrate the main features of the method. Two control variables are studied: throttle and spark advance; for both, step and sweep responses are used. The utility of the method is illustrated by analysing several aspects: motorcycle benchmarking, road surface and velocity sensitivity.

3.1 Introduction

In the past several years, the automotive industry has greatly benefitted from the mass diffusion of closed loop control systems. Closed loop control systems increase efficiency, performance, driving pleasure and most of all safety. The success of these systems has been recognized by legislators: anti-locking systems (ABS) are now compulsory for passenger cars and active stability control systems will soon follow. Most of these technologies were born on the racing track, a highly competitive and well-sponsored environment that acts as a technology showcase demonstrating the potential of these technologies to the public. The development of these technologies for powered two-wheelers (PTW) started after a considerable delay, but has been regaining ground in the past few years. Both the market and the legislator are now aware of the benefits that those systems can bring to PTWs as

---

Modelling, Simulation and Control of Two-Wheeled Vehicles, First Edition.
Edited by Mara Tanelli, Matteo Corno and Sergio M. Savaresi.
© 2014 John Wiley & Sons, Ltd. Published 2014 by John Wiley & Sons, Ltd.

well. In Europe, ABS systems will become compulsory for motorcycles built after 2015. The initial delay is rooted in technological reasons. The dynamics of single track vehicles are far more complex than that of four-wheeled vehicles and consequently also the design of control systems is more complex.

Although the first scientific papers on bicycles and their stability appeared at the end of the 19th century (e.g. Limebeer and Sharp 2006; Rankine 1869; Schwab et al. 2007), only in the past few years has the modelling complexity been systematically addressed with the development of the multibody simulation and modelling approach. Many authors have developed custom-made simulators (Cossalter and Lot 2002; Hauser and Saccon 2006; Lot 2004; Nehaoua et al. 2007; Sharp et al. 2004; Tanelli et al. 2006) and studied motorcycle stability under different driving conditions. (Limebeer et al. 2001) used a simulator to analyse vehicle stability under acceleration and braking, both on a level surface and on a decline or incline. They then improved their simulator, introducing a rider model and focusing on steady turning, stability and design parameter sensitivity in (Limebeer et al. 2002; Sharp and Limebeer 2004). Another state-of-the-art simulator is Cossalter's (documented in Cossalter 2002; Cossalter et al. 1999, 2004a,b). Multibody simulation is well suited to dynamic modal analysis, but not to control system design. These models are too complex and require order reduction techniques to be effectively exploited (see Skogestad and Postlethwaite 2007); and multibody simulators have a large number of parameters that need to be measured or evaluated, which makes validation of the model expensive and time consuming. The black-box perspective (Ljung 1987) represents a complementary approach: being based on input-output data, it does not need the identification of specific physical parameters and, if correctly applied, it automatically returns the minimum complexity model. In this chapter a black-box identification protocol for the engine to rear wheel slip dynamics of a motorcycle during straight running is proposed. The protocol can be employed to identify the relevant dynamics for traction control purposes. The second part of the chapter illustrates the potential of the method by quantitatively analysing several aspects: the comparison of throttle and spark advance, the effect of velocity and road surface and the comparison between two different kinds of motorcycles. The chapter is structured as follows: Section 3.2 presents the experimental setup. In Section 3.4, the identification method is described in detail. In Sections 3.4–3.6 the identified dynamics are analysed from several points of view: comparison of different motorcycles, effects of surface differences and velocity sensitivity. Section 3.7 draws some conclusions.

3.2 Experimental Setup

The reference bike is a superbike championship racing motorcycle. It is equipped with (Figure 3.1):

- An electronic throttle body (ETB), which electronically controls the position of the throttle valve (see Corno et al. 2011).
- An electronic control unit (ECU), which controls the engine spark-advance; the ECU also logs sensors data. The clock frequency of the ECU is 1 kHz.
- Two wheel encoders, which measure the rotational speed of the wheels. The wheel velocity is computed with the $1/\Delta T$ algorithm and filtered with the adaptive notch filter presented in (Corno and Savaresi 2010).

Identification and Analysis of Motorcycle Engine-to-Slip Dynamics

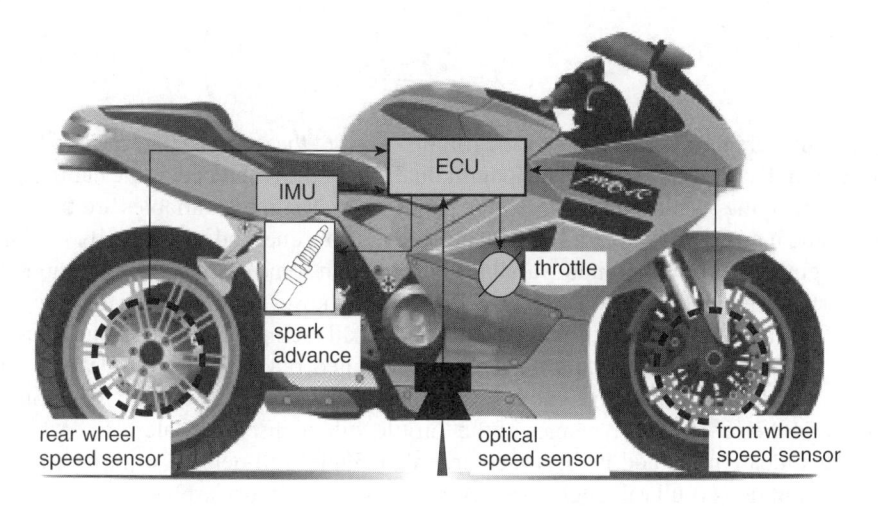

Figure 3.1 Available actuators and sensors

- An optical velocity sensor, which measures the real longitudinal velocity of the vehicle chassis with respect to the road.
- An inertial measurement unit (IMU) located near the vehicle centre of mass, which among other signals, measures the longitudinal acceleration.

The rolling radius of the wheel is estimated through coasting down tests. The test consists of a declutched deceleration from high speed. In these tests, the wheel slip can be assumed to be null; the braking is due to aerodynamic drag and rolling friction. The true velocity, measured with the optical sensor, can be used to calibrate the rolling radii according to the following minimizations:

$$r_f = \underset{r_f}{\operatorname{argmin}} \int_{t_0}^{t_1} |\omega_f r_f - v_{true}| dt$$

$$r_r = \underset{r_r}{\operatorname{argmin}} \int_{t_0}^{t_1} |\omega_r r_r - v_{true}| dt$$

where r_f, r_r are the rolling radii of the front and rear wheel, respectively, ω_f, ω_r are the corresponding wheel rotational speeds and v_{true} is the 'true' vehicle longitudinal speed. The optimization is done over the whole coasting-down manoeuvre. This approach has an advantage over the direct measurement of the rolling radius. The vertical component (lift) of the aerodynamic forces as well as the centrifugal effect on the tyre generate radial forces that deform the tyre at high velocity. The estimation method accounts for these effects.

3.3 Identification of Engine-to-Slip Dynamics

In this section, the problem of estimating the engine-to-slip dynamics is considered. An input/output (I/O) approach is used. The output variable is the longitudinal slip λ of the rear wheel, defined as:

$$\lambda = \frac{\omega r - v}{\omega r} \quad (3.1)$$

where ω and r are the rotational speed and rolling radius of the rear wheel, respectively, and v is the longitudinal vehicle speed. The longitudinal slip of the rear wheel is the natural output variable, since the traction force directly depends on it. The input variables are the throttle position and the spark advance. They both affect engine torque, and can be easily modulated by the engine ECU. Another input variable considered in some cases is the inhibition of the ignition spark in one (or more) cylinders; this variable however provides a very raw torque control. This approach is in fact being abandoned by all racing teams and manufacturers.

When dealing with the throttle, it should be pointed out that the throttle is a servo-controlled mechanism and the control variable is the requested throttle position. The influence of the servo-loop dynamics of the throttle will be discussed later.

The identification method is illustrated around a single well-defined working condition: in-plane conditions (null roll angle), second gear, 14,000 rpm, dry asphalt.

In order to identify the slip dynamics, two kinds of excitations are used: multi-frequency sinusoidal signals (also known as 'frequency sweeps') and step variations. Both tests are performed on a long (3.5 km) straight dry asphalt patch; the test rider brings the motorcycle to a given constant engine speed in a given gear. After steady-state condition is reached, the test is trigged with a button: the ECU completely overrides the driver command, and the excitation signal is applied around the neighbourhood of the initial condition. This experiment is repeatable and provides the real dynamic behaviour of the motorcycle (and rider), on a real test track (on the other hand, test rig experiments are more useful for studying specific components).

In Figure 3.2, two examples of frequency sweep (from 0 to 12 Hz) experiments are displayed. The overall I/O behaviour is significantly nonlinear; this is particularly true when using the spark advance. The nonlinearity of the response is better analysed in Figure 3.3 where the spectrograms (Oppenheim and Schafer 1989) of the two responses are plotted. The spectrograms show the frequency components of the signal as a function of the instantaneous excitation frequency.

From the spectrograms, the nonlinearity of the response shows as high-order harmonics. It is possible to identify three zones:

- **Zone I**, in the range [0–2.5 Hz] the response is essentially linear, only the first harmonic is present in the signal.
- **Zone II**, in the range [2.5–8 Hz] the response is determined by three harmonics. This is the range where the dynamics are mostly nonlinear.
- **Zone III**, in the range [8–12 Hz] the power of the third harmonic diminishes and the response is determined by two harmonics.

The stronger nonlinear behaviour of the spark advance dynamics is also manifest in the greater residual power outside the third harmonic that this response exhibits.

Due to the highly nonlinear nature of the system, the choice of the class of mathematical models to be employed is non-trivial. Here, an extension to higher harmonics of the describing function approach is proposed (see e.g. Gelb and der Velde 1968; Williamson 1976); by 'slicing' the frequency sweep and by looking at a single frequency at a time, it

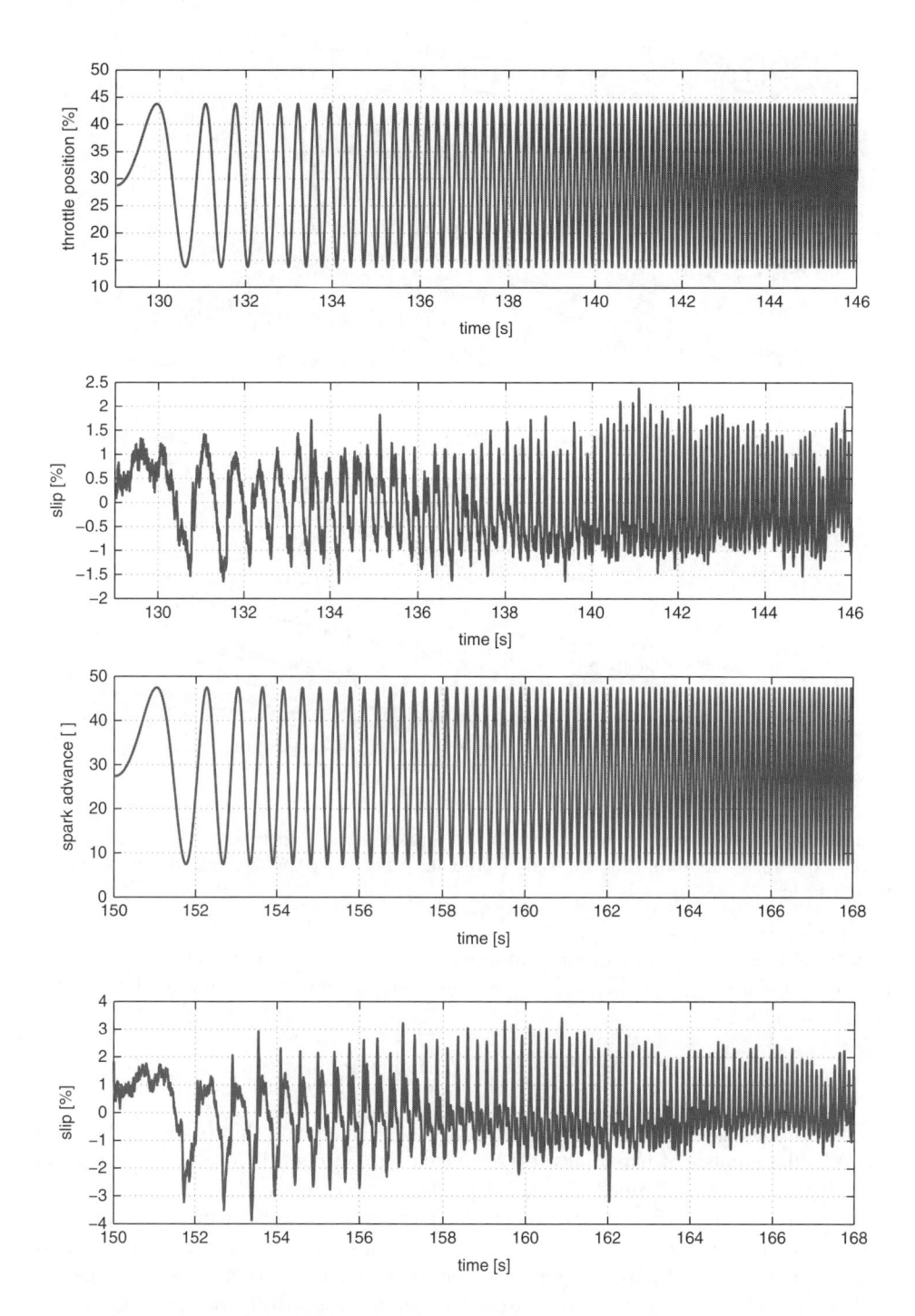

Figure 3.2 Example of a throttle sweep and spark advance sweep experiment. From top to bottom: requested throttle, rear wheel slip, spark advance, rear wheel slip

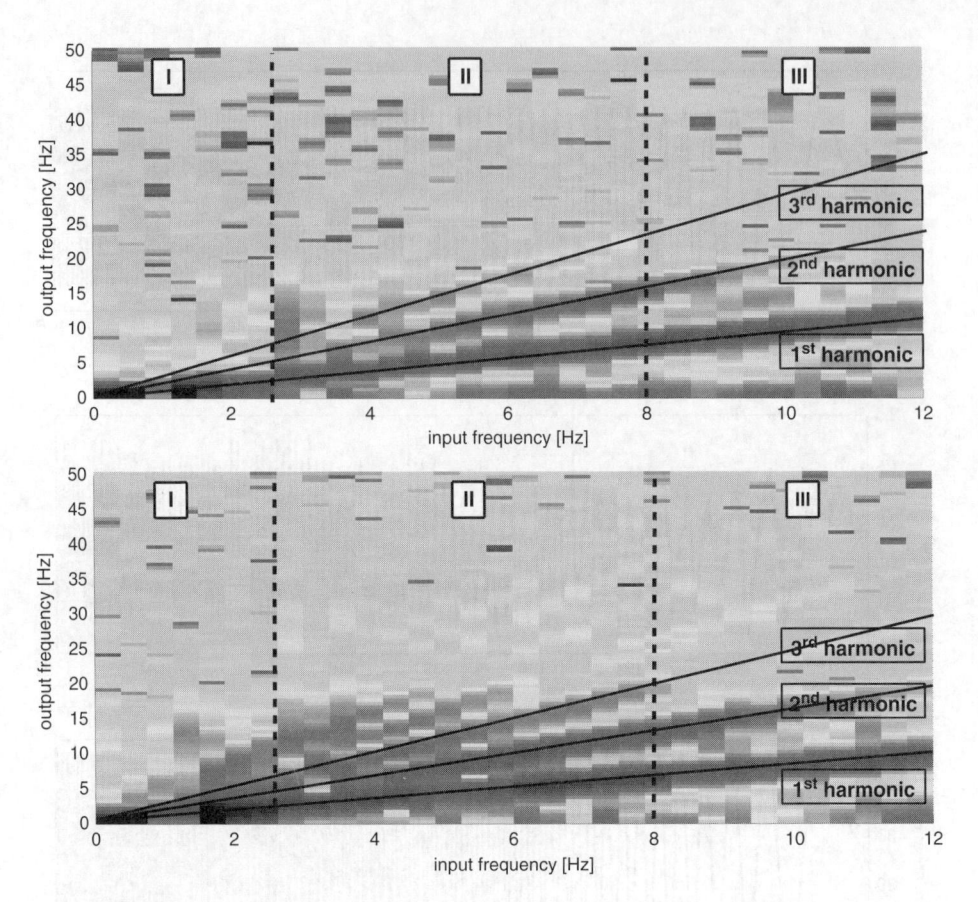

Figure 3.3 Spectrograms of the rear wheel slip throttle sweep experiment (top), spark-advance sweep experiment (bottom). Adapted from (Corno and Savaresi 2010), reproduced with permission from Elsevier

is possible to extract information on amplification and phase shift of each harmonic using Fourier transforms. Essentially, if an input signal $u(t) = \sin(\omega t)$ is considered, the output signal is written as

$$y(t) \approx \sum_{i=1}^{N} A_i(\omega) \sin(i\,\omega t + \psi_i(\omega)) \qquad (3.2)$$

where $A_i(\omega)$ and $\psi_i(\omega)$ are the amplitude amplification and phase shift of the i-th harmonic and N is the number of harmonics that are taken into account. If $N = 1$, then a classical describing function is obtained. Figure 3.4 graphically depicts the idea.

The model is composed of four elements: a pure delay in series with three harmonic generators. The first harmonic generator is a classical linear system with frequency response $G_1(j\omega)$; the second and third harmonic generators are nonlinear systems which, when fed with a sinusoidal input, generate a sinusoidal output at, respectively, twice and three times the input frequency; the amplitude and phase lag of the output signal depend on the input's frequency. Thus, for each input frequency ω_i, the harmonic generators are characterized by

Identification and Analysis of Motorcycle Engine-to-Slip Dynamics

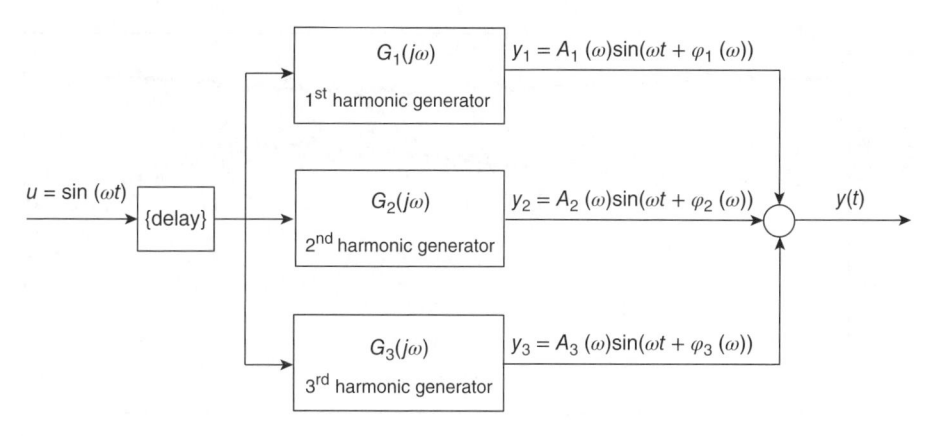

Figure 3.4 Nonlinear slip dynamics model. Adapted from (Corno and Savaresi 2010), reproduced with permission from Elsevier

a complex number representing the amplification and phase shift:

$$G_1(j\omega_i) = \frac{\Lambda_y(j\omega_i)}{\Lambda_u(j\omega_i)} \quad G_2(j\omega_i) = \frac{\Lambda_y(j2\omega_i)}{\Lambda_u(j\omega_i)} \quad G_3(j\omega_i) = \frac{\Lambda_y(j3\omega_i)}{\Lambda_u(j\omega_i)} \quad (3.3)$$

where Λ_y and Λ_u are, respectively, the Fourier transform of the input and the output. $G_1(j\omega)$, $G_2(j\omega)$ and $G_3(j\omega)$ (which can be regarded as higher-order describing functions) can be computed numerically from the experiments described in Figures 3.2 and then represented in a Bode-like diagram. Figure 3.5 depicts the amplification and phase lag for each harmonic, in the case of a throttle sweep experiment (Table 3.1), as a function of the input frequency.

From Figure 3.5 it is possible to draw the following conclusions:

- The first harmonic dominates the others. The three zones mentioned above are seen in this representation. Up to input frequencies of 2.5 Hz the magnitudes of the second and third harmonics are negligible, between 2.5 and 8 Hz both the second and third harmonics have a significant contribution, while above 8 Hz the third harmonic fades.
- All three harmonics have a resonant behaviour. In particular, the first harmonic has a resonance at around 8 Hz, the second at 7 Hz and the third at 4.5 Hz. The third harmonic shows the presence of a second resonance right after the first one. The resonant behaviour of the system is attributed to the elasticity of the transmission; the other resonant phenomena involved in motorcycle dynamics are pitch and heave (see Cossalter 2002).

Table 3.1 Throttle sweep response identified parameters

harmonics	μ	ω_z	ω_{zr}	ξ_{zr}	ω_p	ω_{pra}	ξ_{pra}	ω_{prb}	ξ_{prb}
1st	0.079	4.1 Hz	X	X	1.8 Hz	8.2 Hz	0.27	X	X
2nd	0.009	6 Hz	5.5 Hz	−0.3	X	5.2 Hz	0.22	7 Hz	0.31
3rd	0.008	−4.7 Hz	6 Hz	−0.15	X	4.4 Hz	0.31	6 Hz	0.12

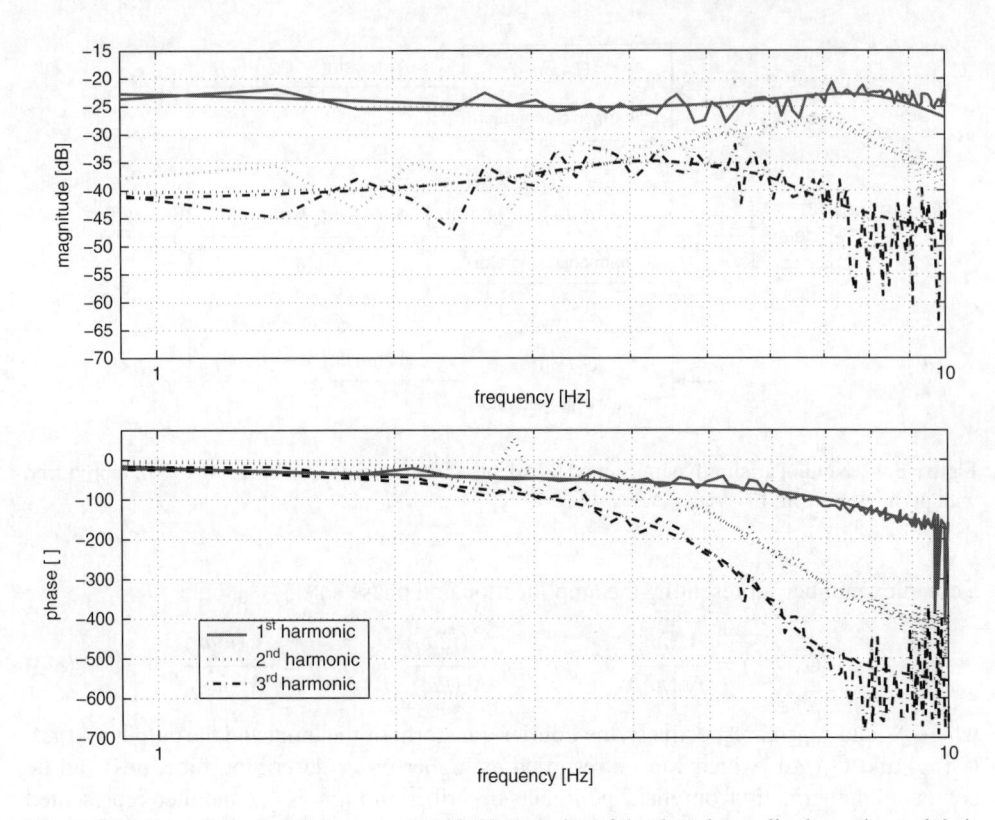

Figure 3.5 Describing functions up to the third harmonics of the throttle-to-slip dynamics and their rational approximations. The plots are reported from u to y, that is, the pure delay is also represented in the phase

The generalized describing functions can be described by rational complex functions of the form:

$$G(j\omega) = \frac{B(j\omega)}{A(j\omega)} = \frac{b_1(j\omega)^n + b_2(j\omega)^{n+1} + \ldots b_{n+1}(j\omega)}{a_1(j\omega)^m + a_2(j\omega)^{m-1} + \ldots a_{m+1}(j\omega)}, \quad (3.4)$$

where $\omega \in \mathbb{R}$. The parameters a_i and b_i are determined by solving the following non-parametric identification problem:

$$\min_{b,a} \sum_{k=1}^{l} w_f(k) \left| h(k) - \frac{B(\omega(k))}{A(\omega(k))} \right|^2, \quad (3.5)$$

where l is the number of available frequencies, $w_f(k)$ is a weight that drives the fitting toward certain frequencies and $h(k)$ is the experimental frequency response. The identified describing functions take the form

$$G_1(j\omega) = \mu_1 \frac{\left(\frac{j\omega}{\omega_{z1}} + 1\right)}{\left(\frac{(j\omega)^2}{\omega_{pr1}^2} + 2\frac{\xi_{pr1}}{\omega_{pr1}}(j\omega) + 1\right)\left(\frac{j\omega}{\omega_{p1}} + 1\right)}$$

$$G_2(j\omega) = \mu_2 \frac{\left(\frac{j\omega}{\omega_{z2}}+1\right)\left(\frac{(j\omega)^2}{\omega_{zr2}^2}+2\frac{\xi_{zr2}}{\omega_{zr2}}(j\omega)+1\right)}{\left(\frac{(j\omega)^2}{\omega_{pr2a}^2}+2\frac{\xi_{pr2a}}{\omega_{pr2a}}(j\omega)+1\right)\left(\frac{(j\omega)^2}{\omega_{pr2b}^2}+2\frac{\xi_{pr2b}}{\omega_{pr2b}}(j\omega)+1\right)}$$

$$G_2(j\omega) = \mu_3 \frac{\left(\frac{j\omega}{\omega_{z3}}+1\right)\left(\frac{(j\omega)^2}{\omega_{zr3}^2}+2\frac{\xi_{zr3}}{\omega_{zr3}}(j\omega)+1\right)}{\left(\frac{(j\omega)^2}{\omega_{pr3a}^2}+2\frac{\xi_{pr3a}}{\omega_{pr3a}}(j\omega)+1\right)\left(\frac{(j\omega)^2}{\omega_{pr2b}^2}+2\frac{\xi_{pr3b}}{\omega_{pr3b}}(j\omega)+1\right)}.$$

Notice that the first harmonic generator, thanks to the linearity hypothesis, can be treated and analysed as a transfer function and can be regarded as a linear approximation of the throttle-to-slip dynamic. A 10 ms pure delay has been introduced to model the air-box dynamics.

Figure 3.5 shows the comparison between the experimental describing functions and the analytical expression in the frequency domain. The identified describing functions are also validated in the time domain by comparing the measured slip at different frequencies with the output simulated according to the nonlinear model depicted in Figure 3.4. This validation is reported in Figure 3.6. The figure exemplifies the advantage of taking into account different harmonics and confirms that for high frequencies a two harmonic approximation yields good fitting, while a three harmonic approximations is needed for Zone II.

By looking at the time domain response, it is clear that the nonlinearities are partly due to the asymmetric behaviour of the system. The slip response to an opening of the throttle is different from the closing one. Figure 3.7 better describes the phenomenon. The differences between the rising phase and the falling phase are clear:

- The falling phase starts almost immediately after the throttle starts closing; the raising phase is delayed. The time span between the zero derivative point of the throttle and the change in sign of the derivative of the slip is four times longer in the rising phase than in the falling phase.
- In the rising phase, the rate of change during the raising phase is greater than in the falling phase.

A physical interpretation can be conjectured. When the slip increases, it does so because of the torque generated by the engine; in the opposite phase the slip decreases because of friction and engine braking. The torque generation is located at the engine, while friction is distributed along the entire transmission and this may motivate the faster reaction to the closing of the throttle.

A switching linear system captures the observed phenomenon. Two linear systems are defined, and the switching between them is driven by the throttle reference first derivative:

$$\begin{aligned}\dot{x}(t) &= A_i x(t) + B_i u(t) \\ y(t) &= Cx(t)\end{aligned} \quad \text{where} \quad \begin{aligned} A_i &= A_{open} \text{ and } B_i = B_{open} \text{ if } \dot{u}(t) > 0 \\ A_i &= A_{close} \text{ and } B_i = B_{close} \text{ if } \dot{u}(t) \leq 0 \end{aligned}$$

where A_{open}, B_{open}, A_{close}, B_{close} and C are linear system matrices in the observability canonical form. The choice of the observability canonical form enables a smooth transition between the two systems simply by keeping the state vector unchanged at the switching. The parameters of the two linear models have been identified from the step response experiments, which

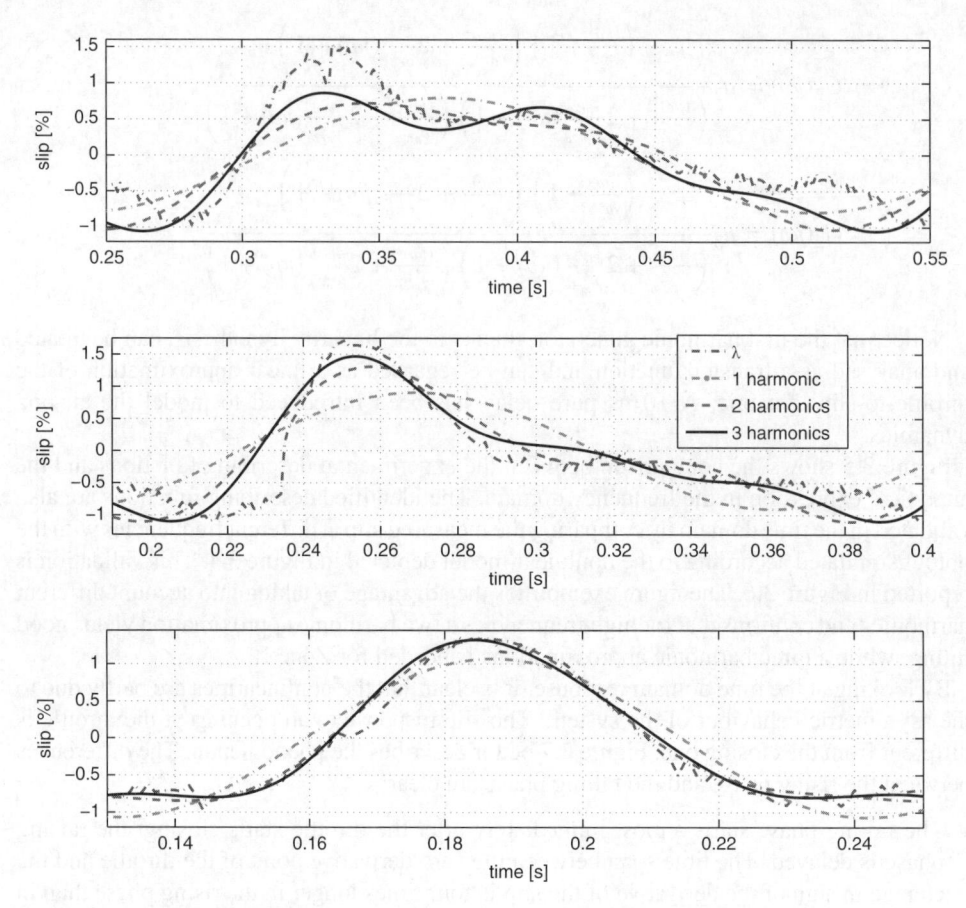

Figure 3.6 Time domain throttle responses at 3.5 Hz (top), 5 Hz (centre) and 9 Hz (bottom). Adapted from (Corno and Savaresi 2010), reproduced with permission from Elsevier

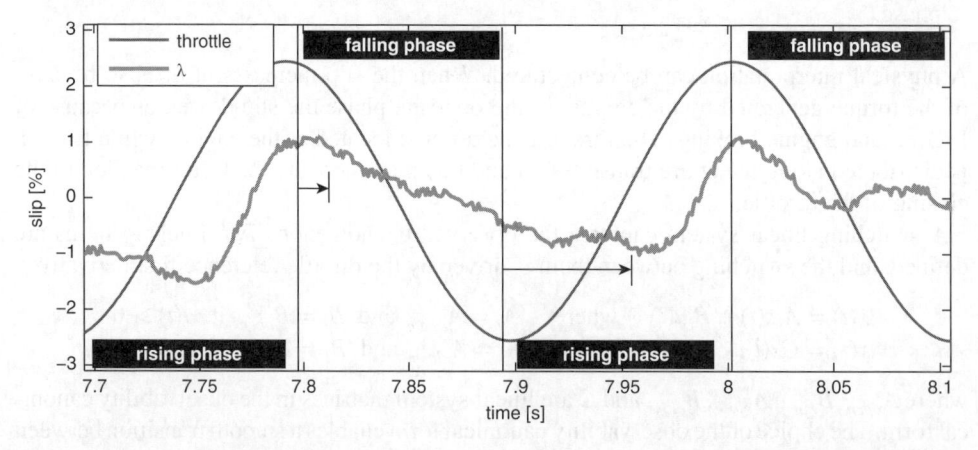

Figure 3.7 Throttle sweep experiment (detail). Adapted from (Corno and Savaresi 2010), reproduced with permission from Elsevier

better isolate the asymmetries. The first harmonic approximation transfer function has been used as the initial point for the following non-convex optimization problem:

$$\min_{\mu,\omega_z,\omega_p} \sum_{k=1}^{N} w_t(k) \left(\frac{y(k) - y_{sim}(\mu,\omega_z,\omega_p)}{N} \right)^2 \qquad (3.6)$$

where $y(k)$ is the measured slip, $y_{sim}(\mu,\omega_z,\omega_p)$ is the slip simulated with the switched system, N is the number of samples in the experiment and w_t is a weight. Note that only the gain, real pole and real zero are subject to optimization. The resonant mode is not changed because it is attributed to the transmission chain dynamics which are assumed symmetric. The obtained values are summarized in the Table 3.2. Figure 3.8 shows the validation of the switched model.

The overall fitting is satisfactory. To put the fitting in perspective, consider that wheel slip is subject to a noise with a peak amplitude of 0.5%. This conclusion is confirmed by the validation test run on the sweep data. Figure 3.9 shows the response of the switched system for three different frequencies. The switched system successfully captures the asymmetric behaviour.

The same identification procedure is applied to the spark advance experiments; the results are briefly commented. Figure 3.10 illustrates the high-order describing functions and their rational approximations.

From Figure 3.10 it is possible to note the following:

- The first harmonic dominates the others. The three zones are still present: the spark-to-slip dynamics is characterized by a faster rise of the second and third harmonics: at 3 Hz the second and third harmonics are 14 dB lower than the first harmonic in the throttle

Table 3.2 Throttle switched systems identified parameters

open	0.1027	2.2 Hz	1.5 Hz	8.2 Hz	0.27
close	0.071	4.1 Hz	1.9 Hz	8.2 Hz	0.27

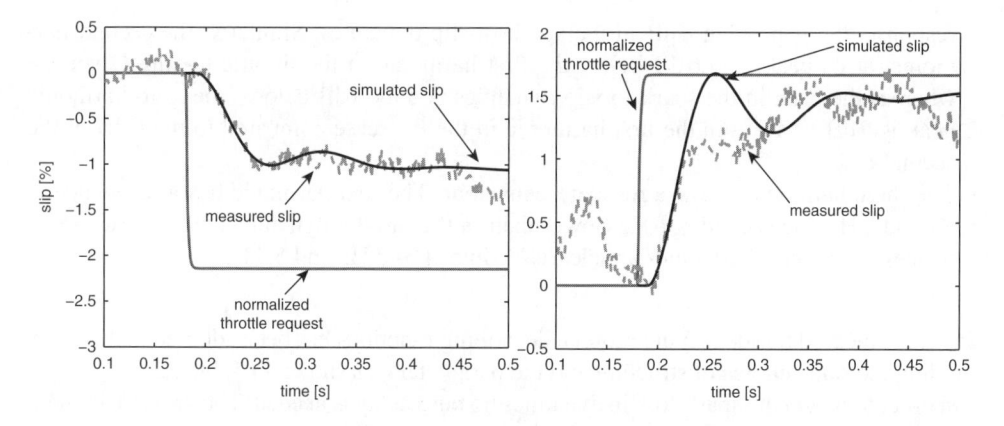

Figure 3.8 Measured and simulated throttle step responses

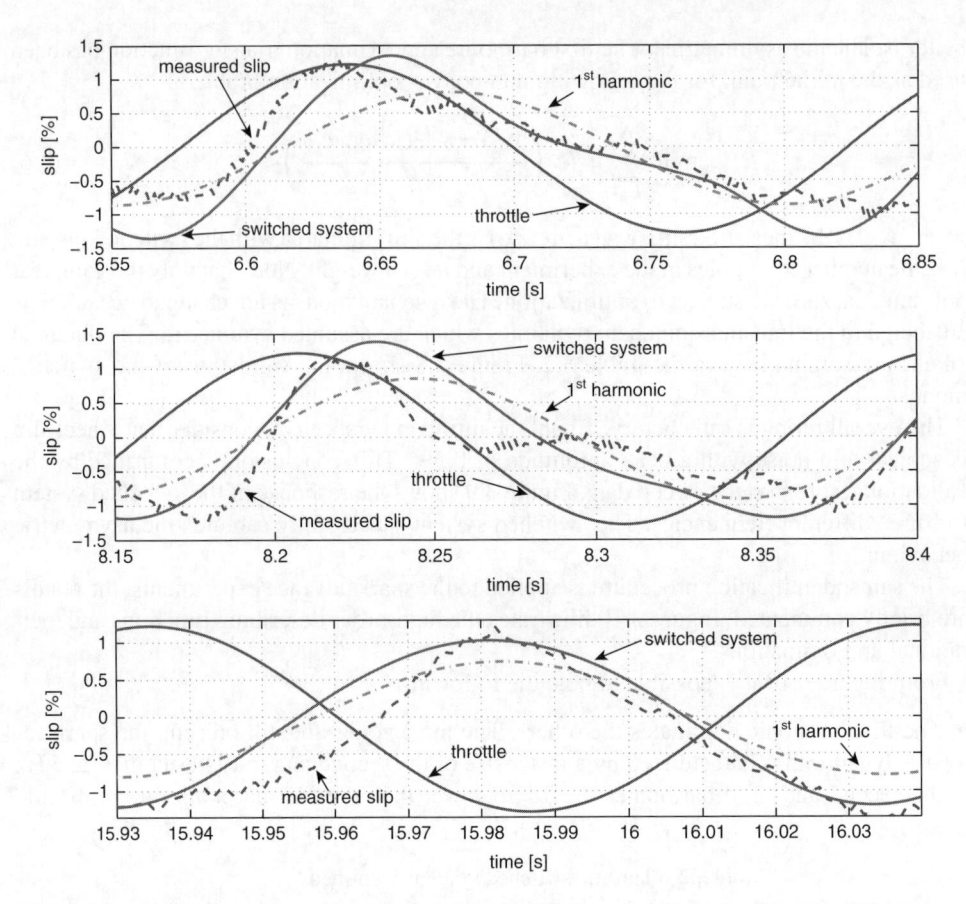

Figure 3.9 Validation of the switched system on a throttle sweep experiment. Detail at 3 Hz (top), at 5 Hz (centre) and 9 Hz (bottom). The plots show the throttle input, the measured slip and two simulated slips (first harmonic approximation and switched system)

response, the gap being 8 dB in the spark-to-slip dynamics. Similarly, the second harmonic, at its peak, is 4.6 dB below the first harmonic in the throttle-to-slip dynamics, whereas the peak in the spark-to-slip dynamics is only 3 dB below. The third harmonic peak is 9 dB away from the first harmonic in the first case compared to the 6 dB in the second case.
- The three harmonics show a resonant behaviour. The first harmonic has a resonance at around 8 Hz, the second at 6 Hz (lower than in the throttle dynamics) and the two resonances of the third harmonics are clearly visible at 3–5 Hz and 5.4 Hz.

The same method for identifying rational describing functions has been adopted in this case, yielding the same choice of structure and the parameters, as shown in Table 3.3.

In order to model the spark-to-slip dynamics no pure delay is needed. The previously mentioned asymmetric behaviour is noted also in the spark-to-slip dynamics; a second switching linear system is derived to describe this behaviour. Table 3.4 shows the identified parameters

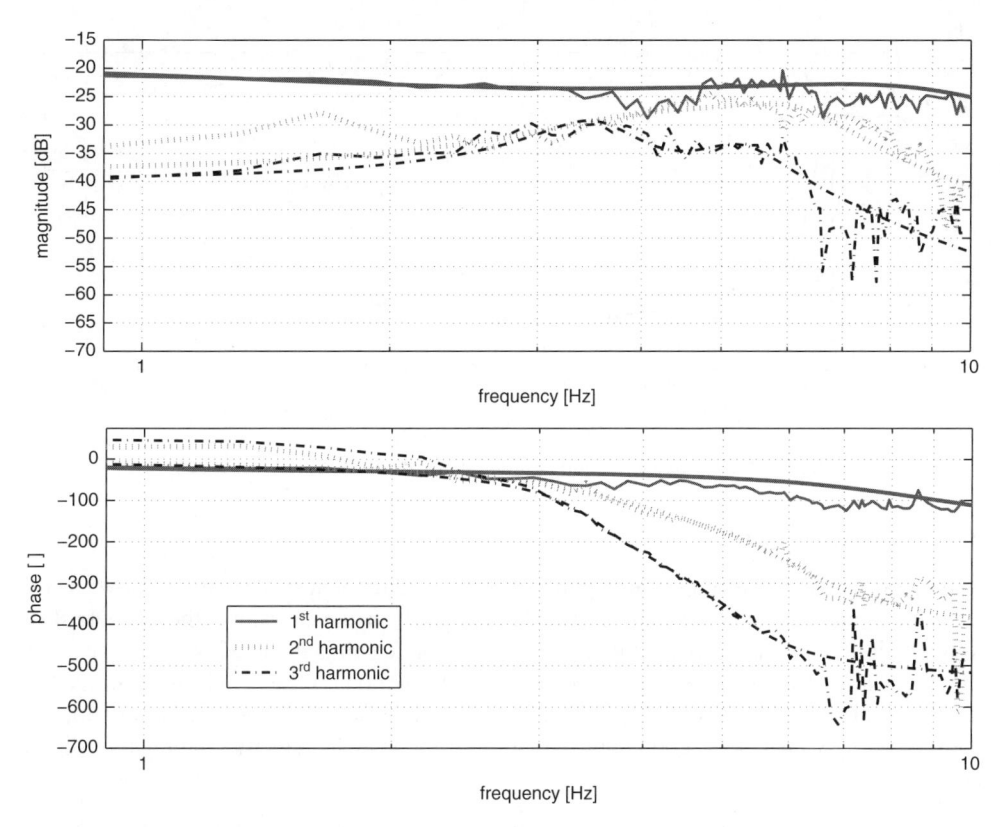

Figure 3.10 Describing functions up to the third harmonics of the spark advance-to-slip dynamics and their rational approximations. The plots are reported from u to y, i.e. the pure delay is also represented in the phase

Table 3.3 Spark advance response identified parameters.

1st	0.1069	2.2 Hz	X	X		1.1 Hz	8.9 Hz	0.42	X	X
2nd	0.0127	4.1 Hz	6.2 Hz	−0.59	X		4.2 Hz	0.34	6.1 Hz	0.13
3rd	0.0104	X	4.4 Hz	−2.25	X		3.5 Hz	0.14	5.4 Hz	0.13

Table 3.4 Throttle switched systems identified parameters.

open	0.0713	4.2 Hz	2.1 Hz	8.9 Hz	0.42
close	0.1069	4.0 Hz	2.0 Hz	8.9 Hz	0.42

and Figure 3.11 plots a comparison between simulation and measured step data. The two dynamics differ mainly in the low frequency gain.

As can be appreciated, the fitting in the time domain is satisfactory, although not as good as in the throttle case.

The identification has been carried out assuming the throttle set point as the input and the absolute rear wheel slip as the output; the following remarks elaborate upon the consequences of these choices.

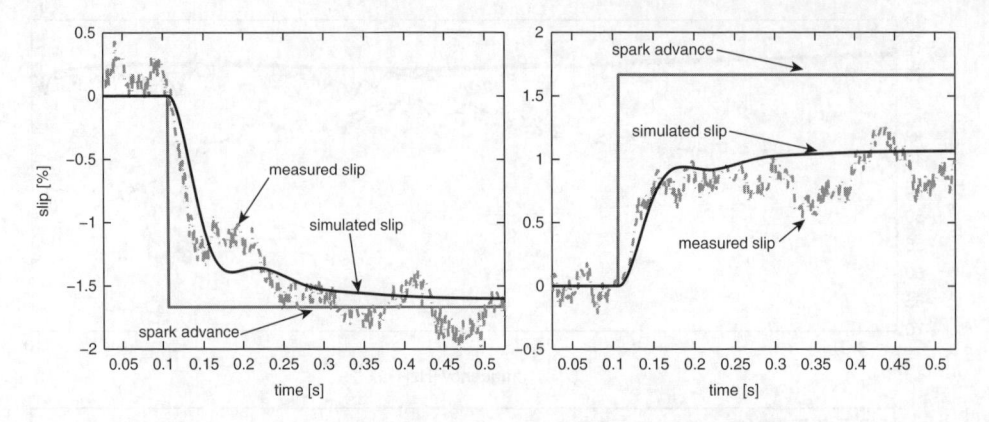

Figure 3.11 Measured and simulated spark advance step responses

3.3.1 Relative Slip

The measurement of absolute slip requires expensive equipment that can be employed only for testing purposes. Traction control systems cannot rely on the measure of the absolute slip; as a solution, the relative wheel slip is used. The relative slip is obtained by comparison of the rear wheel speed with the front wheel speed. This method is based on the hypothesis that during acceleration the longitudinal slip of the front tyre is null. If the tyre is not slipping then its velocity provides a good approximation of the vehicle velocity. The relative slip is computed as:

$$\lambda_r = \frac{\omega_r r_r - \omega_f r_f}{\omega_r r_r}.$$

The effects of using the relative slip are assessed in Figure 3.12, which shows the experimental first harmonic for the two output variables derived from throttle reference sweep experiments. The two dynamics are essentially the same. A slight difference in the frequency range [2, 4] Hz related to the vertical dynamics of the motorcycle is observed. Load variations on the front wheel cause variation of the rolling radius and introduce a bias in the estimation of the velocity.

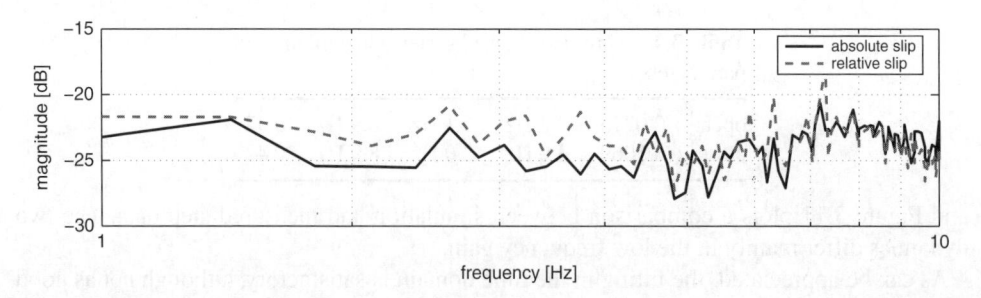

Figure 3.12 First harmonics approximation of the absolute slip (continuous line) and relative slip (dashed line) dynamics

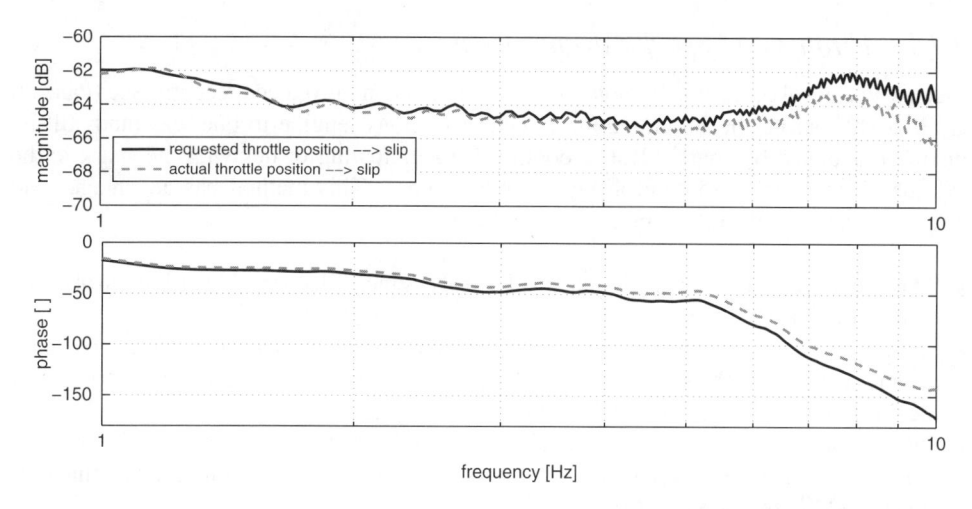

Figure 3.13 First harmonics approximation of the controlled (continuous line) and intrinsic (dashed line) dynamics

3.3.2 Throttle Dynamics

The throttle set point has been used as the input variable because that is the control variable available to a traction control system. Nevertheless, it is interesting to study the intrinsic wheel slip dynamics, i.e. the dynamics from actual throttle position to rear wheel slip. Figure 3.13 shows the first harmonic approximations of the rear wheel slip controlled dynamics (considering the throttle reference as the input) and the rear wheel slip dynamics (the dynamics from the actual throttle position). The throttle servo controller introduces a phase loss in the loop. In particular, at 10 Hz, around 30° of phase loss is due to the servo.

3.4 Engine-to-Slip Dynamics Analysis

The proposed identification method has several benefits: it can be easily performed on any motorcycle equipped with a programmable ECU and does not require specific devices (conversely, white-box modelling requires the accurate measurement of many parameters such as engine map, tyre characteristics, transmission stiffness, etc.). Despite its simplicity, it provides quantitative analysis useful for both designing traction control systems and comparing different motorcycles and scenarios.

In this section, a series of example uses of the proposed method are provided. The following aspects will be considered: an in-depth quantitative comparison of throttle control and spark advance control, an example of how the method can benchmark different motorcycles and an analysis of the effect of different road surfaces and speeds on the engine-to-slip dynamics. Please note that these analyses have been done over a period of years of research. Each analysis has been carried out on a different motorcycle, and possibly under different conditions, so the conclusions cannot be cross-analysed beyond what is explicitly stated.

3.4.1 Throttle and Spark Advance Control

Electronic control of throttle in motorcycles has only been introduced recently, see (Panzani et al. 2013); before the introduction of such technology, engine torque was (and still is) mainly modulated by spark advance control. By anticipating or delaying the spark in the cylinder, it is possible to control the generated torque. This method has advantages and disadvantages over throttle regulation, in particular:

- Spark advance regulation is possible in all petrol engines. Electronic throttle control needs extra hardware.
- Spark advance regulation directly influences the engine torque. The effects of spark advance are seen in the same engine cycle that the variation is applied, while variation of the throttle takes longer to manifest in terms of torque.
- Optimal operation of the engine requires the spark advance to be in a relatively small range; if engine torque is regulated via spark advance for a long period, the engine efficiency and life may be affected.
- Spark advance guarantees a smaller modulability than throttle control; that is, at a given engine speed the torque variation that can be generated by action on spark advance is limited. Throttle control can reach the engine torque limits.
- Spark advance responses are characterized by stronger nonlinearities than throttle responses.

The above considerations are valid, but their quantification is not trivial. In this subsection, the first harmonic approximation method and measured data lead to conclusions relevant to the design of a traction control for motorcycles. In order to have a better vision of the features of the dynamics, the first harmonics approximation for the throttle-to-slip and spark-to-slip dynamics are shown in Figure 3.14, while Figure 3.15 shows the comparison between step responses.

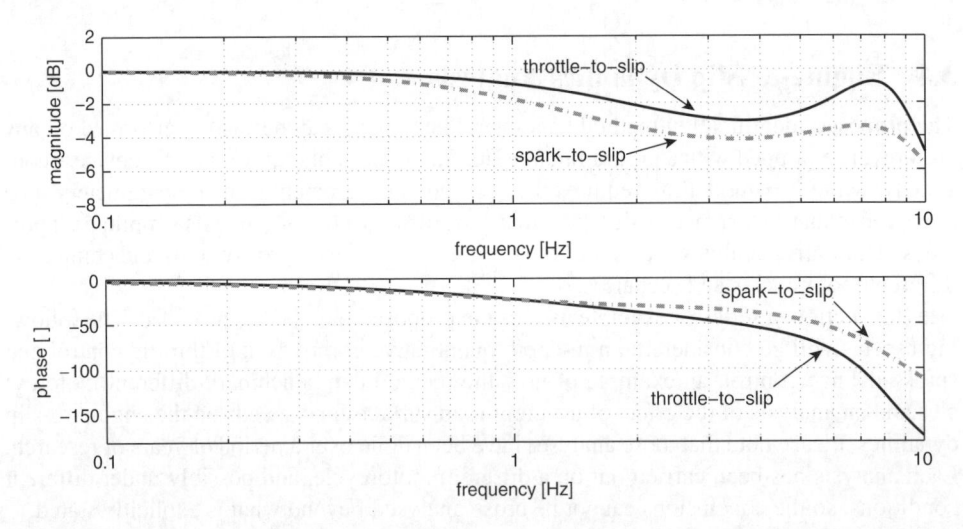

Figure 3.14 Spark advance-to-slip and throttle-to-slip dynamics. For easier comparison, the low frequency gains are normalized to 1

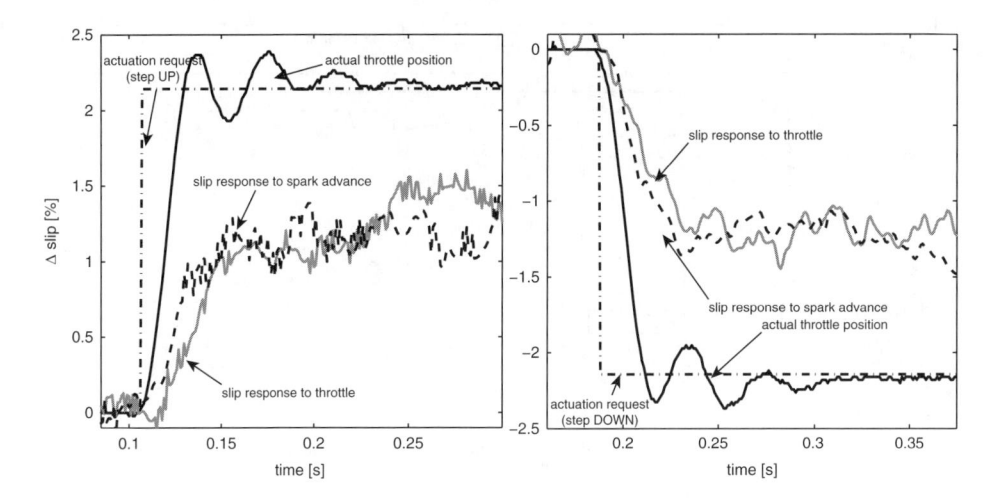

Figure 3.15 Experimental step responses; comparison between different actuations

The figure points to the following conclusions:

- Both dynamics show the transmission resonance at 8 Hz. The resonance in the throttle-to-slip dynamics seems less damped because of a coupling with the throttle control loop. By coincidence, the servo loop cut-off frequency is coupled with the resonance of the transmission and amplifies it.
- Spark advance is 'faster' than throttle action. At 10 Hz, there is a $60°$ difference in phase: half of this loss is due to the servo-loop. This observation proves that although torque control is better achieved by spark advance, the real bottleneck is the transmission. Slip control through throttle control is indeed achievable.
- Although the spark advance provides a slightly faster actuation, it should also be noted that the response of the system to spark advance variation is less linear, and therefore more difficult to model and control.

These conclusions are confirmed by the analysis of the step responses, summarized in Figure 3.15. The response to spark-advance is faster than the response to throttle, but the difference is not critical. This difference is even smaller in closing dynamics which is more relevant to safety-oriented traction control.

3.4.2 Motorcycle Benchmarking

The proposed method can be used to quantitatively compare different motorcycles. Here we show the comparison between a racing bike and a standard road bike. The two motorcycles are different not only for their physical properties but also in terms of electronic system architecture. The racing bike has a single programmable ECU on which the throttle servo loop runs and the hardware that generates the voltage for the spark plug is integrated in the ECU. The road bike electronic system is more modular and based on a CAN-bus. This architecture does not support the direct actuation of a desired throttle position; the engine control unit is torque-based. The desired torque is sent (via CAN at a send rate of 10 ms) to

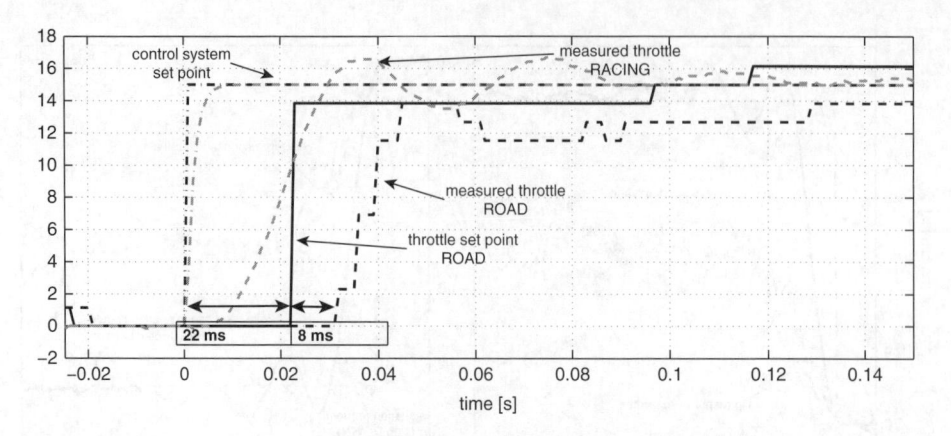

Figure 3.16 Step responses to the control variable

the ECU, specifying when the torque command is actuated through throttle or spark advance (the system documentation refers to these two channels respectively as ASR slow and ASR fast). Once the torque set point has been received by the ECU, a proprietary control algorithm generates the throttle set point (or spark) which is then sent (via CAN at a send rate of 4 ms) to the throttle servo loop (EFI) and architecture is functionally laid out. The CAN bus introduces non-deterministic communication delays; (Tindell et al. 1995). The consequences of the two architectural choices are shown in Figure 3.16 where two responses to step variation of the control variable are shown overlaid. The delays affecting the road bike are marked in the figure; once the torque reference step signal is sent, the throttle reference takes around 22 ms to start rising, and another 8 ms are needed for the throttle to actually start moving. It is easy to predict that these delays will affect the maximum achievable bandwidth of the traction control system.

The road bike is affected by another disadvantage. Commercial bike engines are mapped to meet pollution regulations; this limits the range of variation of the spark advance, to the point where it is practically impossible to perform the protocol to identify the spark advance dynamics.

Some remarks are due:

- The road bike's frequency response is dependable only up to 6 Hz; above that frequency the signal-to-noise ratio greatly degrades.
- The road bike's transfer function has a different structure. It has a couple of complex conjugate poles, a real pole and a couple of complex conjugate zeros at 6 Hz.
- The pure delay affecting the road bike is much longer: 75 ms against 10 ms.
- The road bike's transmission resonance falls outside the aforementioned frequency limit; nevertheless it exhibits a resonance at 4 Hz which is due to the vertical dynamics. Apparently the suspensions are less stiff.

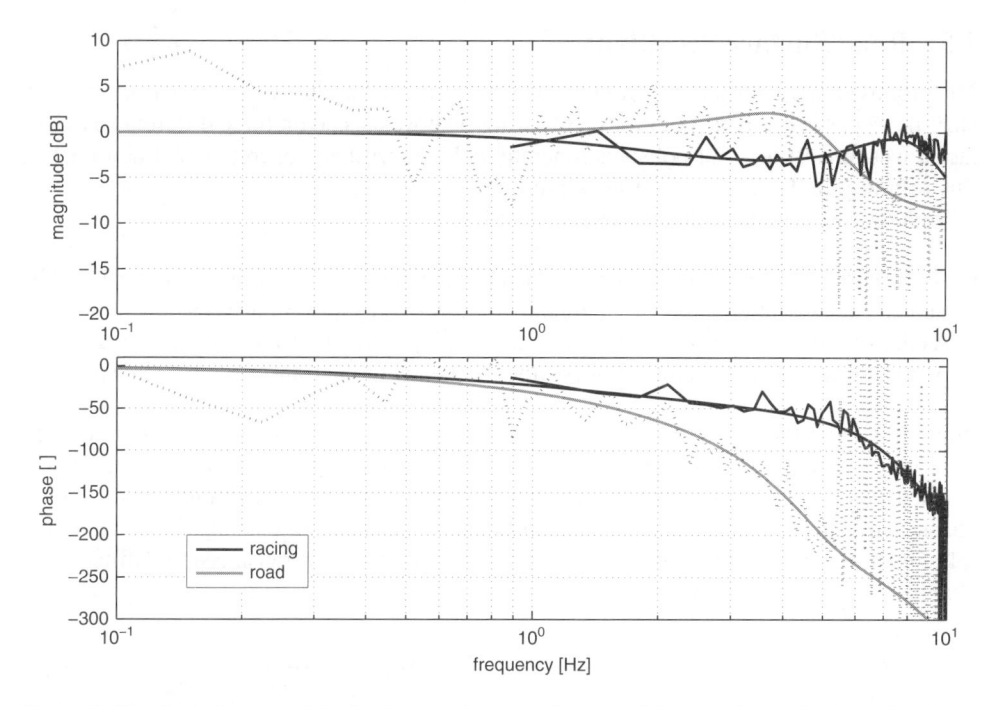

Identification and Analysis of Motorcycle Engine-to-Slip Dynamics

Figure 3.17 Bode diagram of the first harmonic approximation of the throttle-to-slip dynamics. Road bike (dashed), racing bike (solid). The low frequency gain has been normalized to improve readability

Figure 3.18 Test ground with three different surfaces

The road bike is, not surprisingly, of lower performance: it has a slower slip response and it is critically affected by delay (Figure 3.17).

3.5 Road Surface Sensitivity

The proposed method is also useful for quantifying the dynamic effect of other parameters. Here the focus is on road conditions. A touring bike was tested on three different road surfaces: wet basalt tiles, wet polished concrete and wet brushed concrete (see Figure 3.18). The resulting Bode diagrams are shown in Figure 3.19.

From the analysis the following can be noted:

- As the surface becomes more slippery, the DC gain of the transfer function increases. The same torque variation generates a greater wheel slip on slippery surfaces.
- As the surface becomes more slippery, the damping of the transmission mode increases.
- As the surface becomes more slippery, the frequency of the transmission decreases.

These tests offer the possibility of better understanding the role of the transmission dynamics. These experiments shows that the frequency of the transmission resonance is not fixed, but depends on the road surface. Although this may seem counter-intuitive, one should consider that the tyre–road surface dynamics add an extra equivalent stiffness to the transmission. The tyre–road interaction effectively changes the over transmission stiffness thus changing its resonant frequency.

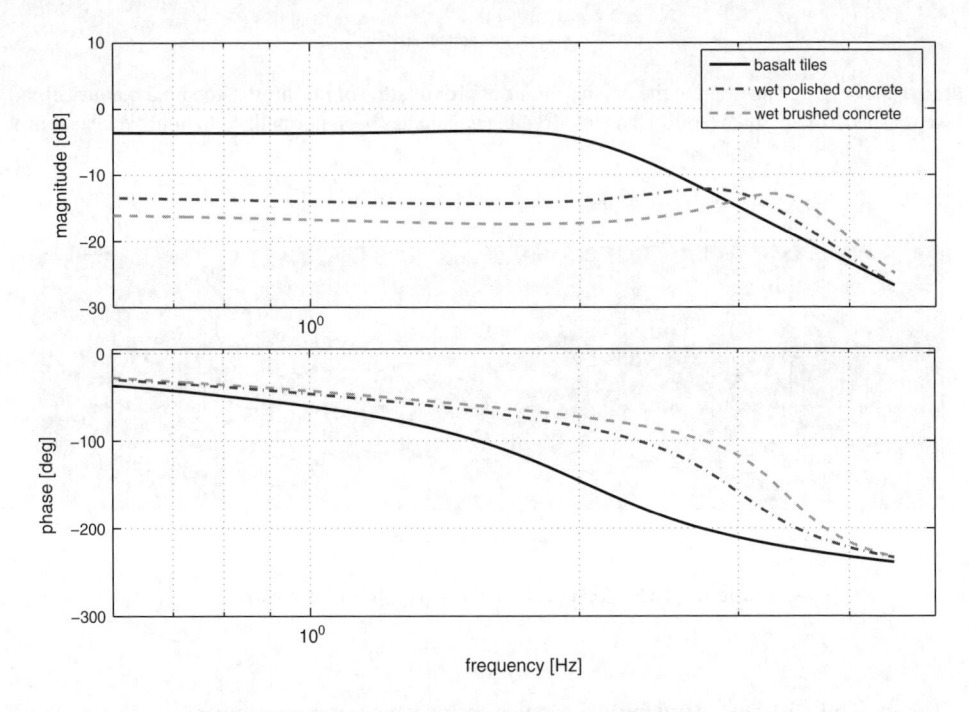

Figure 3.19 Rear slip transfer functions on different surfaces

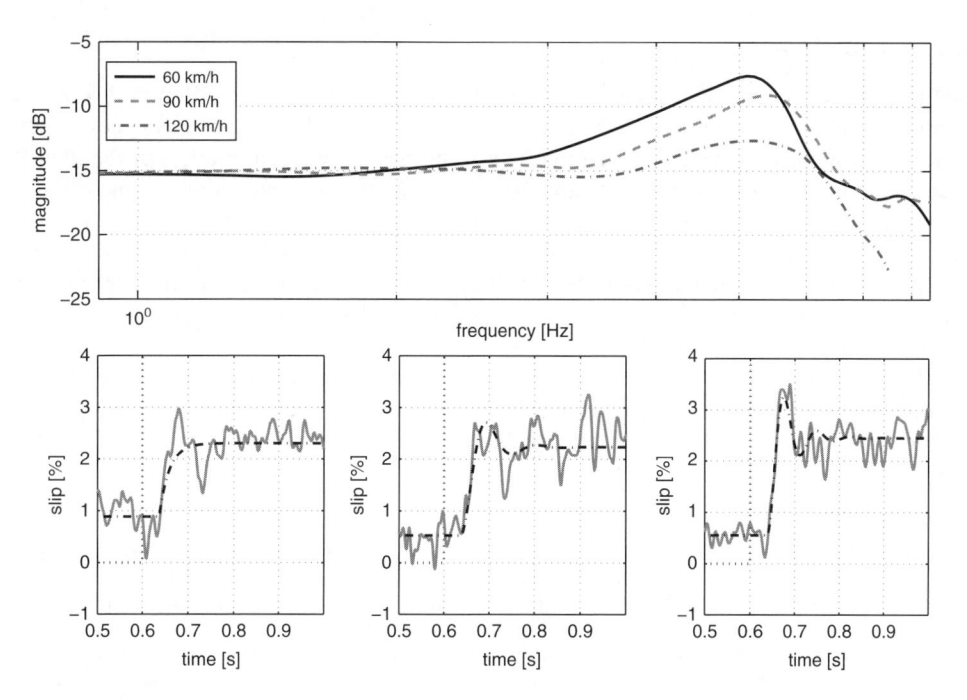

Figure 3.20 Engine-to-rear-slip transfer function (upper) and step responses (lower) performed at different velocities: 120 km/h (left), 90 km/h (centre), 60 km/h (right)

3.6 Velocity Sensitivity

The final analysis relates to the effect of vehicle velocity on the engine-to-slip dynamics. The well-known and often employed single-corner model of wheel slip dynamics shows that the velocity has an effect only on the position of the pole and not on the gain of the transfer function. The tests performed on a racing motorcycle on dry asphalt and shown in Figure 3.20 confirm this analysis and enable other observations:

- The velocity mainly affects the damping of the resonance. It is better damped at high velocity.
- The wheel slip pole variations – that according to the single-corner model should move toward high frequency as the vehicle decelerates – are completely overcome by the dependency of the transmission resonance on velocity. If anything, the position of the complex poles that determine the resonance *increases* with the velocity of the vehicle.

The above considerations underline that the classical single-corner model is not adequate for describing motorcycle traction dynamics, as shown in (Corno et al. 2009) for braking.

3.7 Conclusions

In this chapter an engine-to-wheel slip identification protocol has been proposed. The proposed method is based on tests performed in real conditions. The tests are frequency sweeps and step responses. The data-driven black-box approach yields two models of the dynamics: a simple, first harmonic approximation and a more advanced switched linear model that can model asymmetries in the response. The first harmonic approximation is a powerful method for drawing conclusions useful for the design of traction control systems. The chapter analyses the following aspects: the differences between slip actuation via throttle and spark advance control, the response of different motorcycles and electronic architectures, the effect of different surfaces and velocities.

The proposed analysis opens the way to advanced slip controllers, in which the two control variables are used in coordination to implement advanced strategies that, for example, can achieve fine traction control without impeding engine efficiency.

References

Corno M and Savaresi S 2010 Experimental identification of engine-to-slip dynamics for traction control applications. *European Journal of Control* **16**(1), 88–108.
Corno M, Savaresi S and Balas G 2009 On linear parameter varying (LPV) slip-controller design for two-wheeled vehicles. *International Journal of Robust and Nonlinear Control* **19**(12), 1313–1336.
Corno M, Tanelli M, Savaresi S and Fabbri L 2011 Design and validation of a gain-scheduled controller for the electronic throttle body in ride-by-wire racing motorcycles. *IEEE Transactions on Control Systems Technology* **38**(7), 18–30.
Cossalter V 2002 *Motorcycle Dynamics*. Race Dynamics, Milwaukee, USA.
Cossalter V and Lot R 2002 A motorcycle multi-body model for real time simulations based on the natural coordinates approach. *Vehicle System Dynamics: International Journal of Vehicle Mechanics and Mobility* **37**, 423–447.
Cossalter V, Doria A and Lot R 1999 Steady turning of two-wheeled vehicles. *Vehicle System Dynamics* **31**(3), 157–181.
Cossalter V, Doria A and Lot R 2004a Development and validation of a motorcycle riding simulator *Proceedings of the World Automotive Congress (FISITA)*.
Cossalter V, Lot R and Maggio F 2004b The modal analysis of a motorcycle in straight running and on a curve. *Meccanica* **39**(1), 1–16.
Gelb A and der Velde WV 1968 *Multiple-Input Describing Functions and Nonlinear System Design*. McGraw-Hill, New York.
Hauser J and Saccon A 2006 Motorcycle modelling for high-performance maneuvering. *Control Systems, IEEE* **26**(5), 89–105.
Limebeer D and Sharp R 2006 Bicycles, motorcycles, and models. *Control Systems Magazine, IEEE* **26**(5), 34–61.
Limebeer D, Sharp R and Evangelou S 2001 The stability of motorcycles under acceleration and braking. *Proc. I. Mech. E., Part C, Journal of Mechanical Engineering Science* **215**, 1095–1109.
Limebeer D, Sharp R and Evangelou S 2002 Motorcycle steering oscillations due to road profiling. *Journal of Applied Mechanics* **69**(6), 724–739.
Ljung L 1987 *System identification: theory for the user*. Prentice-Hall Englewood Cliffs, NJ.
Lot R 2004 A motorcycle tire model for dynamic simulations: theoretical and experimental aspects. *Meccanica* **39**(3), 207–220.
Nehaoua L, Hima S, Arioui H, Seguy N and Espie S 2007 Design and modelling of a new motorcycle riding simulator *American Control Conference, 2007. ACC '07*, pp. 176–181 IEEE.
Oppenheim A and Schafer R 1989 *Discrete-Time Signal Processing*. Prentice-Hall, Englewood Cliffs, NJ.
Panzani G, Corno M and Savaresi SM 2013 On adaptive electronic throttle control for sport motorcycles. *Control Engineering Practice* **21**, 42–53.
Rankine W 1869 On the dynamical principles of the motion of velocipedes. *The Engineer* **28**(79), 129.
Schwab A, Meijaard J and Kooijman J 2007 Some recent developments in bicycle dynamics. *Proceedings of the 12th World Congress in Mechanism and Machine Science*.

Sharp R and Limebeer D 2004 On steering wobble oscillations of motorcycles. *Proceedings of the Institution of Mechanical Engineers, Part C: Journal of Mechanical Engineering Science* **218**(12), 1449–1456.

Sharp R, Evangelou S and Limebeer D 2004 Advances in the modelling of motorcycle dynamics. *Multibody System Dynamics* **12**, 251–283.

Skogestad S and Postlethwaite I 2007 *Multivariable feedback control, analysis and design*. Wiley.

Tanelli M, Schiavo F, Savaresi SM and Ferretti G 2006 Object-oriented multibody motorcycle modelling for control systems prototyping *Proceedings of the 2006 IEEE International Symposium on Computer-Aided Control Systems Design (CACSD), Munich, Germany*, pp. 2695–2700.

Tindell K, Burns A and Wellings A 1995 Calculating Controller Area Network (CAN) Mes Response Times. *Control Engineering Practice* **3**(8), 1163–1169.

Williamson D 1976 Describing function analysis and oscillations in nonlinear networks. *International Journal of Control* **24**(2), 283–296.

4

Virtual Rider Design: Optimal Manoeuvre Definition and Tracking

Alessandro Saccon[a], John Hauser[b], and Alessandro Beghi[c]
[a]*Department of Mechanical Engineering, Eindhoven University of Technology, The Netherlands*
[b]*Department of Electrical, Computer, and Energy Engineering, University of Colorado Boulder, USA*
[c]*Department of Information Engineering, University of Padova, Italy*

4.1 Introduction

Modern sports motorcycles possess an impressive combination of power and agility, and are capable of a truly broad range of manoeuvres.

Producing a high-performance motorcycle prototype entails extensive engineering, before skilled riders can evaluate the performance and handling qualities of the physical system. At the engineering stage, *virtual prototyping* can play an important role. During the past two decades, a major effort has been made by researchers to obtain ever more accurate and detailed mathematical models of two-wheeled vehicles. For handling and manoeuvrability studies, a key role is played by multibody modelling: (Cossalter and Lot 2002; Evangelou 2003; Limebeer and Sharp 2006; Lot and Da Lio 2004; Sharp and Limebeer 2001). Models are used to explore the dynamic properties of motorcycles including linearization to study vibration modes under constant-speed and constant-turn-radius conditions (Sharp et al. 2004). Multibody codes and related tools are finding increased use among designers and engineers interested in improving handling qualities, performance and safety.

Simulation studies determine whether the motorcycle can perform manoeuvring tasks such as path following with a specified velocity profile. Although open-loop simulation can be used to analyse automobile handling, a virtual motorcycle rider is needed to stabilize the

---

Modelling, Simulation and Control of Two-Wheeled Vehicles, First Edition.
Edited by Mara Tanelli, Matteo Corno and Sergio M. Savaresi.
© 2014 John Wiley & Sons, Ltd. Published 2014 by John Wiley & Sons, Ltd.

roll mode. Without feedback to provide appropriate steering and throttle/brake inputs, the simulated motorcycle usually falls quickly. The present contribution focuses on the high-performance manoeuvring required in racing competitions, providing a selective literature review and summarizing the authors' contributions in the field.

Simulation models can be used to explore the maximum performance and minimum lap time problem for terrestrial vehicles. Strategies for approximating the solution to the minimum-time problem for cars and motorcycles have been extensively developed. For instance, the quasi-static strategies developed in (Milliken and Milliken 1995) for approximating the minimum lap time performance for racing cars have been used since the 1950s. Although transient behaviour is neglected, the approach of (Milliken and Milliken 1995) allows the use of detailed racing car models. A more direct attack on the minimum-time problem for racing car performance is described in (Casanova et al. 2000), where an optimal control problem involving a seven-degree-of-freedom (7DOF) racing car model is discretized using a parallel shooting method. The resulting nonlinear programme is solved by using a sequential quadratic programming algorithm. This work produces both the race line and the velocity profile, dealing with the dynamic behaviour of the racing car model in a more complete manner. We also note that (Casanova et al. 2000) provide a comprehensive review of maximum performance research for cars.

Due mainly to instability issues, the exploration of maximum performance for motorcycle models is more recent. In (Cossalter et al. 1999), optimal control techniques are used to define and assess a notion of motorcycle manoeuvrability. The cost function uses penalty functions to address constraints such as the width of the road. Using a combination of penalties, (Cossalter et al. 1999) produce plausible approximate race lines. A more direct attack on the maximum performance and minimum lap time problem for motorcycles is presented in (Bertolazzi et al. 2005), which also uses penalty functions to handle inequality constraints. The optimal solution is found by solving a discretized two-point boundary value problem expressing the first-order optimality conditions. Symbolic software is used to develop routines for evaluating the derivatives that occur in the boundary value problem. This strategy is used to compute an optimal trajectory for a complete race track.

One way to reduce the complexity of the maximum performance problem for motorcycles is to seek the optimal velocity profile for a given fixed path. Removing from consideration the selection of the race line, we can focus directly on how various constraints limit the performance of a motorcycle. This is the approach we describe in this contribution.

We have developed over the years a quasi-steady-state technique for approximating the velocity constraint and acceleration limits that are in play for the vehicle at each location along the desired path. With this information to hand, an approximately optimal velocity constraint is constructed and used to build an approximate motorcycle trajectory. With this tool available, engineers have the possibility of quantifying the impact that a parameter modification has on the total time required for the vehicle to traverse a given ground path. Furthermore, such an approximate optimal velocity profile can be used as a reference for exploring the aggressive trajectory space of a given multibody motorcycle model, obtained through the use of a closed-loop control strategy, that is, using a virtual rider.

From a control theory point of view, the manoeuvring control problem for a multibody motorcycle continues to pose a number of interesting challenges. In the study of motorcycle handling, closed-loop strategies are necessary since open-loop manoeuvres, typically used when studying car handling, are not effective.

Theoretical investigations with simplified motorcycle models (Hauser et al. 2004a, 2005a) suggest that one may *uniquely* parameterize every *upright* motorcycle trajectory in the time interval $(-\infty, +\infty)$ by specifying the planar trajectory of the rear wheel contact point. In (Hauser et al. 2004c), an optimal control strategy was developed for finding a state-input trajectory of a non-holonomic motorcycle that (approximately) implements a desired planar trajectory. The developed technique provides a means of parametrizing the trajectories of a non-holonomic motorcycle by those of a non-holonomic car.

Early theoretical studies for tracking a desired ground path with a (simplified) non-holonomic riderless bicycle can be found in (Getz 1994; Getz 1995 and Getz and Marsden 1995). This inspired an approach for controlling a multibody motorcycle model along a specified ground path–see (Frezza and Beghi 2003). The control strategy, detailed in (Saccon et al. 2004), employs the simplified bicycle model from (Getz 1994) and uses a model-based predictive control (MPC) paradigm. The roll angle is used as a virtual control input in place of the steering angle. The strategy has been applied to controlling various multibody models of scooters and motorcycles, as discussed in (Frezza and Beghi 2006).

Recent works on the control of the multibody motorcycle may also be found in (Lot and Cossalter 2006) and (Lot et al. 2007). In the former, a look-ahead strategy is adopted and two separate control loops are employed to regulate speed and lateral deviation errors. In the latter, the proposed rider model is composed of two parts. A path planning procedure, the optimal manoeuvre method (Bertolazzi et al. 2006), is used to compute the reference ground path and speed to be followed. The reference optimal trajectory is then stabilized using independent PID loops to control speed and lateral deviation.

A discrete-time optimal linear quadratic regulator with preview approach is adopted in (Sharp 2007). The reference road information is transformed into a vehicle-based reference frame, so that they represent the rider's view of the road. The control problem is then expressed with respect to this reference frame. Speed control is independent of steering and body lean control and is achieved using a PI regulation. Of interest is the addition of rider upper body lean as a control input, together with the usual throttle/brake and steering torque controls.

A model-based predictive control is also employed in (Massaro and Lot 2010). Arc length and lateral deviation from the desired path provide a notion of tracking error, much as we propose in this work. There are, however, significant differences. The desired state-control curve (the tracking reference) is specified by trimming the vehicle pointwise along the desired ground path as if it were performing a constant-speed, constant-radius turn at each point. The resulting curve is *not* a system trajectory, in contrast to that provided by the dynamic inversion procedure developed in (Saccon et al. 2012). Note also that the control law in (Massaro and Lot 2010) is obtained using the linearization about this non-trajectory. This approach appears to work well for a class of smooth manoeuvres for which that curvature does not vary too rapidly.

The second part of this chapter presents our contribution to this field, describing a nonlinear control system (the virtual rider) with the capability of driving a multibody two-wheeled vehicle along a general user-specified ground path with a desired velocity profile.

The virtual rider system rests on three main pillars: (a) a *dynamic inversion procedure* for computing a state-control trajectory corresponding to a desired manoeuvring task, (b) an *inverse optimal control strategy* for shaping the closed loop dynamics and (c) a *manoeuvre regulation controller* charged with executing the manoeuvre planned in step (a) using the results of step (b) in the computation of the feedback controller.

The numerical strategy for accomplishing (a) was developed in (Saccon et al. 2012), while a detailed description of parts (b) and (c) can be found in (Saccon et al. 2013), together with simulation results of the virtual rider driving a multibody motorcycle model, developed using a commercial multibody software package. Preliminary versions of this work have been presented in (Saccon 2006) and (Saccon et al. 2008).

The virtual rider is based on a simplified motorcycle model, the *sliding plane motorcycle* (SPM) model, which was developed in (Saccon et al. 2012). This rigid motorcycle model captures many important aspects of real motorcycle dynamics including sliding and load transfer, but retains a level of simplicity that makes it suitable for control design purposes. This *rigid motorcycle model* is used in the dynamic inversion procedure to map a desired *planar* trajectory into a corresponding state-input trajectory. We assume that the rider is firmly attached to the main body of the vehicle, leaving the discussion on the important control and configuration effects offered by rider motion for future investigation.

The dynamic inversion procedure, inverse optimal control strategy and manoeuvre regulation controller are all based on this simplified model. The complete multibody motorcycle model, with state dimension typically greater than 20, is accessed through an interface that makes it looks like the SPM model from an input–output perspective. The manoeuvre regulation controller for the SPM model is connected directly to this interface, allowing for closed loop simulations. The SPM model can be viewed as a nonlinear reduced-order model of the multibody motorcycle, with the virtual rider being a reduced-order controller.

4.2 Principles of Minimum Time Trajectory Computation

During a race competition, the goal of the rider is to complete each lap in the minimum time. At each point on the track, the motorcycle is subject to physical constraints that limit the available acceleration and deceleration. The most important constraint is due to the tyres. Since the lateral and longitudinal forces that a tyre can produce are coupled, a large lateral force greatly reduces the available longitudinal force. Additional constraints are mainly due to the engine, aerodynamics and mass distribution.

Consider first the case in which the motorcycle is moving along a straight line. No lateral force is needed in this case, and the longitudinal acceleration is limited at low speed by the maximum longitudinal force that the rear tyre can produce. At high speed, owing to the presence of increased aerodynamic drag, the longitudinal acceleration limit is determined instead by the maximum engine torque. In contrast, during a turn, the lateral acceleration is limited by the maximum lateral force that the tyres can produce. The available engine torque is not normally a limiting factor, however, since the available longitudinal tyre force is greatly reduced due to the coupling of lateral and longitudinal tyre forces.

The transition from straight running to cornering must be approached with care. As the radius of curvature of a turn decreases, the required lateral tyre force increases. Furthermore, the required lateral acceleration (and corresponding lateral tyre forces) is proportional to the square of the velocity. Consequently, it is easy to enter a turn with so much speed that the lateral force required for turning far exceeds what the tyres can produce. In that case, the rider must modify the desired trajectory to avoid losing control of the motorcycle.

4.2.1 Tyre Modelling

To better describe the longitudinal and lateral tyre force coupling, we briefly describe how tyre forces are modelled using Pacejka's *magic formula* (Pacejka 2002b). The magic formula is a set of equations relating load, slip ratio, slip angle and camber angle, denoted by F_z, κ, α and γ, respectively, to the longitudinal force, side force and aligning moment. These equations use a clever composition of trigonometric functions to provide a family of parameterized functions for fitting empirical tyre data. The original formula developed for car tyres has become standard in that context. The extension to motorcycle tyres necessitates substantial changes to accommodate the different roles of sideslip and camber forces in the two cases (Sharp et al. 2004, Lot 2004).

Tyre forces and moments are produced through a combination of geometry and slip. The *camber angle* γ is the angle between the wheel plane and the line perpendicular to the road surface. Owing to the shape of a motorcycle tyre, non-zero camber results in a lateral force called *camber thrust*. Additional tyre forces and moments are produced by *slip* between the tyre and the road surface. Roughly speaking, slip occurs when the velocity vector of the contact point between tyre and road is different from the velocity vector determined by the speed and heading of the wheel. The *slip angle* α is the angle between the wheel's actual direction of travel and the direction toward which it is pointing, while the *slip ratio* κ provides a non-dimensional description of the relative motion between the tyre and the road surface (Pacejka 2002b). The slip ratio is non-zero when the tyre's rotational speed is greater or less than the free-rolling speed.

Magic formula parameter values for a given tyre are used to calculate the steady state force and moment system for realistic operating conditions. Additional features for modelling dynamic effects are developed in (Pacejka 2002b), but are not discussed here. In the magic formula scheme, the cases of pure longitudinal slip and pure lateral slip are treated separately and then combined using loss functions that characterize the reduction of forces in combined slip.

In pure longitudinal slip, the slip angle α and camber angle γ are set to zero so that the tyre does not generate any lateral force. The longitudinal force F_{x0} in pure slip is then a function of the slip ratio κ and the normal load F_z. This function is given by

$$F_{x0}(\kappa, F_z) = D_x \sin\left[C_x \arctan(B_x\kappa - E_x(B_x\kappa - \arctan(B_x\kappa)))\right],$$

where the coefficient functions $B_x = B_x(F_z)$, $D_x = D_x(F_z)$ and $E_x = E_x(F_z, \text{sgn}(\kappa))$, as well as the constant C_x, shape the response. The longitudinal force increases with increasing slip ratio κ up to a maximum longitudinal force, followed by a significant drop. When the tyre is forced to work beyond the peak, the rider experiences a sudden loss of grip as the slip dynamics transition from a stable region with positive slope to an unstable region. Physically, the tyre spins up rapidly under power as the shear force decreases under increasing slip ratio while the engine torque remains nearly constant.

In pure lateral slip, the slip ratio κ is set to zero so that the tyre does not generate any longitudinal force. The lateral force in pure lateral slip, which is a function of the slip angle α, the camber angle γ and the normal load F_z, has the form

$$F_{y0}(\alpha, \gamma, F_z) = D_y \sin\left[C_y \arctan(B_y\alpha - E_y(B_y\alpha - \arctan(B_y\alpha)))\right.$$
$$\left. + C_\gamma \arctan(B_\gamma\gamma - E_\gamma(B_\gamma\gamma - \arctan(B_\gamma\gamma)))\right],$$

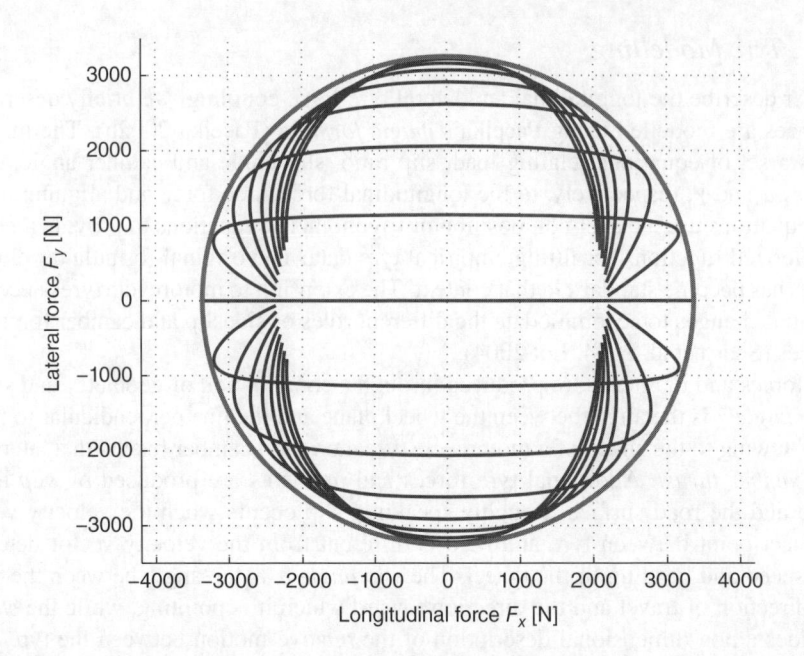

Figure 4.1 The friction ellipse. The envelope of longitudinal and lateral tyre forces, obtained by varying the slip ratio κ and slip angle α, resembles an ellipse, whence the name derives. The shape and position of this ellipse depend on the load and camber angle

where C_y, C_γ and E_γ are constant and $B_y = B_y(F_z, \gamma)$, $D_y = D_y(F_z, \gamma)$, $E_y = E_y(\gamma, \text{sgn}(\alpha))$ and $B_\gamma = B_\gamma(F_z, \gamma)$.

The longitudinal and lateral tyre forces under combined slip are given by (Sharp et al. 2004)

$$F_x = F_{x0}(\kappa, F_z)\, G_{x\alpha}(\alpha, \kappa, F_z)$$

and

$$F_y = F_{y0}(\alpha, \gamma, F_z)\, G_{y\kappa}(\alpha, \kappa, \gamma, F_z),$$

where $G_{x\alpha}(\cdot)$ describes the loss of longitudinal force due to sideslip and $G_{y\kappa}(\cdot)$ describes the loss of lateral force due to longitudinal slip. For further details, see (Pacejka 2002b).

Longitudinal and lateral forces for a given normal load F_z are illustrated in Figure 4.1. The envelope of these curves, representing the maximum available traction and cornering forces, is called the *friction ellipse*. The shape and position of the friction ellipse change according to the normal load and camber angle.

4.2.2 Engine and Drivetrain Modelling

The engine supplies the torque needed for controlling the speed of the vehicle. In a modern motorcycle, the engine torque is transmitted through a chain or a driveshaft to the rear wheel.

The interaction between the rear wheel and the ground produces a shear force causing the vehicle to move.

It is common practice to measure the steady-state engine torque on a test bench. This measurement is performed by setting the throttle valve to a fixed position and then modulating the load torque to achieve a desired engine speed. This gives the steady-state torque for several *load points*, that is, for several combinations of throttle opening position and engine speed. The steady-state torque map is used for simulating the engine in handling and performance analysis.

Engine torque is transmitted to the rear wheel through a set of gears and a chain. A simple mathematical description of the drivetrain takes into account inertial properties of the crankshaft, the main and counter shafts and the rear wheel. The load seen by the engine depends on factors such as the selected gear and the slip ratio.

Figure 4.2 provides an example of how the engine torque, Figure 4.2(a), is transformed to the rear wheel torque, Figure 4.2(b), as the gear setting ranges from first to sixth. The envelope of those curves, given by the bold solid line in Figure 4.2(b), represents maximum wheel torque versus wheel speed. For a more thorough discussion about this transformation, see (Hauser and Saccon 2006).

4.2.3 Brake Modelling

Since acceleration at high speed is limited by available engine torque, we might expect a similar deceleration limitation due to the brakes. In reality, this case does not exist since modern brake disks, pads and hydraulics are dimensioned so that brakes have virtually no capacity problem. On a racing motorcycle, the rider can always apply sufficient brake force to lock the wheels. In high fidelity simulations, brakes are typically modelled as torques applied to the front and rear wheels.

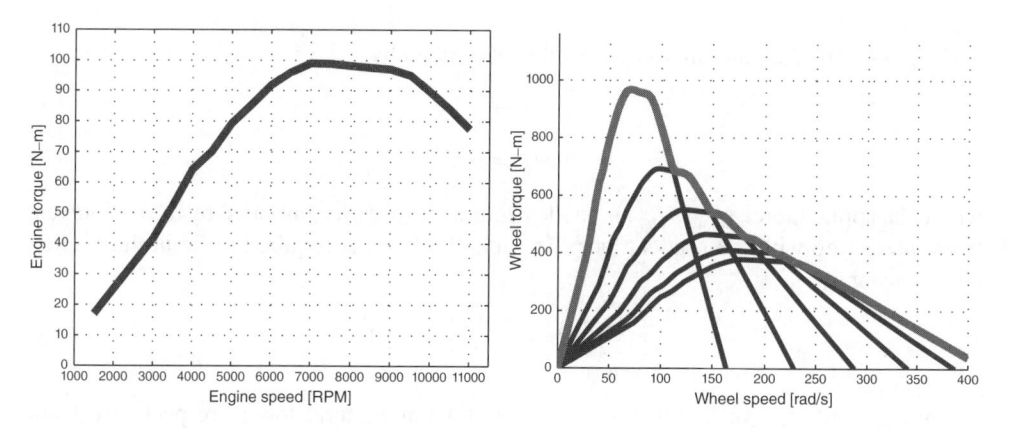

Figure 4.2 Engine and equivalent wheel torque curves. The effect of the gearbox is to modify the torque transmitted to the rear wheel, so that engine power is optimized for different vehicle speeds. The bold solid curve, which depicts the available rear wheel torque, is the envelope of the equivalent engine torque curves corresponding to first to sixth gears for a sport motorcycle

4.2.4 Wheelie and Stoppie

Skilled riders like to show off by performing tricks such as the wheelie and the stoppie in which the front and rear wheels, respectively, come off the ground. Upon passing the *apex* (roughly, the point of maximum curvature of a turn), the racing rider opens the throttle to accelerate out of the turn. The apex of a corner is the place where the chosen race line touches the inside edge of the track. Since modern motorcycles possess engines with considerable power and torque, this exit acceleration can result in the front wheel lifting off the ground. This phenomenon is often observed during professional races such as the Super-Bike and MotoGP championships. Race wheelies, however, are typically short-lived since a sustained wheelie requires precise throttle action to precisely control the pitch motion of the motorcycle. A similar phenomenon, the stoppie, can occur during hard braking when entering a turn. In that situation, if too much torque is applied to the front wheel, the rear wheel can lift off the ground.

4.3 Computing the Optimal Velocity Profile for a Point-Mass Motorcycle

We now describe the basis for estimating the maximum velocity profile for a given path. We start with a detailed description of the algorithm for a point-mass vehicle.

Consider a smooth curve in the plane, which we wish to traverse in minimum time. Here, we view the motorcycle as a point mass moving along the curve with velocity v. Using a moving frame, the accelerations seen by the point motorcycle are given as a tangential or *longitudinal acceleration* $\dot{v}$ and a perpendicular or *lateral acceleration* σv^2. The lateral acceleration depends on the instantaneous *curvature* $\sigma = \pm 1/R$, where R is the radius of the osculating circle, that is the second-order tangent to the curve at the current location. As viewed from above, the sign of σ is positive when the curve is turning right and negative when the curve is turning left. The curvature is zero at points where the curve is straight, that is, $R = \infty$.

The physics of the point motorcycle is thus described by

$$m\dot{v} = f_{long}, \qquad (4.1)$$

$$m\sigma v^2 = f_{lat}, \qquad (4.2)$$

where the applied force (f_{long}, f_{lat}) is an idealization of the force provided by the motorcycle tyres interacting with the road surface. As such, the force is required to lie in the friction ellipse given by

$$\left(\frac{f_{long}}{f_{long}^{max}}\right)^2 + \left(\frac{f_{lat}}{f_{lat}^{max}}\right)^2 \leq 1, \qquad (4.3)$$

where f_{long}^{max} and f_{lat}^{max} are the maximum longitudinal and lateral forces, respectively. Consistent with our curvature definition, we see that f_{long} and f_{lat} act in the forward and right directions, respectively.

Our goal is to find the velocity *profile* v as a function of the arc length s to traverse a given curve in minimum time. Normally, we view the velocity v as a function of time, not arc length. To indicate that arc length rather than time is the independent variable, we use a bar to indicate that a quantity is an explicit function of s. Thus, a curvature profile is given as $\bar{\sigma}(s)$ for $s \in [s_0, s_1]$, whereas the corresponding curvature *trajectory* σ as a function of t is given by $\sigma(t) = \bar{\sigma}(s(t))$, where $s(\cdot)$ is an arc-length trajectory. Restated, our goal is to find a velocity profile $\bar{v}(\cdot)$ that minimizes the time that it takes to traverse the curve given by a curvature profile $\bar{\sigma}(\cdot)$.

Note that the shape of a curve is determined by its curvature profile $\bar{\sigma}(\cdot)$. Indeed, let $(\bar{x}(\cdot), \bar{y}(\cdot))$ be a smooth arc-length-parameterized curve, where the y axis is oriented 90 degrees in the clockwise direction from the x axis, for example, x pointing north and y pointing east. Differentiating with respect to s (denoted by $'$), the orientation of the unit length tangent vector can be specified by a smoothly varying *heading angle* $\bar{\psi}(\cdot)$. That is,

$$\begin{pmatrix} \bar{x}'(s) \\ \bar{y}'(s) \end{pmatrix} = \begin{pmatrix} \cos \bar{\psi}(s) \\ \sin \bar{\psi}(s) \end{pmatrix}. \tag{4.4}$$

Differentiating (4.4), we obtain the arc-length acceleration vector

$$\begin{pmatrix} \bar{x}''(s) \\ \bar{y}''(s) \end{pmatrix} = \bar{\psi}'(s) \begin{pmatrix} -\sin \bar{\psi}(s) \\ \cos \bar{\psi}(s) \end{pmatrix},$$

which is perpendicular to the tangent vector. Fitting the osculating circle to the curve, we obtain

$$\bar{\psi}'(s) = \bar{\sigma}(s), \tag{4.5}$$

so that the curvature is also the rate at which the curve changes direction with respect to arc length. Integrating (4.4) and (4.5) from an initial position $(\bar{x}(0), \bar{y}(0))$ and heading $\bar{\psi}(0)$ shows that there is a three-dimensional family of curves with the same shape. We assume that the curvature profile $\bar{\sigma}(\cdot)$ is continuously differentiable so that $(\bar{x}(\cdot), \bar{y}(\cdot))$ is a C^3 curve, providing a five-dimensional profile $(\bar{x}(\cdot), \bar{y}(\cdot), \bar{\psi}(\cdot), \bar{\sigma}(\cdot), \bar{\sigma}'(\cdot))$ that prescribes a portion of the desired vehicle behaviour.

The motion of the point motorcycle can be described using either a velocity trajectory $v(\cdot)$ or a velocity profile $\bar{v}(\cdot)$ since each is uniquely determined by the other when the velocity is strictly positive. This dependence follows from the fact that $\dot{s}(t) = \bar{v}(s(t)) = v(t)$ yields an arc-length trajectory $t \mapsto s(t)$ that is strictly monotone increasing and hence invertible.

We find it useful to work in the *spatial* domain with the arc length s as the independent variable rather than in the time domain. To this end, using $\dot{v}(t) = \bar{v}'(s(t)) \bar{v}(s(t))$ and $\bar{a}(s) = \bar{f}_{long}(s)/m$, we write the longitudinal dynamics (4.1) as

$$\bar{v}'(s) \bar{v}(s) = \bar{a}(s)$$

or, suppressing the independent variable s,

$$\bar{v}' = \bar{a}/\bar{v}. \tag{4.6}$$

Combining (4.2) and (4.3), it follows that the *input* acceleration is constrained by

$$\bar{a}_{\min}(s, \bar{v}) \leq \bar{a}(s) \leq \bar{a}_{\max}(s, \bar{v}), \tag{4.7}$$

where, in this case,

$$\bar{a}_{\max}(s, \bar{v}) = +\frac{f_{long}^{\max}}{m} \sqrt{1 - \left(\frac{\bar{\sigma}(s) \bar{v}^2}{f_{lat}^{\max}/m} \right)^2}$$

and
$$\bar{a}_{\min}(s,\bar{v}) = -\frac{f_{long}^{\max}}{m}\sqrt{1-\left(\frac{\bar{\sigma}(s)\bar{v}^2}{f_{lat}^{\max}/m}\right)^2}.$$

A dynamic velocity profile satisfying (4.6) must also satisfy
$$\bar{v}(s) \leq \bar{v}_M(s) \tag{4.8}$$

for all s, where $\bar{v}_M(\cdot)$ is the maximum velocity profile corresponding to the curvature profile $\bar{\sigma}(\cdot)$. From (4.2) and (4.3), we see that the maximum velocity at s is given by
$$\bar{v}_M(s) = \sqrt{\frac{f_{lat}^{\max}/m}{|\bar{\sigma}(s)|}}$$

with $\bar{v}_M(s) = +\infty$ (an extended value) whenever $\bar{\sigma}(s) = 0$ to represent the absence of a limit on velocity.

We say that $\bar{v}(\cdot)$ is a *feasible velocity profile* if $\bar{v}(\cdot)$ satisfies the differential equation (4.6) and the constraints (4.7) and (4.8) on the domain of definition of $\bar{\sigma}(\cdot)$. If $\bar{v}(\cdot)$ is feasible and s_c satisfies $\bar{v}(s_c) = \bar{v}_M(s_c)$, then s_c must be a stationary point of $\bar{v}_M(\cdot)$ since $\bar{v}'(s_c) = 0$ (all available force is used for lateral acceleration), and if $\bar{v}'_M(s_c)$ is not zero then $\bar{v}(s) > \bar{v}_M(s)$ for some s near s_c violating (4.8). Note that, since $\bar{\sigma}(\cdot)$ is continuously differentiable, $\bar{v}'_M(s)$ is defined at all s such that $\bar{v}_M(s)$ is finite. In practice, contact points normally occur at local minimizers of $\bar{v}_M(\cdot)$. A local minimum of $\bar{v}_M(\cdot)$ corresponds to local maximum of $|\bar{\sigma}(\cdot)|$, which roughly corresponds to the apex of a turn. A contact point s_c can also occur at a local maximum of $\bar{v}_M(\cdot)$, especially if $v_M(s_c)$ is also a local minimum, that is, $\bar{v}_M(\cdot)$ is constant in the neighbourhood of s_c. A curvature profile exhibiting this possibility can be easily constructed. While the question remains open, it appears that it may be possible to construct a smooth $\bar{v}_M(\cdot)$ allowing a contact point s_c that is an isolated local maximizer of $\bar{v}_M(\cdot)$.

An optimal velocity profile maximizes the velocity at each point along the path while remaining feasible. This property implies, as occurs in various time optimal problems, that every value of the optimal applied longitudinal force is either the maximum or minimum allowed by (4.7) and (4.8), so that the point-mass motorcycle is always either accelerating or braking as much as possible.

To this end, consider the problem of traversing, in clockwise fashion, the curve depicted in Figure 4.3 whose curvature profile is shown in Figure 4.4. Clearly, the optimal velocity profile must touch $\bar{v}_M(\cdot)$ at at least one point since otherwise it would be possible to find a faster velocity profile. Figure 4.5 depicts the maximum velocity profile $\bar{v}_M(\cdot)$ for the path in Figure 4.3 together with the optimal velocity profile $\bar{v}_{opt}(\cdot)$. As noted above, each location s_c such that $\bar{v}_{opt}(s_c) = \bar{v}_M(s_c)$ satisfies the necessary condition $\bar{v}'(s_c) = 0$. Also, when a turn is sufficiently isolated (for example, turns 1 and 2), the approach involves maximum braking, while the departure involves maximum acceleration. This observation provides a strategy for determining the optimal velocity profile.

Suppose that the number of connected regions of local minimizers for $\bar{v}_M(\cdot)$ is finite, with each minimum region defining a *turn*. For each turn, we compute a locally optimal velocity profile as follows. Starting at a local minimizer, we integrate forward with
$$\bar{v}' = \bar{a}_{\max}(s,\bar{v})/\bar{v} \tag{4.9}$$

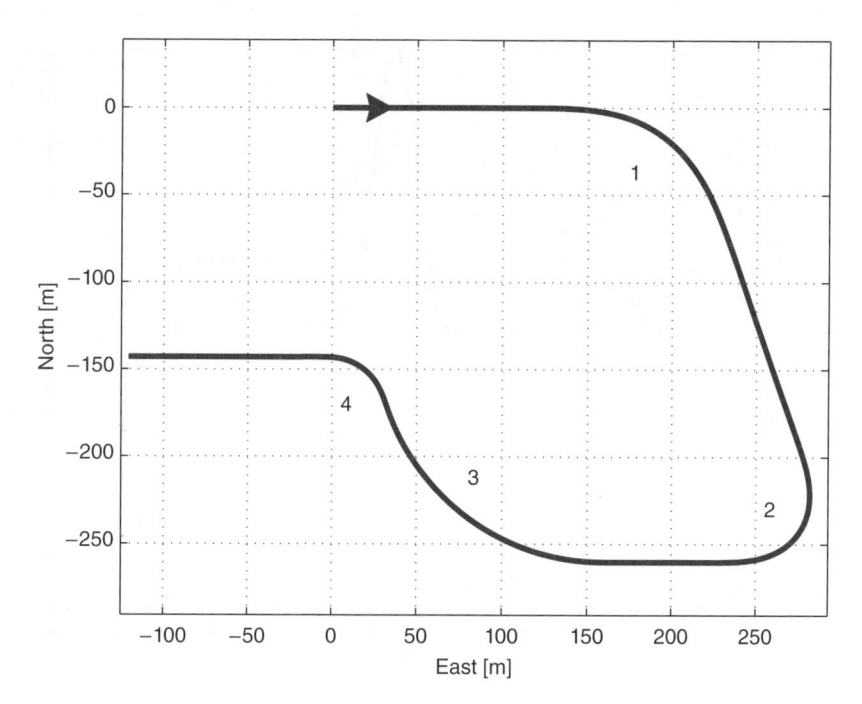

Figure 4.3 An example test track curve. This *x-y* plane curve test track, which is to be followed in clockwise fashion, involves three right turns followed by one left turn

and backward with

$$\bar{v}' = \bar{a}_{\min}(s, \bar{v})/\bar{v} \qquad (4.10)$$

until the maximum velocity constraint (4.8) is violated. The optimal velocity profile $\bar{v}_{opt}(\cdot)$ is then given by the minimum of the local profiles–see Figure 4.5. Switches from maximum acceleration to maximum braking occur at locations where the locally optimal velocity profile that is globally optimal changes.

In our example, switches from maximum acceleration to maximum braking occur at approximately 315 and 548 m. Despite the fact that the track has four well-defined turns, we see in Figure 4.5 that only three locally optimal velocity profiles are used to determine the optimal velocity profile. In this example, it turns out that the locally optimal velocity profile for turn four includes a portion of the locally optimal velocity profile for turn three so that the turn-three profile is not needed (or is redundant). Indeed, as we integrate equation (4.10) backward from the turn-four minimum region of $\bar{v}_M(\cdot)$ (starting at, say, 755 m), the dynamic velocity profile converges to the turn-three minimum region of $\bar{v}_M(\cdot)$ within a finite distance (at approximately 656 m). We then continue until the maximum velocity constraint is violated at approximately 496 m. The turn-four locally optimal velocity profile thus covers both turn three and turn four.

This last feature points to an interesting technical detail. The fact that there is a finite-distance convergence to the turn-three minimum region implies that the vector field at the point of convergence cannot be Lipschitz, and indeed it is not. Furthermore, at that point, we do not have the uniqueness properties that are ensured for a locally Lipschitz vector field.

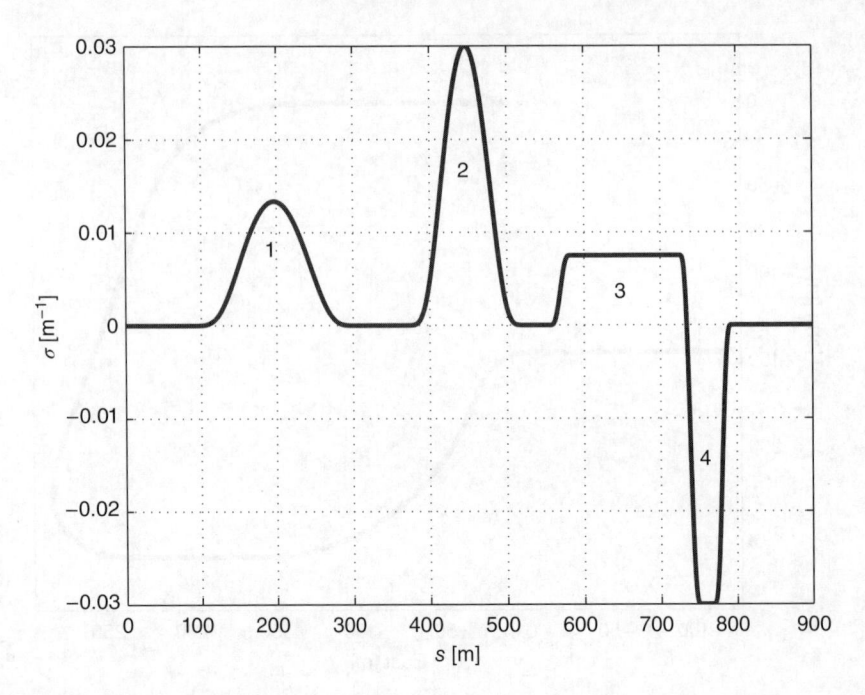

Figure 4.4 Curvature profile of the test track curve in Figure 4.3. The curvature is shown as a function of arc length. The direction of the turns is easily determined by the sign of the curvature (positive for right turns), indicating three right turns followed by one left turn

Non-uniqueness is reflected by the fact that the trajectory can reach the constraint curve at many different points, depending on where the trajectory starts off from the constraint curve. The important point for our purposes is that the curve obtained by integrating (4.10) backward is the unique curve satisfying the given initial condition. Non-uniqueness would be an issue if we needed to begin at a local minimizer of $\bar{v}_M(\cdot)$ and integrate (4.10) forwards. Similar remarks apply to (4.9) with reversed directions.

Suppose now that the point motorcycle is subjected to accelerations of a more general nature, for example, variable aerodynamic drag. In this case, the form of the constrained dynamic system is unchanged, satisfying (4.6), (4.7) and (4.8). That is, a feasible velocity profile satisfies

$$\bar{v}' = \bar{a}/\bar{v},$$

where the available acceleration is constrained according to

$$\bar{a}_{\min}(s, \bar{v}) \leq \bar{a}(s) \leq \bar{a}_{\max}(s, \bar{v}),$$

and the velocity state must satisfy the maximum velocity constraint

$$\bar{v}(s) \leq \bar{v}_M(s).$$

As noted above, the maximum velocity profile $\bar{v}_M(\cdot)$ is continuously differentiable at all points of finite value. The acceleration constraints are continuous in s and $\bar{v}$, $\bar{v} \mapsto \bar{a}_{\min}(s, \bar{v})$

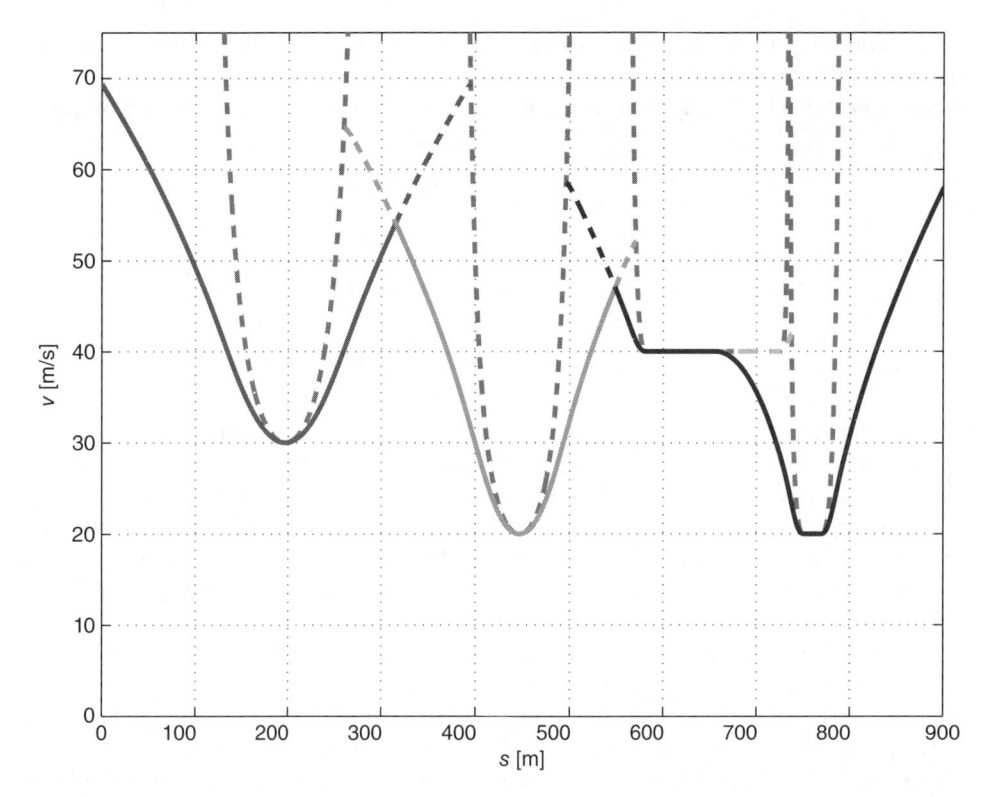

Figure 4.5 Optimal velocity profile. The optimal velocity $\bar{v}_{opt}(\cdot)$ (solid) is shown together with the maximum velocity constraint $\bar{v}_M(\cdot)$ (dashed). The optimal solution touches the maximum velocity profile only at points that locally minimize the maximum velocity profile $\bar{v}_M(\cdot)$. Different shades are used to depict different locally optimal velocity profiles. Switches from maximum acceleration to maximum deceleration occur at approximately 315 and 548 m. The transitions from maximum deceleration to maximum acceleration are, in contrast, smooth

is non-decreasing, $\bar{v} \mapsto \bar{a}_{\max}(s, \bar{v})$ is non-increasing, and $\bar{v} < \bar{v}_M(s)$ implies that $\bar{a}_{\max}(s, \bar{v}) > \bar{a}_{\min}(s, \bar{v})$. We are interested in minimizing

$$J(\bar{v}(\cdot)) = \int_{s_0}^{s_1} \frac{ds}{\bar{v}(s)}, \tag{4.11}$$

subject to the dynamics (4.6) and the constraints (4.6) and (4.7). The cost $J(\bar{v}(\cdot))$, defined in (4.11), is simply the time that it takes to go from s_0 to s_1 using the velocity profile $\bar{v}(\cdot)$.

Now, it is possible to *ride* the velocity constraint in regions where

$$\bar{a}_{\min}\left(s, \bar{v}_M(s)\right) \leq \bar{v}'_M(s)\bar{v}_M(s) \leq \bar{a}_{\max}(s, \bar{v}_M(s))$$

by choosing $\bar{a}(s) = \bar{v}'_M(s)\bar{v}_M(s)$. The condition $\bar{v}'_M(s) = 0$ above is a special case. To obtain the set of locally optimal velocity profiles, we begin in each constraint-riding region and integrate backwards using $\bar{a} = \bar{a}_{\min}(s, \bar{v})$ and forwards using $\bar{a} = \bar{a}_{\max}(s, \bar{v})$ until the maximum velocity constraint is violated. The optimal velocity profile $\bar{v}_{opt}(\cdot)$ is then given by the minimum of the local profiles.

To be concrete, with aerodynamic drag, the acceleration constraints are offset according to

$$\bar{a}_{\max}(s,\bar{v}) = +\frac{f_{long}^{\max}}{m}\sqrt{1-\left(\frac{\bar{\sigma}(s)\bar{v}^2}{f_{lat}^{\max}/m}\right)^2} - b_{\min}\bar{v}^2$$

and

$$\bar{a}_{\min}(s,\bar{v}) = -\frac{f_{long}^{\max}}{m}\sqrt{1-\left(\frac{\bar{\sigma}(s)\bar{v}^2}{f_{lat}^{\max}/m}\right)^2} - b_{\max}\bar{v}^2,$$

where $b_{\min}$ and $b_{\max}$ model the minimum and maximum available drag (for example, $b_{\min} = \rho c_D A/2$, see below). The variable drag coefficient $b \in [b_{\min}, b_{\max}]$ models the extent to which the rider can modulate the drag force through body posture. We thus see that $b_{\min}$ corresponds to the streamlined stance used during acceleration and on high-speed straights, whereas $b_{\max}$ corresponds to a higher drag upright stance used when braking during the approach to a turn. Note that, in this case, a local minimum of $\bar{v}_M(\cdot)$ is no longer a possible contact point for $\bar{v}_{opt}(\cdot)$ and $\bar{v}_M(\cdot)$.

Figure 4.6 shows the optimal velocity profile when variable aerodynamic drag is included. Note the loss of the symmetry that is present when there is no aerodynamic drag. Also, since the presence of aerodynamic drag allows for greater deceleration, the switching points occur further down the track at approximately 327, 553 and 676 m. In this case, all four locally optimal velocity profiles are used to determine the optimal velocity profile. Surprisingly for the drag parameters we use, the time to traverse the 900 m course is slightly shorter than in the no-drag case. Figure 4.7 compares the optimal velocity and acceleration trajectories resulting with and without aerodynamic drag.

When the number of contact regions is finite, the number of acceleration switches is also finite so that $\bar{v}'_{opt}(\cdot)$, hence $\dot{v}_{opt}(\cdot)$, is piecewise continuous. In the above examples, the no-drag case possesses two points of discontinuity, whereas the case with aero drag possesses three, as can be seen in Figures 4.5 and 4.6.

4.3.1 Computing the Optimal Velocity Profile for a Realistic Motorcycle

The algorithm for computing the optimal velocity profile for a point-mass motorcycle provides a framework for estimating optimal velocity profiles for more comprehensive motorcycle models. More comprehensive models possess additional states (including roll angle and rate) and are subject to more complicated constraints than the simple point-mass motorcycle above. We thus need to manage this additional complexity to evaluate the point-wise (in space s) maximum velocity as well as the minimum and maximum acceleration functions. The strategy we have developed is to use a quasi-steady-state approach where the basic idea is to compute quantities as if some of the system states were in steady state.

The algorithm for computing the optimal velocity profile described in the previous subsection can be viewed as providing a quasi-steady-state trajectory for a motorcycle model subject to acceleration constraints that are independent of the roll angle. The quasi-steady-state approach that we discuss in (Hauser and Saccon 2006) is based on the idea that we can

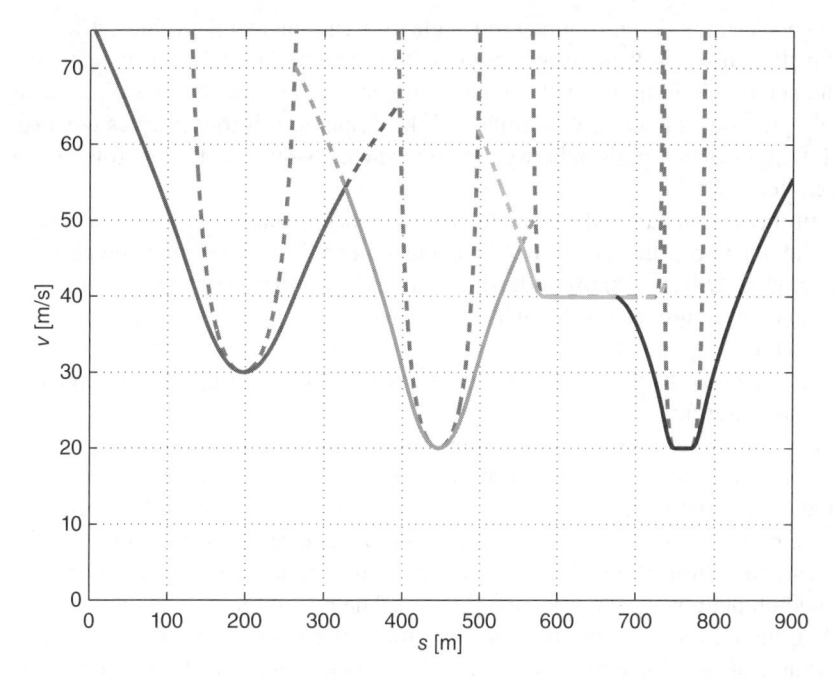

Figure 4.6 Optimal velocity profile with variable aerodynamic drag. The optimal velocity $\bar{v}_{opt}(\cdot)$ (solid), when aerodynamic drag is present, is shown together with the velocity constraint $\bar{v}_M(\cdot)$ (dashed). The loss of symmetry is due to the presence of the aerodynamic drag force. Also, points of contact between the optimal velocity profile and the maximum velocity profile no longer occur at local minimizers of $\bar{v}_M(\cdot)$. As before, different shades are used to depict different locally optimal velocity profiles. Switches from maximum acceleration to maximum deceleration occur at approximately 327, 553 and 676 m

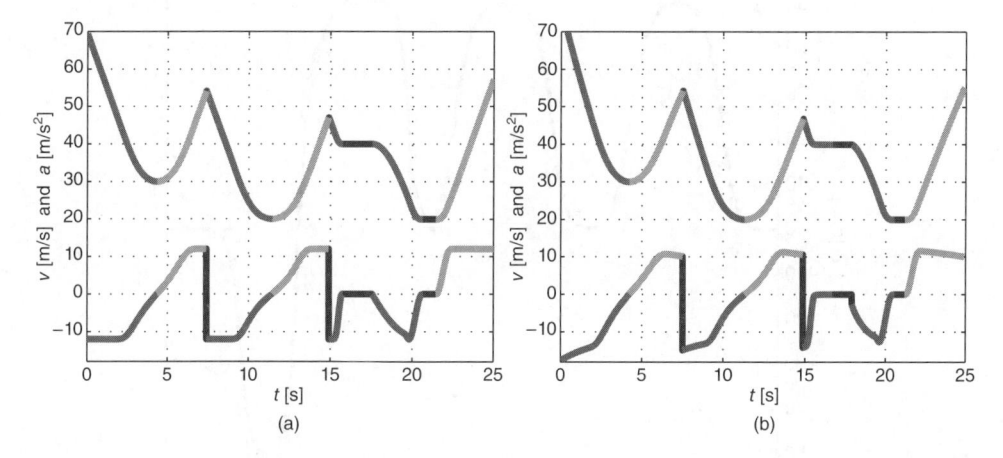

Figure 4.7 Changes in the optimal velocity profile due to aerodynamic drag. (a) shows the longitudinal velocity (upper) and the corresponding acceleration trajectory (lower) for the point-mass vehicle when no aerodynamic drag is present, while (b) shows the longitudinal velocity (upper) and acceleration (lower) for the point-mass vehicle subject to aerodynamic drag. Periods of acceleration and braking are shown with a lighter and darker shade, respectively

obtain a good estimate of the optimal roll angle by assuming that the vehicle is following the desired path with zero roll angle acceleration. It is lear that the quasi-steady-state roll trajectory obtained in this fashion is not a dynamic trajectory, nevertheless the quasi-steady-state roll angle provides a reasonable estimate of the dynamic roll trajectory as our experience suggests. A typical comparison between the quasi-steady-state and actual roll trajectories is presented in Figure 4.8.

Upon further examination, the dynamic roll trajectory appears to be a filtered version of the quasi-steady-state roll trajectory. This relationship can, in fact, be made precise, as shown in (Hauser et al. 2004b, 2005b). Indeed, it can be shown that the dynamic roll trajectory is approximated by a non-causal low-pass filter that depends on the required lateral acceleration trajectory. In particular, the dynamic roll trajectory is close to the quasi-steady-state roll trajectory when the variation of the quasi-steady-state roll trajectory is slow relative to this low-pass filtering.

To determine whether more realistic constraints involving tyre forces, wheelie and so on are satisfied when a manoeuvre is performed by the complex motorcycle model, we develop a method for evaluating the interaction forces between the vehicle and the ground, whose details are described in (Hauser and Saccon 2006). The result is that, for a specific location s, with curvature $\bar{\sigma}(s)$ and velocity v_x, we can compute the maximum and minimum acceleration functions $\bar{a}_{max}(s, v_x)$ and $\bar{a}_{min}(s, v_x)$. The maximum acceleration is obtained by increasing the rear wheel longitudinal force up to the point where the front or rear tyre forces violate the friction ellipse constraint or when the longitudinal force corresponding to the maximum engine torque is reached. The minimum acceleration function is computed in a similar fashion, assuming the rear wheel longitudinal force to be zero and increasing the braking force while checking the tyre and stoppie constraints. As previously mentioned, the

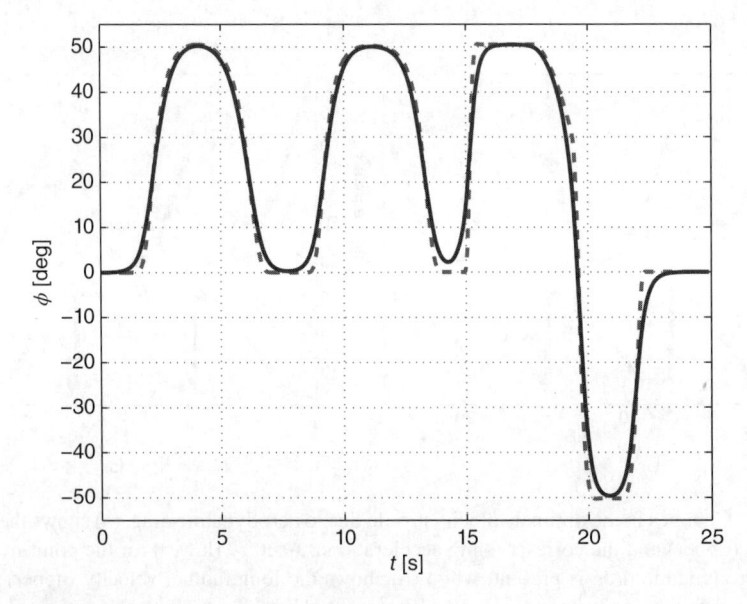

Figure 4.8 Approximation of the roll trajectory. The quasi-steady-state roll trajectory (dashed) approximates the dynamic roll trajectory (solid) satisfying the equations of motion

maximum breaking force is not a real limitation for a modern racing motorcycle, since the rider can always apply enough braking torque to lock the front wheel.

The computation of the optimal velocity profile is accomplished in much the same way as for the point mass vehicle. As before, the integration is performed with respect to the arclength s rather than time. A finite number of points $s_1, \ldots, s_N$ on the track are chosen.

The algorithm then has the following form:

1. At the current position s_k, the velocity is $\bar{v} = \bar{v}(s_k)$ and the curvature is $\bar{\sigma} = \bar{\sigma}(s_k)$.
2. Trim the vehicle at constant velocity $\bar{v}$ and curvature $\bar{\sigma}$ obtaining the steady-state roll angle $\bar{\varphi}_{ss}$ together with the interaction forces between the tyres and ground.
3. Holding φ at $\bar{\varphi}_{ss}$, let $\overline{F}^r_T(s_k)$ be the largest longitudinal rear force F^r_T satisfying:
   (a) tyre constraints: the longitudinal and lateral forces of the front and rear wheels remain within the respective friction ellipses;
   (b) engine constraint: the longitudinal force to be produced by the rear tyre can be generated by the engine;
   (c) wheelie constraint: normal load on the front tyre has to be greater than or equal to zero.
4. Define $\bar{a}_{\max}(s_k, \bar{v}) := \dot{v}_T$, where $\dot{v}_T$ is the longitudinal acceleration corresponding to the maximum lateral force $\overline{F}^r_T(s_k)$. Propagate the velocity profile using

$$\bar{v}(s_{k+1}) = \bar{v}(s_k) + (s_{k+1} - s_k) \cdot \bar{a}_{\max}(s_k, \bar{v}(s_k))/\bar{v}(s_k).$$

5. If the maximum velocity curve is reached (or exceeded) proceed to Step 6. Otherwise, set $k = k + 1$ and go to Step 1.
6. Find the next local minimum of the maximum velocity curve, occurring at the location s_j. Set $\bar{v}(s_j) = \bar{v}_{\max}(s_j)$. Also, set $k = j$ for future use.
7. At the current position s_j, the velocity is $\bar{v} = \bar{v}(s_j)$ and the curvature is $\bar{\sigma} = \bar{\sigma}(s_j)$.
8. Trim the vehicle at constant velocity $\bar{v}$ and curvature $\bar{\sigma}$, obtaining the steady-state roll angle $\bar{\varphi}_{ss}$ together with the interaction forces between tyres and ground.
9. Holding φ at $\bar{\varphi}_{ss}$, let $-\overline{F}^f_T(s_j)$ be the largest decelerating longitudinal front force $-F^f_T$ satisfying:
   (a) tyre constraints: the longitudinal and lateral forces of the front and rear wheels remain within their respective friction ellipses;
   (b) stoppie constraint: normal load on the rear tyre has to be non-negative.
10. Define $\bar{a}_{\min}(s_j, \bar{v}) := \dot{v}_T$, where $\dot{v}_T$ is the longitudinal acceleration corresponding to the decelerating lateral force $\overline{F}^f_T(s_j)$. Propagate the velocity profile backwards using $\bar{v}(s_{j-1}) = \bar{v}(s_j) - (s_j - s_{j-1}) \cdot \bar{a}_{\min}(s_j, \bar{v}(s_j))/\bar{v}(s_j)$, saving for comparison all previously computed values of $\bar{v}(s_{j-1})$.
11. If the presently determined velocity $\bar{v}(s_{j-1})$ exceeds a previously computed velocity (signifying intersection), take $\bar{v}(s_{j-1})$ to be the lower of the two and proceed to step 1 (using the k set in Step 6). Otherwise, set $j = j - 1$ and go to Step 7.

Note that we have omitted the obvious exits or jumps that would be taken whenever we run off the end of the track in the forward or reverse directions. As for the point-mass motorcycle, the optimal velocity profile (estimate) is given by the minimum of the local optimal velocity profiles.

4.3.2 Application to a Realistic Motorcycle Model

For the sake of illustration, consider a track composed by a simple chicane. The curvature of the chicane used in our example is shown Figure 4.9, where a change of sign in σ can be seen. In the same figure, we report the velocity profile computed with the quasi-steady-state method as well as the velocity profile followed by a multibody motorcycle model driven by the virtual rider (dashed line) that we will describe in the second part of this chapter. The lateral acceleration reaches 1 g at the apex of the turns with a corresponding roll angle of 45°.

There is good agreement between the roll angles obtained during dynamic simulation and the quasi-steady-state method. The roll angles are shown in Figure 4.9, together with the longitudinal acceleration profiles.

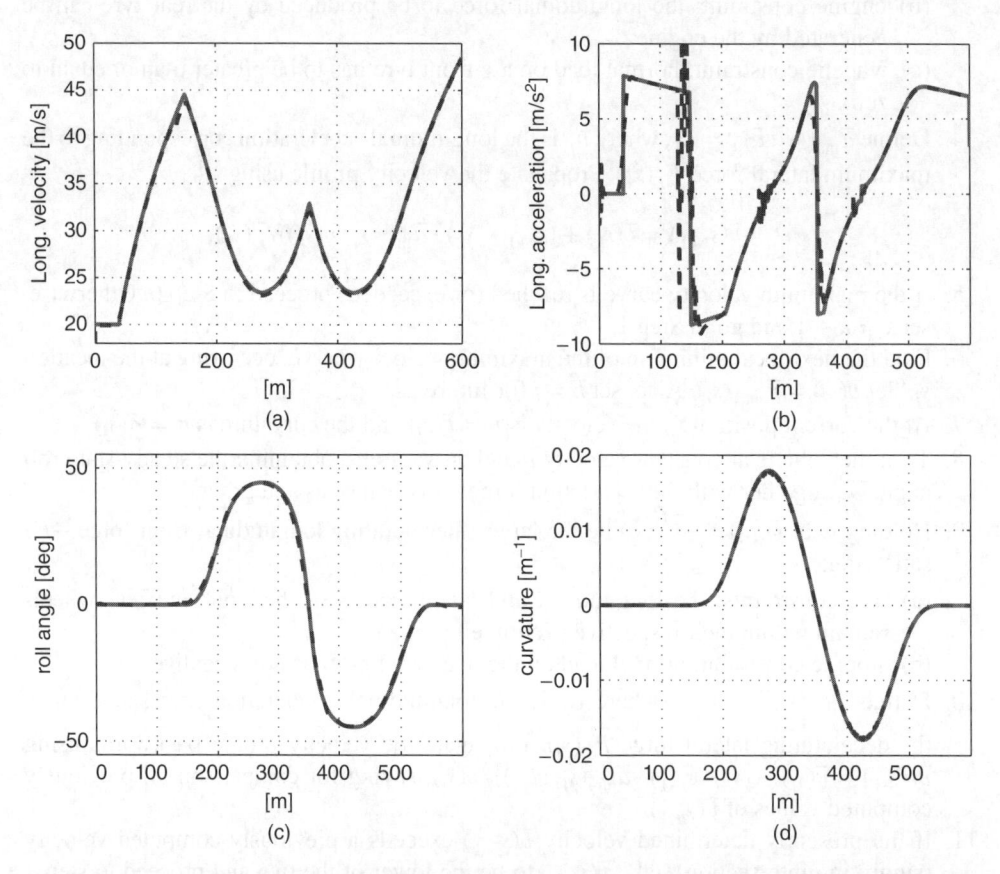

Figure 4.9 Comparison between quasi-steady-state and dynamic simulation through a chicane manoeuvre. (a) velocity profile computed with the quasi-steady-state method (solid) and dynamic simulation with the virtual rider (dashed); (b) acceleration profiles obtained with the quasi-steady-state method (solid) and simulation (dashed); (c) roll angle profile obtained with the quasi-steady-state method (solid) and simulation (dashed); and (d) required curvature profile given as input to the quasi-steady-state method (solid) and that produced by the simulated multibody motorcycle (dashed)

When using the quasi-steady-state method, it is possible to transition instantaneously from acceleration to braking. In contrast, when working with a model that includes the suspension dynamics, it takes time to transition from an acceleration posture in which the rear suspension is more compressed to a deceleration posture in which the front suspension is more compressed. During this load-transfer transient, the normal force at the front tyre may be such that the available braking force is temporarily less than the ideal force predicted by the quasi-steady-state method. Race pilots use a combination of suspension setup and rider skill to manage such load-transfer effects.

The interaction forces between tyres and ground predicted by the quasi-steady-state method and obtained through a dynamic simulation are shown in Figure 4.10. As in the case of Figure 4.9, there is a good agreement between predicted and simulated forces.

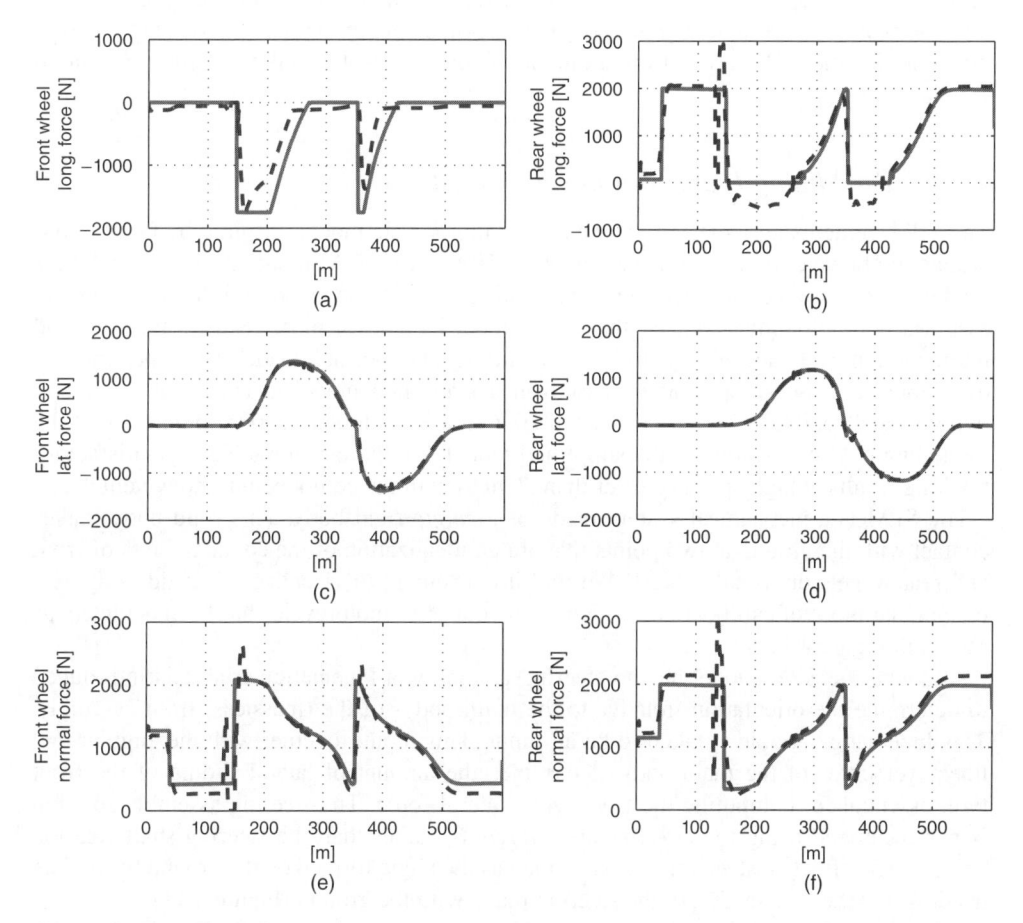

Figure 4.10 Quasi-steady-state and dynamic tyre forces. (a) and (b) show the longitudinal tyre forces computed with the quasi-steady-state method (solid) and those obtained by simulation with the virtual rider (dashed); (c) and (d) show the quasi-steady-state (solid) and dynamic (dashed) lateral tyre forces; (e) and (f) show, for the same manoeuvre, the normal forces computed with the quasi-steady-state method (solid) and those obtained by dynamic simulation (dashed)

Observable differences occur during the transition from acceleration to braking due to suspension dynamics that are neglected in the quasi-steady-state computation.

4.4 The Virtual Rider

The remainder of this chapter is dedicated to describing the basic principles on which the virtual rider used in the previous section is based. Further details can be found in (Saccon et al. 2012; 2013) and references therein.

The virtual rider is based on a simplified motorcycle model, the *sliding plane motorcycle* (SPM) model. This *rigid motorcycle model* is used in a dynamic inversion procedure to map a desired *planar* trajectory into a corresponding state-input trajectory. Given such a state-input trajectory a manoeuvre regulation controller for the SPM model is designed with desired closed loop response. The manoeuvre regulation controller for the SPM model is then connected to a complete multibody motorcycle model through an interface that makes this look like the SPM model from an input–output perspective, allowing for closed-loop simulations.

4.4.1 The Sliding Plane Motorcycle Model

The SPM model is an idealized reduced-order motorcycle model, inspired by Getz's non-holonomic bicycle model (Getz and Marsden 1995). The SPM model differs from Getz's model in the way contact between tyres and ground is modelled. While Getz's model assumes non-holonomic contact, the SPM model includes a more realistic tyre–ground interaction model, accounting for lateral sliding and normal load. Our experience in developing a closed-loop strategy based on a simplified model to control a multibody two-wheeled vehicle, reported in (Saccon 2006 and Saccon et al. 2008), indicates that the modelling of lateral sliding in the simplified model is important for achieving satisfactory tracking results at high speed (greater than 30 m/s) with the complex multibody vehicle.

The SPM is a mechanical system made of a *single* rigid body. This rigid body makes contact with the ground at two points that are an idealization of the contact points of front and rear wheels on a real vehicle. We find it convenient to visualize this rigid body as a rectangular piece of cardboard on which the layout of a "motorcycle" has been depicted, as shown in Figure 4.11.

Although the SPM model has no steering assembly, a kinematic variable is introduced to determine the orientation–relative to the main body–of the (massless) front assembly. This kinematic variable is taken to be an input. This input, together with the angular and linear velocities of the main body, determines the amount of lateral sliding of the front tyre, essential for computing the front tyre's lateral force. The steering assembly rotation is parameterized using the *effective steer angle* δ_f rather than the steering-shaft rotation angle δ. The effective steer angle δ_f is defined as the angle formed on the ground by vehicle heading and the intersection of the steering plane with the ground (Figure 4.11).

We suppose that the longitudinal forces on the front and rear tyres can be directly specified, and are taken to be model inputs. In contrast, the lateral forces are determined using a *linear* tyre model, so that the lateral shear force depends linearly on the *sideslip angle* α and the *(wheel) roll angle* φ. The *normal load* F_z also appears in a linear fashion so that the lateral

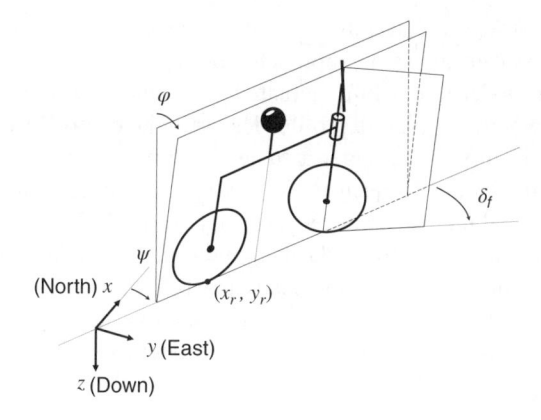

Figure 4.11 Generalized coordinates and steering input for the SPM model

tyre force is given by
$$F_y = (C_\alpha \alpha + C_\gamma \varphi) F_z, \qquad (4.12)$$

for given *cornering stiffness* C_α and *camber stiffness* C_γ.

Considering Figure 4.11, the generalized coordinates describing the configuration of the vehicle are the position of the point of contact of rear wheel (x_r, y_r), roll angle φ, and yaw angle ψ. The control inputs to the model are two. They are the thrust force F_x and effective steer angle δ_f. The generalized coordinates and inputs for the SPM model are written in short as $\mathbf{q} = (x_r, y_r, \varphi, \psi)$, and $\mathbf{u} = (F_x, \delta_f)$, respectively. We obtain a second-order control system of the form $\ddot{\mathbf{q}} = f(\mathbf{q}, \dot{\mathbf{q}}, \mathbf{u})$.

The SPM is a *constrained* mechanical system. Two holonomic constraints are enforced to maintain the two contact points at ground height. Those two holonomic constraints are *non-ideal* as they restrict the configuration space and they *do* work, dissipating or increasing the system's energy. To our knowledge, *non-ideal* holonomic constraints are not typically discussed in classical mechanics books, although they often lead to compact mathematical models that are suitable for control purposes.

4.5 Dynamic Inversion: from Flatland to State-Input Trajectories

A typical motorcycle manoeuvring task is to traverse a specified *ground path* using a desired *velocity profile*. Together, these determine a *desired flatland trajectory* $(x_d, y_d)(t), t \in [0, T]$, with associated velocity $v_d(t)$, course angle $\chi_d(t)$, acceleration $\dot{v}_d(t)$, curvature $\sigma_d(t)$ and curvature rate $\dot{\sigma}_d(t)$.

Our goal is to find an upright SPM trajectory that accomplishes the task of following a desired flatland trajectory. The passage from flatland to state-input trajectory will be denoted *lifting*, as we 'lift' a planar trajectory to a full state-input trajectory (including *roll*) of the dynamic control system.

In (Hauser et al. 2004c), we have discussed a dynamic inversion procedure for a non-holonomic motorcycle model based on the *embedding* of its dynamics into an extended control system with extra (artificial) inputs that make the system easy to control. In (Saccon et al. 2012), we have taken the same approach, adding extra control inputs to extend the under-actuated SPM model into a fully actuated mechanical system that can be made to follow any desired velocity–curvature profile. The artificial control effort is then optimized away, leading to a trajectory of the under-actuated rigid motorcycle model that is still consistent with the specified planar trajectory. By saying 'optimized away', we mean that the magnitude of the artificial inputs is reduced to nearly zero by an iterative optimization procedure that modifies, in a manner to be clarified, the state and input curves.

To provide exact accomplishment of the desired task, we seek roll and heading angle trajectories, $\varphi(\cdot)$ and $\psi(\cdot)$, and appropriate thrust and effective steering angle inputs, $F(\cdot)$ and $\delta(\cdot)$, to satisfy the SPM dynamics along the desired trajectory,

$$\begin{bmatrix} \dot{v}_d(t) \\ \dot{\chi}_d(t) - \dot{\psi}(t) \\ \ddot{\varphi}(t) \\ \ddot{\psi}(t) \end{bmatrix} = f_v \left(\begin{bmatrix} x_d(t) \\ y_d(t) \\ \varphi(t) \\ \psi(t) \end{bmatrix}, \begin{bmatrix} v_d(t) \\ \chi_d(t) - \psi(t) \\ \dot{\varphi}(t) \\ \dot{\psi}(t) \end{bmatrix}, \begin{bmatrix} F(t) \\ \delta(t) \end{bmatrix} \right), \qquad (4.13)$$

with $|\varphi(t)| < \pi/2$ and $|\beta(t)| = |\chi_d(t) - \psi(t)|$ small. Note that, within this section alone, we will use F and δ to refer to the control inputs F_x and δ_f. In particular, δ will refer to the *effective* steer angle rather than the steering shaft rotation angle.

4.5.1 Quasi-Static Motorcycle Trajectory

Our strategy for lifting is to use pointwise or nearly pointwise calculations to get an idea of what the lifted trajectory might look like, and then use trajectory optimization to transform the idealized into an actual trajectory. The intuition is that when a motorcycle manoeuvres in a smooth, slowly varying manner, the resulting trajectory stays *close* to the equilibrium manifold, and that even when there are rapid transitions, the trajectory may still be reasonably close *in function space*.

First note that, when the desired ground trajectory is a constant-speed circle, the lifted trajectory is the *trim* trajectory, $\varphi(t) \equiv \varphi_e$, $\psi(t) = \chi_d(t) - \beta_e$, $F(t) \equiv F_e$, and $\delta(t) \equiv \delta_e$, obtained by solving

$$\begin{bmatrix} 0 \\ 0 \\ 0 \\ 0 \end{bmatrix} = f_v \left(\begin{bmatrix} 0 \\ 0 \\ \varphi_e \\ 0 \end{bmatrix}, \begin{bmatrix} v_d \\ \beta_e \\ 0 \\ \sigma_d v_d \end{bmatrix}, \begin{bmatrix} F_e \\ \delta_e \end{bmatrix} \right), \qquad (4.14)$$

for the unknowns φ_e, β_e, F_e and δ_e. When $v_d(t)$ and $\sigma_d(t)$ are 'slowly' varying, one can imagine that there is a corresponding slowly varying motorcycle trajectory that is close to the pointwise (time-frozen) equilibrium curve, $\varphi(t) \approx \varphi_e(t)$ and so on.

We seek curves $\varphi(\cdot)$, $\psi(\cdot)$, $F(\cdot)$ and $\delta(\cdot)$ that approximately satisfy (4.13) for more *aggressive* ground trajectories. We begin by finding *quasi-static* roll and sideslip angle trajectories

$\varphi_{qs}(\cdot)$ and $\beta_{qs}(\cdot)$ that satisfy the *accelerated trim* condition

$$\begin{bmatrix} \dot{v}_d(t) \\ 0 \\ 0 \\ 0 \end{bmatrix} = f_v \left(\begin{bmatrix} 0 \\ 0 \\ \varphi_{qs}(t) \\ 0 \end{bmatrix}, \begin{bmatrix} v_d(t) \\ \beta_{qs}(t) \\ 0 \\ \dot{\chi}_d(t) \end{bmatrix}, \begin{bmatrix} F_{qs}^{(0)}(t) \\ \delta_{qs}^{(0)}(t) \end{bmatrix} \right). \tag{4.15}$$

Assuming that the desired curve is sufficiently regular ensures that (4.15) can be solved for its four unknowns provided $\dot{v}_d(t)$ is not too large. These regularity conditions are discussed in detail in (Saccon et al. 2012). In practice, using realistic motorcycle and tyre parameters, we have found no difficulties in computing solutions for longitudinal accelerations $|\dot{v}_d|$ up to about 1.6 g (exceeding the capabilities of high performance street bikes!).

Using $\varphi_{qs}(\cdot)$ and $\beta_{qs}(\cdot)$, we define the *quasi-static state trajectory* $(\mathbf{q}_{qs}(t), \mathbf{v}_{qs}(t)), t \in [0, T]$, as

$$\mathbf{q}_{qs}(t) = (x_d(t), y_d(t), \varphi_{qs}(t), \psi_{qs}(t))^T,$$

$$\mathbf{v}_{qs}(t) = (v_d(t), \beta_{qs}(t), \dot{\varphi}_{qs}(t), \dot{\psi}_{qs}(t))^T,$$

where $\psi_{qs}(\cdot)$ and $\dot{\psi}_{qs}(\cdot)$ are defined to be

$$\psi_{qs}(t) = \chi_d(t) - \beta_{qs}(t), \qquad \dot{\psi}_{qs}(t) = \dot{\chi}_d(t) - \dot{\beta}_{qs}(t).$$

In practice, $\dot{\varphi}_{qs}(\cdot)$ and $\dot{\beta}_{qs}(\cdot)$ are obtained by (numerical) differentiation of $\varphi_{qs}(\cdot)$ and $\beta_{qs}(\cdot)$, although they may, in principle, be computed using the linearization of (4.15).

At this point, we have little reason to expect that $(\mathbf{q}_{qs}(\cdot), \mathbf{v}_{qs}(\cdot))$ is the state portion of a trajectory of the SPM motorcycle, as there are four constraints and only two control inputs. To remedy this deficit, we will simply *add* two *artificial* control inputs. In this manner, we are *embedding* the SPM dynamics into the *extended* SPM system

$$\begin{bmatrix} \dot{v} \\ \dot{\beta} \\ \ddot{\varphi} \\ \ddot{\psi} \end{bmatrix} = f_v \left(\begin{bmatrix} x_r \\ y_r \\ \varphi \\ \psi \end{bmatrix}, \begin{bmatrix} v \\ \beta \\ \dot{\varphi} \\ \dot{\psi} \end{bmatrix}, \begin{bmatrix} F \\ \delta \end{bmatrix} \right) + \begin{bmatrix} 0 \\ 0 \\ u_\varphi \\ u_\psi \end{bmatrix}, \tag{4.16}$$

where the non-physical control inputs u_φ and u_ψ have been added to facilitate direct manipulation of $\ddot{\varphi}$ and $\ddot{\psi}$. Now, since F and δ affect $\dot{v}$ and $\dot{\beta}$ in a rather direct way, we can expect to find a quasi-static trajectory of augmented inputs $F_{qs}(\cdot), \delta_{qs}(\cdot), u_\varphi^{qs}(\cdot), u_\psi^{qs}(\cdot)$, that, together with $(\mathbf{q}_{qs}(\cdot), \mathbf{v}_{qs}(\cdot))$, constitutes a trajectory of the extended system (4.16). Indeed, for *each time*, one solves the first two equations of

$$\begin{bmatrix} \dot{v}_d(t) \\ \dot{\beta}_{qs}(t) \\ \ddot{\varphi}_{qs}(t) \\ \ddot{\psi}_{qs}(t) \end{bmatrix} = f_v \left(\begin{bmatrix} x_d(t) \\ y_d(t) \\ \varphi_{qs}(t) \\ \psi_{qs}(t) \end{bmatrix}, \begin{bmatrix} v_d(t) \\ \beta_{qs}(t) \\ \dot{\varphi}_{qs}(t) \\ \dot{\psi}_{qs}(t) \end{bmatrix}, \begin{bmatrix} F_{qs}(t) \\ \delta_{qs}(t) \end{bmatrix} \right) + \begin{bmatrix} 0 \\ 0 \\ u_\varphi^{qs}(t) \\ u_\psi^{qs}(t) \end{bmatrix} \tag{4.17}$$

for the quasi-static force and steering, $F_{qs}(t)$ and $\delta_{qs}(t)$, followed by the (trivial) specification of $u_\varphi^{qs}(t)$ and $u_\psi^{qs}(t)$ using the last two equations. The artificial inputs, $u_\varphi^{qs}(t)$ and $u_\psi^{qs}(t)$, make up for the difference between the (now specified) quasi-static angular accelerations, $\ddot{\varphi}_{qs}(t)$ and $\ddot{\psi}_{qs}(t)$, and the angular accelerations, $\ddot{\varphi}$ and $\ddot{\psi}$, that would result from applying the input $(F_{qs}(t), \delta_{qs}(t))^T$ at the state $(\mathbf{q}_{qs}(t), \mathbf{v}_{qs}(t))$.

4.5.2 Approximate Inversion by Trajectory Optimization

Recall that our goal is to find curves $\varphi(\cdot)$, $\psi(\cdot)$, $F(\cdot)$ and $\delta(\cdot)$ that satisfy (4.13). We attack the (approximate) solution of this time-varying system of differential-algebraic equations in the following manner.

First, we embed the dynamic portion of (4.13) into the extended, second-order system

$$\begin{bmatrix} \ddot{\varphi} \\ \ddot{\psi} \end{bmatrix} = f_v^{34} \left(\begin{bmatrix} 0 \\ 0 \\ \varphi \\ \psi \end{bmatrix}, \begin{bmatrix} v_d(t) \\ \chi_d(t) - \psi \\ \dot{\varphi} \\ \dot{\psi} \end{bmatrix}, \begin{bmatrix} F \\ \delta \end{bmatrix} \right) + \begin{bmatrix} u_\varphi \\ u_\psi \end{bmatrix}, \qquad (4.18)$$

where f_v^{34} indicates the vector containing the third and fourth elements of f_v, and we've ignored the position curve $(x_d(t), y_d(t))$ since f_v does not depend on it. Using state $x = (\varphi, \psi, \dot{\varphi}, \dot{\psi})^T$ and inputs $u = (F, \delta)^T$ and $u_{hg} = (u_\varphi, u_\psi)^T$, this can be written as the time-dependent control system

$$\dot{x} = f(x, u, u_{hg}, t).$$

Then, encoding the constraint portion of (4.13) into the function

$$h(x, u, t) = \begin{bmatrix} \dot{v}_d(t) \\ \dot{\chi}_d(t) - x_4 \end{bmatrix} - f_v^{12} \left(\begin{bmatrix} 0 \\ 0 \\ x_1 \\ x_2 \end{bmatrix}, \begin{bmatrix} v_d(t) \\ \chi_d(t) - x_2 \\ x_3 \\ x_4 \end{bmatrix}, \begin{bmatrix} u_1 \\ u_2 \end{bmatrix} \right),$$

we see that (4.13) is satisfied by (the components of) $(x(\cdot), u(\cdot))$ if and only if

$$\dot{x}(t) = f(x(t), u(t), u_{hg}(t), t)$$
$$0 = h(x(t), u(t), t)$$
$$0 = u_{hg}(t)$$

for (almost all) $t \in [0, T]$. The quasi-static trajectory $x_{qs}(t) = (\varphi_{qs}(t), \psi_{qs}(t), \dot{\varphi}_{qs}(t), \dot{\psi}_{qs}(t))^T$, $u_{qs}(t) = (F_{qs}(t), \delta_{qs}(t))^T$ and $u_{hg}^{qs}(t) = (u_\varphi^{qs}(t), u_\psi^{qs}(t))^T$, constructed above, satisfies both the dynamics and the $h \equiv 0$ constraint, but does not satisfy $u_{hg}(t) \equiv 0$.

To find a trajectory $(x(\cdot), u(\cdot), u_{hg}(\cdot))$ that satisfies the dynamics and approximately satisfies the constraints, we will use a trajectory optimization approach. Indeed, consider the optimal control problem

$$\min_{(x, u, u_{hg})(\cdot)} \int_0^T l(x(\tau), u(\tau), u_{hg}(\tau), \tau) \, d\tau + m(x(T))$$
$$\text{s.t.} \quad \dot{x} = f(x, u, u_{hg}, t), \qquad x(0) = x_0,$$

(4.19)

where the incremental

$$l(x, u, u_{hg}, t) = \tfrac{1}{2}\|(x,u) - (x_{qs}(t), u_{qs}(t))\|_{Q,R}^2$$
$$+ \tfrac{\rho}{2}[\,\|u_{hg}\|^2 + \|h(x,u,t)\|^2\,] \qquad (4.20)$$

and terminal

$$m(x) = \tfrac{1}{2}\|x - x_{qs}(T)\|_P^2 \qquad (4.21)$$

cost functions, with suitable symmetric, positive definite weights Q, R and P, and a suitable penalty $\rho > 0$, are used to discourage the violation of the equality constraints ($h \equiv 0$ and $u_{hg} \equiv 0$) while expressing the belief that there is a suitable trajectory $(x(\cdot), u(\cdot))$ near $(x_{qs}(\cdot), u_{qs}(\cdot))$. Taking $x_0 = x_{qs}(0)$, we see that $(x_{qs}(\cdot), u_{qs}(\cdot), u_{hg}^{qs}(\cdot))$ provides the required initial trajectory.

Given that the dynamics described by f are nonlinear and that the convexity properties of the incremental cost l are complicated due to the presence of h, it is difficult to say much in advance regarding the nature of solutions to the ρ-parametrized optimal control problem. Within our practical experience, the posed optimal control problem appears to be relatively easy to solve when the desired flatland trajectory is within the capabilities of the SPM model.

With an approximate solution $(\tilde{x}(\cdot), \tilde{u}(\cdot))$ to the lifting problem (4.19) in hand, we obtain an actual trajectory of the SPM model by tracking the approximate state-input trajectory

$$\tilde{\mathbf{q}}(t) = (x_d(t),\ y_d(t),\ \tilde{\varphi}(t),\ \tilde{\psi}(t))^T$$
$$\tilde{\mathbf{v}}(t) = (v_d(t),\ \chi_d(t) - \tilde{\psi}(t),\ \dot{\tilde{\varphi}}(t),\ \dot{\tilde{\psi}}(t))^T \qquad (4.22)$$
$$\tilde{\mathbf{u}}(t) = (\tilde{F}(t),\ \tilde{\delta}(t))^T$$

using a time varying trajectory tracking controller. That is, we use a projection operator to *project* the *curve* $(\tilde{\mathbf{q}}(\cdot), \tilde{\mathbf{v}}(\cdot), \tilde{\mathbf{u}}(\cdot))$ onto the SPM trajectory manifold to obtain a *trajectory* $(\mathbf{q}(\cdot), \mathbf{v}(\cdot), \mathbf{u}(\cdot))$ that approximately follows the desired flatland trajectory. A suitable time-varying feedback for the trajectory tracking controller can be obtained by solving a time-varying LQR problem for the linearization of the system about the approximate (state-input) trajectory (4.22). At this point, the artificial controls u_φ and u_ψ are no longer present, ensuring a true SPM trajectory.

4.6 Closed-Loop Control: Executing the Planned Trajectory

4.6.1 Manoeuvre Regulation

Given a desired manoeuvre regulation task

$$y_\xi(t) = (x_{r\xi}(t), y_{r\xi}(t)),$$

$t \geq 0$, the lifted state-control trajectory $\xi(t) = (x_\xi(t), u_\xi(t))$, $t \geq 0$, for the SPM model, obtained by using the lifting strategy discussed in the previous section, will be written in expanded form as

$$x_\xi(t) = (x_{r\xi}, y_{r\xi}, \varphi_\xi, \psi_\xi, v_\xi, \beta_\xi, \dot{\varphi}_\xi, \dot{\psi}_\xi)^T(t),$$
$$u_\xi(t) = (F_\xi, \delta_\xi)^T(t).$$

The virtual rider we have developed is based on a manoeuvre regulation control strategy (Hauser and Hindman 1995, Hindman 1999). With a manoeuvre regulation controller, the manoeuvring error is defined using an appropriate 'distance' between the current state $x(t)$ and the *entire* desired state curve $x_\xi(\cdot)$, rather than just the desired state $x_\xi(t)$ at time t as it would be for a trajectory tracking controller. As a consequence, in the case that $x(t) = x_\xi(t - \tau)$, for some $\tau > 0$, the controller simply uses the delayed version of the desired input $u_\xi(t - \tau)$, avoiding the potentially dangerous speed-up phenomenon. In the following, we present some important aspects of this theory, referring the reader to (Banaszuk and Hauser 1995; Hauser and Chung 1994; Hauser and Hindman 1995; Hindman 1999 and Saccon 2006) for further details and results.

In the control of nonlinear systems, it is common and useful to seek out a coordinate system in which the nature of the problem is clarified and for which the problem is, perhaps, somewhat easier to solve. For manoeuvre regulation, we will make use of a *transverse* coordinate system that is adapted to the desired manoeuvring task.

The first step in obtaining a manoeuvre regulation controller is to specify a *longitudinal* parametrization for the desired task. For vehicles, it is natural to use an *arc length* parametrization

$$s_\xi(t) = \int_0^t \sqrt{\dot{x}_{r\xi}^2(\tau) + \dot{y}_{r\xi}^2(\tau)}\, d\tau \ .$$

For tasks for which the desired velocity of the vehicle is bounded away from zero, the mapping $t \mapsto s_\xi(t)$ is invertible. Using the inverse $\bar{t}_\xi(s)$, we obtain the s parametrized curves for the state $\bar{x}_\xi(s) = x_\xi(\bar{t}(s))$, input $\bar{u}_\xi(s) = u_\xi(\bar{t}(s))$ and output (task) $\bar{y}_\xi(s) = y_\xi(\bar{t}(s))$. We see that these quantities are related according to

$$\begin{aligned}\bar{x}_\xi'(s)\bar{v}_\xi(s) &= f(\bar{x}_\xi(s), \bar{u}_\xi(s)) \\ \bar{y}_\xi(s) &= h(\bar{x}_\xi(s)),\end{aligned} \quad (4.23)$$

where f is the SPM model dynamics, $\bar{v}_\xi(s) = v_\xi(\bar{t}_\xi(s))$ and $v_\xi(t) = \dot{s}_\xi(t)$, and the *prime* in (4.23) indicates differentiation with respect to s. Note that, since we are using an arc length parametrization, the desired path velocity $\bar{v}_\xi(s)$ is equal to the vehicle velocity component $\bar{v}_\xi(s)$ of the desired state task $\bar{x}_\xi(s)$. Exploiting this apparent coincidence and with the aim of reducing the possibility of confusion that may arise from the use of v as the *transverse* form control input (that we will introduce shortly), we will use v, $\bar{v}_\xi(s)$ and so on, to refer to vehicle velocity objects below.

The next step in the design of a manoeuvre regulation controller is to construct a local change of coordinates

$$x \mapsto \Psi(x) = (s, w_1, \ldots, w_{n-1})$$

that is valid on a (tubular) neighbourhood of the state-space curve $\bar{x}_\xi(\cdot)$ and that satisfies $\Psi(\bar{x}_\xi(s)) = (s, 0, \ldots, 0)$. We call s the *longitudinal coordinate* and $w = (w_1, \ldots, w_{n-1})$ the *transverse coordinates*. It is useful to partition the coordinate change accordingly, defining the mapping $W : \mathbb{R}^n \to \mathbb{R}^{n-1}$ so that

$$\begin{bmatrix} s \\ w \end{bmatrix} = \begin{bmatrix} \pi(x) \\ W(x) \end{bmatrix} = \Psi(x) \ . \quad (4.24)$$

Note that n is the state dimension and, for the SPM model, we have $n = 8$.

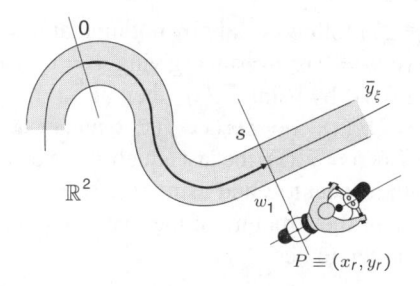

Virtual Rider Design: Optimal Manoeuvre Definition and Tracking

Figure 4.12 Local coordinates around the path. The arclength s of the point having minimum distance from (x_r, y_r) and the (signed) distance w_1 of the point $P = (x_r, y_r)$ from the path can be used as coordinates to indicate the position of rear wheel contact point about the desired path $\bar{y}_\xi(\cdot)$

For the SPM model, in a neighbourhood of the desired output manoeuvre, $\bar{y}_\xi(\cdot)$, we can parametrize points using

$$\begin{bmatrix} x_r \\ y_r \end{bmatrix} = \begin{bmatrix} \bar{x}_{r\xi}(s) \\ \bar{y}_{r\xi}(s) \end{bmatrix} + \begin{bmatrix} -\sin\bar{\chi}_\xi(s) \\ \cos\bar{\chi}_\xi(s) \end{bmatrix} w_1 = \Omega(s, w_1), \tag{4.25}$$

as depicted in Figure 4.12. We have taken the (transverse) displacement to be orthogonal to the (unit) tangent vector $(\bar{x}'_\xi(s), \bar{y}'_\xi(s)) = (\cos\bar{\chi}_\xi(s), \sin\bar{\chi}_\xi(s))$ with course heading $\bar{\chi}_\xi(s) = \bar{\psi}_\xi(s) + \bar{\beta}_\xi(s)$. The inverse $(s, w_1) = \Phi(x_r, y_r)$ of this map is locally well-defined around points (s, w_1) satisfying $w_1 \bar{\sigma}_\xi(s) < 1$ where $\bar{\sigma}_\xi(s) = \bar{\chi}'_\xi(s)$ is the *curvature* of the path at $(\bar{x}_{r\xi}(s), \bar{y}_{r\xi}(s))$. Note that w_1 is the signed distance from (x_r, y_r) to the (locally) *nearest* point (which occurs at location s along the manoeuvre) and represents the lateral displacement of the rear wheel contact point from the desired ground path.

For the SPM model, the local diffeomorphism $(s, w) = \Psi(x)$ can be constructed using the inverse maps of Equation (4.25) and taking, as remaining components of the transverse coordinates, the mapping

$$\begin{aligned} w_2 &= \varphi - \bar{\varphi}_\xi(s), & w_5 &= \beta - \bar{\beta}_\xi(s), \\ w_3 &= \psi - \bar{\psi}_\xi(s), & w_6 &= \dot{\varphi} - \dot{\bar{\varphi}}_\xi(s), \\ w_4 &= v - \bar{v}_\xi(s), & w_7 &= \dot{\psi} - \dot{\bar{\psi}}_\xi(s), \end{aligned} \tag{4.26}$$

with $s = \pi(x)$. The transverse coordinates measure the difference between the current state and desired state along the desired path at position $s = \pi(x)$, recalling that $\pi(x)$ is the first component of the diffeomorphism $(s, w) = \Psi(x)$.

Let $x = \Gamma(s, w)$ denote the inverse of $(s, w) = \Psi(x)$. Using the state-dependent input transformation $u = \bar{u}_\xi(s) + v$, the system dynamics can be written in (s, w) coordinates as (cf. Hauser and Chung 1994)

$$\begin{aligned} \dot{s} &= \bar{v}_\xi(s) + f_1(s, w, v) \\ \dot{w} &= A(s)w + B(s)v + f_2(s, w, v) \end{aligned} \tag{4.27}$$

where $f_1(s, 0, 0) \equiv 0$ and $f_2(s, w, v)$ is higher order in (w, v). We call (4.27) the *transverse form* of the systems dynamics.

The second equation in (4.27) follows easily by noting that $w = 0$ if and only if $x \in \bar{x}_\xi(\cdot)$ and that the state-space curve $w = 0$ is invariant (giving $\dot{w} = 0$) under the flow of the system when $v = 0$ (i.e. we stay on $\bar{x}_\xi(\cdot)$ by using $\bar{u}_\xi(s)$ when at $\bar{x}_\xi(s)$). For the first equation, differentiate $s = \pi(x)$ to get $\dot{s} = D\pi(x) \cdot \dot{x} = D\pi(x) \cdot f(x, u)$ and evaluate with $(w, v) = (0, 0)$ to get $\dot{s} = D\pi(\bar{x}_\xi(s)) \cdot f(\bar{x}_\xi(s), \bar{u}_\xi(s)) = \bar{v}_\xi(s)$; the last equality can be obtained by differentiating the identity $s = \pi(\bar{x}_\xi(s))$ with respect to s and using (4.23).

A closer look reveals some of the structure of the transverse form. Using $\dot{w} = DW(x) \cdot \dot{x}$, the nonlinear $\dot{w}$ dynamics is seen to be

$$\dot{w} = DW(\Gamma(s, w)) \cdot f(\Gamma(s, w), \bar{u}_\xi(s) + w),$$

so that (Saccon 2006)

$$A(s)\,w = \mathbf{D}W(\bar{x}_\xi(s)) \cdot \mathbf{D}_1 f(\bar{x}_\xi(s), \bar{u}_\xi(s)) \cdot Z(s)\,w$$
$$+ \mathbf{D}^2 W(\bar{x}_\xi(s)) \cdot (f(\bar{x}_\xi(s), \bar{u}_\xi(s)), Z(s)\,w), \tag{4.28}$$

$$B(s)\,v = \mathbf{D}W(\bar{x}_\xi(s)) \cdot \mathbf{D}_2 f(\bar{x}_\xi(s), \bar{u}_\xi(s)) \cdot v, \tag{4.29}$$

where $Z(s)\,w = D_2\Gamma(s, 0) \cdot w$. Note the requirement for the *second* derivative of the coordinate change. The terms $\mathbf{D}_1 f(\bar{x}_\xi(s), \bar{u}_\xi(s)))$ and $\mathbf{D}_2 f(\bar{x}_\xi(s), \bar{u}_\xi(s)))$ in the above expression are those occurring in the standard linearization of the system, evaluated at $(x, u) = (\bar{x}_\xi(s), \bar{u}_\xi(s))$. In (Saccon et al. 2013), the matrices $A(s)$ and $B(s)$ for the SPM model (and other planar vehicles) are computed explicitly.

One key idea in manoeuvre regulation, is to restrict the attention to inputs that do not depend explicitly on time, so that one can eliminate t (in a neighbourhood of the task in state space) obtaining

$$\bar{w}' = A_T(s)\bar{w} + B_T(s)\bar{v} + f_T(s, \bar{w}, \bar{v}), \tag{4.30}$$

where $A_T(s) = A(s)/\bar{v}_\xi(s)$, $B_T(s) = B(s)/\bar{v}_\xi(s)$ and $f_T(s, \bar{w}, \bar{v})$ is higher order in $(\bar{w}, \bar{v})$. The systems (4.27) and (4.30) are *trajectory equivalent* in the sense that relevant trajectories of each system can be mapped to trajectories of the other. Indeed, each bounded trajectory $(\bar{w}(s), \bar{v}(s))$, $s \geq s_0$ of (4.30) gives rise to a (4.27) trajectory $(s(t), w(t), v(t))$, $t \geq 0$, with $(s(0), w(0)) = (s_0, \bar{w}(s_0))$; just integrate the scalar differential equation $\dot{s} = \bar{v}_\xi(s) + f_1(s, \bar{w}(s), \bar{v}(s))$ to determine the strictly increasing s trajectory $s(t)$, $t \geq 0$, and write $w(t) = \bar{w}(s(t))$ and $v(t) = \bar{v}(s(t))$. Conversely, *every* bounded trajectory $(s(t), w(t), v(t))$, $t \geq 0$ of (4.27) with strictly positive and bounded $\dot{s}(\cdot)$ (and regardless of how $v(\cdot)$ was determined!) gives rise to the (4.30) trajectory $(\bar{w}(s), \bar{v}(s)) = (w(\bar{t}(s)), v(\bar{t}(s)))$, where $\bar{t}(s)$ is the inverse function of $s(t) : \bar{t}(s(t)) = t$, $t \geq 0$.

Expressing the transverse dynamics as a differential equation where the longitudinal state s becomes the independent variable enables the development of a *time-invariant* control law for regulating the transverse states w to zero. If, for example, the *transverse linearization*

$$\bar{w}' = A_T(s)\bar{w} + B_T(s)\bar{v} \tag{4.31}$$

is exponentially stabilized by a (s-varying) linear state feedback $\bar{v} = -K(s)\bar{w}$, then the time-invariant *nonlinear* state feedback $v = -K(s)w$ will exponentially stabilize the manoeuvre

$\bar{x}_\xi(\cdot)$ for (4.27), (Hauser and Chung 1994). Written in the original coordinates, the nonlinear feedback

$$u = k(x) = \bar{u}_\xi(\pi(x)) - K(\pi(x))W(x) \tag{4.32}$$

exponentially stabilizes the manoeuvre $\bar{x}_\xi(\cdot)$ for $\dot{x} = f(x, u)$. Such a $K(\cdot)$ might be developed using an LQR strategy if the transverse linearization is, for example, uniformly or instantaneously controllable.

A manoeuvre regulation control law can be obtained if a linear feedback $\bar{v}(s) = -K(s)\bar{w}(s)$ that exponentially stabilizes the transverse linearization (4.30) can be found. Generally speaking, stability concepts are not applicable on finite intervals, as is the case here where the path to be followed has length $L < \infty$. However, if we extend the path using constant curvature and the task using constant velocity to one of infinite extent, then a stabilizing controller can be designed. In practice, it is sufficient that the constant-velocity, constant-curvature tail of the task be long enough that the components (neglecting (x_r, y_r) and ψ) of the lifted state-control task $(\bar{x}_\xi(\cdot), \bar{u}_\xi(\cdot))$ exhibit near-steady-state values as s approaches L. In that case, $A_T(L) \approx A_T^{(\sigma,v)}$ and $B_T(L) \approx B_T^{(\sigma,v)}$ where $\sigma = \bar{\sigma}_\xi(L)$ and $v = \bar{v}_\xi(L)$ are the constant terminal curvature and velocity.

For such a task, we can obtain the desired stabilizing $K(\cdot)$ by solving the finite horizon, linear quadratic optimal control problem

$$\text{minimize} \int_{s_0}^{L} \bar{w}(s)^T Q\, \bar{w}(s)/2 + \bar{v}(s)^T R\, \bar{v}(s)/2 \, ds + \bar{w}(L)^T P_L \bar{w}(L)/2 \tag{4.33}$$

$$\text{subject to} \quad \bar{w}' = A_T(s)\bar{w} + B_T(s)\bar{v}, \quad \bar{w}(s_0) = w_0$$

for (the limiting case) $s_0 = 0$. Here $Q = Q^T$ and $R = R^T$ are *arbitrary* positive definite matrices (and possibly time-varying) and P_L is the positive definite algebraic Riccati equation solution associated with the time-invariant LQR problem for $A_T^{(\sigma,v)}, B_T^{(\sigma,v)}, Q, R$. The *feedback* solution to (4.33) provides the (space-varying) gain matrix

$$K(s) = R^{-1} B_T^T(s) P(s), \tag{4.34}$$

where $P(s)$, $s \in [0, L]$, is the solution of the *differential Riccati equation* (DRE)

$$-P' = A_T^T(s)P + PA_T(s) - PB_T(s)R^{-1}B_T^T(s)P + Q, \quad P(L) = P_L. \tag{4.35}$$

The minimum value (or cost to go) (Anderson and Moore 1989) of (4.33) is $V(s_0, w_0) = w_0^T P(s_0)\, w_0/2$. Since the cost in (4.33) is strongly positive definite on the linear space of homogenous trajectories of the linear dynamics, it is clear that the Riccati solution $P(s)$ is positive definite for each $s \in [0, L]$. Extending $P(\cdot)$ to the infinite horizon using $P(s) = P_L$, $s > L$, we obtain a $K(\cdot)$ that exponentially stabilizes (the extended version of) the transverse linearization (4.30) as well as the manoeuvring task $w \equiv 0$ for the nonlinear system in transverse form (4.27).

We emphasize again that the exponential stability property is obtained for *every* bounded, uniformly positive definite choice of $Q(\cdot)$ and $R(\cdot)$. In this fashion, we obtain a feedback controller where the *gain* is *scheduled* (according to the longitudinal state of the system) and *stability* is guaranteed. Note, however, that the *quality* of regulation for the obtained $K(\cdot)$ is strongly affected by the choice of Q and R. One possible way to choose the Q and R matrices for the SPM model is discussed in the following.

4.6.2 Shaping the Closed-Loop Response

Our task is to choose the linear quadratic regulator weights (or weighting functions) Q and R in such a manner that the (space-varying) closed loop system

$$\overline{w}' = \left[A_T(s) - B_T(s)\, K(s) \right] \overline{w} \tag{4.36}$$

has a satisfactory dynamic response (for a desired class of admissible tasks). For instance, it is essential that the response from reasonable initial conditions $\overline{w}(s_0) = w_0$ converges to zero in a manner that is neither too fast nor too slow and that does not result in too large a response in, for example, the roll φ. Many of these (and other) considerations are driven by the fact that the system is a mechanical system that cannot be made to move rapidly without exerting forces that are larger than what can be produced by tyre–road interations. The roll (lean) angle is limited by similar considerations. Finally, since we are interested in controlling a more complex multibody motorcycle model, the feedback $\overline{v} = -K(s)\overline{w}$ should provide a *regulator* that is somewhat robust to modelling errors including parameter mismatch and unmodelled dynamics such as the intentionally neglected suspension.

Denote with $A_T^{(\sigma,v)}$, $B_T^{(\sigma,v)}$ the transverse linearization of the SPM model along the trim trajectory consisting of a circular path of curvature σ traversed at constant speed v. In (Saccon et al. 2013), we have shown that the transverse linearization $A_T^{(\sigma,v)}$, $B_T^{(\sigma,v)}$ are constant matrices due to the particular choice of the transverse coordinates.

Our approach is to find, if possible, a (single) positive definite pair Q, R that results in a satisfactory (linearized LQR) dynamic response over a desired class of constant operating conditions (trim trajectories). For each constant (σ, v) in the desired set, we want the dynamic response of the linear system

$$\overline{w}' = \left[A_T^{(\sigma,v)} - B_T^{(\sigma,v)} K_{(Q,R)}^{(\sigma,v)} \right] \overline{w} \tag{4.37}$$

to satisfy the chosen performance measures, where $K_{(Q,R)}^{(\sigma,v)}$ is the constant LQR gain associated with $A_T^{(\sigma,v)}$, $B_T^{(\sigma,v)}$, Q, R. This Q, R pair is then used to construct $K(s)$ for the chosen task using (4.34), (4.35) above. Although we do not have a formal justification, our experience seems to indicate that the s-varying system (4.36) inherits many of the performance characteristics of the family of constant systems (4.37).

Due to space limitations, we refer the interested reader to (Saccon et al. 2013) for the details of the inverse optimal control strategy that we have adopted to obtain the Q and R matrices. The strategy been successful in designing Q and R matrices that are suitable for a wide range of velocities and lateral accelerations (hence curvatures). As an example, in Figure 4.13, we report the *locus* of open and closed loop poles for a sport motorcycle as the lateral acceleration is varied from zero to 1.2 g, while holding the velocity constant at $v = 30$ m/s. For the entire range of lateral accelerations, the closed loop eigenvalues remain inside a rather small region of the complex plane, with sufficient damping. A similar root locus plot with varying velocity shows a similar dynamic response grouping. We remark that this design heuristic has been used thus far in the development of manoeuvre regulation controllers for three racing motorcycles and a couple of sport bikes.

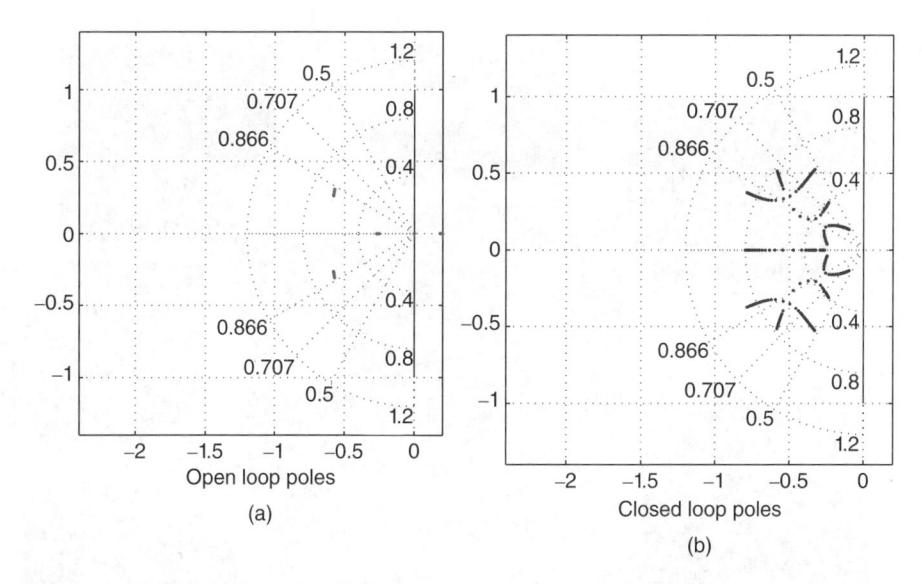

Figure 4.13 Open and closed loop poles using the same Q and R. Open loop poles are reported in (a), and closed loop in (b). The velocity is held constant at $v = 30$ m/s while the lateral acceleration is varied from zero to 1.2 g

4.6.3 Interfacing the Maneuver Regulation Controller with the Multibody Motorycle Model

This subsection describes how the manoeuvre regulation controller can be used to control a multibody model of a fully articulated motorcycle. To this end, we have used the commercial multibody code Adams/Car together with the VI-Motorcycle add-on devoted to motorcycle dynamics simulation (Figure 4.14).

A modern sport motorcycle is composed of a centre subsystem (including frame, engine, gear box and fuel tank), a rear subsystem (including swing arm and brake callipers), an upper steering subsystem (including handlebars and upper fork), a bottom steering subsystem (including lower fork and brake callipers), and front and rear wheel subsystems (including tyres, rims, and brake disks). Tyre–ground interaction is modelled using the Pacejka magic formula tyre model (PAC-MC 1.1). The tyre model (Pacejka 2002a) includes relaxation length, providing phase lag between kinematic slip and tyre force (for both longitudinal slip and sideslip angle).

The aerodynamics is modelled as a drag force parallel to the ground acting at the centre of pressure in the opposite direction of vehicle's motion. The motorcycle model includes a telescopic fork suspension at the front and a monoshock (single linkage) suspension at the rear. The engine is modelled as a massive disk rotating about the crankshaft axis, which is orthogonal to the vehicle plane of symmetry. Depending on its spinning rate (RPM) and the opening of the (virtual) throttle valve, the engine provides a torque that is counteracted by the frame. A motorcycle gearbox is modelled and connected to the engine. The gearbox includes primary drive gear ratio, first to sixth gear ratios, and secondary drive gear ratio,

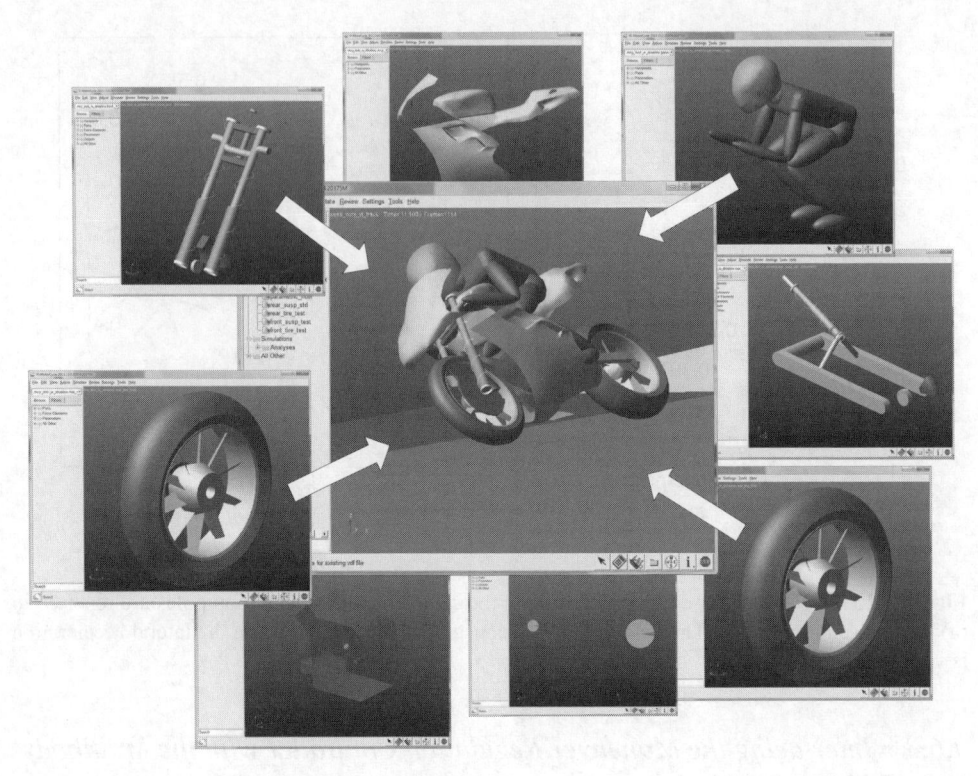

Figure 4.14 The (standard) VI-Motorcycle motorcycle model. The motorcycle is described using different assemblies stored in a database that can be connected to form a complete vehicle, ready for dynamic simulations

this last specified by drive and wheel sprocket radii. The secondary drive is composed of a massless chain, modelled as two nonlinear 'springs' attached at the bottom and top of the drive and wheel sprockets. The rider is modelled as a rigid body that is firmly attached to the centre assembly: no rider motion is used in this work.

Because the states and inputs of the sliding plane motorcycle (the *control design model*) are different from those of the multibody motorcycle (the *plant*), we need to develop a suitable interface between the controller and the plant in order for the former to act on the latter as if it were the control design model.

The input to the controller is obtained by taking measurements of the multibody plant: roll angle and rate, yaw angle and rate, and rear then the wheel contact point position and velocity from the multibody vehicle are passed to the controller without further processing. The rear wheel contact point for the multibody vehicle is the centre of the contact patch that the (torus-shaped) rear tyre makes with the ground.

The maneouvre regulation controller produces, as output, the thrust force and the effective steering angle. These are transformed into the plant inputs, steering angle (at the handle bar) and throttle and brake commands, as follows. To provide a steering command, the effective steering angle (output) from the controller are transformed using the approximation $\cos \varphi \tan \delta_f = \tan \delta$ (Saccon et al. 2012, Saccon and Hauser 2009).

The maximum traction force available depends primarily on the current gear ratio and the engine RPM. During dynamic simulation, the thrust force demand F from the maneouvre regulation controller is either positive or negative. If the required amount can be generated by the engine, then the throttle valve is opened accordingly, up to the wide open position, at which point it saturates. Saturations may also occur in the other direction (as the throttle command is rolled off) when the desired (negative) braking force F cannot be fully obtained using engine braking torque. In that condition, brakes are used to provide the additional required braking force. For simplicity, the braking torque is always applied on the front wheel. Further details and numerical simulations can be found in (Saccon et al. 2012; 2013).

4.7 Conclusions

We have presented an effective means for exploring aggressive motorcycle trajectories. In the first part of this chapter, we considered the problem of generating an optimal speed profile, taking into account the constraints imposed to the maximum acceleration and deceleration by the tyres, engine and the possible lifting of the front wheel (wheelie) and rear wheel (stoppie). In the second part of this chapter, we detailed a reduced-order manoeuvre regulation controller for controlling a multibody motorcycle model along the obtained optimal speed profile.

The manoeuvre regulation controller is based on a simplified motorcycle model, the sliding plane motorcycle model. Dynamics inversion for this simplified model is obtained by embedding its dynamics into an extended control system by adding two non-physical control inputs. By optimizing away this additional control effort, we were able to lift a flatland trajectory into a full state-input trajectory. This trajectory of the sliding plane motorcycle model is then used in a manoeuvre regulation control system that can drive a multibody motorcycle model along the desired curvature–velocity profile.

The proposed control strategy has been found to perform well in the presence of unmodelled dynamics such as that due to the suspension system, gear shifting and even short-lived wheelies. This control architecture has been chosen for several reasons. First, the virtual rider had to be interfaced with commercial multibody software and needed to be applicable to different types of two-wheeled vehicles (e.g. scooters or motorcycles) with quite different suspension and drive train systems, for which standard parametric models are not available. Second, we are sceptical of the notion that excellent tracking results can only be obtained with a virtual rider based on a full multibody model. It is our opinion that human riders base their actions (and predictions) on a mental model (of different kinds, depending on experience and skill) that neglects many model details (including, for example, tyre relaxation length, vehicle mass distribution or tyre profiles). Also, our previous experience in designing a virtual rider (Saccon et al. 2004), suggests that it is possible to obtain quite remarkable tracking performance, for relatively low speeds (less that 30 m/s), using only a simple *non-holonomic* vehicle model combined with a simple look-ahead strategy.

The proposed virtual rider provides quite satisfying tracking results for aggressive manoeuvres (such as that experienced, in racing applications). A detailed discussion of such numerical experiments, however, goes beyond the scope of this contribution, where the aim has been to expose the key ideas underlying the virtual rider and to show that the feedback strategy has been successfully applied to a multibody (code) model. We are confident that this control paradigm can be extended to include rider motion (thought of as an input) and paths with banking and elevation changes, which are of interest when simulating high performance motorcycle manoeuvres for racing applications. We refer the reader to (Hauser and Saccon 2006; Saccon et al. 2012; 2013) for further discussion on the generation of optimal velocity profiles and the development of the virtual rider.

4.8 Acknowledgements

We owe a great debt to Professor Ruggero Frezza for invaluable insights, advice and contributions to this project. We also express our sincere gratitude to Diego Minen and his research and development team from VI-Grade for fruitful collaboration and support during this project. Finally, we would like to thank Professor Vittore Cossalter and his group for providing some of the numerical data used to characterize the multibody motorcycle model used in the presented numerical simulations. This research was supported in part by MSC.Software Corporation and VI-grade.

References

Anderson BDO and Moore JB 1989 *Optimal Control: Linear Quadratic Methods*. Prentice-Hall International, Inc.
Banaszuk A and Hauser J 1995 Feedback linearization of transverse dynamics for periodic orbits. *Systems and Control Letters* **26**, 95–105.
Bertolazzi E, Biral F and Da Lio M 2005 Symbolic-numeric indirect method for solving optimal control problems for large multibody systems: The time-optimal racing vehicle example. *Multibody System Dynamics* **13**(2), 233–252.
Bertolazzi E, Biral F and Da Lio M 2006 Symbolic-numeric efficient solution of optimal control problems for multibody systems. *Journal of Computational and Applied Mathematics* **185**, 404–421.
Casanova D, Sharp R and Symonds P 2000 Minimum time manoeuvring: The significance of yaw inertia. *Vehicle System Dynamics* **2**(34), 77–115.
Cossalter V and Lot R 2002 A motorcycle multi-body model for real time simulations based on the natural coordinates approach. *Vehicle System Dynamics*.
Cossalter V, Da Lio M, Lot R and Fabbri L 1999 A general method for the evaluation of vehicle manoeuvrability with special emphasis on motorcycles. *Vehicle System Dynamics* **31**(2), 113–135.
Evangelou S 2003 *The control and stability analysis of two-wheeled road vehicles*. PhD thesis. Imperial College London.
Frezza R and Beghi A 2003 *New Trends in Nonlinear Dynamics and Control and their Applications* Springer Berlin / Heidelberg chapter Simulating a Motorcycle Driver, pp. 175–186.
Frezza R and Beghi A 2006 A virtual motorcycle driver for closed-loop simulation. *IEEE Control Systems Magazine* **26**(5), 62–77.
Getz N 1994 Control of balance for a nonlinear nonholonomic non-minimum phase model of a bicycle *Proceedings of the American Control Conference (ACC), Baltimore*, vol. 1, pp. 148–151.
Getz N 1995 Internal equilibrium control of a bicycle *Proceedings of the 34th IEEE Conference on Decision and Control*, vol. 4, pp. 4285–4287.
Getz N and Marsden J 1995 Control for an autonomous bicycle *Proceedings of the IEEE International Conference on Robotics and Automation (ICRA)*, vol. 2, pp. 1397–1402.
Hauser J and Chung CC 1994 Converse Lyapunov functions for exponentially stable periodic orbits. *Systems and Control Letters* **23**, 27–34.
Hauser J and Hindman R 1995 Maneuver regulation from trajectory tracking: Feedback linearizable systems *Proceedings of the IFAC Symposium on Nonlinear Control Systems (NOLCOS), Lake Tahoe, CA, USA*.
Hauser J and Saccon A 2006 Motorcycle modelling for high-performance maneuvering. *IEEE Control Systems Magazine* **26**(5), 89–105.
Hauser J, Saccon A and Frezza R 2004a Achievable motorcycle trajectories *Proceedings of the 43rd IEEE Conference on Decision and Control (CDC)*, vol. 4, pp. 3944–3949.

Hauser J, Saccon A and Frezza R 2004b Achievable motorcycle trajectories *43rd IEEE Conference on Decision and Control*, pp. 3944–3949–, Paradise Island, Bahamas.

Hauser J, Saccon A and Frezza R 2004c Aggressive motorcycle trajectories *Proceedings of the 6th Symposium on Nonlinear Control Systems (NOLCOS)*, Stuttgart, Germany.

Hauser J, Saccon A and Frezza R 2005a On the driven inverted pendulum *Proceedings of the 44th IEEE Conference on Decision and Control (CDC), Seville, Spain.*

Hauser J, Saccon A and Frezza R 2005b On the driven inverted pendulum *44th IEEE Conference on Decision and Control*, pp. 6176–6180–, Seville, Spain.

Hindman R 1999 *A Nonlinear Control Design Methodology Using Maneuvering Ideas* PhD thesis University of Colorado at Boulder.

Limebeer D and Sharp R 2006 Single-track vehicle modelling and control, bicycle, motorcycles, and models. *IEEE Control Systems Magazine* **26**(5), 34–61.

Lot R 2004 A motorcycle tire model for dynamic simulations: Theoretical and experimental aspects. *Meccanica* **39**(3), 207–220.

Lot R and Cossalter V 2006 A nonlinear rider model for motorcycles *FISITA 2006 World Automotive Congress, Yokohama, Japan.*

Lot R and Da Lio M 2004 A symbolic approach for automatic generation of the equations of motion of multibody systems. *Multibody System Dynamics*.

Lot R, Massaro M and Sartori R 2007 Advanced motorcycle virtual rider *Proceedings of the 20th IAVSD Symposium, Dynamics of Vehicles on Roads and Tracks, Berkeley, California.*

Massaro M and Lot R 2010 A virtual rider for two-wheeled vehicles *Proceedings of the 49th IEEE Conference on Decision and Control*, pp. 5586–5591.

Milliken W and Milliken D 1995 *Race Car Vehicle Dynamics*. Society of Automotive Engineers (SAE).

Pacejka H 2002a *Tire and Vehicle Dynamics* 2nd edn. Butterworth-Heinemann, Oxford.

Pacejka H 2002b *Tyre and Vehicle Dynamics*. Butterworth Heinemann, Oxford.

Saccon A 2006 *Maneuver Regulation of Nonlinear Systems: The Challenge of Motorcycle Control* PhD thesis University of Padova, Italy.

Saccon A and Hauser J 2009 An efficient Newton method for general motorcycle kinematics. *Vehicle System Dynamics* **47**(2), 221–241.

Saccon A, Hauser J and Beghi A 2008 A virtual rider for motorcycles: An approach based on optimal control and maneuver regulation *Proceedings of the 3rd IEEE International Symposium on Communications, Control and Signal Processing (ISCCSP), St. Julians, Malta.*

Saccon A, Hauser J and Beghi A 2012 Trajectory exploration of a rigid motorcycle model. *IEEE Transactions on Control Systems Technology* **20**(2), 424–437.

Saccon A, Hauser J and Beghi A 2013 A virtual rider for motorcycles: Maneuver regulation of a multi-body vehicle model. *IEEE Transactions on Control Systems Technology* **21**(2), 332–346.

Saccon A, Hauser J and Frezza R 2004 Control of a bicycle using model predictive control strategy *Proceedings of the 6th IFAC Symposium on Nonlinear Control Systems (NOLCOS), Stuttgart, Germany.*

Sharp R 2007 Motorcycle steering control by road preview. *Journal of Dynamic Systems, Measurement, and Control* **129**, 373–381.

Sharp R and Limebeer D 2001 A motorcycle model for stability and control analysis. *Multibody System Dynamics* **6**, 123–142.

Sharp R, Evangelou S and Limebeer D 2004 Advances in the modelling of motorcycle dynamics. *Multibody System Dynamics* **12**(3), 251–283.

5

The Optimal Manoeuvre

Francesco Biral, Enrico Bertolazzi, and Mauro Da Lio
Department of Industrial Engineering, University of Trento, Italy

Motorcycle riding mostly owes its popularity to the sense of freedom that users experience (Broughton et al. 2009; Sexton et al. 2004). Thanks to the comparable mass between the human and the machine, motorcyclists use their upper body movements as an additional input enhancing the sensation of control over the vehicle and giving the impression of being part of the machine entity. Those additional degrees of freedom let riders express themselves with a personal riding style that is not possible to such an extends in other motorised vehicles. Moreover, compared to cars, due to the natural vehicle instabilities such as wobble and weave (Cossalter et al. 2004; Limebeer et al. 2001), motorcycles provide more of an opportunity to challenge the rider's control skills, manoeuvring around obstacles in a variety of road scenarios (e.g., rural and/or mountain roads or off roads). Sensation seeking, which is one of the main factors for riding, is provided by the perception of longitudinal acceleration, roll rate and speed (Watson et al. 2007; Sexton et al. 2004). The more motorcyclists practice and improve their riding experience the more they push the vehicle to its limits seeking fun and coming close to the intrinsic machine instabilities. Riding experience and skill plays an important role in the efficiency of the control actions and repeatability of the manoeuvres (Rice 1978). The above considerations make it evident that the rider has a major influence on the machine performance and dynamic behaviour. Therefore, it is more true than for other ground vehicles that the assessment of two-wheeled vehicles cannot be done without the inclusion of the rider in the evaluation process. Assessing motorcycle performance is quite a difficult task due to a large variability of riding style, thanks to the number of inputs that can contribute to the control strategy, and the central role of the subjective perception of the riding experience. Despite the difficulty, the objective evaluation of motorcycle performance is a fundamental mission of the motorcycle designer for both commercial and racing machines. On the other hand the results of the assessment process, both from numerical simulations and experimental tests, depend on how the vehicle is manoeuvred by the rider

---

Modelling, Simulation and Control of Two-Wheeled Vehicles, First Edition.
Edited by Mara Tanelli, Matteo Corno and Sergio M. Savaresi.
© 2014 John Wiley & Sons, Ltd. Published 2014 by John Wiley & Sons, Ltd.

to accomplish a given task or goal such as a standardised test, or a manoeuvre to avoid an obstacle or to complete a lap in the minimum time. Thus the fundamental question in the evaluation process is: given a specific motorcycle, is there a manner in which to ride the machine in the most efficient way according to a desired goal? If such a manner exists, we will call it the *optimal manoeuvre* and it can be used as a reference to compare and rank other similar motorcycles. If the goal is an objective definition of the absolute performance, then the optimal manoeuvre is a measure, or baseline, of the best intrinsic performance that can be achieved with the machine, regardless of the human rider's skills. On the other hand, as we have said above, the rider's skills and experience must be included somewhat in the evaluation process. Therefore the optimal manoeuvre can be searched among the set of manoeuvres that also satisfy the human's riding skills and preferences. This manoeuvre is still optimal, but only for the specific group of riders whose skills and preferences were objectively defined. The optimal manoeuvre that satisfies the riding preferences in some measure models how a real human rider of the selected group will ride to accomplish the given task. For this reason, later on we will also call it the *reference manoeuvre*. Thus assessing the rider–motorcycle performance is also a matter of defining the limitations and riding preferences of a selected group of users, which is not always an easy task.

The reference manoeuvre may also be a central concept for the design of intelligent riding support systems to help riders to safely manoeuvre. Motorcyclists ride their machines more for pleasure than for mobility reasons and they say that they really enjoy riding fast (Broughton et al. 2009). Enjoyment of the riding experience contributes to the tendency to engage in dangerous behaviour that, combined with overconfidence of proper riding skills, downplays hazardous conditions (Chen and Chen 2011), which leads to loss of control with dramatic consequences for riders' health (injuries and death) due to their physical vulnerability. Therefore the development of riding support systems that help the motorcyclist to correctly assess the risk level of the road scenario and the manoeuvre they are executing is expected to have great potential for reducing motorcycle crashes. Nevertheless, their development is a challenging task since, from one side they should interfere as little as possible with the riding enjoyment and must take into account the riding style and skills, and on the other side they must understand correctly the level of risk of the actual manoeuvre – to avoid false or missed alarms. Therefore, to efficiently support the rider, the system must be able to evaluate the safest manoeuvre that a particular group of users (i.e., novice or experienced) should do in a certain situation and provide suitable warning and intervention if they deviate from the identified correct behaviour. An example of such a system is presented in (Biral et al. 2010; 2012 and Huth et al. 2011).

The essential idea proposed here is to develop an unique approach both to objectively measuring the motorcycle-rider performance and also to generate realistic rider control behaviour by calculating optimal manoeuvres according to a desired goal and the machine role.

The approach proposed here may also have a relevant impact on the definition and architecture of virtual rider models. Especially in its wider definition, the reference manoeuvre can be adopted as a target optimal plan that is used to model riders' planning capabilities and introduced into a receding horizon control scheme. This aspect is out of the scope of the chapter but a short discussion will be reported in the conclusions paragraph.

5.1 The Optimal Manoeuvre Concept: Manoeuvrability and Handling

Many researchers have faced the problem of assessing the motorcycle performance from both theoretical and experimental points of view (Kooijman and Schwab 2011). To serve this purpose a number of indices that quantify motorcycle performance and how easily they are controlled have been proposed in the literature and some of them experimentally compared for a lane change manoeuvre (Cossalter and Sadauckas 2006). Most of those indices are combined with the subjective opinion of the riders in order to evaluate the machine–rider system as a whole. Alternatively, here the intent is to propose an objective evaluation method based on quantitative metrics that also include the rider's behaviour and characteristics. To succeed in achieving this goal it is necessary to map subjective opinion or preferences to measurable quantities. On the other hand, objectifying personal sensations, perceptions or the workload is a hard task and the work done here does not pretend to be complete. Nevertheless, a general framework is derived, and some examples of how to mathematically transpose the rider's skills into the proposed formulation are shown.

We start by recalling that, in the early 1950s, the aeronautics and aerospace fields faced the problem of evaluating the best intrinsic performance of an aeroplane which can be realistically obtained by a human pilot. Despite the differences between the two fields (a well-trained pilot behaves in a standardized manner, aircrafts are designed to behave similarly for each role, etc.) there are also some points of contact with the motorcycle dynamics such as the combination of roll and yaw motion and high impact of pilot behaviour on aircraft performance.

(Paranjape and Ananthkrishnan 2006) reviewed the performance metric for combat aircraft, which are called *agility metrics*. The agility metrics are measures of merit used to quantify the manoeuvring capability of aircraft to execute certain tasks. When these manoeuvres are complex and on a long timescale (i.e. greater than 10–20s) they are called functional agility metrics (e.g. combat cycle time is the time to complete the manoeuvre that has the goal to accomplish a 180° turn heading change and then return to the same Mach number).

In the aeronautics fields the agility is considered as the combination of *manoeuvrability* and *controllability*. The first concept measures the highest values of some parameters (e.g. peak velocities and turn rates) that the aircraft can achieve during a certain manoeuvre and they form the so-called *manoeuvrability envelope*. The second concept quantifies how easy it is to control precisely those parameters (Blaye 2002). Manoeuvrability and controllability are usually contradictory objectives since a precise control of the aircraft flight parameters is more easily attained when the peak values of the parameters are limited, which means manoeuvring the vehicle far from the manoeuvrability envelope. Therefore agility relates both to aircraft performance and handling qualities which are defined as 'those qualities or characteristics of an aircraft that govern the ease and precision with which a pilot is able to perform the tasks required in support of an aircraft role' (Harper and G.E 1986).

As a result we can conclude that *handling* is intimately linked to the characteristic dynamic response of the vehicle and human pilot acting together to perform, at best, a certain task, whereas the *manoeuvrability* is an expression of the maximum performance that can be achieved by the machine for the same task, regardless of the influence of the pilot's ability.

In other words, handling is the result of the ability of the pilot to exploit the controllability of an aircraft to flight close to the manoeuvrability envelope. Both manoeuvrability and handling are simultaneously affected by several variables such as the dynamics system's parameters and environmental characteristics. Additionally, both metrics are task dependent: what is preferred for one task may be less desirable for another.

The above concepts are applicable to all vehicles, and in particular they are here developed for two-wheeled vehicles. It is evident that the manoeuvre that generates the maximum performance of the motorcycle represents the most efficient way to ride the vehicle and therefore can be defined as *optimal*. This consideration links the manoeuvrability metric with the *optimal manoeuvre* which is an ideal manoeuvre that does not involve the rider's ability, and it makes it possible to evaluate the metric itself. For instance, if the metric is the minimum time to complete a lane change, then the optimal manoeuvre is the motorcycle trajectory, velocity and so on, along with the history of the rider's controls that actually generates the fastest manoeuvre. When the metric is the minimum time, the path of the optimal manoeuvre traced in the state space of the motorcycle is the so-called manoeuvrability envelope, which is defined by the time evolution of the combination of peak values of the state variables when the motorcycle accomplishes a desired manoeuvre (e.g., lane change, curve entrance, obstacle avoidance or minimum lap time). As a consequence, for the same machine, the manoeuvrability envelope differs from task to task but in any case encloses the set of all possible feasible manoeuvres (or time evolution of the states) that the machine could execute to successfully complete the task. If we include the rider influence in the evaluation, that is we want to assess the motorcycle handling, we need to define some constraints in the manoeuvre execution to comply with the human rider's limitations, riding skills and preferences. Therefore, the optimal manoeuvre that is searched for in this case is a subset of manoeuvres contained in the manoeuvrability envelop. Similarly it happens for other metrics, such as the safest or a low consumption manoeuvres. Therefore the manoeuvrability envelope of a manoeuvre outlines the best possible performance out of *every motorcycle* and contains the set of all admissible states.

Given a machine type (e.g., sport, racing etc.), for *every motorcycle* we mean every possible layout or setup obtained by changing the design parameters. From the discussion above, it emerges that the generation of an optimal manoeuvre involves an optimisation process that also human riders, in a similar fashion, adopt to learn how to ride at best their motorcycle. The concept of the optimal manoeuvre to evaluate the performance of a motorcycle as a solution of an optimisation problem was first introduced by (Cossalter et al. 1999). They formulated an optimisation problem to minimize a goal function which is the mathematical representation of the manoeuvrability/handling metric. They associate the minimum value of the metric (i.e. a unique scalar) to the best manoeuvre in order to compare different motorcycles. They had in mind the problem of the optimal design of racing motorcycles. However, the concept can be extended and shift the focus from the assessment of two-wheeled vehicle handling qualities to the evaluation of the manoeuvre itself. In other words, given the motorcycle as it is, with its own handling qualities, the objective is to find the best way to ride the motorcycle in a specific situation trying to meet a desired goal (Bertolazzi et al. 2009; Biral et al. 2005). Developing further this concept, provided a real-time and robust solver, the optimal manoeuvre can be used in a receding horizon scheme to control a real or virtual motorcycle that calculates not only the optimal controls but also the reference manoeuvre to accomplish.

5.1.1 Optimal Manoeuvre Mathematically Formalised

There are four essential ingredients of the optimal manoeuvre concept. Each of these ingredients has a mathematical representation that is used to formulate the optimisation problem that yields the optimal manoeuvre as a solution.

The first ingredient is the two-wheeled vehicle with its handling qualities and peculiarities. The vehicle is characterised by its dynamics which is often modelled with a set of ordinary differential equations (or differential algebraic):

$$x'(t) = f(x(t), u(t), t, p) \tag{5.1}$$

where $x(t) \in \mathbb{R}^n$ are the states of the dynamical system, $u(t) \in \mathbb{R}^m$ are the controls and $p \in \mathbb{R}^k$ are the parameters that identify different systems. It is worth noting that some rider characteristics maybe modelled as differential equations that can be added to (5.1).

The second ingredient is the manoeuvre to be accomplished in a time horizon T starting from initial time t_i. The manoeuvre may be generated in a short timescale (e.g., lane change, obstacle avoidance) or on a long timescale (e.g. complete a lap of a circuit, follow a piece of rural road). Usually the manoeuvre definition is also completed by the specification of an initial and a final state: in some cases the manoeuvre is actually a transition between these two states.

$$x(t_i) = x_{\text{init}} \tag{5.2a}$$

$$x(t_i + T) = x_{\text{fin}} \tag{5.2b}$$

where t_i is the initial time and $t_i + T$ is the final time. The constraints are the third ingredient. They help to better define the manoeuvre: for instance, they restrict the space where the manoeuvre can be executed. Sometimes the constraints are part of the goal definition: for example they may implement some safety aspects such as keeping a certain margin from maximum tyre adherence or clearance from an obstacle. Finally, the rider's preferences and skills are represented as a range of values that can be assumed by both the control and the state variables. All these constraints might be of equality or inequality type and they might be active at a single point (position in space or time, e.g. pass a location with a predetermined speed) and/or path constraints.

$$C(x(t), u(t)) \leq 0 \tag{5.3}$$

The final ingredient is the metric or the goal (also called the cost function) that defines the concept of optimal. As stated it can be a measure of the vehicle qualities or a way to specify the properties of the manoeuvre that we want to generate (safest, fastest, etc.). A general mathematical definition of the goal is:

$$\mathcal{F}(u) = \Phi(x(t_i), x(t_i + T)) + \int_{t_i}^{t_i+T} L(x(t), u(t), t, p) dt, \tag{5.4}$$

where L is an integral merit function whose value must be minimized/maximized over the horizon T (e.g. the energy consumption) and Φ is a terminal term which measures the metric at the initial and/or final states (e.g. condition of steady-state or stable condition).

The concept of the optimal manoeuvre has its natural counterpart in the formulation of an optimal control problem (OCP) based on the above ingredients. Constrained optimal control problems lead naturally to the concepts of feasible control and feasible manoeuvre.

Definition 5.1.1 Admissible Control, Feasible Manoeuvre *An admissible control $\bar{u}(t) \in \mathcal{U}$ is said to be feasible, provided that (i) the response $\bar{x}(t)$ is defined on the entire interval $t_i \leq t \leq t_i + T$, and (ii) $\bar{u}(t)$ and $\bar{x}(t)$ satisfy all the constraints (5.1), (5.2) in the time interval. The pair $\xi(t) = (\bar{u}(t), \bar{x}(t))$ is called the feasible manoeuvre. The set of feasible manoeuvres Ξ is defined as:*

$$\Xi := \{\xi(\cdot) : \xi \text{ feasible}\}$$

Similarly, the set of feasible controls $\mathcal{U}$ is defined as:

$$\mathcal{U} := \{u(\cdot) : \text{ there exist } x(\cdot) \text{ such that } (u(\cdot), x(\cdot)) \in \Xi\}$$

Having defined the merit function (5.4), the set of constraints to be satisfied (5.2), (5.3), and the set of admissible controls and feasible manoeuvres, we can state the optimal control problem as follows: *find an admissible control $u \in \mathcal{U}$ which satisfies the constraints in such a manner that the cost functional $\mathcal{F}(u)$ has a minimum value when the dynamic system (5.1) transits from the initial to the final states (5.2a), (5.2b)*. The optimal manoeuvre is that special feasible manoeuvre that is associated to the minimum value of the cost functional. Therefore we can define what an optimal manoeuvre is:

Definition 5.1.2 Optimal Manoeuvre *A feasible manoeuvre $\xi^o \in \Xi$ is called optimal if it satisfies the constraints in such a manner that the cost functional $\mathcal{F}(u)$ has a minimum value when the dynamic system transits from the initial to the final states.*

If the rider's actuation limits and abilities are not included in the problem formulation the optimal manoeuvre $\xi^o(t) = (u^o(t), x^o(t))$ is the best intrinsic manoeuvre that can be extracted from the machine and it defines the manoeuvrability envelope for the given task. If the rider's influence is in some way included then the optimal manoeuvre qualifies the motorcycle properties in relation to the human rider point of view. Finally, if the focus is the quality of the manoeuvre then the optimal manoeuvre is the manoeuvre that best fits the specifications defined by the goal function and the constraints. For instance, it can be the safest manoeuvre that a group of human riders can produce when entering into a rural road curve.

5.1.2 The Optimal Manoeuvre Explained with Linearized Motorcycle Models

To exemplify the concepts discussed in Section 5.1 we use a set of linearized models of the motorcycle lateral dynamics which are described in (Astrom et al. 2005; Limebeer and Sharp 2006). In Section 5.3 a more complex model and example will be considered.

As a first approximation, the rider controls the motorcycle direction by generating lateral forces via the steering system. If we initially assume that there is no sideslipping of the

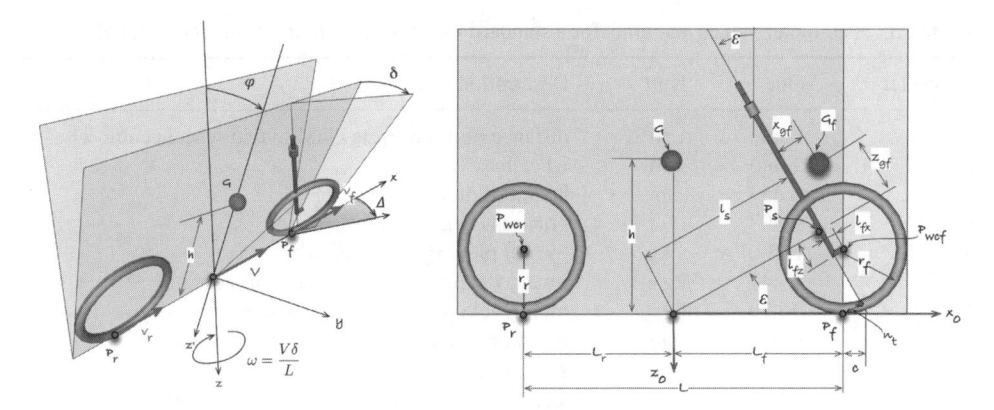

Figure 5.1 Perspective view of inverted pendulum model of a motorcycle with fixed rider (left), side view of the motorcycle model with front frame (right). G is the overall centre of mass of machine and human together. The picture on the right shows the main vehicle model parameters such as the inclination of the steering axis ε and the front frame and overall centre of mass positions in their respective frames

vehicle tyres and thus the rolling is non-holonomic, we can assimilate the linearized lateral dynamics of a motorcycle to that of an inverted pendulum on a kart. Despite its simplicity, the model captures the main dynamic behaviour of a motorcycle entering into a stationary corner at constant speed for small roll angles. Although the model does not use the rider's actual control (i.e., the steering torque) the results and observations obtained have a general validity (and can be extended to more general problems and more complex models).

The inverted pendulum has one degree of freedom: the roll angle ϕ and being the lateral velocity and yaw rate constrained by the no-slipping condition on the tyre contact point. ϕ_{dot} represents the roll rate and the linearized lateral dynamics is expressed by the following set of first-order differential equations:

$$\phi'(t) = \phi_{\text{dot}}(t), \tag{5.5a}$$

$$I_{xx_0}\phi'_{\text{dot}}(t) = m\,g\,h\,\phi(t) - \frac{m\,h\,V^2}{L}\delta(t), \tag{5.5b}$$

The overall moment of inertia with respect to the roll axis I_{xx_0} is the sum motorcycle moment of inertia about roll axis I_x through the centre of mass and the Stainer term $m\,h^2$. The model parameters are described in Table 5.1. In reality, a human rider moves the handlebar to generate the lateral forces F_y on the tyres and control the motorcycle roll angle and vehicle directionality. In this model, we neglect the steering system dynamics and assume that the rider directly controls the lateral force and the motorcycle directionality by defining the trajectory curvature radius with the steering angle $\delta(t)$. Of course, this assumption is valid as far as the tyre slipping condition is held true and the steering axis is vertical (i.e. $\varepsilon = 0$).

We now want to find the fastest manoeuvre to move the motorcycle from the straight running to steady-state cornering with a given target small roll angle (i.e., curve entrance

Table 5.1 Parameter values and units for a standard machine used in the numerical examples

Parameter	Value	Unit	Description
L_r	0.606	m	distance of motorcycle CoM to rear wheel centre
L	1.443	m	wheelbase
h	0.654	m	height of motorcycle CoM
g	9.807	m	gravity acceleration
m	273.0	kg	overall motorcycle mass
I_{xz}	−5.47	kg m^2	mixed x-y motorcycle moment of inertia
I_x	40.79	kg m^2	motorcycle roll moment of inertia
r_f	0.292	m	rear wheel radius
r_r	0.317	m	front wheel radius
ε	24.5	°	steering axis inclination
m_f	27.75	kg	front frame mass
I_{x_f}	2.487	kg m^2	front frame roll moment of inertia
I_{y_f}	0.143	kg m^2	front frame pitch moment of inertia
I_{z_f}	0.484	kg m^2	front frame steering axis moment of inertia
x_{g_f}	0.024	m	front frame centre of mass x position
z_{g_f}	−0.189	m	front frame centre of mass z position
l_{f_z}	0.082	m	distance between steering system origin and wheel centre
l_{f_x}	0.028	m	front frame offset

CoM stands for centre of mass and $L_f = L - L_r$

manoeuvre). The associated OCP that yields the fastest manoeuvre can be formulated as follows:

$$\text{minimize:} \quad T \tag{5.6a}$$

$$\text{subject to:} \quad \text{ODE} \quad (5.5) \tag{5.6b}$$

$$\text{and boundary conditions:} \quad \phi(0) = \phi_{\text{dot}}(0) = 0, \quad \phi(T) = \phi_1, \phi_{\text{dot}}(T) = 0, \tag{5.6c}$$

$$\text{and control constraint:} \quad -\delta^{\max} \leq \delta(t) \leq \delta^{\max} \tag{5.6d}$$

where $[-\delta^{\max}, \delta^{\max}]$ is the set of admissible control steering angles, T is the minimum manoeuvring time. Since the rider influence is not taken into account, the goal of the formulated problem is the assessment of the motorcycle manoeuvrability, and the corresponding metric is the minimum time. The analytical and numerical solution of a general optimal control problem is studied in some detail in Section 5.2. Here we use the analytical solution of the problem (5.6) that is derived in Section 5.2.3 to understand the manoeuvrability concept. Therefore, from the analytical solution, the minimum time manoeuvre is obtained, applying the following control history:

$$\delta(t) = \text{sign}(\phi_1)\delta^{\max} \begin{cases} -1 & \text{for } 0 \leq t < t_s, \\ +1 & \text{for } t_s \leq t \leq T, \end{cases}$$

where t_s and T are, respectively, the switching time and the manoeuvre time given by

$$f = -|\phi_1|\frac{gL}{V^2\delta^{max}}, \quad \alpha = \frac{\sqrt{mgh}}{\sqrt{I_{xx_0}}}, \quad w = 1 - f - \frac{1}{2}f^2 + \frac{1}{2}\sqrt{f(f+4)(f^2-4)}$$

$$T = \frac{\ln(w) - \ln(1+f)}{\alpha}, \quad t_s = \frac{\ln(1+w) - \ln(2)}{\ln(w) - \ln(1+f)}, \quad (5.7)$$

that yields the solution for roll angle and roll rate:

$$\phi(t) = \text{sign}(\phi_1)\frac{\delta^{max}V^2}{gL}\begin{cases}\cosh(t\alpha) - 1 & \text{for } t < t_s, \\ \cosh(t\alpha) + 1 - 2\cosh((t-t_s)\alpha) & \text{for } t \geq t_s,\end{cases} \quad (5.8a)$$

$$\omega(t) = \alpha\,\text{sign}(\phi_1)\frac{\delta^{max}V^2}{gL}\begin{cases}\sinh(t\alpha) & \text{for } t < t_s, \\ \sinh(t\alpha) - 2\sinh((t-t_s)\alpha) & \text{for } t \geq t_s,\end{cases} \quad (5.8b)$$

The solution is made of two parts related to the steering angle that switches between the minimum and maximum value at instant time t_s (i.e. switching time). This problem is known as bang-bang control. Notice that a solution exists provided that $-1 \leq f < 0$, that is if

$$|\phi_1| < \frac{\delta^{max}V^2}{gL}. \quad (5.9)$$

Equation (5.9) shows that the optimal control exists only if the inequality is satisfied. In other words, not all states can be reached, and this depends on the forward velocity V and the motorcycle design parameters. The chart on the left of Figure 5.2 shows, in the state space, the roll angle ϕ and roll rate ϕ_{dot} associated to the minimum time manoeuvre to move from the initial state (steady-state straight running) to the final state (steady-state cornering). The fastest manoeuvre – the *manoeuvrability envelope* – encloses the set of admissible and reachable states for this type of manoeuvre. In other words, all other suboptimal manoeuvres are included in this region (shaded area in Figure 5.2). The size of the set of reachable states determines the degree of manoeuvrability and the larger the area the greater the manoeuvrability of the machine. However, note that the shape and size of the manoeuvrability envelope depends on a specific task or manoeuvre (e.g., a lane change has a different manoeuvrability envelope compared to an obstacle avoidance manoeuvre). The analytical solution of Equation (5.7), (5.8) shows that optimal manoeuvre and the minimum time T are functions of the motorcycle design parameters. Therefore, if we change the machine design parameters in order to improve its manoeuvrability, the set of reachable states is expected to be larger. Indeed, this is what happens. For example, given the same control force history the machine with a lower centre of mass completes the manoeuvre in less time, yielding larger roll rates, as shown in Figure 5.2. But a machine with higher centre of mass needs more time to complete the manoeuvre and is characterized by a smaller set of admissible states (i.e. it reaches lower roll rates). As a first conclusion, *the degree of manoeuvrability is an intrinsic characteristic of the machine, which defines the best achievable performance for a desired manoeuvre* and the optimal manoeuvre extracts the best manoeuvre from the set of the admissible ones. Figure 5.2b shows the initial counter-steering manoeuvre: the rider firstly turns the handlebar on the opposite side to the direction he/she intends to steer

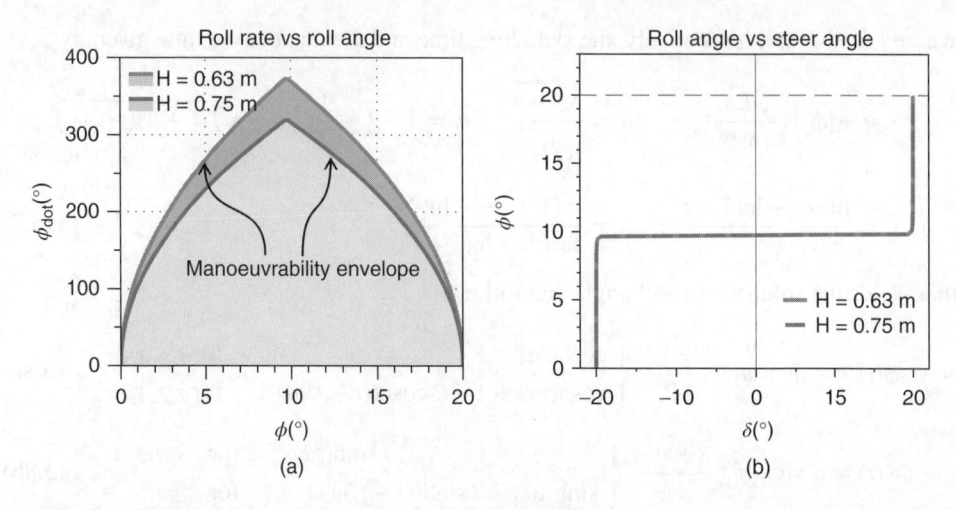

Figure 5.2 Minimum time optimal manoeuvre with bang-bang control steering angle. The optimal manoeuvre has been calculated with two values of motorcycle centre of mass height. The phase portrait and the control angle are respectively shown on the left and on the right graphs. The plot compares two manoeuvres with different height h of centre of mass: $T = 0.107s$ for $h = 0.634$ m and $T = 0.125$ s for $h = 0.750$ m. For this simulation, the roll moment of inertia is $I_{xx_0} = m\,h^2$ and the mixed moment of inertia $I_{xz} = 0$.

and then the steer angle goes in the same direction, to stabilize the machine at the desired roll angle.

The optimal manoeuvre calculated with the previous model corresponds to the maximum performance achievable with the given machine. However, it is not possible for a human rider to change the steering angle in zero time, therefore the steering rate has to be limited. An additional equation is added to (5.5) with a new inequality constraint to limit the steering rate:

$$\delta'(t) = \delta_{\text{dot}}(t) \quad -\delta_{\text{dot}}^{\max} \leq \delta_{\text{dot}}(t) \leq \delta_{\text{dot}}^{\max}$$

With the new optimal control problem the resulting manoeuvre is significantly slower, and the set of reachable states is a subset of the bang-bang solution on steering angle (Figure 5.3). Since the rider's limits are taken into account, the set of reachable states now defines the *handling envelope*, which measures the overall performance of machine and rider together. We could have carried out parametric analysis on design variables to understand their effect on handling. However, the rider's main control input is the steering torque, being leaning (which applies a roll moment to the rear frame) 'too sluggish to be practical' in fast manoeuvres, such as emergency situations (Limebeer and Sharp 2006). Therefore, to include the steering torque in the model, the linearized dynamics of the steering system is considered. To this end an additional body – the front frame – is constrained with a hinge to rotate with respect to the rear frame about a steering axis. The steering axis is inclined at an angle ε with respect to the axis z, orthogonal to the road plane. The front wheel is attached to the front frame with a horizontal offset l_{f_x} with respect to the steering axis. The offset has a direct contribution to the normal trail value n_t (Figure 5.1), which in turn affects motorcycle stability in straight

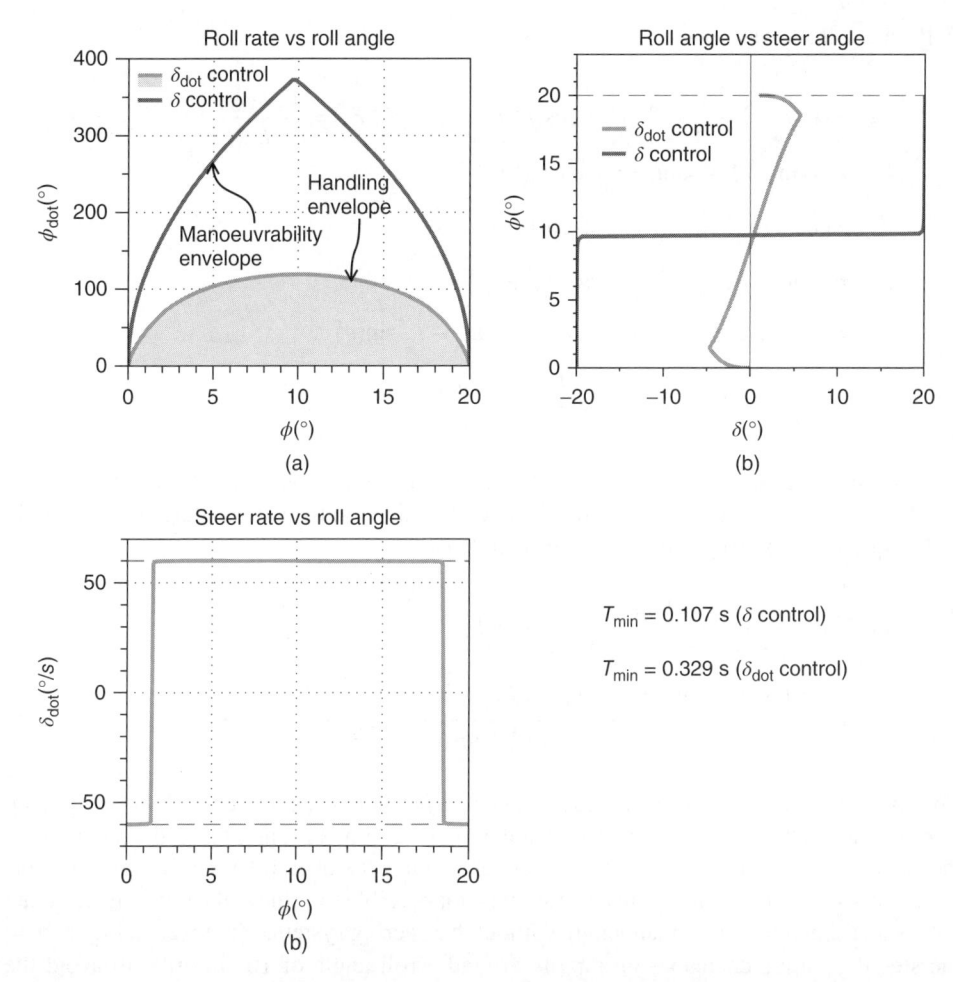

Figure 5.3 Comparison between the minimum time optimal manoeuvre, with bang-bang steering angle and the minimum time with limited steering torque δ_{dot}. The figure shows (a) the phase portrait between roll angle ϕ and roll rate δ_{dot}, (b) the steering angle versus roll angle and (c) steering torque versus roll angle. For this simulation, the roll moment of inertia is $I_{xx_0} = m\,h^2$ and the mixed moment of inertia $I_{xz} = 0$

running and the rider effort via the steering torque. The new model becomes:

$$I_{xx_0}\phi'_{dot}(t) = I_1\delta'_{dot}(t) + I_2\,V\delta_{dot}(t) + N_1\delta(t) + m\,g\,h\,\phi(t) - \frac{m\,h\,V^2}{L}\Delta(t), \quad (5.10a)$$

$$I_{zz_{f0}}\delta'_{dot}(t) = N_1\sin(\varepsilon)\delta(t) + F_{z_f}\,n_t\phi(t) + \tau(t), \quad (5.10b)$$

$$\phi'(t) = \phi_{dot}(t), \quad \delta'(t) = \delta_{dot}(t), \quad \tau'(t) = \tau_{dot}(t) \quad (5.10c)$$

where $\Delta(t) = \cos(\varepsilon)\delta(t)$ is the steering angle on the road plane and $\tau_{dot}(t)$ is the steering torque rate, which is the limited control: $-\tau_{dot}^{max} \leq \tau_{dot}(t) \leq \tau_{dot}^{max}$. The other parameters in

(5.10) are as follows:

$$n_t = \sin(\varepsilon) \, r_f - l_{f_x}, \qquad l_s = \cos(\varepsilon) \, L_f + n_t, \qquad F_{z_f} = \frac{L_r \, m \, g}{L}$$

$$h_{g_f} = -\sin(\varepsilon) \, l_s - \sin(\varepsilon) \, x_{g_f} + \cos(\varepsilon) \, z_{g_f}$$

$$l_{g_f} = \cos(\varepsilon) \, x_{g_f} + \sin(\varepsilon) \, z_{g_f} + \cos(\varepsilon) \, l_s$$

$$I_{xx_0} = h^2 \, m + I_x, \qquad I_{zz_{f0}} = m_f \, x_{g_f}^2 + I_{z_f}$$

$$I_{xz_0} = I_{xz} - L_r \, h \, m, \qquad I_{xz_{fG0}} = h_{g_f} \, m_f \, x_{g_f} - I_{z_f} \sin(\varepsilon)$$

$$I_1 = \frac{I_{xz_0} \, n_t}{L} + I_{xz_{fG0}}, \qquad I_2 = \frac{I_{xz_0}}{L} - \frac{m \, h \, n_t}{L}, \qquad N_1 = n_t \, F_{z_f} + g \, m_f \, x_{g_f}$$

The optimal manoeuvre being searched is again the transition from straight running to steady-turning at constant speed. This requires to calculate the steady-state steering angle and torque from (5.10) given the final roll angle $\phi(T)$:

$$\delta(T) = -\frac{g \, h \, m \, L}{(-h \, m \, V^2 \, \cos(\varepsilon) + L \, N_1)} \phi(T) \qquad (5.11\text{a})$$

$$\tau(T) = \frac{(h \, m \, V^2 \, \cos(\varepsilon) - L \, N_1) \, n_t \, F_{z_f} + L \, \sin(\varepsilon) \, g \, h \, m \, N_1}{L \, N_1 - h \, m \, V^2 \, \cos(\varepsilon)} \phi(T) \qquad (5.11\text{b})$$

The solution of the bang-bang steering torque rate problem is compared with the bang-bang steering rate solution. Despite the fact that the steering torque is bang-bang, the inclusion of the steering system dynamics in the model deteriorates the overall lateral dynamics performance. As shown in Figures 5.4a and 5.4b it is not possible to achieve the same steering rate values that characterize the manoeuvre without the steering system. Moreover, in Figure 5.4d the steering torque changes sign rapidly around a roll angle of 10° in order to avoid the steering rate overcoming the limit of 60°/s. From the results, we can state that the steering system reduces the manoeuvrability performance of the motorcycle. On the other hand, an accurate design of the steering system may improve the motorcycle stability and reduce the rider effort, in other words improving the machine handling. One of the key parameters is the normal trail n_t which, as shown in (5.11b), has an effect on the steering torque value in steady-state conditions and is expected to also have an effect in transient phase (Biral et al. 2003a). In order to exemplify the effect of the steering system parameters on the machine performance we have compared the optimal manoeuvres obtained with three motorcycles that differ from each others in terms of the normal trail only. The simulation results are reported in Figures 5.5 and 5.6. One major difference with the manoeuvre obtained in Figure 5.4 is that the phase portrait now has a different shape close to the final roll angle. The reason is the higher roll moment of inertia, which is now included in the simulations (not present in simulation of Figure 5.4). The roll dynamics is slower and a roll angle overshoot is necessary to stabilize the machine at the desired steady-state value. The phase portraits of Figures 5.5a–c show that there is no significant difference in the three manoeuvres except in the optimal control τ_{dot} shown in Figure 5.5d. However, the steering torques are different

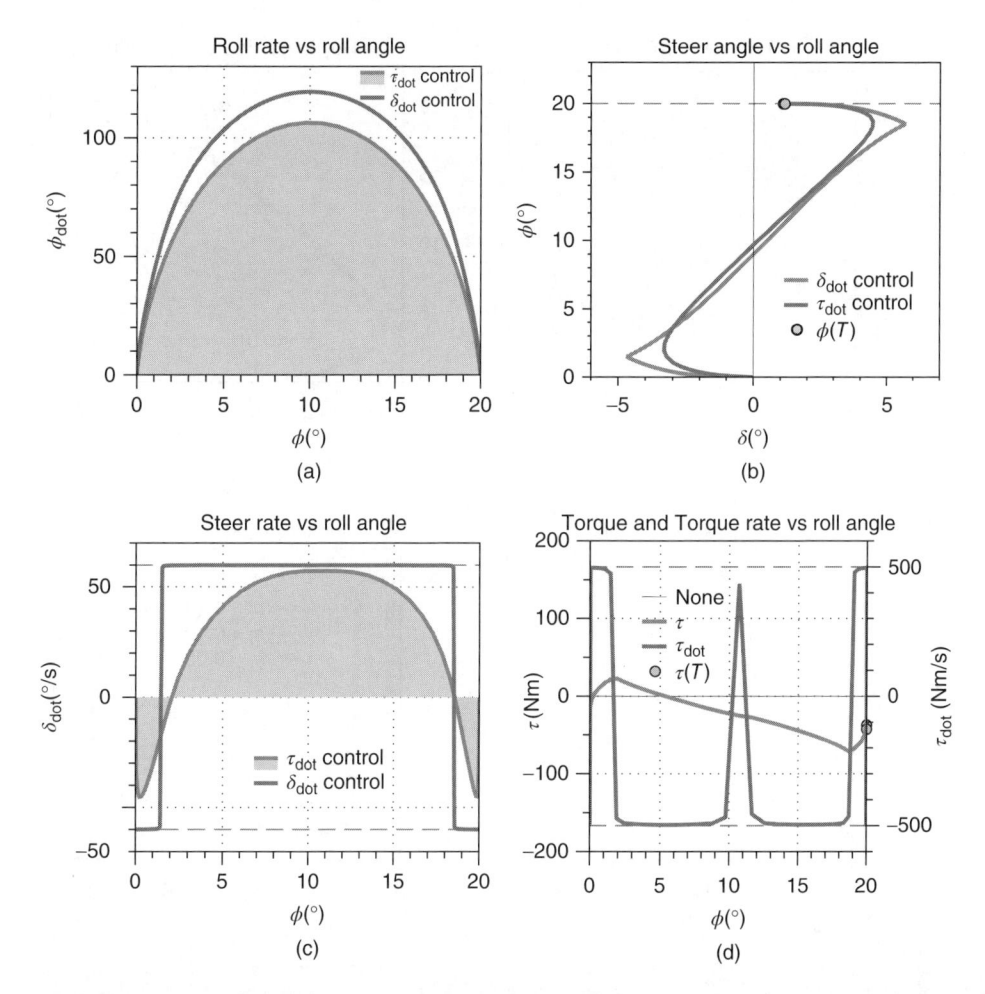

Figure 5.4 Comparison between the minimum time optimal manoeuvre with bang-bang steering torque the and minimum time with bang-bang steering torque rate. The figure shows (a) the phase portrait between roll angle ϕ and roll rate δ_{dot}, (b) the steering angle versus roll angle, (c) steering rate versus roll angle and (d) the steering torque and steering rate versus the roll angle. For this simulations, the roll moment of inertia is $I_{xx_0} = m\,h^2$ and the mixed moment of inertia $I_{xz} = 0$

in the three cases, as shown in Figure 5.6a. In particular, the steering torque is lower for smaller normal trail than the reference one ($n_t = 0.1$ m). In other words, despite the fact that the reduction of the normal trail does not yield a better manoeuvring time, which is substantially the same for all three manoeuvres, the rider's effort was reduced. Figure 5.6b gives more insight on the benefit derived by the reduction of the normal trail. The graph compares the steering power and the steering energy in the three cases that are respectively defined as:

$$P(t) = \delta_{\text{dot}}(t)\,\tau(t), \qquad E(t) = \int_0^T |P(t)|\,\mathrm{d}t.$$

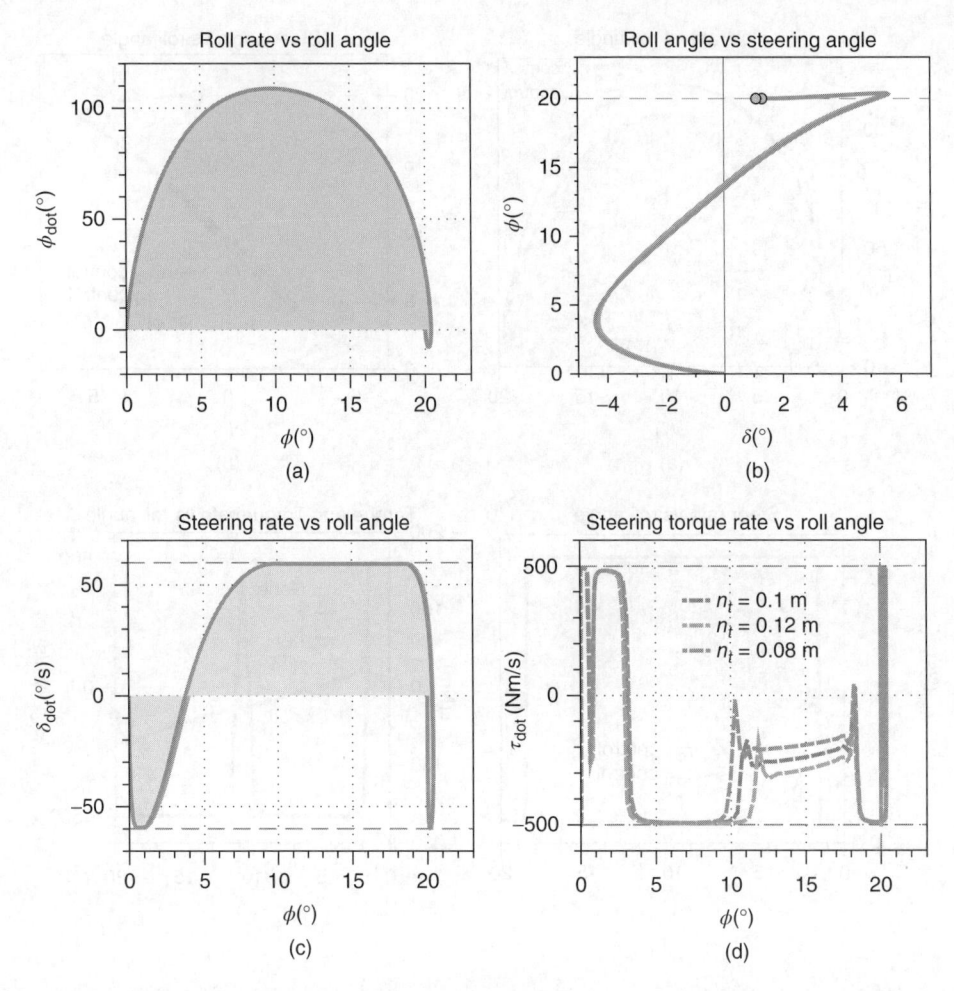

Figure 5.5 Comparison between the minimum time optimal manoeuvres with bang-bang steering torque with three different normal trail values. The figure shows (a) the phase portrait between roll angle ϕ and roll rate δ_{dot}, (b) the steering angle versus roll angle and (c) the steering rate versus roll angle and (d) the steering torque rate versus roll angle. For this simulation, the roll moment of inertia is $I_{xx_0} = I_x + m\,h^2$ and the mixed moment of inertia $I_{xz} \neq 0$

Reducing the normal trail requires a lower steering torque to complete the manoeuvre in the same time and indeed reduces the steering energy and therefore the rider's manoeuvring effort. The reader may also note that the steering power is also negative, since the rider has to slow down the roll motion to zero roll rate. This example clearly shows that it is possible to improve the motorcycle handling without deteriorating the machines' intrinsic performance limit (i.e., the manoeuvrability). Therefore, with linearized models of motorcycle lateral dynamics, we have shown that handling and manoeuvrability are two distinct concepts that can be objectively assessed by posing them as the same optimal control problem but with different metrics. The solutions adapt the input's time evolution for riding a motorcycle to

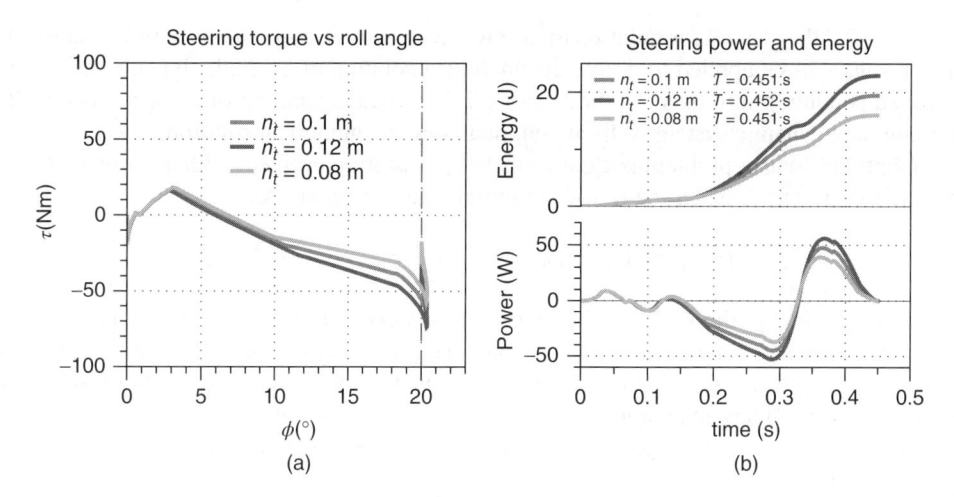

Figure 5.6 Graph (a) compares the steering torque for three different values of normal trail. Graph (b) compares the steering power and energy for the same three normal trail values. Top part graph also reports the minimum manoeuvring time T

accomplish a desired manoeuvre (such as curve entrance, lane change, minimum time lap etc.) in a way that minimizes a given metric. In particular, for handling, the metric that measures the rider's effort has an effect on the solution and the use of the rider's controls. Comparing the results obtained by solving the same problem (i.e., transit from straight running to steady-state cornering), we get more complex manoeuvres that reflect the additional complexity added in the motorcycle dynamic models. Usually this complexity restricts the set of reachable states and therefore the machine manoeuvrability. Those concepts were largely applied to investigate the effect of the power train (Cossalter et al. 2010; Biral et al. 2003b), tyre characteristics (Da Lio et al. 1999), the position of the centre of mass and other design parameters (Bobbo et al. 2009; Biral et al. 2002) of racing motorcycles. It was also used to understand the influence of the rider's upper torso movements on the manoeuvre effectiveness (Biral et al. 2006; Bertolazzi et al. 2007b).

5.2 Optimal Manoeuvre as a Solution of an Optimal Control Problem

The analytic solution of an optimal control problem (OCP) is possible only for very simple cases, but when the motorcycle model is nonlinear and subject to constraints of different nature (i.e. equality, non-linear inequalities, etc.) a numerical solution is the only viable method. This section explains the derivation of the necessary equations of the optimal control problem. The approach used is to start from unconstrained problems, then consider limitations on control input only, and finally extend to the general case of constrained OCP. Initially an unconstrained OCP is formulated as an optimization problem of an integral function subject to a feasible manoeuvre. Via the calculus of variation the necessary conditions are derived, including the equations for the optimal controls. In Section 5.2.1 the necessary OCP equations are derived for the case with limited controls using an novel approach that is easier to understand compared to the original derivation of the principle of Pontryagin. In

Section 5.2.2 the general formulation for an OCP with constraints only on controls is derived. The results will be applied to derive the analytical solution of a simple dynamic model of inverted pendulum introduced in Section 5.1.2. Finally, the solution of a constrained OCP for complex dynamic system is discussed, including the numerical solution.

An optimal control problem is a constrained optimization problem with a *dynamical system* as constraint. Consider the following initial value problem (IVP):

$$x'(t) = f(x(t), u(t), t), \qquad x(t_i) = x_{\text{init}}, \qquad t \in (t_i, t_f), \tag{5.12}$$

where $x(t) \in \mathbb{R}^n$ are the *states* of the *dynamical system* and $u(t) \in \mathbb{R}^m$ are the *controls*. If the controls $u(t)$ are known and regular enough (e.g. piecewise continuous) and $f(x, u(t), t)$ is regular enough as a function of (x, t) (e.g. piecewise continuous in (x, t) and Lipschitz in x) then IVP (5.12) has an unique solution. This is sometimes called a *controlled dynamical system*. Now we introduce a payoff or performance index functional $\mathcal{F}(u)$:

$$\mathcal{F}(u) = \Phi(x(t_i), x(t_f)) + \int_{t_i}^{t_f} L(x(t), u(t), t) dt, \tag{5.13}$$

which returns a scalar where $x(t)$ satisfies (5.12) (i.e., is a feasible manoeuvre) given a control history $u(t)$. The term $\Phi(x(t_i), x(t_f))$ in (5.12) is called the *initial/final payoff* while $L(x(t), u(t), t)$ in (5.13) is called the *running payoff*. An optimal control is a control history $u(t)$ which determines $x(t)$ by the dynamical system (5.12) that minimizes the (total) payoff functional (5.13). The optimal control problem is classified depending on the structure of the performance index. If it contains only the terminal payoff, it is called the Mayer problem, if it contains only the running payoff it is called the Lagrange problem, otherwise, in general it is called a Bolza problem.

A necessary conditions for a solution of problem (5.12) with (5.13) is that the first variation of the functional $\mathcal{J}(x, \lambda, u)$ is zero where:

$$\mathcal{J}(x, \lambda, u) = \Phi(x(t_i), x(t_f)) + \int_{t_i}^{t_f} H(x, \lambda, u, t) - \lambda(t) \cdot x'(t) dt, \tag{5.14a}$$

$$H(x, \lambda, u, t) = L(x, u, t) + \lambda \cdot f(x, u, t), \tag{5.14b}$$

and (5.14b) is called the *Hamiltonian*. The first variation of the functional is denoted as $\delta \mathcal{J}$ and is evaluated as the following derivative (the Gâteaux variations),

$$\delta \mathcal{J} = \frac{d}{d\alpha} \mathcal{J}(x + \alpha \delta x, \lambda + \alpha \delta \lambda, u + \alpha \delta u)|_{\alpha=0} = - \int_{t_i}^{t_f} \lambda(t) \cdot \delta x' + \delta \lambda(t) \cdot x'(t) dt$$

$$+ \int_{t_i}^{t_f} [\partial_x H(x, \lambda, u, t) \delta x + \partial_\lambda H(x, \lambda, u, t) \delta \lambda + \partial_u H(x, \lambda, u, t) \delta u] dt$$

$$+ \partial_{x_i} \Phi(x(t_i), x(t_f)) \delta x(t_i) + \partial_{x_f} \Phi(x(t_i), x(t_f)) \delta x(t_f),$$

where $\lambda(t)$ are Lagrange multipliers or *costates*, used to satisfy constraint (5.12). The functions $\delta x(t)$, $\delta u(t)$, $\delta \lambda(t)$ are the *variations* and can be interpreted as the direction of the derivative in the infinite-dimensional function space. The function $\delta x(t)$, $\delta u(t)$, $\delta \lambda(t)$ can be arbitrarily chosen except for $\delta x(t)$, which at initial point must satisfy $\delta x(t_i) = 0$ to match initial boundary condition. Function $\delta x'(t)$ is the derivative of $\delta x(t)$ so that can be integrated by parts

$$-\int_{t_i}^{t_f} \lambda(t) \cdot \delta x'(t) dt = \lambda(t_i) \cdot \delta x(t_i) - \lambda(t_f) \cdot \delta x(t_f) + \int_{t_i}^{t_f} \lambda'(t) \cdot \delta x(t) dt. \tag{5.15}$$

Collecting the terms involving the variations (and noticing that $\delta x(t_i) = 0$) the following equality is obtained:

$$0 = \frac{d\mathcal{J}}{d\alpha}\bigg|_{\alpha=0} = \int_{t_i}^{t_f} [A \cdot \delta x + B \cdot \delta u + C \cdot \delta \lambda] dt + D \cdot \delta x(t_f), \tag{5.16}$$

where

$$\begin{aligned} A &= \partial_x^T H(x, \lambda, u, t) + \lambda'(t), & B &= \partial_u^T H(x, \lambda, u, t), \\ C &= \partial_\lambda H(x, \lambda, u, t) - x'(t), & D &= \partial_{x(t_f)}^T \Phi(x(t_i), x(t_f)) - \lambda(t_f). \end{aligned} \tag{5.17}$$

Equation (5.16) must be satisfied for any admissible *variation* $\delta x(s)$, $\delta \lambda(s)$ and $\delta u(s)$. The following Lemma is a simplified version of Du Bois–Reymond lemma and is used to build an equivalent boundary value problem (BVP) from (5.16).

Lemma 5.2.1 Du Bois–Reymond *If $h(t)$ is a continuous function on $[a, b]$ and for all $v(t)$ continuous with $v(a) = v(b) = 0$ we have $\int_a^b h(t)v(t)dt = 0$, then $h \equiv 0$.*

From Lemma 5.2.1, it follows that $A = B = C = 0$; moreover from the arbitrariness of $\delta x(t_f)$ it follows that $D = 0$. Thus the solution of OCP (5.12) with (5.13) must satisfy the BVP:

$$\text{equations:} \quad x' = \partial_\lambda^T H(x, \lambda, u, t), \tag{5.18a}$$

$$\text{co-equations:} \quad \lambda' = -\partial_x^T H(x, \lambda, u, t), \tag{5.18b}$$

$$\text{BC:} \quad x(t_i) = x_i, \quad \lambda(t_f) = \partial_{x_f}^T \Phi(x(t_i), x(t_f)), \tag{5.18c}$$

Notice that (5.18a) is the original ODE (5.12), and (5.18a) are a new set of differential equations called *co-equations*. The last equation in (5.18c) is called the *transversality condition*. Finally, $u(t)$ must satisfy the algebraic equation

$$\partial_u^T H(x, \lambda, u, t) = 0. \tag{5.19}$$

If $x(t)$, $\lambda(t)$, $u(t)$ are solution of (5.18) with (5.19) and this solution is a minimum then the second variation of $\mathcal{J}$ must be non-negative:

$$\delta^2 \mathcal{J} = \frac{d^2}{d\alpha^2} \mathcal{J}(x + \alpha \delta x, \lambda + \alpha \delta \lambda, u + \alpha \delta u)\bigg|_{\alpha=0} \geq 0 \tag{5.20}$$

Choosing $\delta x(t) = \delta \lambda(t) = \mathbf{0}$ the condition on second variation reduces to

$$\int_{t_i}^{t_f} \delta u^T [\partial_u^2 H(x, \lambda, u, t)] \delta u \, dt \geq 0, \tag{5.21}$$

and because $\delta u(t)$ is an arbitrary function it follows that $\partial_u^2 H(x, \lambda, u, t)$ is a semi-definite positive matrix. This suggests that (5.19) is a stationary point for the function $H(x, \lambda, u, t)$ with x, λ, t fixed and (5.21) imply that this stationary point is a local minimum. Hence $u(t)$ is the solution of the more general problem

$$u(t) = \operatorname{argmin}\{\mathbf{v} \in \mathbb{R}^m \mid H(x(t), \lambda(t), \mathbf{v}, t)\}, \tag{5.22}$$

and thus the solution of OCP (5.12) with (5.13) must satisfy the following BVP:

$$\begin{aligned}
\text{equations:} \quad & x' = \partial_\lambda^T H(x, \lambda, u, t), \\
\text{co-equations:} \quad & \lambda' = -\partial_x^T H(x, \lambda, u, t), \\
\text{BC:} \quad & x(t_i) = x_i, \quad \lambda(t_f) = \partial_{x_f}^T \Phi(x(t_i), x(t_f)), \\
\text{CTRL:} \quad & u(t) = \operatorname{argmin}\{\mathbf{v} \in \mathbb{R}^m \mid H(x(t), \lambda(t), \mathbf{v}, t)\}.
\end{aligned} \tag{5.23}$$

Notice that the boundary conditions are derived by selecting the variation $\delta x(t)$ which satisfy $\delta x(t_i) = \mathbf{0}$; alternatively, it is possible to choose $\delta x(t)$ freely and impose the condition $x(t_i) + \alpha \delta x(t_i) = x_i$. This additional constraint can be forced by using a multiplier $\boldsymbol{\mu}$ yielding a new functional:

$$\tilde{\mathcal{J}}(x, \lambda, u, \boldsymbol{\mu}) = \mathcal{J}(x, \lambda, u) + \boldsymbol{\mu} \cdot (x(t_i) + \alpha \delta x(t_i) - x_i).$$

Minimization of $\tilde{\mathcal{J}}$ results in the following variation associated to the optimal control problem (5.12) with (5.13)

$$\begin{aligned}
0 = \left.\frac{d\tilde{\mathcal{J}}}{d\alpha}\right|_{\alpha=0} &= \left.\frac{d\mathcal{J}}{d\alpha}\right|_{\alpha=0} + \delta\boldsymbol{\mu} \cdot (x(t_i) - x_i) + \boldsymbol{\mu} \cdot \delta x(t_i) \\
&= \int_{t_i}^{t_f} [A \cdot \delta x + B \cdot \delta u + C \cdot \delta \lambda] dt \\
&\quad + D \cdot \delta x(t_f) + (E + \mu) \cdot \delta x(t_i) + \delta\boldsymbol{\mu} \cdot (x(t_i) - x_i),
\end{aligned}$$

where A, B, C, D are defined in (5.17) and E is defined as

$$E = \partial_{x_i}^T \Phi(x(t_i), x(t_f)) + \lambda(t_i). \tag{5.24}$$

Again from Lemma 5.2.1 it follows that $A = B = C = \mathbf{0}$ and from the arbitrariness of $\delta x(t_i)$, $\delta x(t_f)$ and $\delta \mu$ it follows that $D = \mathbf{0}$, $E + \mu = \mathbf{0}$ and $x(t_i) = x_i$. And thus we obtain (5.18) with (5.19). Notice that μ appears only in $E + \mu = \mathbf{0}$ and thus E can take any values which will be compensated by μ, thus, equations $E + \mu = \mathbf{0}$ and multiplier μ can be eliminated, and we get BVP (5.23). Using an appropriate multiplier it is possible to impose initial and final values for $x(t)$.

5.2.1 The Pontryagin Minimum Principle

More often the controls $u(t)$ are bounded in a compact set $\mathcal{U}$, that is,

$$u(t) \in \mathcal{U}, \tag{5.25}$$

then $\delta u(t)$ cannot be arbitrary but must satisfy $u(t) + \alpha \delta u(t) \in \mathcal{U}$. This complicates the derivation of a necessary condition using only variations. A simple heuristic to deduce necessary conditions is provided by using barrier function:

$$p(u, \varepsilon) = -\varepsilon \log \ \text{dist}(u, \mathbb{R}^m \setminus \mathcal{U}), \quad \text{where} \quad \text{dist}(u, \mathcal{A}) = \inf \{|u - v|, v \in \mathcal{A}\}, \quad (5.26)$$

which is small and positive for $u \in \mathcal{U}$ and is ∞ for $u \notin \mathcal{U}$. Adding this barrier function to the performance index (5.13) results in:

$$\tilde{\mathcal{F}}(u) = \Phi(x(t_i), x(t_f)) + \int_{t_i}^{t_f} L(x(t), u(t), t) + p(u(t), \varepsilon) dt. \quad (5.27)$$

Repeating the derivation of the BVP (5.23) now with *unconstrained* $u(t)$ results in the same ODE and BC but CTRL modifies to:

$$\text{CTRL}': \quad u(t) = \text{argmin}\{v \in \mathbb{R}^m \mid H(x(t), \lambda(t), v, t) + p(u(t), \varepsilon)\}. \quad (5.28)$$

From the definition of $p(u, \varepsilon)$ it follows that $u(t) \in \mathcal{U}$. Moreover, as ε becomes small the problem (5.28) approximates the following minimization

$$\text{CTRL}'': \quad u(t) = \text{argmin}\{v \in \mathcal{U} \mid H(x(t), \lambda(t), v, t)\}. \quad (5.29)$$

Thus, the solution of OCP (5.12) with (5.13) and controls bounded by (5.25) is equivalent to BVP (5.23) with the CTRL equation substituted with CTRL''. Equation (5.29) is a part of the Pontryagin minimum principle (PMP) which yields an equivalent BVP from an OCP. For a detailed and rigorous derivation of the PMP see (Pontryagin et al. 1962) or (Kirk and Kreider 1970) or (Ross 2009).

5.2.2 General Formulation of Unconstrained Optimal Control

The optimal control problem can be formulated in slightly more general form as follows:

$$\text{minimize:} \quad \mathcal{F}(u) = \Phi(x(t_i), x(t_f)) + \int_{t_i}^{t_f} L(x(t), u(t), t) dt, \quad (5.30a)$$

$$\text{ODE constraint:} \quad \mathbf{A}(x(t), t) x'(t) = f(x(t), u(t), t), \quad (5.30b)$$

$$\text{boundary Conditions:} \quad \mathbf{b}(x(t_i), x(t_f)) = \mathbf{0}, \quad (5.30c)$$

$$\text{controls constraints:} \quad u(t) \in \mathcal{U}. \quad (5.30d)$$

$\mathbf{A}(x, t)$ is a non-singular matrix with continuous and piecewise differentiable entries and it corresponds to the *mass matrix* of the multibody model considered in many problems.

Denote with $\tilde{\mathcal{J}}(x, \lambda, u, \mu)$ the functional:

$$\tilde{\mathcal{J}}(x, \lambda, u, \mu) = \mathcal{B}(x(t_i), x(t_f), \mu) + \int_{t_i}^{t_f} H(x, \lambda, u, t) - \lambda(t) \cdot (\mathbf{A}(x, t) x'(t)) dt,$$

$$\mathcal{B}(x_i, x_f, \mu) = \Phi(x_i, x_f) + \mu \cdot \mathbf{b}(x_i, x_f), \quad (5.31)$$

$$H(x, \lambda, u, t) = L(x, u, t) + \lambda \cdot f(x, u, t).$$

Denote with $\boldsymbol{\eta}(\boldsymbol{x}, \boldsymbol{\lambda}, t) = \mathbf{A}(\boldsymbol{x}, t)^T \boldsymbol{\lambda}$ so that the variation of the term $\boldsymbol{\lambda}^T \mathbf{A}\, \boldsymbol{x}'$ takes the form:

$$\delta(\boldsymbol{\lambda}^T \mathbf{A}(\boldsymbol{x},t)\boldsymbol{x}') = \frac{d}{d\alpha}\left(\boldsymbol{\eta}(\boldsymbol{x}+\alpha\delta\boldsymbol{x}, \boldsymbol{\lambda}+\alpha\delta\boldsymbol{\lambda}) \cdot (\boldsymbol{x}'+\alpha\delta\boldsymbol{x}')\right)\Big|_{\alpha=0},$$

$$= \boldsymbol{\eta}(\boldsymbol{x},\boldsymbol{\lambda},t) \cdot \delta\boldsymbol{x}' + \boldsymbol{x}' \cdot \left[\partial_x \boldsymbol{\eta}(\boldsymbol{x},\boldsymbol{\lambda},t)\delta\boldsymbol{x} + \partial_\lambda \boldsymbol{\eta}(\boldsymbol{x},\boldsymbol{\lambda},t)\delta\boldsymbol{\lambda}\right].$$

Integrating by parts and noticing that $\frac{\partial \boldsymbol{\lambda}}{\partial \boldsymbol{\eta}} = \mathbf{A}(\boldsymbol{x},t)^T$, the following equality is obtained:

$$\delta(\boldsymbol{\lambda}^T \mathbf{A}(\boldsymbol{x},t)\boldsymbol{x}') = \frac{d}{dt}(\boldsymbol{\eta}\cdot\delta\boldsymbol{x}) - \frac{d\boldsymbol{\eta}}{dt}\cdot\delta\boldsymbol{x} + \boldsymbol{x}'\cdot\left[\partial_x\boldsymbol{\eta}(\boldsymbol{x},\boldsymbol{\lambda},t)\delta\boldsymbol{x} + \partial_\lambda\boldsymbol{\eta}(\boldsymbol{x},\boldsymbol{\lambda},t)\delta\boldsymbol{\lambda}\right],$$

$$= \frac{d}{dt}(\boldsymbol{\eta}\cdot\delta\boldsymbol{x}) + \left[\mathbf{N}(\boldsymbol{x},\boldsymbol{\lambda},t)\boldsymbol{x}' - \mathbf{A}(\boldsymbol{x},t)^T\boldsymbol{\lambda}'\right]\cdot\delta\boldsymbol{x} + \left[\mathbf{A}(\boldsymbol{x},t)\boldsymbol{x}'\right]\cdot\delta\boldsymbol{\lambda},$$

where

$$\mathbf{N}(\boldsymbol{x},\boldsymbol{\lambda},t) = [\partial_x \boldsymbol{\eta}(\boldsymbol{x},\boldsymbol{\lambda},t)]^T - \partial_x\boldsymbol{\eta}(\boldsymbol{x},\boldsymbol{\lambda},t). \tag{5.32}$$

Because $\boldsymbol{u}(t)$ is determined by the PMP, we consider the first variation of (5.31) only for $\delta\boldsymbol{x}(t)$, $\delta\boldsymbol{\lambda}(t)$ and $\delta\boldsymbol{\mu}(t)$:

$$\delta\tilde{\mathcal{J}} = \int_{t_i}^{t_f}\left[\tilde{A}\cdot\delta\boldsymbol{x} + \tilde{B}\cdot\delta\boldsymbol{\lambda}\right]dt + \tilde{C}\cdot\delta\boldsymbol{x}(t_f) + \tilde{D}\cdot\delta\boldsymbol{x}(t_i) + \tilde{E}\cdot\delta\boldsymbol{\mu}, \tag{5.33}$$

where (omitting some parameter dependencies)

$$\tilde{A} = \partial_x^T H + \mathbf{A}^T \boldsymbol{\lambda}' - \mathbf{N}\boldsymbol{x}', \qquad \tilde{B} = \partial_\lambda^T H - \mathbf{A}\boldsymbol{x}',$$
$$\tilde{C} = \partial_{x_f}^T B - \boldsymbol{\eta}(\boldsymbol{x}(t_f),\boldsymbol{\lambda}(t_f),t_f), \qquad \tilde{D} = \partial_{x_i}^T B + \boldsymbol{\eta}(\boldsymbol{x}(t_i),\boldsymbol{\lambda}(t_i),t_i), \tag{5.34}$$
$$\tilde{E} = \partial_\mu^T B.$$

From the Du Bois–Reymond Lemma 5.2.1, it follows that $\tilde{A} = \tilde{B} = \mathbf{0}$; moreover, from the arbitrariness of $\delta\boldsymbol{x}(t_i)$, $\delta\boldsymbol{x}(t_f)$ and $\delta\boldsymbol{\mu}$ it follows that $\tilde{C} = \tilde{D} = \tilde{E} = \mathbf{0}$. Then the solution of OCP (5.30) must satisfy the following BVP:

$$\text{ODE:} \quad \mathbf{S}(\boldsymbol{x},\boldsymbol{\lambda},t)\begin{pmatrix}\boldsymbol{x}'\\\boldsymbol{\lambda}'\end{pmatrix} = \begin{pmatrix}\partial_x^T H(\boldsymbol{x},\boldsymbol{u},t)\\\partial_\lambda^T H(\boldsymbol{x},\boldsymbol{u},t)\end{pmatrix}, \quad t\in(t_i,t_f) \tag{5.35a}$$

$$\text{BC:} \qquad \mathbf{0} = \mathbf{B}(\boldsymbol{x}(t_i),\boldsymbol{\lambda}(t_i),\boldsymbol{x}(t_f),\boldsymbol{\lambda}(t_f),\boldsymbol{\mu}), \tag{5.35b}$$

$$\text{PMP:} \qquad \boldsymbol{u}(t) = \text{argmin}\{\boldsymbol{v}\in\mathcal{U} \mid H(\boldsymbol{x}(t),\boldsymbol{\lambda}(t),\boldsymbol{v},t)\}, \tag{5.35c}$$

where

$$\mathbf{S}(\boldsymbol{x},\boldsymbol{\lambda},t) = \begin{pmatrix}\mathbf{N}(\boldsymbol{x},\boldsymbol{\lambda},t) & -\mathbf{A}(\boldsymbol{x},t)^T\\ \mathbf{A}(\boldsymbol{x},t) & 0\end{pmatrix},$$

$$\mathbf{B}(\boldsymbol{x}_i,\boldsymbol{\lambda}_i,\boldsymbol{x}_f,\boldsymbol{\lambda}_f,\boldsymbol{\mu}) = \begin{pmatrix}\partial_{x_i}\mathcal{B}(\boldsymbol{x}_i,\boldsymbol{x}_f,\boldsymbol{\mu}) + \mathbf{A}(\boldsymbol{x}_i,t_i)^T\boldsymbol{\lambda}_i\\ \partial_{x_f}\mathcal{B}(\boldsymbol{x}_i,\boldsymbol{x}_f,\boldsymbol{\mu}) - \mathbf{A}(\boldsymbol{x}_f,t_f)^T\boldsymbol{\lambda}_f\\ \partial_\mu \mathcal{B}(\boldsymbol{x}_i,\boldsymbol{x}_f,\boldsymbol{\mu})\end{pmatrix},$$

and

$$\partial_x^T H(x,u,t) = \partial_x^T J(x,u,t) + \partial_x^T f(x,u,t)\lambda,$$
$$\partial_\lambda^T H(x,u,t) = f(x,u,t),$$
$$\partial_{x_i}^T B(x_i,x_f,\mu) = \partial_{x_i}^T \Phi(x_i,x_f) + \partial_{x_i}^T \mathbf{b}(x_i,x_f)\mu,$$
$$\partial_{x_f}^T B(x_i,x_f,\mu) = \partial_{x_f}^T \Phi(x_i,x_f) + \partial_{x_f}^T \mathbf{b}(x_i,x_f)\mu,$$
$$\partial_\mu^T B(x_i,x_f,\mu) = \mathbf{b}(x_i,x_f).$$

5.2.3 Exact Solution of a Linearized Motorcycle Model

Let us make use of the method so far described to derive the analytical solution of an optimal manoeuvre for the linearized model (5.6) that transits the machine from steady-state straight running to steady-state turning at constant velocity V. It is more convenient to make a change of variable for (5.5) to reformulate the problem (5.6) into one with a fixed domain by setting $t = \zeta T$. If we define $\hat{x}(\zeta) = x(\zeta T)$ for a generic state variable x then it follows that $\hat{x}'(\zeta) = x'(\zeta T)T$, which yields the new formulation of the optimal control problem (5.6):

minimize: T, (with $T > 0$)

subject to: $\hat{\phi}'(\zeta) = T\,\hat{\phi}'_{\text{dot}}(\zeta),\ \hat{\phi}'_{\text{dot}}(\zeta) = T\dfrac{h}{I_{xx_0}}\left(m\,g\hat{\phi}(\zeta) - \dfrac{m\,h\,V^2}{L}\hat{\delta}(\zeta)\right),$ (5.36)

$\hat{\phi}(0) = \hat{\phi}_{\text{dot}}(0) = \hat{\phi}_{\text{dot}}(1) = 0,\ \hat{\phi}(1) = \phi_1,\ \left|\hat{\delta}(\zeta)\right| \leq \delta^{\max}.$

The optimal control (5.36) contains the parameter T to be optimized. The general formulation of optimal control of Section 5.2.2 does not consider parameter optimization. Although the formulation can be extended to include this case, for simplicity we recast the problem to fit the formulation considered in Section 5.2.2.

The new constant function $\hat{T}(\zeta)$, (that satisfies $\hat{T}'(\zeta) = 0$) is added, and the problem (5.36) becomes:

minimize $\hat{T}(1)$, (with $\hat{T}(\zeta) > 0$)

subject to: $\hat{\phi}'(\zeta) = \hat{T}(\zeta)\,\hat{\phi}'_{\text{dot}}(\zeta),\ \hat{\omega}'(\zeta) = \hat{T}(\zeta)\dfrac{h}{I_x}\left(mg\hat{\phi}(\zeta) - \dfrac{m\,h\,V^2}{L}\hat{\delta}(\zeta)\right),$

(5.37)

$\boxed{\hat{T}'(\zeta) = 0},\ \hat{\phi}(0) = \hat{\phi}_{\text{dot}}(0) = \hat{\phi}_{\text{dot}}(1) = 0,\ \hat{\phi}(1) = \phi_1,\ |\hat{\delta}(\zeta)| \leq \delta^{\max}.$

Now the problem is formulated as in (5.30) with the following correspondence

$$x = (\hat{\phi},\hat{\phi}_{\text{dot}},\hat{T})^T,\quad \eta(x,\lambda,t) = \lambda = (\lambda_1,\lambda_2,\lambda_3)^T,\quad \Phi(x(0),x(1)) = \hat{T}(1),$$
$$L(x,u,t) = 0,\quad A(x,t) = \mathbf{I},\quad N(x,\lambda,t) = \mathbf{0},\quad \mathcal{U} = \left[-\delta^{\max},\delta^{\max}\right],$$

$$f(x,u,t) = \hat{T}\begin{pmatrix}\hat{\phi}_{\text{dot}} \\ \dfrac{h}{I_{xx_0}}\left(mg\hat{\phi} - \dfrac{m\,h\,V^2}{L}\hat{\delta}\right) \\ 0\end{pmatrix},\quad \mathbf{b}(x(0),x(1)) = \begin{pmatrix}\hat{\phi}(0) \\ \hat{\phi}_{\text{dot}}(0) \\ \hat{\phi}(1) - \phi_1 \\ \hat{\phi}_{\text{dot}}(1)\end{pmatrix}.$$

After a few manipulations (5.37) becomes

equations: $\hat{\phi}' = \hat{T}\hat{\phi}_{dot}, \quad I_{xx_0}\hat{\phi}'_{dot} = \hat{T}h\left(mg\hat{\phi} - \frac{m\,h\,V^2}{L}\hat{\delta}\right), \quad \hat{T}' = 0,$

co-equations: $I_{xx_0}\lambda'_1 = -mgh\hat{T}\lambda_2, \quad \lambda'_2 = -\hat{T}\lambda_1,$

$$I_{xx_0}\lambda'_3 = -\lambda_1 I_x \hat{\omega} - \lambda_2 h\left(mg\hat{\phi} - \frac{m\,h\,V^2}{L}\hat{\delta}\right);$$

BC: $\hat{\phi}(0) = \hat{\phi}_{dot}(0) = \hat{\phi}_{dot}(1) = 0, \quad \hat{\phi}(1) = \phi_1,$

$\mu_1 + \lambda_1(0) = 0, \quad \mu_2 + \lambda_2(0) = 0, \quad \lambda_3(0) = 0,$

$\mu_3 - \lambda_1(1) = 0, \quad \mu_4 - \lambda_2(1) = 0, \quad \lambda_3(1) = 1;$

PMP: $\hat{\delta}(\zeta) = \underset{\delta\in[-\delta^{max},\delta^{max}]}{\operatorname{argmin}} \hat{T}\left[\lambda_1\hat{\omega} + \lambda_2(h/I_{xx_0})\left(mg\hat{\phi} - \frac{m\,h\,V^2}{L}\delta\right)\right].$ (5.38)

This is a BVP with unknowns $\hat{\phi}, \hat{\phi}_{dot}, \hat{T}, \lambda_1, \lambda_2$ and λ_3. From PMP and the physical condition $\hat{T}(\zeta) > 0$ we can solve the control $\hat{\delta}(\zeta) = \operatorname{sign}(\lambda_2(\zeta))\delta^{max}$. This means that the control $\hat{\delta}(\zeta)$, apart from the special case when $\lambda_2(\zeta) = 0$, switches between the extrema values $\pm\delta^{max}$. In this case the solution (or the control) is called bang-bang.

Notice that $\lambda_1(0), \lambda_2(0), \lambda_1(1), \lambda_2(1)$ can be chosen freely because we have four multipliers μ that can be used to satisfy the corresponding boundary condition. Thus, the multiplier μ and the boundary condition for λ_1 and λ_2 can be eliminated. The general solution of the first two co-equations with $\hat{T}(\zeta) = T$ constant is:

$$\lambda_1(\zeta) = C_1 e^{\zeta T \alpha} + C_2 e^{-\zeta T \alpha},$$

$$\lambda_2(\zeta) = \frac{C_2 e^{-\zeta T \alpha} - C_1 e^{\zeta T \alpha}}{\alpha} h\,m\,g,$$

$$\alpha = \frac{\sqrt{m\,g\,h}}{\sqrt{I_{xx_0}}}.$$
(5.39)

From the PMP we know that the control $\hat{\delta}(\zeta)$ switches at the point where $\lambda_2(\zeta) = 0$, that is when

$$C_2 e^{-\zeta T\alpha} = C_1 e^{\zeta T\alpha} \quad\Rightarrow\quad \frac{C_2}{C_1} = e^{2\zeta T\alpha} \quad\Rightarrow\quad \zeta_s = \frac{1}{2T\alpha}\log\frac{C_2}{C_1}. \quad (5.40)$$

and, thus, there is at most one *switching* point ζ_s only when C_1 and C_2 are different from 0 and with equal sign; moreover, if the switching point is on $\zeta_s > 0$ then $|C_2| > |C_1|$.

Using this information we build the general solution of the first two state equations using the constant functions

$$\hat{T}(\zeta) = T \quad \text{and} \quad \hat{\delta}(\zeta) = \frac{g\,L}{h\,V^2}F, \quad T \text{ and } F \text{ constants}$$

before the switching point ζ_s, thus, for $\zeta \leq \zeta_s$:

$$\hat{\phi}(\zeta) = \frac{1}{\alpha}(C_3 e^{\zeta T\alpha} - C_4 e^{-\zeta T\alpha}) + F,$$
$$\hat{\omega}(\zeta) = C_3 e^{\zeta T\alpha} + C_4 e^{-\zeta T\alpha}. \tag{5.41}$$

The general solution of the first two state equations with the functions $\hat{T}(\zeta) = T$ and $\hat{\delta}(\zeta) = -\frac{g}{h}\frac{L}{V^2}F$ constants after the switching point ζ_s is for $\zeta \geq \zeta_s$:

$$\hat{\phi}(\zeta) = \frac{1}{\alpha}(C_5 e^{(\zeta-\zeta_s)T\alpha} - C_6 e^{-(\zeta-\zeta_s)T\alpha}) - F,$$
$$\hat{\omega}(\zeta) = C_5 e^{(\zeta-\zeta_s)T\alpha} + C_6 e^{-(\zeta-\zeta_s)T\alpha}. \tag{5.42}$$

The constants C_1, C_2, C_3 and C_4 must satisfy

$$2C_1 = \lambda_1(0) - \alpha\lambda_2(0), \qquad 2C_2 = \lambda_1(0) + \alpha\lambda_2(0),$$
$$2C_3 = \hat{\omega}(0) + \alpha\hat{\phi}(0) - \alpha F = -\alpha F, \qquad 2C_4 = \hat{\omega}(0) - \alpha\hat{\phi}(0) + \alpha F = \alpha F,$$

and thus they can be determined from initial values $\hat{\phi}(0) = \hat{\omega}(0) = 0$, $\lambda_1(0)$ and $\lambda_2(0)$; we can rewrite (5.39) and (5.41) as

$$\lambda_1(\zeta) = \lambda_1(0)\cosh(\zeta T\alpha) - \alpha\lambda_2(0)\sinh(\zeta T\alpha), \qquad \hat{\phi}(\zeta) = F(1 - \cosh(\zeta T\alpha)),$$
$$\lambda_2(\zeta) = -\frac{\lambda_1(0)}{\alpha}\sinh(\zeta T\alpha) + \lambda_2(0)\cosh(\zeta T\alpha), \qquad \hat{\omega}(\zeta) = -\alpha F\sinh(\zeta T\alpha). \tag{5.43}$$

The constants C_5 and C_6 must satisfy

$$2C_5 = \hat{\omega}(\zeta_s) + \alpha\hat{\phi}(\zeta_s) + \alpha F, \qquad 2C_6 = \hat{\omega}(\zeta_s) - \alpha\hat{\phi}(\zeta_s) - \alpha F, \tag{5.44}$$

and thus they can be determined from values at switching point ζ_s obtained from the initial piecewise solution:

$$\hat{\phi}(\zeta_s) = F(1 - \cosh(\zeta_s T\alpha)), \qquad \hat{\omega}(\zeta_s) = -\alpha F\sinh(\zeta_s T\alpha), \tag{5.45}$$

and (5.42) become:

$$\hat{\phi}(\zeta + \zeta_s) = \frac{\hat{\omega}(\zeta_s)}{\alpha}\sinh(\zeta T\alpha) + \hat{\phi}(\zeta_s)\cosh(\zeta T\alpha) - F(1 - \cosh(\zeta T\alpha)),$$
$$\hat{\omega}(\zeta + \zeta_s) = \hat{\omega}(\zeta_s)\cosh(\zeta T\alpha) + \alpha\hat{\phi}(\zeta_s)\sinh(\zeta T\alpha) + \alpha F\sinh(\zeta T\alpha). \tag{5.46}$$

Using the identity for hyperbolic functions:

$$\cosh(x + y) = \sinh x \sinh y + \cosh x \cosh y,$$
$$\sinh(x + y) = \cosh x \sinh y + \sinh x \cosh y,$$

in (5.46) with (5.45), after some manipulation:

$$\hat{\phi}(\zeta + \zeta_s) = F(2\cosh((\zeta + \zeta_s)T\alpha) - \cosh(\zeta T\alpha) - 1),$$
$$\hat{\omega}(\zeta + \zeta_s) = \alpha F(2\sinh((\zeta + \zeta_s)T\alpha) - \sinh(\zeta T\alpha)). \tag{5.47}$$

Using the identity $\zeta T = t$ and $\zeta_s T = t_s$ in (5.43) and (5.47) we can write exact solution as follows:

$$\phi(t) = F \begin{cases} 1 - \cosh(t\alpha) & 0 \le t < t_s, \\ 2\cosh((t - t_s)\alpha) - \cosh(t\alpha) - 1 & t_s \le t \le T, \end{cases}$$

$$\omega(t) = \alpha F \begin{cases} -\sinh(t\alpha) & 0 \le t < t_s, \\ 2\sinh((t - t_s)\alpha) - \sinh(t\alpha) & t_s \le t \le T, \end{cases} \quad (5.48)$$

where t_s is the switching time and

$$F = \pm \frac{h\, V^2}{g\, L} \delta^{max}$$

where the sign depends on the boundary condition. To compute switching time t_s and manoeuvre time T we must solve the nonlinear system:

$$\phi_1 = \phi(T) = F(2\cosh((T - t_s)\alpha) - \cosh(T\alpha) - 1),$$
$$0 = \omega(T) = \alpha F(2\sinh((T - t_s)\alpha) - \sinh(T\alpha)),$$

whose solution is

$$T = \frac{\ln(w) - \ln(1+f)}{\alpha}, \qquad t_s = \frac{\ln(1+w) - \ln(2)}{\alpha}, \quad (5.49)$$

$$w = 1 + f - \frac{1}{2}f^2 + \frac{1}{2}\sqrt{f(f+4)(f-2)(f+2)}, \qquad f = -\frac{\phi_1}{F}.$$

Notice that real solutions for T and t_s are obtained for $f \in (-1, 0]$. This means that

$$F = -\text{sign}(\phi_1) \frac{h\, V^2}{g\, L} \delta^{max}$$

and not all the angles ϕ_1 are reachable but only the ones that satisfy:

$$|\phi_1| < \frac{h\, V^2}{g\, L} \delta^{max}.$$

5.2.4 Numerical Solution and Approximate Pontryagin

The exact solution of steering problem (5.6) formulated as the BVP (5.35) uses the Pontryagin minimum principle (5.35c) to obtain the control law $F(t)$. However, the solution of this problem for multiple controls and a complex domain $\mathcal{U}$ may be difficult. Also for a simpler problem like the curve initiation problem the solution may be discontinuous. In real applications exact solution of the optimal control problem is a rare event so that numerical solution of the BVP (5.35) is mandatory. Discontinuity of the control is a problem for the convergence of a numerical method for the BVP (5.35). To simplify the numerical solution we can *smooth* the problem and obtain the solution of an approximate problem (and thus, a suboptimum). A solution based on barrier or penalty functions is the following. Let $p(x)$ be a function such that $p(x) = \infty$ for x outside $\mathcal{V}$ and $p(x) = 0$ inside $\mathcal{V}$; this is an ideal barrier function which can be used to transform the problem (5.30) to the equivalent one:

$$\text{minimize:} \quad F(u) = \Phi(x(t_i), x(t_f)) + \int_{t_i}^{t_f} L(x(t), u(t), t) + p(u(t)) dt, \quad (5.50a)$$

$$\text{ODE:} \quad A(x(t), t) x'(t) = f(x(t), u(t), t), \quad (5.50b)$$

$$\text{BC:} \quad b(x(t_i), x(t_f)) = 0, \quad (5.50c)$$

where the constraint (5.30d) is substituted with the barrier function $p(x)$ in equation (5.50a). This problem is now approximated by using a smooth barrier $p_\varepsilon(x)$ which has the property that $p_\varepsilon(x)$ is regular and small for x internal points of $\mathcal{V}$. For example, for $\mathcal{V} = [-\delta^{\max}, \delta^{\max}]$ a possible barrier function is

$$p_\varepsilon(x) = -\varepsilon \log \cos\left(\frac{\pi}{2} \frac{x}{\delta^{\max}}\right). \quad (5.51)$$

The effect of ε and the shape of $p_\varepsilon(x)$ is shown in Figure 5.7. Using barrier (5.51) in problem (5.38) after some manipulations results in:

$$\text{minimize} \quad \hat{T}(1) - \varepsilon \hat{T}(1) \int_0^1 \log \cos\left(\frac{\pi}{2} \frac{\hat{\delta}(\zeta)}{\delta^{\max}}\right) \zeta, \quad (\text{with } \hat{T}(\zeta) > 0)$$

$$\text{subject to:} \quad \hat{\phi}'(\zeta) = \hat{T}(\zeta) \hat{\omega}(\zeta), \quad \hat{\omega}'(\zeta) = \hat{T}(\zeta) \frac{m h}{I_x}\left(g\hat{\phi}(\zeta) - \frac{h V^2}{L} \hat{\delta}(\zeta)\right), \quad (5.52)$$

$$\hat{T}'(\zeta) = 0, \quad \hat{\phi}(0) = \hat{\omega}(0) = \hat{\omega}(1) = 0, \quad \hat{\phi}(1) = \phi_1.$$

and after few manipulation from (5.52) and PMP the control $\hat{\delta}(\zeta)$ satisfies:

$$\hat{\delta}(\zeta) = \underset{\delta \in \mathbb{R}}{\arg\min} H = \underset{\delta \in \mathbb{R}}{\arg\min} \hat{T}\left[\lambda_1 \hat{\omega} + \lambda_2 \frac{m h}{I_x}\left(g\hat{\phi} - \frac{h V^2}{L} \delta\right) - \varepsilon \log \cos\left(\frac{\pi}{2} \frac{\delta}{\delta^{\max}}\right)\right].$$

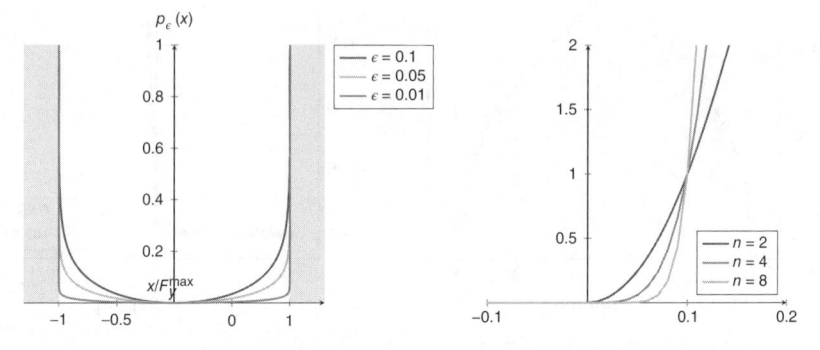

Figure 5.7 On the left is shown the shape of barrier function (5.51) for $\varepsilon = 0.1$, $\varepsilon = 0.05$ and $\varepsilon = 0.01$, and on the right, the shape of the penalty function (5.56) for $n = 2$, $n = 4$ and $n = 8$

The minimum can be found by searching for stationary points of the Hamiltonian function

$$0 = \partial_\delta H = \hat{T}\left[-\lambda_2 \frac{m\,h^2\,V^2}{I_x\,L} + \frac{\varepsilon}{\delta^{\max}}\frac{\pi}{2}\tan\left(\frac{\pi}{2}\frac{\delta}{\delta^{\max}}\right)\right],$$

and thus

$$\hat{\delta}(\zeta) = \delta^{\max}\left\{\frac{2}{\pi}\arctan\left(\lambda_2(\zeta)\frac{2\delta^{\max}m\,h^2\,V^2}{\pi\,\varepsilon\,I_x\,L}\right)\right\}.$$

Notice that $\frac{2}{\pi}\arctan\left(\frac{x}{\varepsilon}\right) \approx \mathrm{sign}(x)$ and thus this solution is an approximation of the true optimal control. This control is now smooth, and any numerical method can be used to approximate the solution of the BVP derived from problem (5.52). For example, in (Bertolazzi et al. 2005; 2006, 2007a) second-order finite difference is used for the discretization. In Figure 5.8 the results using a second-order discretization with $\varepsilon = 10^{-4}$ on a mesh of 40 points is shown.

A second-order finite difference for the BVP (5.35) is the following:

$$\frac{\mathbf{S}(\mathbf{x}_{k+1/2}, \lambda_{k+1/2}, t_{k+1/2})}{\Delta T}\begin{pmatrix}\mathbf{x}_{k+1} - \mathbf{x}_k \\ \lambda_{k+1} - \lambda_k\end{pmatrix} = \begin{pmatrix}\partial_x^T H(\mathbf{x}_{k+1/2}, \mathbf{u}_{k+1/2}, t_{k+1/2}) \\ \partial_\lambda^T H(\mathbf{x}_{k+1/2}, \mathbf{u}_{k+1/2}, t_{k+1/2})\end{pmatrix}, \tag{5.53}$$

$$\mathbf{0} = \mathbf{B}(\mathbf{x}_0, \lambda_0, \mathbf{x}_N, \lambda_N, \mu), \tag{5.54}$$

$$\mathbf{u}_{k+1/2} = U(\mathbf{x}_{k+1/2}, \lambda_{k+1/2}, t_{k+1/2}), \tag{5.55}$$

where $U(\mathbf{x}, \lambda, t)$ is a function assumed known or computable which solves the PMP:

$$U(\mathbf{x}, \lambda, t) = \mathrm{argmin}\{\mathbf{v} \in \mathcal{U} \mid H(\mathbf{x}(t), \lambda(t), \mathbf{v}, t)\}.$$

Here $\Delta T = (t_f - t_i)/N$ and $t_k = t_i + k\,\Delta T$. Moreover $\mathbf{x}_k \approx \mathbf{x}(t_k)$ and $\lambda_k \approx \lambda(t_k)$. Half integer such as $k+1/2$ means averaging, for example $\lambda_{k+1/2} = (\lambda_k + \lambda_{k+1})/2$. Discretized BVP (5.53–5.54–5.55) constitutes a non-linear system which must be solved. This non-linear system is difficult to solve and a robust nonlinear system solver must be used like TEN-SOLVE (Bouaricha and Schnabel, 1997) or affine invariant Newton solver (Deuflhard 2011).

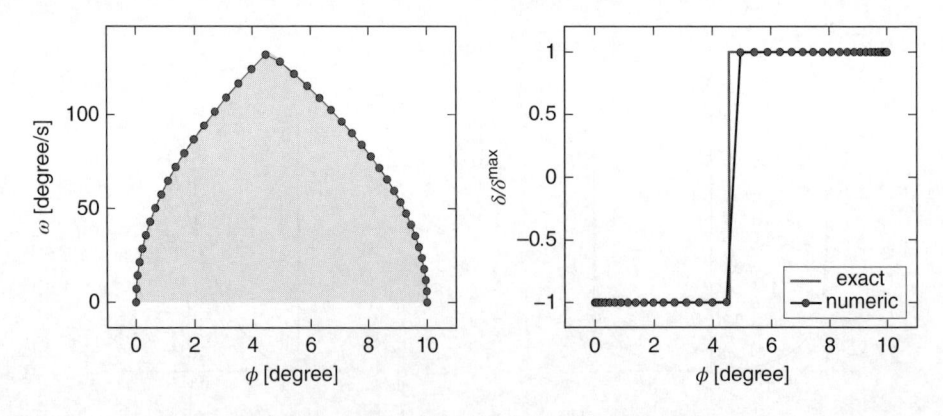

Figure 5.8 Exact and numerical solution of inverted pendulum bicycle model with $\varepsilon = 10^{-4}$, $m = 250$, $g = 9.81$, $F_y^{\max} = mg$, $I_x = 45$, $\phi_1 = 10°$, $h = 0.6$

(Bertolazzi et al. 2007a) describes a variant of the affine invariant Newton solver for the solution of the nonlinear systems as a result of the discretization of the optimal control problems.

5.2.4.1 Optimal Control with Unilateral Constraints

In many applications not only the controls are subject to unilateral constraints but also the states, so we have inequalities like:

$$\mathbf{C}(x(t), u(t)) \leq \mathbf{0}, \quad \text{where} \quad \mathbf{C}(x, u) = \begin{pmatrix} C_1(x, u) \\ C_2(x, u) \\ \vdots \\ C_m(x, u) \end{pmatrix},$$

which are added to problem (5.30). The introduction of inequality constraints complicates the variational approach, but the use of barrier functions allow the problem to be reduced again to the form of (5.30). A barrier function is a continuous and positive function over the feasible set which goes to ∞ as x approaches the boundary. A barrier function for $C_1(x, u)$ may be $b(C_1(x, u))$ where:

$$b(x) = -\varepsilon \log x, \quad \text{or} \quad b(x) = \frac{1}{x}.$$

This kind of barrier produces a nonlinear system that is extremely difficult to solve. An alternative is to use penalty functions which are a regularized version of the function $[x]^+ = \max(x, 0)$ and take the form:

$$p(x; h, n) = \begin{cases} 0 & x < 0, \\ \left(\frac{x}{h}\right)^n & x \geq 0. \end{cases} \quad (5.56)$$

The behaviour of (5.56) for $n = 2, 4, 8$ is shown by the plot on the left of Figure 5.7. This penalty is equal to zero for non-positive x and increases very rapidly for $x > h$ and positive n. The parameter h determines the degree of influence of the penalty. See (Bertolazzi et al. 2007a) for a deeper discussion on penalties and the choice of the parameters.

5.3 Applications of Optimal Manoeuvre to Motorcycle Dynamics

The concepts of manoeuvrability and handling are mathematically formalized in Section 5.1.1 as the solution of an optimal control problem that minimises a desired metric. In Section 5.1.2 examples have been introduced that practically explains the above concepts by means of motorcycle linearized models to transit from steady-state straight running to steady-state cornering. It has been shown how the manoeuvrability is an intrinsic characteristic of a machine, therefore it depends only on the design parameters, while handling involves the rider's skills and how the machine may fit those skills and limitations.

In this section the above concepts are further extended by modelling and introducing the rider's riding capability and the corresponding limitations in the optimal control formulation to complete minimum time manoeuvres. The machine model used in the simulation of this section includes the main non-linearities (e.g. tyre and engine characteristics, and large roll angles) as described in (Bertolazzi et al. 2006). Comparisons of simulations and experimental data show that in such cases the optimal manoeuvres retain the same characteristics as real ones.

5.3.1 Modelling Riders' Skills and Preferences with the Optimal Manoeuvre

The way that an expert human rides depends on what is his/her final goal and personal riding skills and experience.

For example, the study of (Davoodi et al. 2012) found that most riders are capable of responding to an unexpected object along the roadway in 2.5 s or less. Expert riders show superior control of the machine compared to first-time riders, but training leads to an improvement in the riding skills of novice riders, improving their capacity to adapt the speed to the situation, reducing trajectory-corrective movements, and changing their pattern of gaze exploration (Stasi et al. 2011). Again, (Rice 1978) has experimentally shown that expert riders efficiently move their upper body to precisely control the motorcycle. In particular, in some situations they lean in such a way as to reduce the time necessary to enter into a curve, eliminating the counter-steering manoeuvre. Most of the above aspects, and others, can be engineered in the optimal control formulation with additional differential equations that model the band width of the human rider control actions (Schwab and Meijaard 2013) (e.g., body part movements or steering torque or throttle rotation) or with inequality constraints on some variables such as maximum steering torque and steering rate. One complex aspect in motorcycle dynamics is the concurrent use of the four inputs (steering torque, front brake, rear brake and throttle). In particular, the rear brake is quite critical in cornering manoeuvres since its incorrect use can lead to a rider falling. Only few top riders are able to correctly use the rear tyre braking force, because this causes a lack of rear lateral support which results in an open-loop unstable state (Kelly 2008): for F1 drivers and (Corno et al. 2008) for motorcycles. One possible way to take into account the way riders use the longitudinal and lateral controls is to properly shape the pattern of longitudinal and lateral accelerations generated in the g-g plots. In fact, it has been experimentally shown (Lot and Biral 2009) that the envelope of longitudinal and lateral accelerations of a large set of manoeuvres of a rider shows how efficiently the rider is able to act on the motorcycle controls. On the other hand, it is also a measure of the set of accelerations that the rider considers comfortable or reachable, given the uncertain road surface conditions and the machine dynamic properties.

Figure 5.9 shows the g-g plot of two riders, with different skills and experience, riding the same motorcycle on the same circuit. As shown, the two pattern are different, and in neither case does it completely covers the corresponding adherence ellipse. In particular, rider 1, who is a less aggressive rider, even if with long riding experience, used the rear brake much less in cornering compared to rider 2. Nevertheless, rider 2 is still not using all available lateral acceleration in the braking phase. The pattern of the acceleration envelope can be properly introduced in the OCP formulation with a specific inequality constraint for longitudinal acceleration as function of lateral acceleration. One possible implementation is the following:

$$a_x(t) \geq -a_{y0} - \left| \frac{a_y(t)}{a_{y_L}} \right|^{n_B}, \qquad (5.57)$$

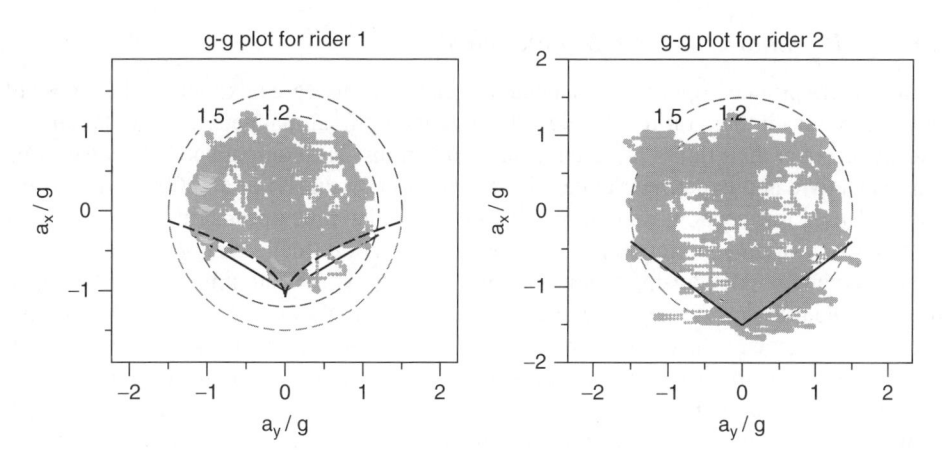

Figure 5.9 Distribution of normalized lateral and longitudinal accelerations for two riders on the same race track and the same machine. Lighter points mean more frequent value for the pair $(a_x/g, a_y/g)$. Figure also shows circle of estimated roll angle. Data 2002 courtesy of MDRG of University of Padova

where a_{y0} and n_B are parameters that define the shape of the envelope in the braking phase and can be estimated from the acceleration envelope of a single or group of riders to be modelled. When $n_B = 1$, the inequality reduces to a straight line, and for $n_B < 1$ it is a curved line as shown in Figure 5.9 for rider 1.

Another interesting piece of information that helps in designing an OCP problem that produces human-like manoeuvres is based on the nature of human motor strategies. A number of studies have been carried out to try to understand what is the criteria that produce human motion-like in general (Liu and Todorov 2007; Viviani and Flash 1995; Breteler et al. 2002, and for the specific domain of driving Yamakado and Abe 2008). Minimum jerk is one of the most important movement rules and is regarded as a principle of maximum smoothness. Drivers' bodies are sensitive to jerk, and jerk reflects the changing forces applied to the vehicle. For example, car drivers exploit this information to keep total tyre force within a certain range by a coordinated control between longitudinal and lateral accelerations. For example, in driving, minimum jerk trajectories are those that minimize the root mean square (rms) value of driver control. One example of implementation is the following:

$$\mathcal{L}_j(j_x, j_y) = \left(\frac{j_x}{j_{x0}}\right)^2 + \left(\frac{j_y}{j_{y0}}\right)^2, \tag{5.58}$$

where j_x, j_y are the generic definition for the longitudinal and lateral jerk of the vehicles, which can be mapped into the corresponding derivatives of the lateral and longitudinal controls. In a similar way other rider characteristics can be objectified and introduced in the OCP formulation forming a sort of weighted sum of different criteria. In fact, riders' control strategies often arise from a trade-off of several other criteria which allows us to model a variety of motor strategies. For example, a cost function combining minimum jerk and minimum time can model a variety of human behaviours under different time pressures. In Section 5.3.2, as an application example of all the above concepts, the minimum lap time problem of motorcycles is discussed.

5.3.2 Minimum Lap Time Manoeuvres

Racing riders produce optimized manoeuvres that in some cases extract the best possible performance from the motorcycle. Nevertheless, the optimal manoeuvre formulation must take into account their riding preferences and limitations. The comparison between the optimal manoeuvre and experimental data is reported in (Bertolazzi et al. 2006 and Cossalter et al. 2010). In this section, we compare the minimum lap time optimal manoeuvre with no limitation on combined longitudinal and lateral accelerations (called *reference*) and the optimal manoeuvre obtained using the limitations imposed by (5.57) to mimic the behaviour of rider 1 of the previous section. The circuit used in the simulation is located in Adria (Italy) since it is the same as the experimental data of Figure 5.9. The motorcycle dynamic model is nonlinear and is the one explained in (Bertolazzi et al. 2006). A quite convenient way to formulate the corresponding OCP is to change the coordinates from cartesian $x(t), y(t), \psi(t)$ to curvilinear $\zeta(t), n(t), \alpha(t)$, since it is easy to model the road shape and it is straightforward to impose the path constraints. The curvilinear coordinates are the position of the motorcycle along the road middle line ζ, the orientation of the motorcycle with respect to the tangent to the middle line $\alpha(t)$ and the lateral displacement $n(t)$.

The relationship between cartesian coordinates and curvilinear coordinates is established on the basis of their derivatives (i.e. the velocities) as follows:

$$\frac{d\zeta(t)}{dt} = \frac{\cos(\alpha(t))u(t) - \sin(\alpha(t))v(t)}{s_n(t)\kappa(\zeta(t)) - 1}, \tag{5.59a}$$

$$\frac{ds_n\zeta(t)}{dt} = \cos(\alpha(t))v(t) - \sin(\alpha(t))u(t), \tag{5.59b}$$

$$\frac{d\alpha(t)}{dt} = \dot{\psi}(t) + \frac{(\sin(\alpha(t))v(t) - \cos(\alpha(t))u(t))\kappa(\zeta(t))}{s_n(t)\kappa(\zeta(t)) - 1}, \tag{5.59c}$$

where $u(t)$ and $v(t)$ are the components of the motorcycle reference point velocity vector projected in the moving frame (Figure 5.10). To reformulate Equations (5.1) in space domain, let us consider the set of coordinates $x(\zeta)$ as a function of ζ instead of time t and compute the total derivative with respect to ζ:

$$\frac{dx(t(\zeta))}{d\zeta} = \frac{dx(t)}{dt}\frac{dt}{d\zeta}. \tag{5.60}$$

Let us call $\zeta_v = \frac{d\zeta(t)}{dt}$, which can be obtained by solving (5.59a). Using (5.60) to compute the derivatives in the dynamic system model of (5.1) and changing variable t to variable ζ we get:

$$\frac{dx(\zeta)}{d\zeta}\zeta_v = f(x(\zeta), u(\zeta), \zeta, p). \tag{5.61}$$

In this new set of coordinates the path constraints that force the motorcycle to stay within the road geometry are:

$$W_L(\zeta) \le s_n(\zeta) \le W_R(\zeta), \tag{5.62}$$

where $W_L(\zeta)$ and $W_R(\zeta)$ represent respectively the left and right widths of the road, which can be different along the road profile. The reader may also note that the domain of integration is now fixed and corresponds to the circuit length L, which is an advantage for the

The Optimal Manoeuvre

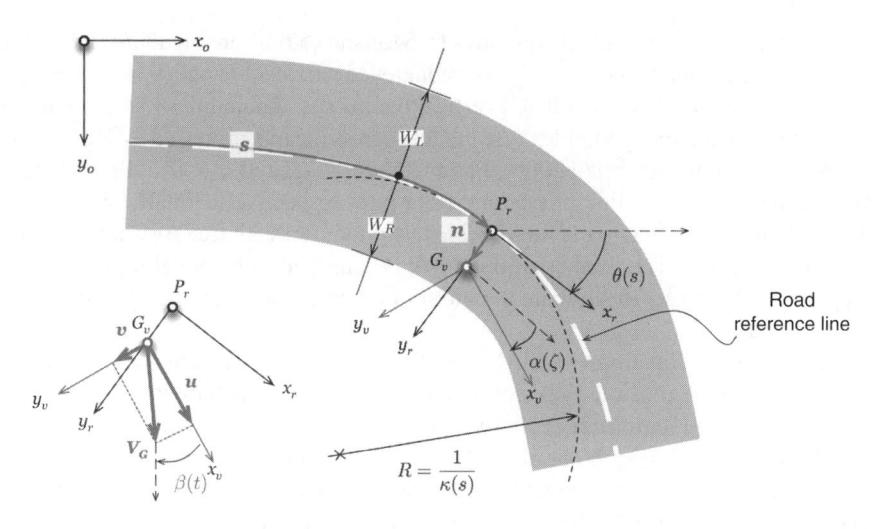

Figure 5.10 Meaning of the curvilinear coordinates. The road is defined by its reference line which is described with a sequence of segments of given length L and curvature κ and left W_L and right W_R width. P_r projection of the vehicle point G_v on the reference line. n is the distance of G_v from reference line and α is the orientation of the vehicle with respect to reference line tangent in P_r.

numerical method's convergence. The goal function becomes:

$$L(x(\zeta), u(\zeta)) = \int_0^L \left[\frac{w_T}{\zeta_v(\zeta)} + \left(\frac{j_\tau(\zeta)}{j_{\tau_{max}}}\right)^2 + \left(\frac{j_{S_r}(\zeta)}{j_{S_{r_{max}}}}\right)^2 + \left(\frac{j_{S_f}(\zeta)}{j_{S_{f_{max}}}}\right)^2 \right] d\zeta, \quad (5.63)$$

and in combination with the inequality constraint (5.57) it models a rider who wants to complete a manoeuvre in minimum time combining to some degree the longitudinal and lateral forces in the braking phase. Other trajectory constraints are the following:

$$|\tau(\zeta)| \leq \tau^{max} \quad (5.64a)$$

$$|\delta_{\text{dot}}(\zeta)| \leq \delta_{\text{dot}}^{max} \quad (5.64b)$$

$$S_r(\zeta) \leq Sr_{max}(u(\zeta)) \quad (5.64c)$$

$$S_f(\zeta) \leq 0 \quad (5.64d)$$

$$\left(\frac{S_r(\zeta)}{\mu_{Xr} N_r(\zeta)}\right)^2 + \left(\frac{F_r(\zeta)}{\mu_{Yr} N_r(\zeta)}\right)^2 \leq 1 \quad (5.64e)$$

$$\left(\frac{S_f(\zeta)}{\mu_{Xf} N_f(\zeta)}\right)^2 + \left(\frac{F_f(\zeta)}{\mu_{Yf} N_f(\zeta)}\right)^2 \leq 1. \quad (5.64f)$$

Inequalities (5.64a) and (5.64b) limit the maximum steering torque and steering rate which are due to human rider's limitations. Inequality (5.64c) limits the maximum traction at rear wheel so as not to exceed the engine torque envelope. Inequality (5.64d) prevents the front wheel producing tractive forces. Inequalities (5.64e) and (5.64f) are an approximate implementation of the rear and front tyre ellipse of adherence where S_i, F_i and N_i, with $i = r, f$, are respectively the longitudinal, lateral and vertical tyre forces. Inequalities (5.64e) and (5.64f) could be avoided using more complex tyre model such as the one proposed by Pacejka which uses the Magic Formula (Pacejka 2006). Parameters μ_{Xr}, μ_{Yr} and μ_{Xf}, μ_{Yf} are the longitudinal and lateral adherence limits for rear and front tyres respectively. Finally, we impose the initial conditions to be equal to the motorcycle in straight run with free forward velocity and rear longitudinal force. The final conditions will be imposed to be equal to initial one (i.e. *cyclic* conditions) in order to simulate the motorcycle performing a circuit lap starting with maximum speed.

The optimal control equations which derived from (5.61)–(5.64) yield a highly non-linear boundary value problem of the type defined by (5.35) that is numerically solved with the software described in (Bertolazzi et al. 2007a).

Figure 5.11 reports the g-g plots of the two simulated minimum manoeuvres combined with the rear and front tyre engagements. The plots show the typical pattern of Figure 5.9. The acceleration points do not fill completely the friction circle in the upper part (i.e. for positive accelerations) because the front wheel load almost reached zero. For rider 1, the negative accelerations are not optimally combined with lateral ones and therefore the friction ellipse is not fully filled with points which means that tyres are not used at their limits. On the other hand the maximum use of tyre forces would be possible by properly distributing the braking tyre effort between front and rear. The simulation results were obtained with a constraints on center of mass accelerations but we could get almost the same result by putting the limit on the rear brake only. A minimum use of the rear braking force when cornering has a clear effect on the longitudinal velocity evolution. The top graph of Figure 5.12 highlights

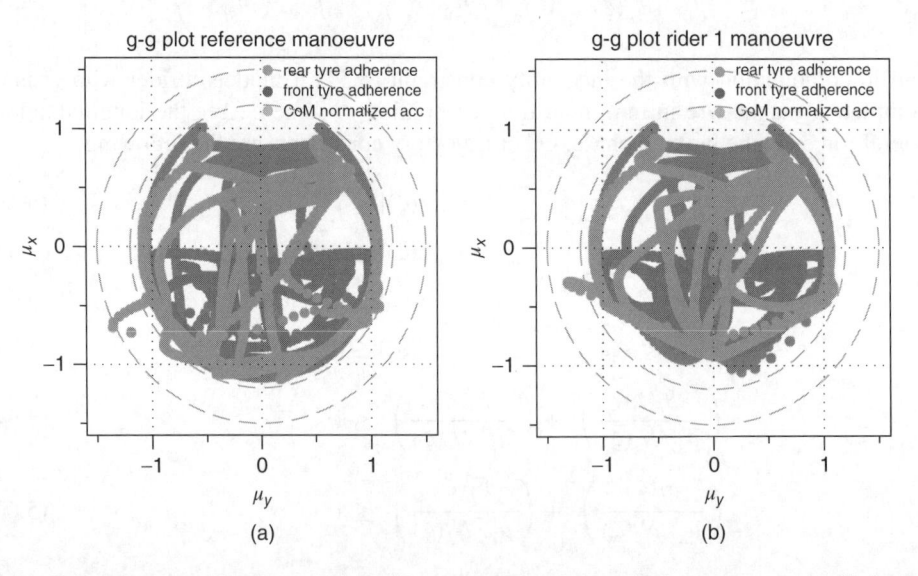

Figure 5.11 g-g plots of reference optimal manoeuvre and of optimal manoeuvre with limitation of combined longitudinal and lateral accelerations. The plots show rear and front tyre adherence and acceleration of the centre of mass projected on the ground

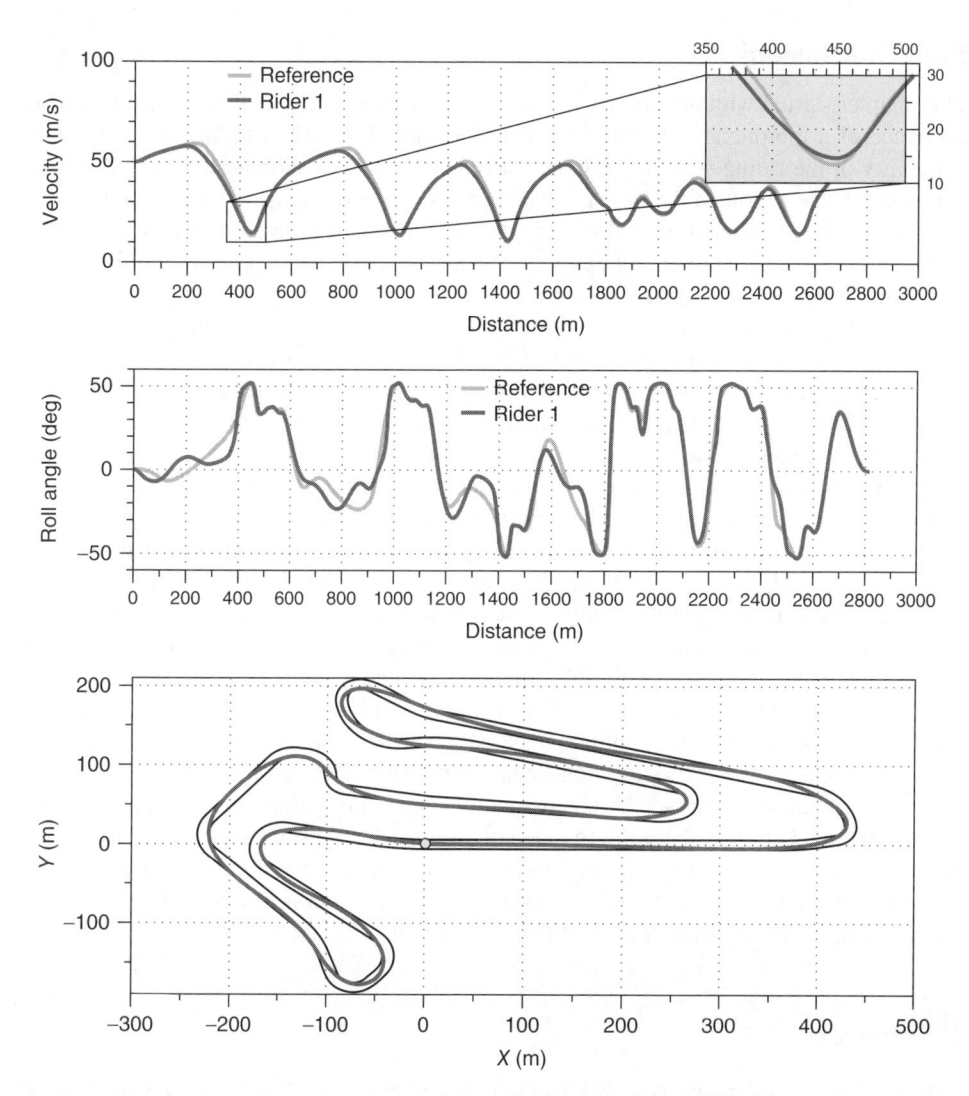

Figure 5.12 Comparison between reference optimal manoeuvre and optimal manoeuvre with limitation of the combined longitudinal and lateral accelerations. The top plot shows forward velocity with zoom on the first corner. The middle plot shows the roll angle. The bottom plot is the trajectory for the reference manoeuvre

that the reference manoeuvre yields a more prolonged and stronger deceleration despite the higher roll angle. On the contrary the rider 1's manoeuvre has to anticipate the braking before rolling the motorcycle into the corner in order to reduce the speed since he brakes less due to the constraint (5.58). This behaviour is more evident for greater decelerations and at the transition between fast straight pace and sharp corners. The simulated lap time difference is about 1.6s (the reference manoeuvre being the fastest) which is due to only a suboptimal use of the rear brake when leaning the motorcycle. The results clearly show the effectiveness of the proposed method to model the rider's skill and preferences.

5.4 Conclusions

This chapter started with the question of assessing in an objective manner the intrinsic performance of a motorcycle or combined with its rider. The task is tough due to the large variability of the riding style, the number of input that may contribute to the control strategy, and the central role of the subjective perception of the riding experience. The answer is found in the formulation of the task to complete a manoeuvre as an optimal control problem that minimizes a desired metric of the performance and possibly includes riding limitations, skills and preferences. Similarly to aircraft engineering, if we only consider the machine characteristics we evaluate the manoeuvrability, otherwise we assess handling. In both cases, either we consider the rider's limitations or not, the solution is optimal and is a baseline, or a reference manoeuvre, for others obtained by changing the design parameters. The final value of the metric is the performance index which evaluates the machine and can be used to compare the performance of different vehicles. With the aid of linearized models of motorcycle lateral dynamics we have shown that handling and manoeuvrability are two distinct concepts that measure the performance of the machine alone or with the rider in the loop. We have also shown that manoeuvrability is restricted by the machine's subsystem dynamic properties but it is possible to design the machine in order to reduce the effort and substantially keep the same manoeuvrability performance. In the last part of the chapter we proposed a method to include the human rider's experience or attitude limitations in order to obtain an suboptimal manoeuvre which resembles that of a particular category of riders.

Despite the benefits that may derive from using the optimal control approach to solve the above problems the solution of optimal manoeuvres with complex motorcycle models is not trivial. The reasons are the highly non-linear relationship that links the optimal inputs and the state variables that must satisfy the inequality constraints. Additionally, as proved for the linearized model of (5.5), the solution may not even exist since some states are not reachable due to the control limitations. To this end, we have proposed an indirect method based on an approximation of the Pontryagin's minimum principle that is numerically efficient and can also handle complex motorcycle models (Cossalter et al. 2010).

References

Astrom K, Klein R and Lennartsson A 2005 Bicycle dynamics and control: adapted bicycles for education and research. *Control Systems, IEEE* **25**(4), 26–47.

Bertolazzi E, Biral F and Da Lio M 2005 Symbolic–Numeric Indirect Method for Solving Optimal Control Problems for Large Multibody. The racing vehicle example. *Journal of Multibody System Dynamics* **13**, 233–252.

Bertolazzi E, Biral F and Da Lio M 2006 Symbolic–Numeric Efficient Solution of Optimal Control Problems for Multibody Systems. *Journal of Computational and Applied Mathematics* **185**(2), 404–421.

Bertolazzi E, Biral F and Lio MD 2007a Real-time motion planning for multibody systems. *Multibody System Dynamics* **17**, 119–139.

Bertolazzi E, Biral F, Lio MD and Cossalter V 2007b Influence of rider's upper body motions on motorcycle minimum time *Multibody Dynamics 2007: ECCOMAS Thematic Conference: Milano, 25–28 June 2007*. Dipartimento di Ingegneria Aerospaziale, Politecnico, Milano.

Bertolazzi E, Biral F, Lio MD, Saroldi A and Tango F 2009 Supporting drivers in keeping safe speed and safe distance: The Saspence Subproject Within the European Framework Programme 6 Integrating Project PReVENT. *IEEE Transactions on Intelligent Transportation Systems* **2009**, 1–14.

Biral F, Bortoluzzi D, Cossalter V and Da Lio M 2003a Experimental study of motorcycle transfer functions for evaluating handling. *Vehicle System Dynamics* **39**(1), 1–25, 2003.

Biral F, Da Lio M and Bertolazzi E 2005 Combining safety margins and user preferences into a driving criterion for optimal control-based computation of reference maneuvers for an ADAS of the next generation *Intelligent Vehicles Symposium, 2005*, pp. 36–41 number 94 in *Proceedings. IEEE*.

Biral F, Da Lio M and Bertolazzi E 2006 Motion planning algorithms based on optimal control for motorcycle rider system *FISITA 2006 World Automotive Congress*. JSAE. Paper Code Yokohama2006/F2006V20.

Biral F, Da Lio M and Maggio F 2003b How gear ratio influences lap time and driving style. An analysis based on time-optimal maneuvers In *Small Engine Technology Conference & Exhibition* (ed. International S) SAE. Paper Number 2003-32-0056.

Biral F, Da Lio M, Lot R and Sartori R 2010 An intelligent curve warning system for powered two-wheel vehicles. *European Transport Research Review* pp. 1–10.

Biral F, Lot R and Garbin S 2002 Enhancing the performance of high powered motorcycles by a proper definition of geometry and mass distribution *Motorsports Engineering Conference & Exhibition* SAE International, Indianapolis IN, USA.

Biral F, Lot R, Rota S, Fontana M and Huth V 2012 Intersection support system for powered two-wheeled vehicles: Threat assessment based on a receding horizon approach. *Intelligent Transportation Systems, IEEE Transactions on* **PP**(99), 1–12. in press.

Blaye PL 2002 Agility: History, definitions and basic concepts *RTO HFM-052 Lecture Series on "Human Consequences of Agile Aircraft"* NATO Research & Technology Organization. published in RTO-EN-12.

Bobbo S, Cossalter V, Massaro M and Peretto M 2009 Application of the 'optimal maneuver method' for enhancing racing motorcycle performance. *SAE International Journal of Passenger Cars–Electronic and Electrical Systems* **1**(1), 1311–1318.

Bouaricha A and Schnabel RB 1997 Algorithm 768: Tensolve: a software package for solving systems of nonlinear equations and nonlinear least-squares problems using tensor methods. *ACM Trans. Math. Softw.* **23**(2), 174–195.

Breteler M, Meulenbroek R and Gielen S 2002 An evaluation of the minimum-jerk and minimum-torque-change principles at the path, trajectory, and movement-cost levels. *Motor Control* **6**, 69–83.

Broughton P, Fuller R, Stradling S, Gormley M, Kinnear N, O'Dolan C and Hannigan B 2009 Conditions for speeding behaviour: a comparison of car drivers and powered two- wheeled riders. *Transportation Research Part F: Traffic Psychology* **12**, 417–427.

Chen CF and Chen CW 2011 Speeding for fun? Exploring the speeding behavior of riders of heavy motorcycles using the theory of planned behavior and psychological flow theory. *Accident Analysis & Prevention* **43**(3), 983–990.

Corno M, Savaresi SM, Tanelli M and Fabbri L 2008 On optimal motorcycle braking. *Control Engineering Practice* **16**, 644–657.

Cossalter V and Sadauckas J 2006 Elaboration and quantitative assessment of manoeuvrability for motorcycle lane change. *Vehicle System Dynamics* **44**(12), 903–920.

Cossalter V, Da Lio M, Lot R and Fabbri L 1999 A general method for the evaluation of vehicle manoeuvrability with special emphasis on motorcycles. *Vehicle System Dynamics* **31**(2), 113–135.

Cossalter V, Lot R and Maggio F 2004 The modal analysis of a motorcycle in straight running and on a curve. *Meccanica* **39**(1), 1–16.

Cossalter V, Peretto M and Bobbo S 2010 Investigation of the influences of tyre–road friction and engine power on motorcycle racing performance by means of the optimal manoeuvre method. *Proceedings of the Institution of Mechanical Engineers, Part D: Journal of Automobile Engineering* **224**(4), 503–519.

Da Lio M, Cossalter V, Lot R and Fabbri L 1999 The influence of tyre characteristics on motorcycle manoeuvrability *European Automotive Congress, Conference II: Vehicle Dynamics and Active Safety*, Barcelona, Spain.

Davoodi SR, Hamid H, Pazhouhanfar M and Muttart JW 2012 Motorcyclist perception response time in stopping sight distance situations. *Safety Science* **50**(3), 371–377.

Deuflhard P 2011 *Newton Methods for Nonlinear Problems: Affine Invariance and Adaptive Algorithms* Springer Series in Computational Mathematics. Springer.

Harper R and Cooper GE 1986 Handling qualities and pilot evaluation. *Journal of Guidance, Control, and Dynamics* **9**(5), 515–529.

Huth V, Biral F, Martin O and Lot R 2011 Comparison of two warning concepts of an intelligent curve warning system for motorcyclists in a simulator study. *Accident Analysis and Prevention.* **44**(1), 18–25, 2012.

Kelly PD 2008 Lap Time Simulation with Transient Vehicle and Tyre Dynamics PhD thesis Cranfield University School of Engineering. Automotive Studies Grou.

Kirk D and Kreider D 1970 *Optimal Control Theory: An Introduction* Prentice-Hall Electrical Engineering Series. Prentice-Hall.

Kooijman JDG and Schwab AL 2011 A review on handling aspects in bicycle and motorcycle control In *Proceedings of the ASME 2011 International Design Engineering Technical Conferences & Computers and Information in Engineering Conference* (ed. ASME) ASME. paper number DETC2011/MSNDC-47963.

Limebeer D and Sharp R 2006 Bicycles, motorcycles, and models. *Control Systems, IEEE* **26**(5), 34–61.

Limebeer D, Sharp R and Evangelou S 2001 The stability of motorcycles under acceleration and braking. *J. Mech. Eng. Sci* **215**(9), 1095–1109.

Liu D and Todorov E 2007 Evidence for the flexible sensorimotor strategies predicted by optimal feedback control. *J. Neuroscience* **27**(35), 9354–9368.

Lot R and Biral F 2009 An interpretative model of g-g diagrams of racing motorcycle *Proceedings of the 3rd ICMEM International Conference on Mechanical Engineering and Mechanics*, Beijing, P. R. China.

Pacejka, H.B. and Society of Automotive Engineers Series SAE-R 2006 Tire and Vehicle Dynamics. Society of Automotive Engineers, Incorporated.

Paranjape AA and Ananthkrishnan N 2006 Combat aircraft agility metrics – a review. *Time* **58**(2), 143–154.

Pontryagin LS, Boltyanskii VG, Gamkrelidze RV and Mishchenko E 1962 *The mathematical theory of optimal processes (International series of monographs in pure and applied mathematics)*. Interscience Publishers.

Rice RS 1978 Rider skill influences on motorcycle manoeuvering *SAE Technical Papers* SAE International. paper number 78312.

Ross I 2009 *A Primer on Pontryagin's Principle in Optimal Control*. Collegiate Publishers.

Schwab AL and Meijaard JP 2013 A review on bicycle dynamics and rider control. *Vehicle System Dynamics: International Journal of Vehicle Mechanics and Mobility*.

Sexton B, Baughan C, Elliot M and Maycock G 2004 The accident risk of motorcyclists TRL Report no. 607. Technical Report, Transport Research Laboratory, Crowthorne, England.

Stasi LD, Contreras D, Cándido A, Cañas J and Catena A 2011 Behavioral and eye-movement measures to track improvements in driving skills of vulnerable road users: First-time motorcycle riders. *Transportation Research Part F: Traffic Psychology and Behaviour* **14**(1), 26–35.

Viviani P and Flash T 1995 Minimum-jerk, two-thirds power law, and isochrony: converging approaches to movement planning. *J. Exp. Psychology* **21**, 32–53.

Watson B, Tunnicliff D, White K, Schonfeld C and Wishart D 2007 Psychological and social factors influencing motorcycle rider intentions and behaviour. Technical Report RSRG 2007-04, Centre for Accident Research and Road Safety Queensland University of Technology.

Yamakado M and Abe M 2008 An experimentally confirmed driver longitudinal acceleration control model combined with vehicle lateral motion. *Vehicle System Dynamics* **46**(sup1), 129–149.

6

Active Biomechanical Rider Model for Motorcycle Simulation

Valentin Keppler
Biomotion Solutions GbR and Department of Sports Science, University of Tübingen, Germany

Compared with other vehicles such as cars, the mechanical interaction between rider and motorcycle is much closer. Today the dynamics of motorcycles is well understood (Limebeer and Sharp 2006) and even complex multibody models of motorcycles have been realized (e.g., Berritta et al. 2000). But models that consider the influence of the rider on the dynamics of the driver–vehicle system are mostly restricted to upper body lean and steering torques acting between frame and steering head. On the other hand, it is well known that experienced riders are able to damp out weave mode by relaxing their arms, and that things get worse when riders try to stiffen up in order to get control of the motorcycle again. So there is evidently a strong influence from the mechanical coupling between rider and motorcycle on the ride stability. Therefore, to enable the analysis of issues such as motorcycle ride comfort, ride safety and unstable ride modes, we have developed a biomechanical rider model which is able to steer the motorcycle by pushing the handlebars. This approach describes the rider–motorcycle interaction realistically by means of a physical model of an active human rider.

At the beginning of this chapter we describe some basic principles of biomechanical full-body models, human motor control principles and how they can be applied in multibody models to simulate moving virtual humans. In the next section we describe the biomechanical active rider model (Keppler 2010, 2012) which has been developed by Biomotion Solutions and which is available as a control element for SIMPACK, a state-of-the-art multibody system (MBS) simulation package.

Finally, some simulation results are presented to demonstrate how the biomechanical rider model (Figure 6.1) is used to identify parameters of rider vibration characteristics, to investigate the influence of the rider's biomechanical parameters on the dynamic ride stability and to perform some ride simulations with road excitations to calculate the ride comfort. As the

---

Modelling, Simulation and Control of Two-Wheeled Vehicles, First Edition.
Edited by Mara Tanelli, Matteo Corno and Sergio M. Savaresi.
© 2014 John Wiley & Sons, Ltd. Published 2014 by John Wiley & Sons, Ltd.

Figure 6.1 A biomechanical rider model steering a motorcycle by arm movements

biomechanical rider model is an alternate approach compared to the artificial steering-torque approach, we discuss the differences between *steer by torque* and *steer by moving the handlebars*. Both approaches will be compared and the pros and cons will be discussed.

6.1 Human Biomechanics and Motor Control

Biomechanical models of the human operator can be used in a broad variety of application fields. The best known are whole body simulation, models for car crash simulation, in which the exposure and influence of external forces on the human driver can be analysed. As the human body mass and the forces applied from the human to the car are small in comparison to the car, the man–machine interaction is neglected in most cases. On the other hand it has been shown that, surprisingly, the coupling of a pilot and his F-16 fighter is too strong to be neglected. The so-called roll ratcheting is caused by the high amplification of the fighter pilot's steering commands to the aircraft as it is fly-by-wire controlled. Pilot-induced oscillations have been described in Smith and Montgomery (1996). A stopped roll-motion of the plane excited soft tissue oscillations in the pilot's upper extremities are amplified by the fly-by-wire controller which finally lead to unexpected plane oscillations. Clearly, it has to be expected that the rider's motions can not be neglected in the case of rider–motorcycle interaction, so the use of biomechanical full body models will gain more and more importance in the future. The motivation was to develop an active biomechanical rider model that can be used in a broad range of applications. Therefore it should be able to perform standard riding tasks like crossover manoeuvres or cycle passing.

To make the model capable of pushing the handlebars by arm movements a bridge has to be built between virtual rider control algorithms such as the steer by road preview and human motor control which has been used in the past to simulate fast targeted arm movements or the simulation of standing or walking (e.g., Henze 2002; Mergner et al. 2003).

6.1.1 Biomechanics

The mechanical principles of human motion have been the subject of scientific research for a long time. So Giovanni Alfonso Borelli who lived in Italy from 1608 till 1678 described in his book *De motu animalium** how muscles and bones act as systems of lever arms (Figure 6.2), deflection pulleys and ropes, which finally lead to the observed motions of animals and man. Since then a lot of research in the field of biomechanics has been done. With the development of computer simulation techniques biomechanical models of humans have been developed to analyse human motion.

6.1.1.1 The Human Body as a Multibody System

It is commonly accepted to describe the human body as a marionette consisting of 17 rigid bodies for MBS simulations. There are two principal ways to parametrize the body properties for each segment. Of course, parameters like mass, inertia and geometric properties can be determined for one specific subject, which might be of interest for individual computer-aided surgery planning or performance analysis of athletes. But more common are human body

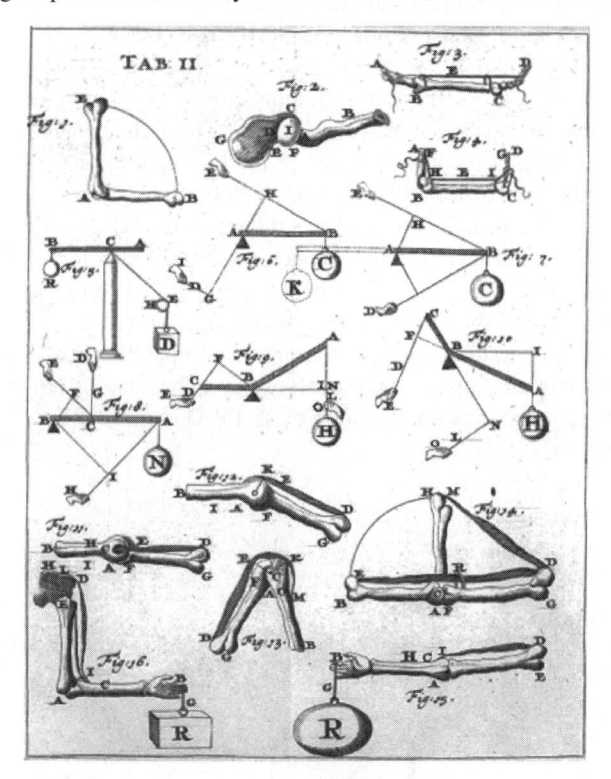

Figure 6.2 One of the oldest biomechanical publications by (Borelli 1680) explains biological motion as a result of muscles and skeleton. Source: Giovanni Alfonso Borelli, De Motu Animalium [Public domain], via Wikimedia Commons

---

* N.B. *De Motu Animalium* (Movement of Animals) is also the title of a text by Aristotle (384–322 BC) about the general principles of animal motion.

models which are based on statistical data. There are some established sets of data which can be used, such as those of (Chandler et al. 1975; Clauser et al. 1969 and NASA 1978).

6.1.1.2 Muscles and Skeleton

As there are no servo motors acting in human limbs, active motion results from contraction of muscles which act between two or more joints. As an additional complication there is no such thing as a pushing muscle, so actuated joints need more than one muscle to enable controlled motion, and this leads to the classification of muscles as agonist or antagonist. Segment motion and dynamically adjusted joint stiffness is realized by fine-tuned muscle contractions of all involved muscles. The interested reader might look at the work of (van Soest 1992), who describes how muscle-skeletal models can be helpful in understanding human movement and how the 'hill type' muscles (Hill 1938), the de facto standard in muscle simulation, can be modelled in biomechanical simulation.

By a defined contraction of agonist and antagonist muscles (Figure 6.3) the segment's position can be held in a stable position and joint stiffness can be modulated by increasing or decreasing of the co-contraction level. Increased co-contraction results in an inherent stability of the body segments against spontaneous disturbances such as impacts during ground contact whilst walking, and this is known as a *preflex*. More information on the major influence of optimally tuned joint stiffness and damping can be found, for example, in (Blickhan et al. 2007). It seems to be expected that a motorcycle rider will adapt co-contraction during manoeuvres or if expecting disturbances due to road condition. At least for car drivers it has been experimentally shown by (Pick and Cole 2006) that the muscle co-contraction is increased during lane change manoeuvres. Even if it might be possible to model the complete body as a musculo-skeletal model, this approach might be too complex as it would entail building up a motor controller which is capable of handling this complex system (consisting of more than 100 joints and about 750 muscles). Co-contraction results in a net joint stiffness at a desired equilibrium point and we are looking at comparable small motion amplitudes around the equilibrium point, so it is a reasonable approach to use a simplified model for our purposes that models the co-contracted muscles as a PID-controller for closed loop control of a joint position.

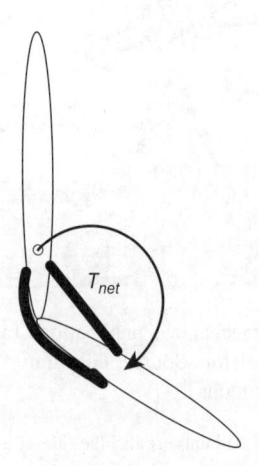

Figure 6.3 Net joint torque resulting of co-contraction of agonist and antagonist muscles

6.1.1.3 The Motion of Soft Tissue

The human body mass contains a significant fraction of soft tissue like muscles, viscera and adipose, it treating as a system of rigid bodies might lead to false simulation results. From analyses of the soft tissue motion during running, for example, it has been shown that wobbling mass oscillations range from 1 to 30 Hz (Schmitt 2011). So, especially if MBS models of humans are used to simulate impacts (e.g. drop jumps) or vibration analyses (hand–arm vibration, ride comfort) pure rigid body models lack accuracy. To solve this problem the so-called *wobbling mass model* was introduced in (Denoth et al. 1985 and Gruber et al. 1987). The wobbling mass model has been shown to be important in the simulation of impact dynamics (e.g. Gruber et al. 1998; Keppler and Günther 2006) and in ride comfort analysis (Mutschler et al. 2004).

6.1.2 Motor Control

Having modelled the 17-segment-body with joint actuators alone is not sufficient to perform forward dynamic simulation of movements. The model also has to implement control structures, which makes the calculation of appropriate joint torques possible, and this in turn leads to the desired motion pattern. The way humans control their motions is the subject of the research of human motor control. In the following, some basic aspects of how the human body processes sensory input in order to find a suitable muscle activation pattern to perform desired motions, we will discuss.

6.1.2.1 The Input

As everyone knows our senses can be dazzled. One obvious example is a subject standing on one leg in front of a vertically patterned wall. If this pattern is suddenly shifted sideways the sensory signals are erroneous and the subject tends to fall over. If the subject closes their eyes the overall stability is a little bit poorer but upright standing is still possible. This shows that sensory input stimulus is redundant. Even simple human motions such as standing upright can be hard to model, but there are simulation models that deal with human sensor fusion and motor control (Maurer et al. 2006; Mergner et al. 2003).

An interesting early work on the simulation of a bicycle rider model was published by (Doyle 1987) who investigated blindfolded bicycle riding (Figure 6.4). He recorded the bicycle's motion by use of gyroscopes and the steering angle changes resulting from the rider's input. He then reproduced the rider's steering input in a computer model and described the steering pattern of the rider as a combination of delayed repeat of the roll rate and short intermittent ballistic accelerations. As this rider model just uses the vestibular and proprioceptive sensor channels steering responses might be smoother if the experiment were reproduced with the same bike but with eyes open because the vestibular input is based on accelerations, so slow low amplitude movements with respect to the reference system can not be detected properly.

Doyle (personal communication) states, 'The main thing the visual system provides is a clear indication of the vertical angle which the vestibular system is not good at. If the rider can see that the vertical is being approached he can anticipate with a counter correction. Blindfolded he either tends to go through the vertical and only picks up confirmation as the acceleration builds up on the other side or the current input fails to drive the bike as far as the upright and it starts to fall back on the same side.'

Figure 6.4 Cyclist riding blindfolded. Source: Doyle 1987. Reproduced with permission from Anthony Doyle

6.1.2.2 The Output

The next question is how we process the combined and weighted sensor signals to muscle stimulus patterns. There are a lot of muscles that have to be stimulated in a coordinated pattern to shift a single joint position, so we are dealing with highly overdetermined systems. The same problem occurs at the level of multi joint movements. In order to control the steering angle, the motorcycle rider needs to change about 14 joint angles, so the question is how the human handles this problem. One approach to explain how human motor control deals with overdetermined problems is the λ-model of the equilibrium point hypothesis which was published by (Feldman 1974). Basically it describes the idea that for any given joint space configuration there exists a set of muscle lengths in which this position is at a mechanical equilibrium point. So by giving a desired motion it should be possible to calculate the corresponding target muscle sets and stimulation patterns.

It has been said that there is a need of more detailed internal models (van Ingen Schenau et al. 1995) which are probably located in the cerebellum. So (Pellionisz and Llinas 1985), for example, postulated that the cerebellum can be seen as a metric tensor which transforms between different referent systems. In reality a motorcycle rider uses vision, the vestibular system,[†] the proprioceptive channels and the sensations that come from the machine to hands and buttocks to calculate the actual motion state of his motorcycle and then he uses this information by means of internal models to interpolate into the future in order to initiate suitable roll motions at the right time by counter steering.

For our motorcycle rider simulation we generalized the ideas about an internal model of body metrics and the λ-model of the equilibrium point hypothesis and postulated the

---

[†] For the details of the mechanics of the human vestibular system and a description of otholites see for example (Jäger and Haslwanter 2004).

joint space model (JSM) which handles the sensor fusion and copes with redundancy to calculate desired joint trajectories for a given motion target. This includes the idea that for every steering angle there exists a posture vector in human joint space which fulfils the boundary conditions. We also decided to use as input a fusion of visual input (road preview) and vestibular input (roll angle) to calculate target steering angles as output from the motor control unit.

6.2 The Model

Motorcycle dynamics is well understood by analytical models (e.g. Limebeer and Sharp 2006) which, due to the chosen analysis method, require stringent model reduction. Nowadays, it is even possible to simulate complex motorcycle models based on real existing designs (e.g., Berritta et al. 2000). By using one of the available industry standard multibody simulation packages even complex three-dimensional simulations can be handled comfortably. The underlying multibody package used in our simulation is SIMPACK (SIMPACK AG, Gilching Germany) which is often used in automotive or wheel–rail simulation. The simulation is based on three models: the human body, the motorcycle and the motion control model.

6.2.1 The Human Body Model

Based on anthropometrical data (e.g., NASA 1978) we have implemented a 17-segment full body model. The model consists of two legs (foot, shank and thigh), a three-part trunk, neck, head and two arms (hand, fore arm, upper arm) as depicted in Figure 6.5. Furthermore, each of these rigid bodies is coupled with a so-called wobbling mass, which takes into account that human body tissue is not a rigid material (see Figure 6.6). The consideration of wobbling masses is important for ride comfort simulations (Keppler and Günther 2006; Mutschler et al. 2004).

Figure 6.5 Human body model consisting of 17 segments. Automatic model generation allows for variation of anthropometric parameters such as stature, body weight and gender

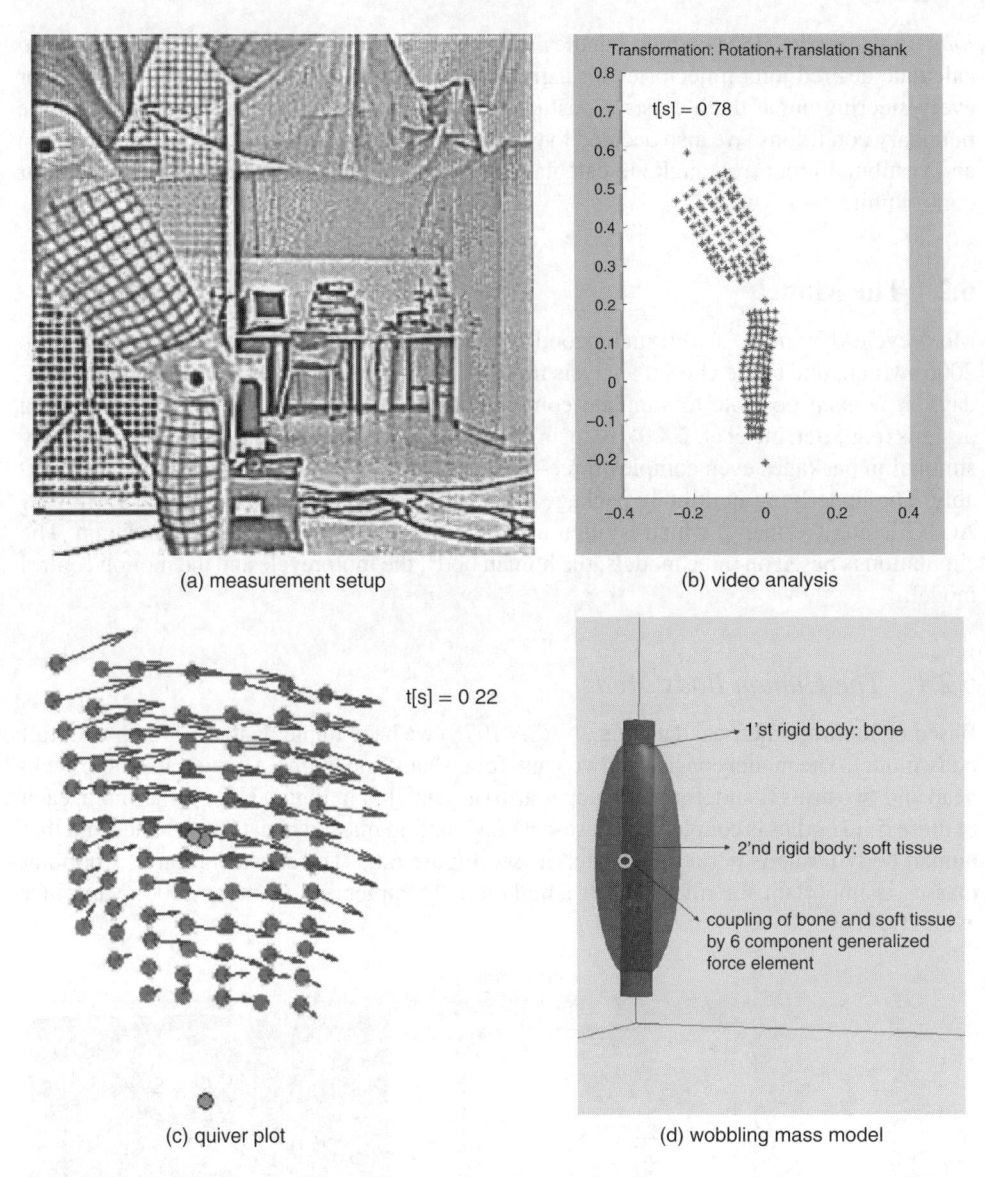

Figure 6.6 Wobbling mass model: A second rigid body, which simulates the soft tissue, is coupled to the bone. The wobbling mass can rotate and translate with respect to coupling marker and is connected by a nonlinear force element (d). The pictures show a measurement setup (a) which was used to visualize and quantify the motion and frequencies of wobbling masses as presented by (Keppler and Günther 2006). The depicted grids on the shank and thigh have been filmed at high speed and the coordinates have been tracked. The motion of the grid points is visualized as a quiver plot (c). (Source: Keppler V and Günther M 2006. Reproduced with permission from Elsevier)

A human body model can be built up from scratch, piece by piece using the SIMPACK preprocessor, while each anthropometric parameter like mass or inertia is calculated by means of regression equations. But due to the large number of model parameters, building a human body model is an elaborate and error-prone task. It might be interesting to generate a range of different rider anthropometrics so it is of benefit to use an automatic model wizard (e.g., Varibody, Biomotion Solutions figure as shown in Figure 6.5) to generate the human body model based on user input of gender, stature and body weight.

6.2.1.1 Wobbling Masses

$$Fel\,(r_{i,\ i\in\{x,y,z\}}) = A_c * (Cnl_i * \Delta r_i^{Se_{ct}} + C_i * \Delta r_i) \tag{6.1}$$

$$Fdiss\,(r_{i,\ i\in\{x,y,z\}}) = A_c * \left(Dnl_i * \frac{d}{dt}(r_i)^{Se_{dt}} + D_i * \frac{d}{dt}(r_i)\right)$$

$$Tel\,(r_{i,\ i\in\{al,be,ga\}}) = A_c * (Cnl_i * \Delta r_i^{Se_{cr}} + C_i * \Delta r_i)$$

$$Tdiss\,(r_{i,\ i\in\{al,be,ga\}}) = A_c * \left(Dnl_i * \frac{d}{dt}(r_i)^{Se_{dr}} + D_i * \frac{d}{dt}(r_i)\right)$$

$$F\,(r_{i,\ i\in\{x,y,z\}}) = Fel_i + Fdiss_i$$

$$T\,(r_{i,\ i\in\{al,be,ga\}}) = Tel_i + Tdiss_i$$

Parameter symbols are explained in Table 6.1.

Table 6.1 Symbols for wobbling mass force law

Symbol	Unit	Description
F_i	N	Force acting in i
T_i	Nm	Torque acting about i
Fel_i	N	Force elastic in i
$Fdis_i$	N	Force dissipative in i
Tel_i	Nm	Torque elastic in i
$Tdis_i$	Nm	Torque dissipative in i
Δr_i	m	Delta translation in r_i $i \in \{x,y,z\}$
$\frac{d}{dt}(r_i)$	m/s	Velocity translation in r_i
Δr_i	rad	Delta rotation r_i $i \in \{al,be,ga\}$
$\frac{d}{dt}(r_i)$	rad/s	Velocity rotation in r_i $i \in \{al,be,ga\}$
A_c	m^2	Cross-sectional area scaling
C_i	N/m	Stiffness linear in i
Cnl_i	N/m^{Se}	Stiffness non-linear in i
Se		Exponent nonlinear term
D_i	Ns/m	Damping linear in i
Dnl_i	Ns/m^{Se}	Damping nonlinear in i
C_a	Nm/rad	Stiffness linear rotational
Cnl_a	Nm/rad^{Sr}	Stiffness nonlinear rotational

Table 6.2 Symbols for joint actuators

Symbol	Description
K_P	proportional gain
K_D	derivative gain
K_I	integral gain

6.2.1.2 Joint Actuators

To enable both the passive posture maintenance and the active motion generation, actuators have to be implemented at each joint of the human body model. The torques which are acting between the two coupling points at the segments are calculated by Equation 6.2, and the actuator works at three different modes:

- passive torque for posture maintenance
- active motion defined by motorcycle rider controller
- torque/force by motorcycle rider controller

In the "defined motion" case, PID controllers are used to let the joint "degree of freedom (DOF)" follow set values. The formula for the PID controllers are as follows: The set point (SP) is the value to be reached by the controller, the process variable (PV) is the measured value. The only variable used by the controller is $\Delta(t) = SP - PV$.

$$P_{out}(t) = K_P * \Delta(t) + K_D * \frac{d}{dt}(\Delta(t)) + K_I * \int_{-\inf}^{t} \Delta(t') dt' \qquad (6.2)$$

The parameters are explained in Table 6.2, and suitable parameter values have been identified by comparison with measurement data found in literature.

6.2.1.3 Grip Force

There are grip force models described in literature (e.g., Dong et al. 2007) which are mainly used to calculate vibration issues. We implemented a simplified grip force model as a three-component force and torque model as described in Equation 6.3 and parameter symbols are declared in Table 6.3. Parameter values for the grip force model have been taken in accordance with literature data (Dong et al. 2007).

$$F_i\left(r_i, \frac{d}{dt}(r_i)\right)_{i \in \{x,y,z\}} = C_t * (r_i - r_i^{offset}) + D_t * \frac{d}{dt}(r_i)$$

$$T_i\left(r_i, \frac{d}{dt}(r_i)\right)_{i \in \{al,be,ga\}} = C_r * (r_i - r_i^{offset}) + D_r * \frac{d}{dt}(r_i) \qquad (6.3)$$

6.2.1.4 Seating Force

The seat contact force has been modelled as a simplified sphere-to-plane contact, as described in Equation 6.4. To ensure accurate results, the force element uses the root

Table 6.3 Symbols for grip force law

Symbol	Unit	Description
F_i	N	Force acting in i
T_i	Nm	Torque acting about i
Δr_i	m	Delta translation in r_i, $i \in \{x, y, z\}$
$\frac{d}{dt}(r_i)$	m/s	Velocity translation in r_i
Δr_i	rad	Delta rotation r_i, $i \in \{al, be, ga\}$
$\frac{d}{dt}(r_i)$	rad/s	Velocity rotation in r_i, $i \in \{al, be, ga\}$
C_i	N/m	Stiffness linear in i
D_i	Ns/m	Damping linear in i
C_a	Nm/rad	Stiffness linear rotational
D_a	Nm/rad	Damping linear rotational

switching functionality from the SODASRT solver from SIMPACK. Tangential forces are calculated with a velocity regularization friction model.

$$F_z\left(z, \frac{d}{dt}(z)\right) = \begin{cases} z \geq 0 : 0 \\ z < 0 : C_z * z + Fd_z \end{cases} \tag{6.4}$$

$$v_{\tan} = \sqrt{(\sum \frac{d}{dt}(r_i)^2)}_{|i \in \{x,y\}}$$

$$F_{\tan}\left(r_i, \frac{d}{dt}(r_i)\right)_{i \in \{x,y\}} = \begin{cases} F_z \geq 0 & : 0 \\ F_z < 0 \wedge v_{\tan} > v_{reg} & : \mu * F_z \\ F_z < 0 \wedge v_{\tan} \leq v_{reg} & : \mu * F_z * (v_{\tan}/v_{reg}) * (2 - v_{\tan}/v_{reg}) \end{cases}$$

$$F_i\left(r_i, \frac{d}{dt}(r_i)\right)_{i \in \{x,y\}} = F_{\tan}\left(r_i, \frac{d}{dt}(r_i)\right) * \left(\frac{d}{dt}(r_i)/v_{\tan}\right)$$

$$Fd_z\left(z, \frac{d}{dt}(z)\right) = \begin{cases} \frac{d}{dt}(z) \geq 0 : 0 \\ \frac{d}{dt}(z) < 0 : D_z * \frac{d}{dt}(z) * \begin{cases} z \geq l_{dt} : 1 \\ z < l_{dt} : z/l_{dt} \end{cases} \end{cases}$$

6.2.2 The Motorcycle Model

The motorcycle model is built as a 3D multibody system using SIMPACK as simulation platform. The bike consists of a fork-mounted front wheel connected to the steering axis by a one-DOF prismatic joint. The steering is connected by a one-DOF hinge joint to the frame. The swing-arm-mounted rear wheel also has one DOF of rotation (see Figure 6.7). Parameters for inertia and masses were chosen in accordance with commonly used literature. Suspension systems were modelled as linear spring damper force elements.

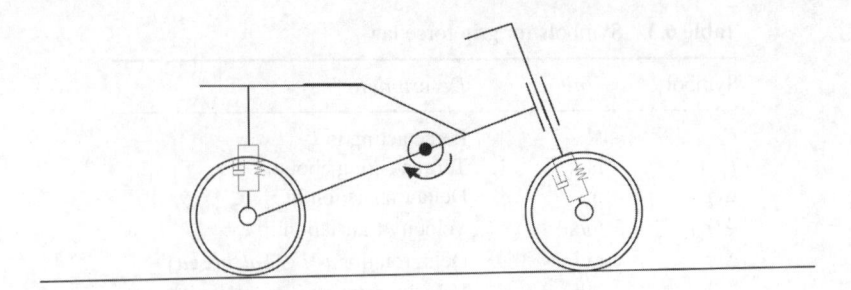

Figure 6.7 Motorcycle model consisting of six rigid bodies

6.2.3 Steering the Motorcycle

It has been shown that a motorcycle model can be enabled to follow a track or even perform optimal manoeuvres which result in minimized lap time. The common approach is to control the roll angle by a torque that acts between the frame and the steering. This is a phenomenological model which has to be restricted to fit real human riders' abilities such as physical strength and reaction times. So, commonly torques are saturated to a maximum torque which is in agreement with human riders. As the roll angle to torque transformation is almost instantaneous there is also a need to low-pass filter the steering torque to a value of about 5–8 Hz cut-off. Generally, there are two different approaches to get the desired roll angle which has to be set to perform ride simulations. The first approach is to let the virtual rider 'look' at a point in front of its actual position and to control the roll angle with respect to the expected lane position error. This method has the benefit that it is quite flexible and needs no prior 'optimal ride' calculation. The second variant results in optimal manoeuvre with respect to a given objective function (lowest fuel consumption or shortest lap time). This method requires calculating the roll angle and speed for each position on the track prior to the MBS simulation by an optimal control approach, such the H_∞.

We decided to use the *road preview approach* as the target for our rider model was to analyse the mechanical interaction between rider and motorcycle. We don't want to reproduce a minimum lap time, but are interested in how the rider–handlebar coupling influences the weave and wobble modes. For this approach, the road preview approach is suitable.

6.2.3.1 JSM–Joint Space Model

We have implemented passive and active actuators in this model. The lower extremities, the trunk and head–neck, are stabilized by passive impedances whose parameters are chosen according to literature values (e.g., Keppler 2003). To enable the rider model to control the handlebars, shoulders and elbows are actuated via muscle torque generators. Muscle torques are generated by PID controllers in the joints, which take the desired joint space configuration as a set value. Their input is provided by the Biomotion motorcycle rider model controller. The controller layout follows the scheme in Figure 6.8 and extends roll-angle control, as described, for example, by (Cossalter and Lot 2006), to a joint space model controller. For every desired steering angle a $R^1 \Rightarrow R^n$ transformation has to be provided by the controller (where n gives the number of controlled segment angles). As discussed earlier,

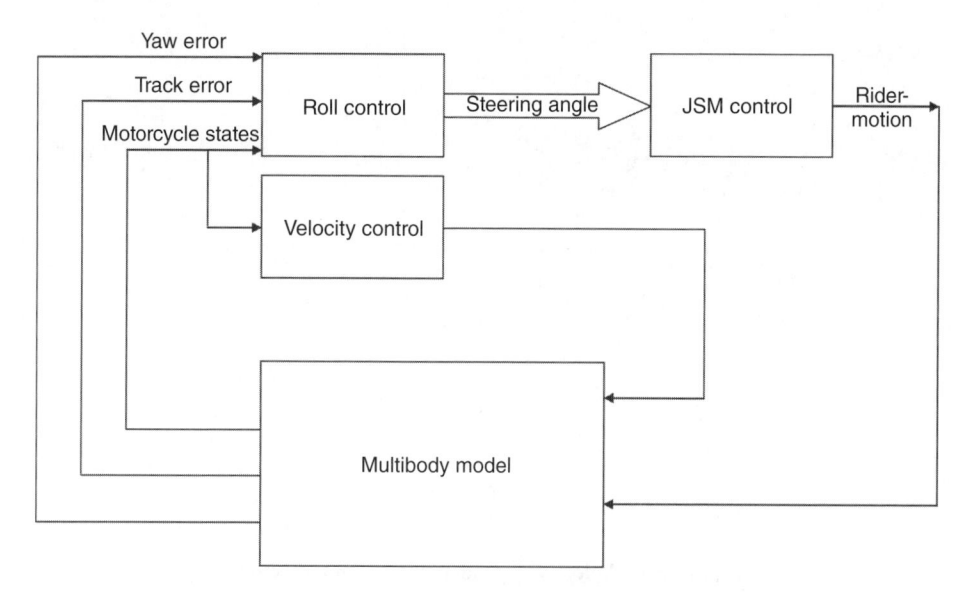

Figure 6.8 Human body model consisting of 17 segments

the human brain is capable of performing coordinated transformations, and manages internal mechanic models, so we can see that the JSM as a kind of low-level model of the cerebellum.

To avoid any issues which may result on a co-simulation approach, we implemented the complete rider controller as a SIMPACK control loop element, which is commercially available as a SIMPACK add-on module. Appropriate roll angles can be computed by PID controllers using as input the lateral track error of the bike's position and the yaw-angle, which describes the difference between the bike's direction and the curvature of the road trajectory. The model uses a speed-dependent road preview (SIM 2011) with a proportional preview time $t_{preview}$ of 1–2 seconds. The calculation of the preview point follows Equation 6.5 and is explained by Figure 6.9. PID parameters have to be chosen carefully to ensure stable operation. Due to the fact that in heavier motorcycles the influence of the (counter) steering is the dominant mode (Sharp 2007) we have not yet implemented a lean control.

$$s_{preview} = s_{motorcycle} + s_0 + t_{preview} * \dot{s}_{vehicle} \qquad (6.5)$$

Speed control is currently implemented as a PID control which allows the model to maintain the predefined set value by applying an appropriate torque at the rear wheel.

6.3 Simulations and Results

With the rider–motorcycle model we conducted a set of simulations. As a first step we identified the joint impedances for a motorcycle rider by comparing the frequency response function (FRF) of the rider with measured data from literature (Cossalter et al. 2006). With this validated model we performed a variation of biomechanical factors and analysed their influence on the ride stability and ride comfort.

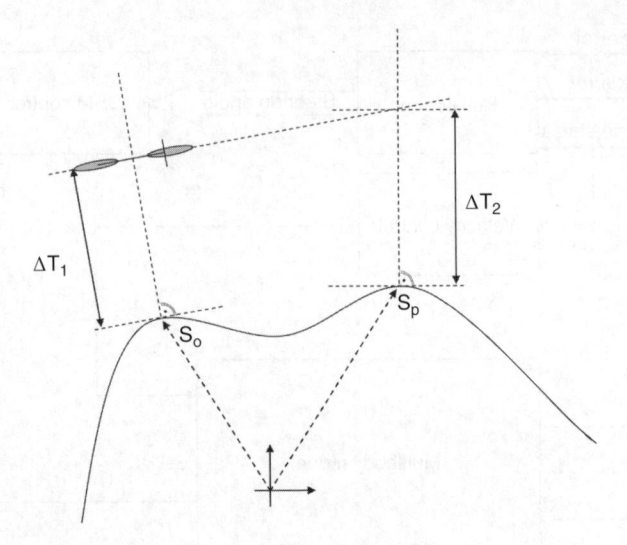

Figure 6.9 Velocity-dependent track preview sensor

6.3.1 Rider's Vibration Response

Values for joint stiffness and damping vary over a broad range in the biomechanical literature. In literature (Cossalter et al. 2006) measurement data describing the rider's FRF (cross spectrum from steering axis acceleration and excitation torque) to steering axis excitation can be found. Cossalter (2006) matched the measured data using a reduced rider model and optimized the parameters for inertia, stiffness and damping. For the whole-body model we reproduced this measurement of the FRF in the simulation of the biomechanical rider model. We performed some parameter variations to evaluate the influence of segment stiffness of trunk and hand–arm system (Figures 6.10 and 6.11). The biomechanical rider model showed a good accordance to the measurements. The FRF plots show that the stiffness of the hand–arm system influences the location of the first maximum of the FRF between 1 and 4 Hz, a frequency which is near the wobble mode of motorcycles. Thus it has to be expected, that the muscular co-contraction has a significant influence on motorcycle weave mode.

6.3.1.1 Comparing FRF to the Steer by Torque

We wanted to take a first look at the differences between the steer by torque and the biomotion model, so we performed the mock-up simulation with a torque-controlled excitation of the handlebars in order to calculate the FRF by cross-spectrum of the excitation torque and angular acceleration of the steering head. As the literature values for the PD gains of a torque control vary over a broad range, we varied them in the simulation too. As expected, the torque controller shows a simpler structure (Figure 6.13) in the FRF compared to the FRF of the biomechanical model (Figure 6.12). By using high gains in the torque controller it is possible to shift the resonance out of the range of wobble and weave. As the stiffness, mass and inertia properties of the human body model are restricted to a physiologically reasonable range, the critical resonances in the biomotion model are given explicitly by a physical model.

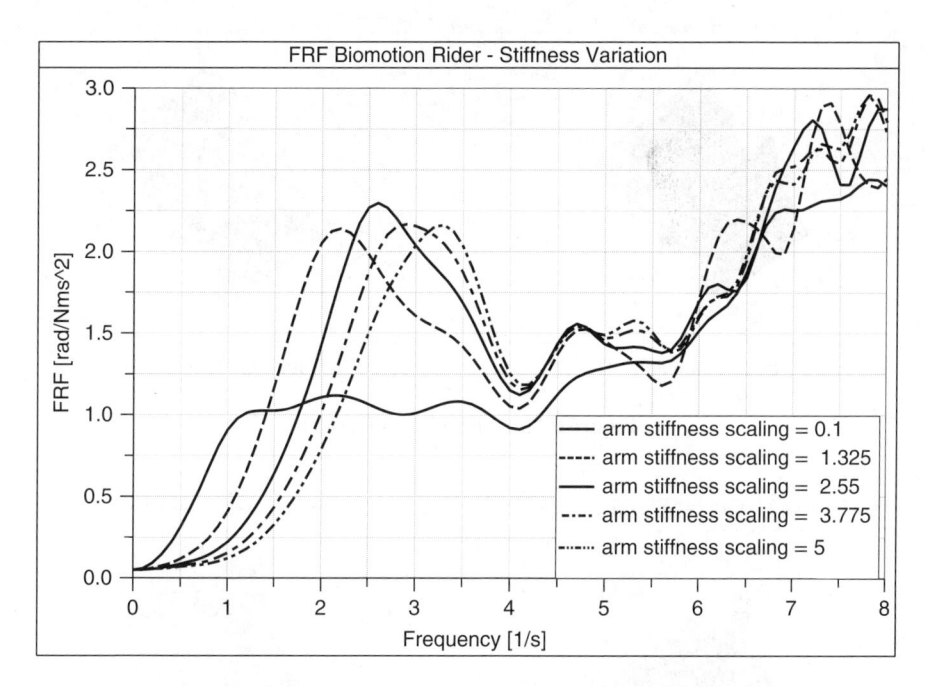

Figure 6.10 Parameter variation of the arm joint stiffness

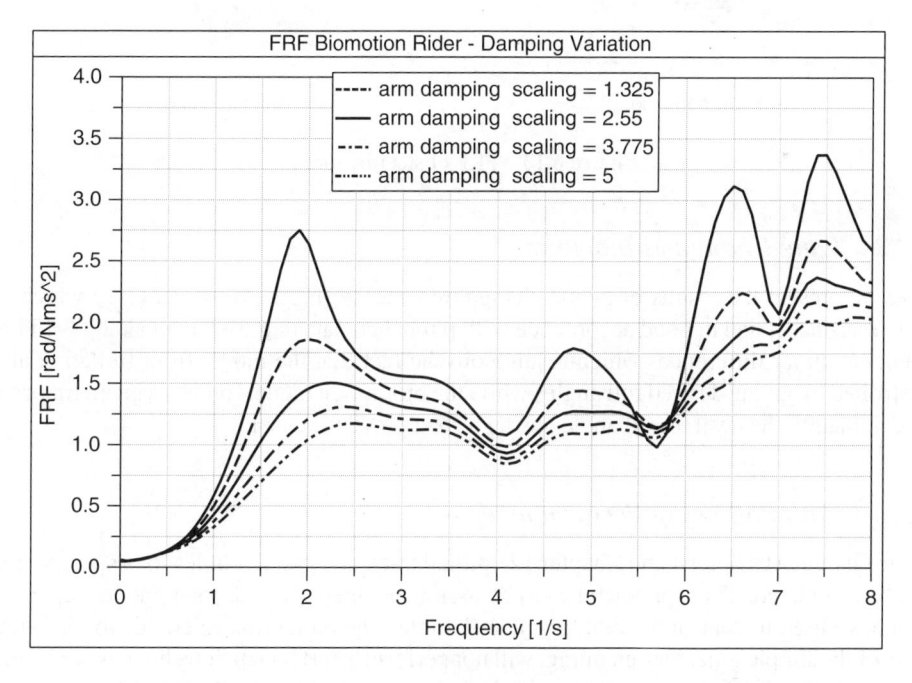

Figure 6.11 Parameter variation of the arm joint damping

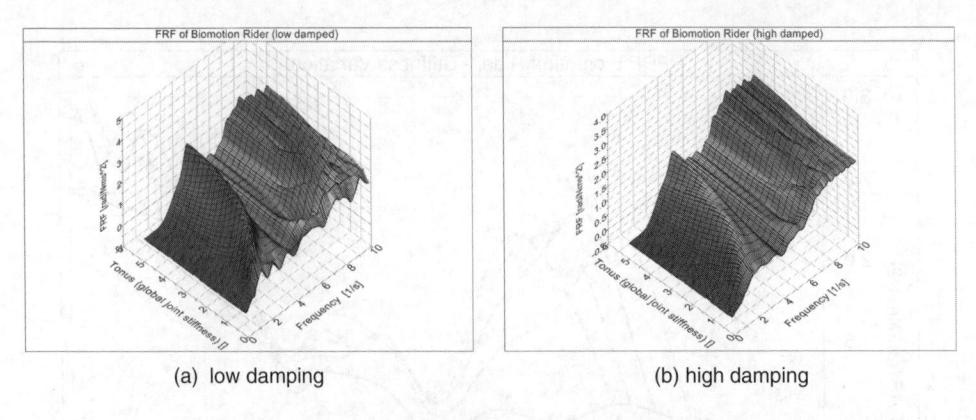

Figure 6.12 FRF biomotion rider

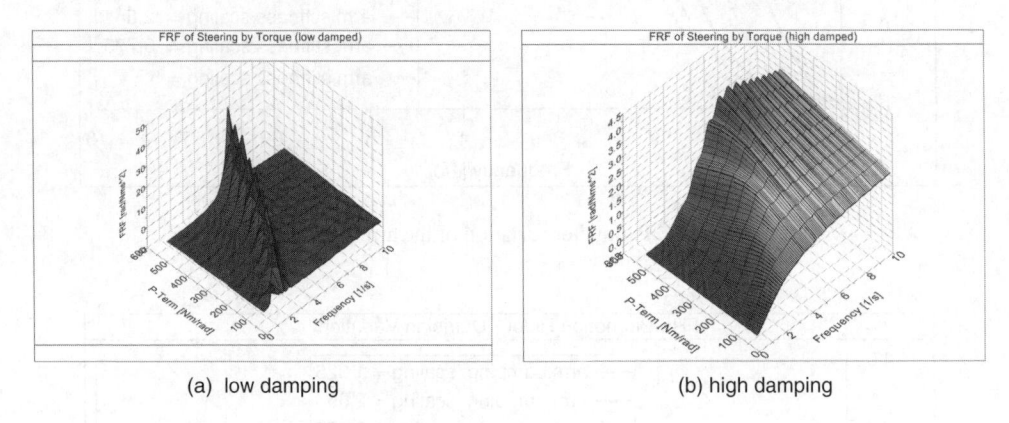

Figure 6.13 FRF of steer by torque

6.3.2 Lane Change Manoeuvre

First we present the results of a lane change manoeuvre (Figure 6.14 and 6.15) which are quite reasonable and in good accordance with published data (e.g. Bellati et al. 2003). There are some minor differences which result from the fact that the model from Bellati et al. is controlled by giving desired roll angles while the biomotion model follows a given trajectory by calculating the desired roll angles using a road preview sensor.

6.3.3 Path Following Performance

As discussed in Chapter 4 and Chapter 12 optimal track and speed profiles for a specific track can be calculated. This approach has to be assumed superior to the road preview approach which we used to control the motorcycle roll angle. The biomotion rider, due to the limitations of the simple trajectory planning, will not perform a perfect lap time, but it is considered to be useful to simulate average riders analysing vehicle handling, ride stability or ride comfort. To get an impression of the limitations and the capabilities of the biomotion rider with

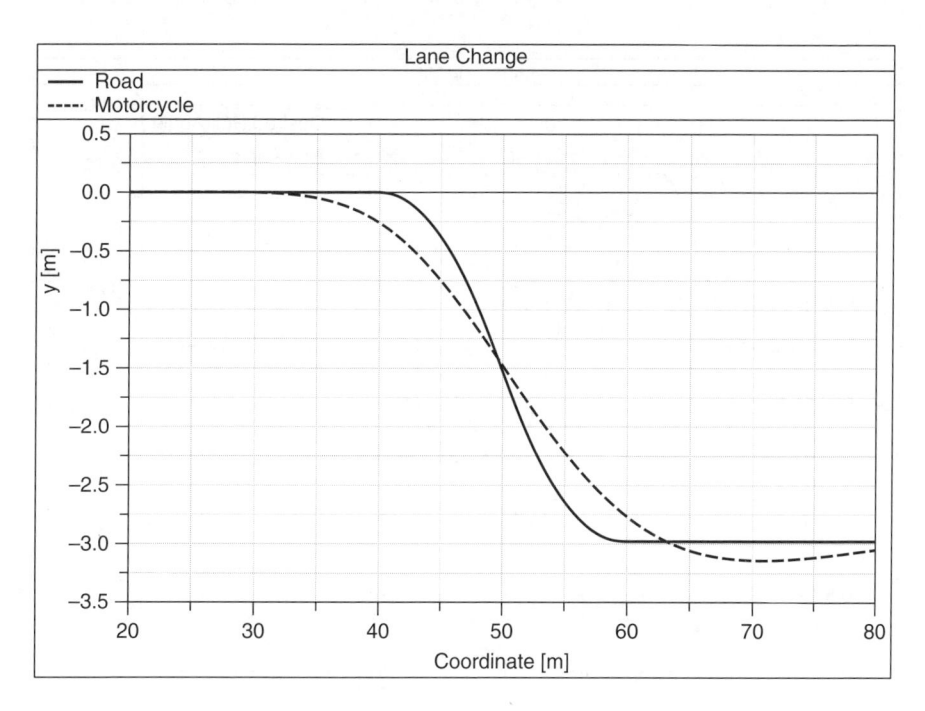

Figure 6.14 The position of the motorcycle during a lane change manoeuvre and the given track

road preview we did some comparisons with the results of optimal manoeuvre methods on two different test tracks (Figure 6.16).

While the performance of the biomotion rider model with simple road preview on test track 1 is acceptable (Figures 6.17 to 6.19, the result shows a minor problem at position 825 m. The motorcycle is too fast, which leads to a peak in roll angle. The tyres reached the physical limits probably because the deceleration wasn't initiated soon enough or the distribution of braking torque between front and rear wheel has been suboptimal. Test track 2 (the course at Adria) seems to be more demanding for the speed and track strategy, so the road preview control only could perform at 60% of optimal speed on this track (Figures 6.20 to 6.22. Note that we used the optimal manoeuvre track as input for the controller. The reason for the limitation to 60% speed on the second test track was that the goal was to perform the complete track. As the biomotion model had the most problems with a chicane near the 1400 m point the speed factor had to be reduced for this track to the value of 60%.

In particular, the calculation of an appropriate speed profile seems to be crucial for the simulation. The current implementation of the speed control of the biomotion rider is a combination of linear roll angle dependency, pitch control to prevent wheelie and stoppie and a fixed braking torque distribution between front and rear wheel. This approach is too simple to show good performance in lap time simulation, especially because the linear roll dependency has the effect that the rider starts braking too near to the corners, which often leads to unstable riding. It can be expected that taking into account more than one preview point would help to enhance lap time performance.

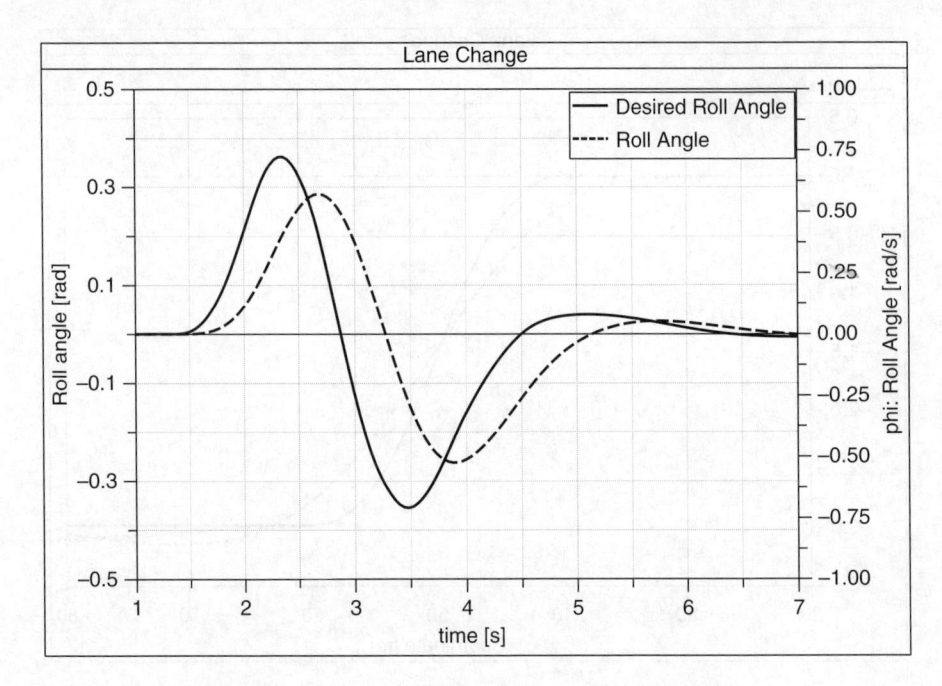

Figure 6.15 The motorcycle roll angle during lane change

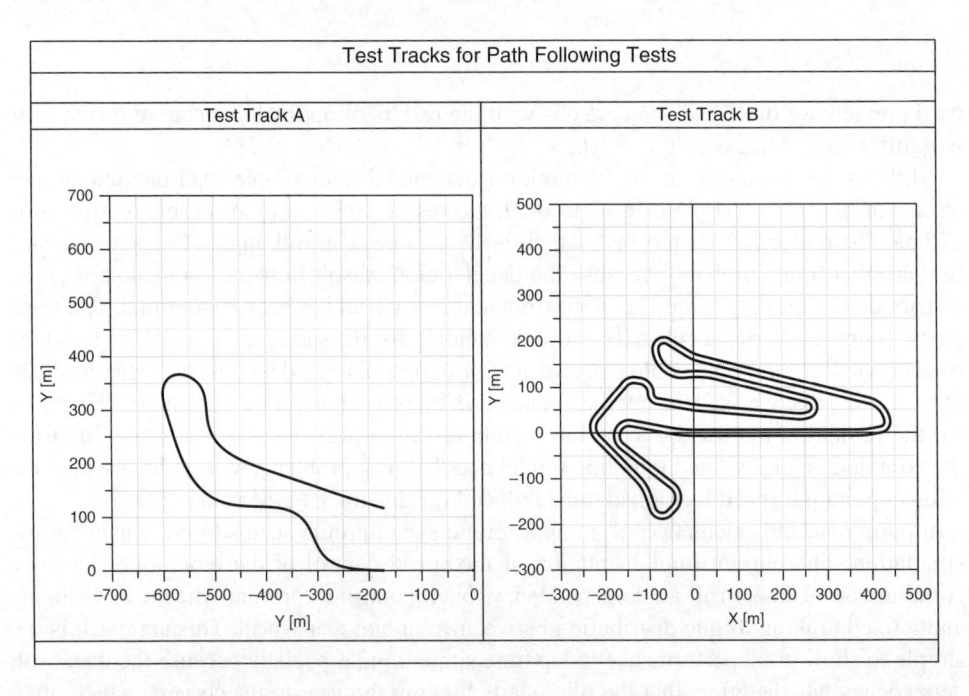

Figure 6.16 Two different tracks have been used to compare the biomotion rider with optimal manoeuvre methods

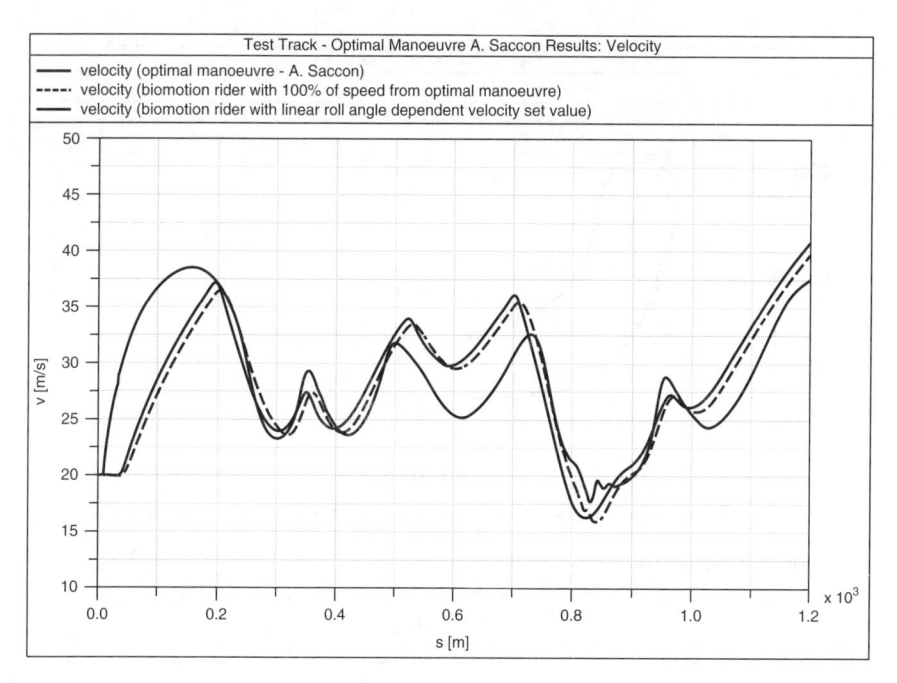

Figure 6.17 The velocity profile of the biomotion rider compared to the optimal manoeuvre results as described in Chapter 4

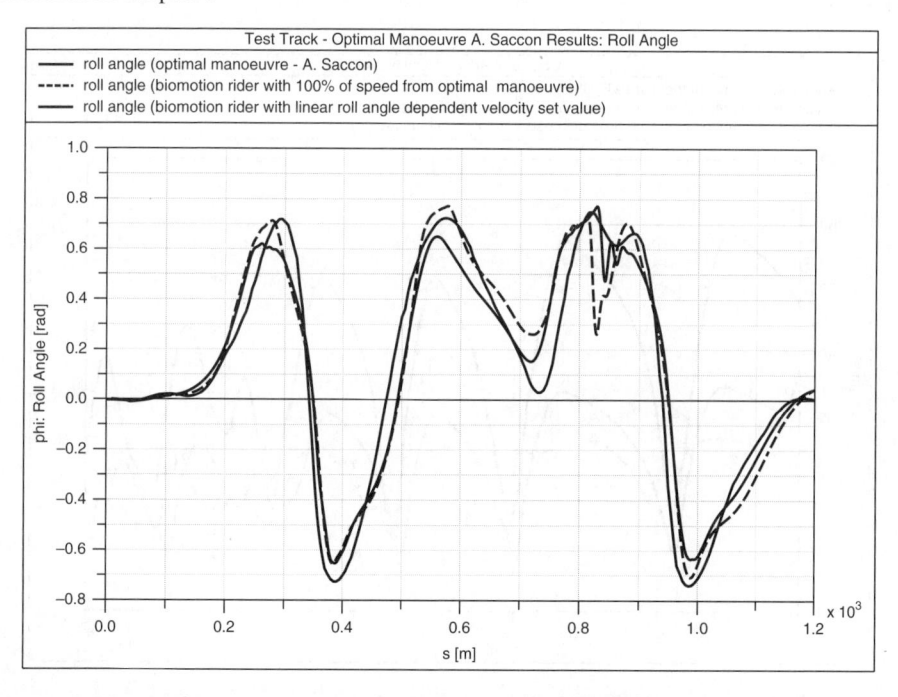

Figure 6.18 The roll angle profile of the biomotion rider compared to the optimal manoeuvre results as described in Chapter 4

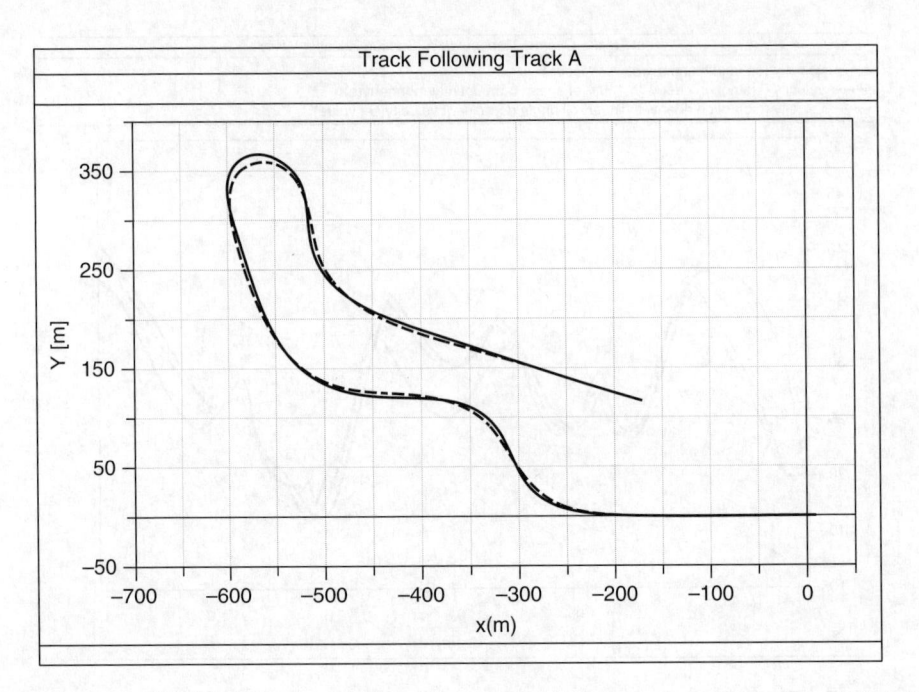

Figure 6.19 Comparison of the performed track and the optimal manoeuvre results described in Chapter 4

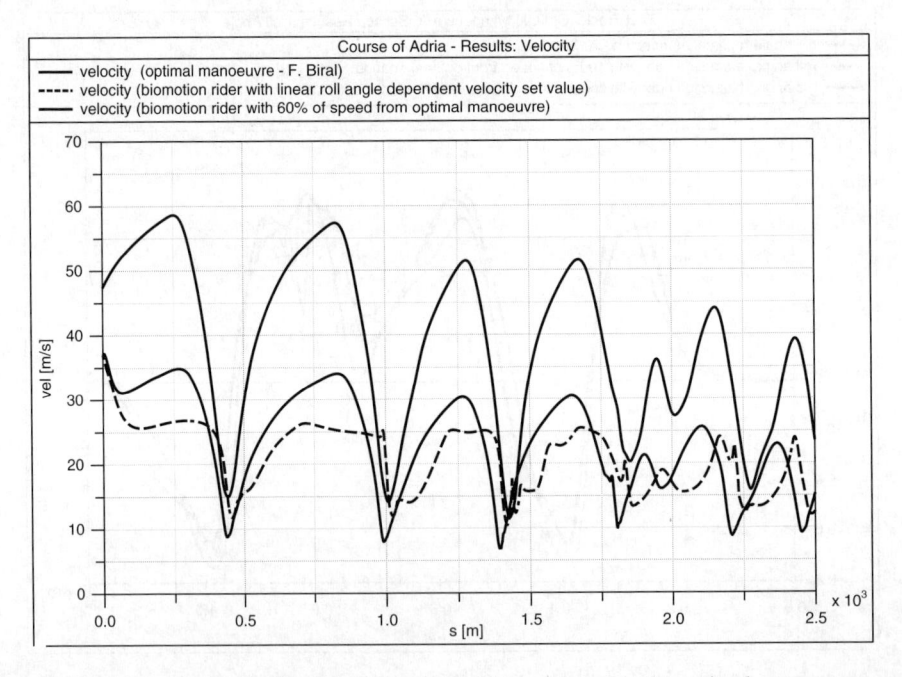

Figure 6.20 The velocity profile of the biomotion rider compared to the optimal manoeuvre results as described in Chapter 12

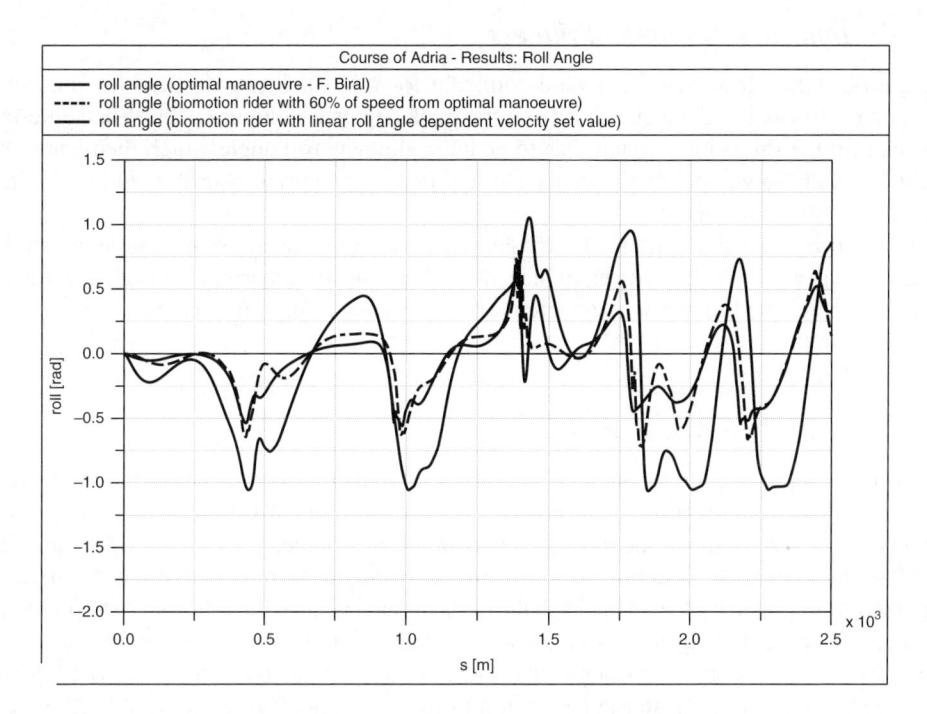

Active Biomechanical Rider Model for Motorcycle Simulation 175

Figure 6.21 The roll angle profile of the biomotion rider compared to the optimal manoeuvre results

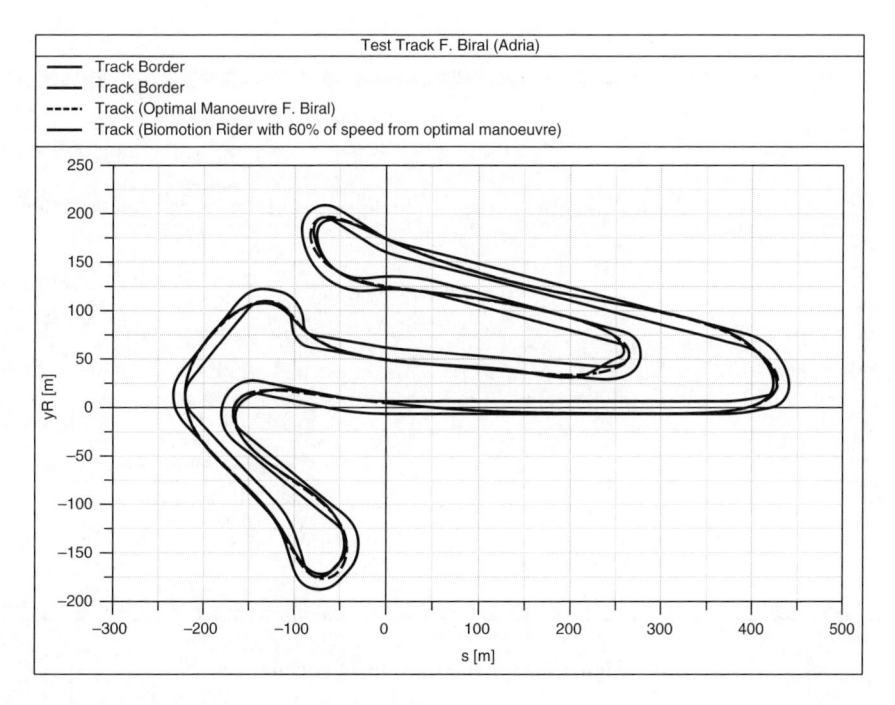

Figure 6.22 Comparison of the performed track to the optimal manoeuvre results using 60% scaled speed profile from optimal manoeuvre

6.3.4 Influence of Physical Fitness

To gain insight into which factors are dominant for ride stability we varied the muscular tension of the rider under a circuit ride on the track depicted in Figure 6.23. When entering the circuit, the rider–bike system has to adapt to the new roll angle which then has to be held constant. So we took the parameters identified in the previous section and multiplied them by a 'strength' factor.

As can been seen in Figure 6.24 the rider–bike system tends to oscillate more when the rider's physical strength is lower (which may also indicate a correlation with the rider's age). We performed the same parameter variation with a crossover manoeuvre confirming this tendency.

6.3.5 Analysing Weave Mode

Motorcycle and rider are a coupled system whose dynamics emerge from the interaction between them. Experienced riders report the possibility of provoking or damping down highly dangerous unstable ride modes such as weave or wobble. We used our rider model to analyse the rider–bike system near weave mode. Riding at 60 m/s, a transient disturbance moment was applied by a sudden hip shake of the rider. The bike then showed a short latency in which a negative damped oscillation showed up, which finally led to exponential rising amplitude in yaw angle and to crashing. Furthermore, we analysed the sensitivity of the weave phenomenon to the muscular tension of the rider's body as depicted in Figure 6.25. The results predict strong influence of biomechanical factors on ride dynamics.

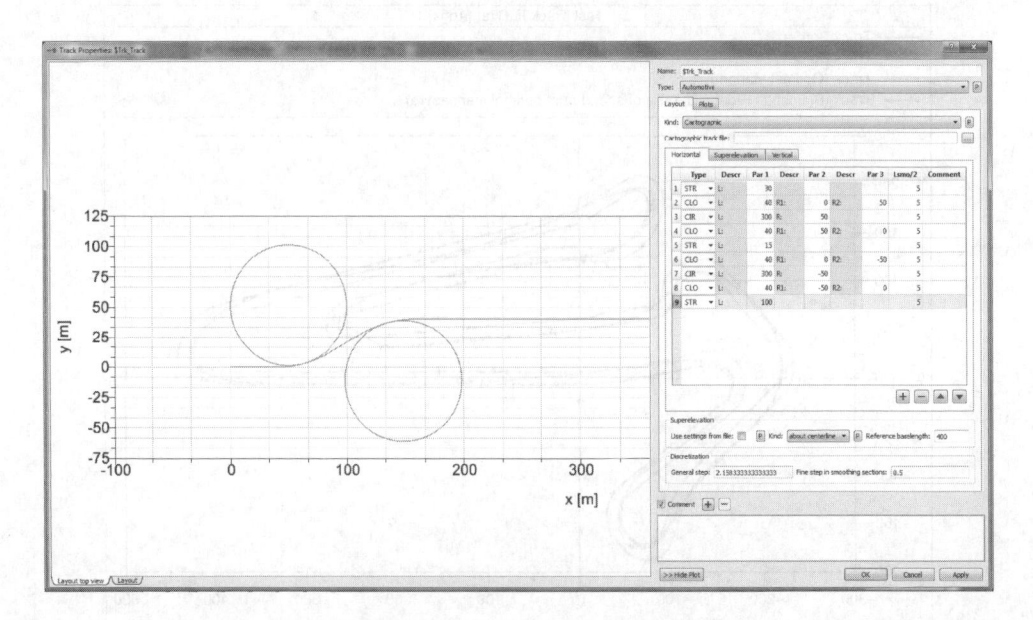

Figure 6.23 Track profile of circuit

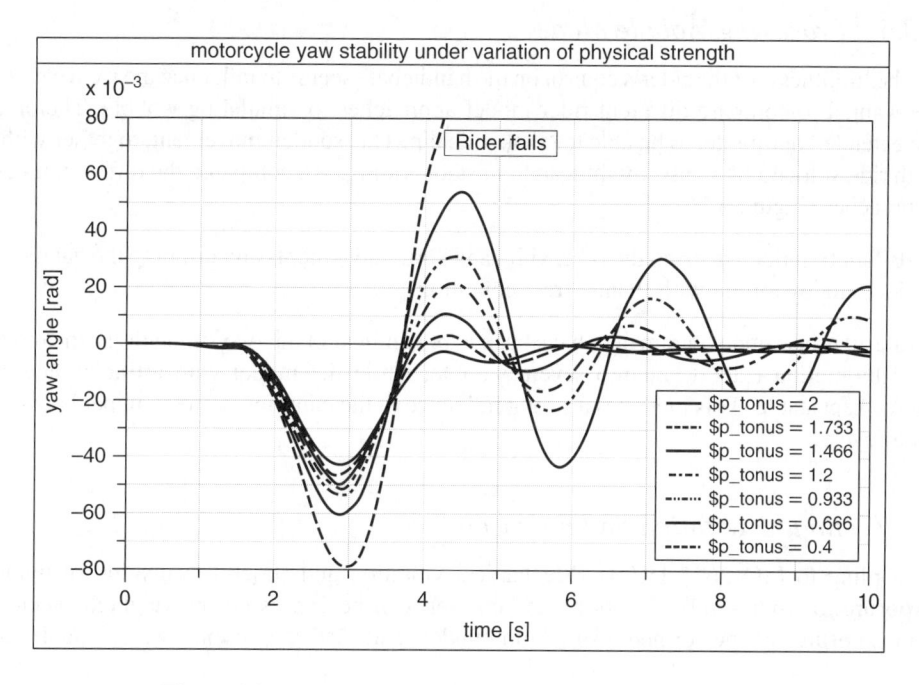

Figure 6.24 Circuit ride under variation of muscular tension

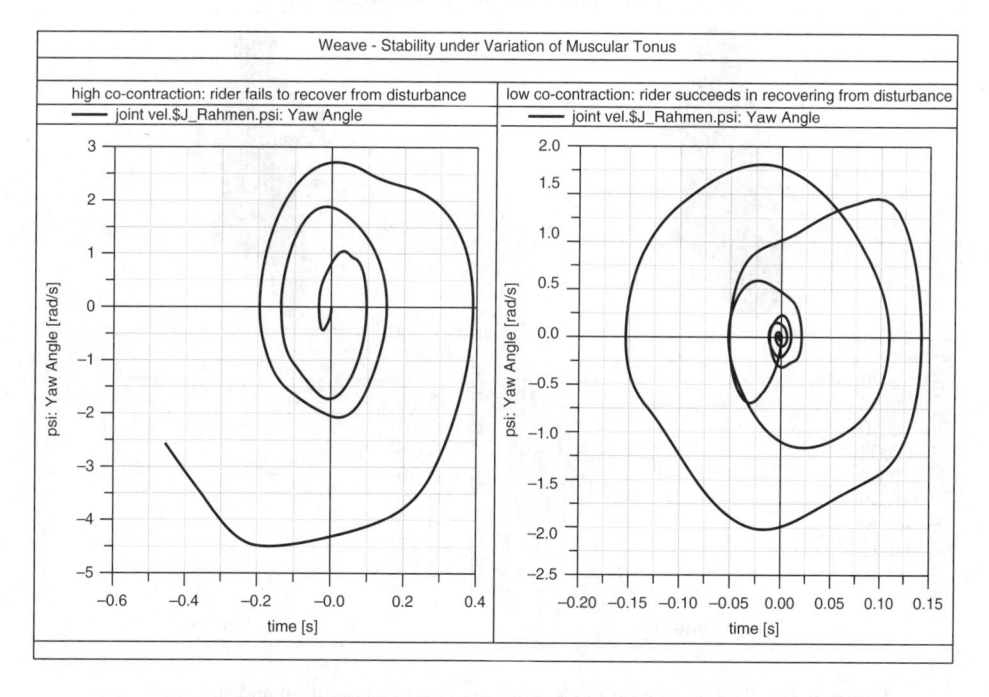

Figure 6.25 Influence of muscular tension on weave

6.3.6 Provoking Wobble Mode

As the 'tightness' of the rider's control on the handlebars seems to influence the weave mode, we wanted to compare different rider model approaches to simulating wobble. Therefore we extended our model to be able to 'shake' its hip. This sudden movement, together with a high ride velocity of 55 m/s might lead to unstable riding. We compared the different model approaches (Figure 6.26):

1. Biomotion rider steering the bike with hand–arm movements and gripping the handlebar
2. Steering by torque with 'hands-free' condition

While all other model parameters have been kept identical in the two simulations, the model with biomotion rider went into severe wobble, while the model with artificial steering torque kept stable. This indicates a strong influence of the coupling between upper body and motorcycle.

6.3.7 Road Excitation and Ride Comfort

According to ISO 2631-1:1997 (Mechanical vibration and shock – evaluation of human exposure to whole-body vibration) ride comfort can be expressed by weighted acceleration exposure. We performed a simulation under 'bad road' conditions, which enables the

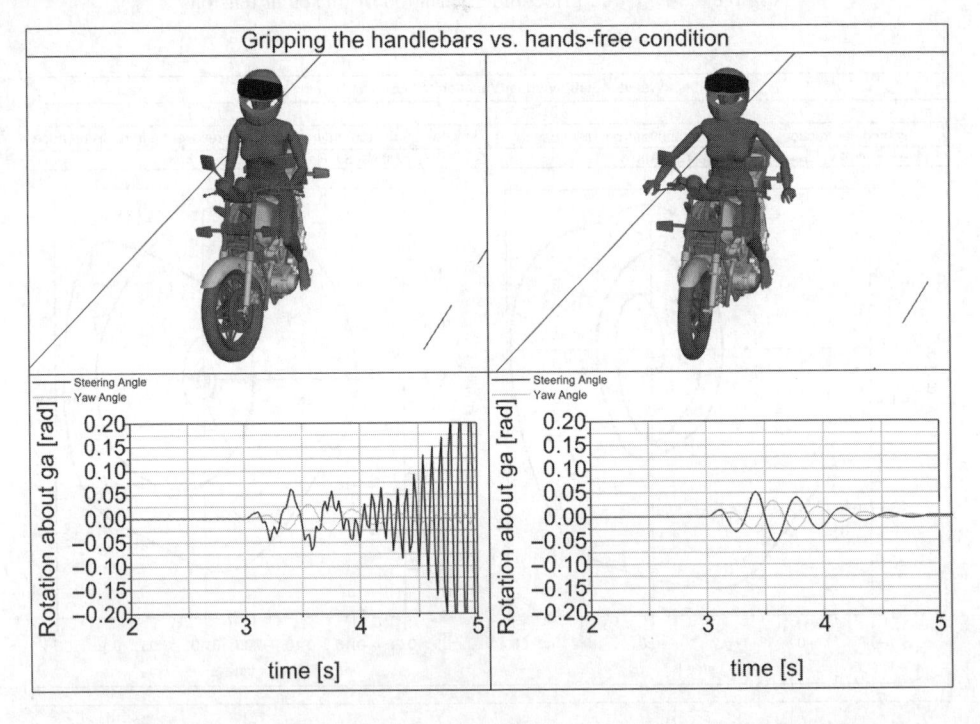

Figure 6.26 The biomotion model with 'hands-free' condition and steer by torque

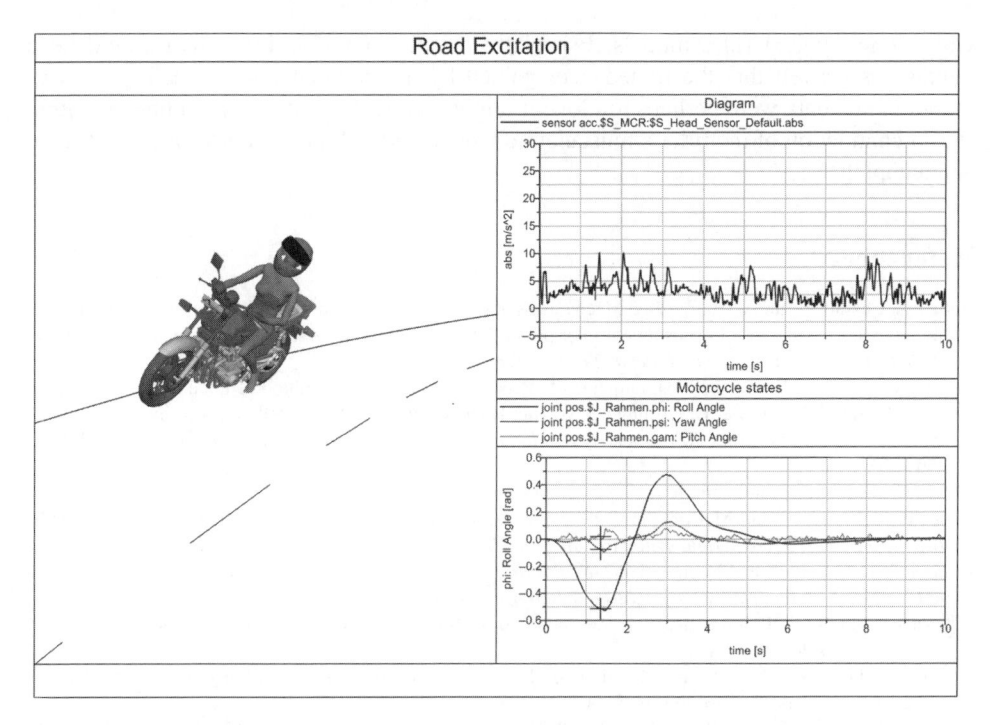

Figure 6.27 The anthropometric model enables ride comfort analysis (e.g. ISO 2631) responding to realistic road excitations

calculation of the rider's vibration response to road excitations (Figure 6.27). Having a whole-body model allows for the estimation of medical or psychophysical ride comfort measures for the head or the wearing comfort of helmets, which may correlate with forces and torques at the cervical spine. In most cases a subjective ride comfort value is a superposition of different ride comfort measures and it is still an open question in vehicle dynamics simulation as to which values correlate with a common subjective ride comfort perception.

6.4 Conclusions

The results of the simulations have shown that it is possible to simulate stable track-following by steering the motorcycle with the handlebars, which allows for taking the rider–bike interaction into account. Simulation results gained by this coupled system show a strong influence of the rider's biomechanics to the dynamics of rider-bike systems. Taking the human factor into account during the engineering design task may help to accelerate the development process and to enhance efficiency in prototyping and testing. By the use of state-of-the art multibody simulation systems it is possible to run a broad variety of motorcycle rider simulations even on notebooks in only a few seconds. We showed that it is helpful to use realistic biomechanical models to understand and interpret measured data. As a prospect for further research we see many open fields which can be worked with biomechanical rider models. Especially measurements during riding would be of interest, as we felt that the impedances gained by the parameter identification from the measurement data were perhaps too low. It might be possible that under 'riding condition' the co-contraction of the rider's musculature would lead to higher values for joint stiffness and damping.

References

Bellati, A, Cossalter V and Garbin S 2003 Mechanisms of steering control of motorcycles (meccanismi di controllo dei motocicli: lo sterzo e lo spostamento del pilota *9th High-Tech Cars And Engines 'Automobili E Motori High-Tech' Mostra-Convegno. Modena, 29–30 maggio.*

Berritta R, Biral F and Garbin S 2000 Evaluation of motorcycle handling with multibody modelling and simulation.

Blickhan R, Seyfarth A, Geyer H, Grimmer S, Wagner H and Günther M 2007 Intelligence by mechanics. *Philosophical Transactions of the Royal Society A: Mathematical, Physical and Engineering Sciences* **365**(1850), 199–220.

Borelli G 1680 De motu animalium.

Chandler R, Clauser R and McConville C 1975 Investigation of inertial properties of the human body. *AMRL Technical Report, NASA Wright-Patterson Air Force Base.*

Clauser C, McConville J and Young J 1969 Weight, volume and center of mass of segments of the human body. *AMRL Technical Report, NASA Wright-Patterson Air Force Base.*

Cossalter V and Lot R 2006 A non-linear rider model for motorcycles. *FISITA 2006, World Automotive Congress 22–27 October 2006 Yokohama).*

Cossalter V, Doria A, Fabris D and Maso M 2006 Measurement and identification of the vibration characteristics of motorcycle riders. *Proceedings of ISMA 2006–dinamoto.it* p. 1793.

Denoth J, Gruber K, Keppler M and Ruder H 1985 Forces and torques during sport activities with high accelerations. *Biomechanics: Current Interdisciplinary Research (Martinus Nijhoff).*

Dong R, Dong J, Wu J and Rakheja S 2007 Modeling of biodynamic responses distributed at the fingers and the palm of the human hand–arm system. *Journal of Biomechanics.*

Doyle A 1987 *The Skill of Bicycle Riding* PhD thesis University of Sheffield.

Feldman A 1974 Control of the length of the muscle. *Biophysics* **19**, 766–771.

Gruber K, Denoth J, Stuessi E and Ruder H 1987 The wobbling mass model. *International Series on Biomechanics, Biomechanics X-B, Human Kinetics Publishers Champaign* pp. 1905–1100.

Gruber K, Ruder H, Denoth J and Schneider K 1998 A comparative study of impact dynamics: wobbling mass model versus rigid body models. *Journal of Biomechanics* **31**, issue 5, 439–444.

Henze A 2002 *Three-dimensional biomechanical modeling and the development of a controller for the simulation of bipedal walking* PhD thesis Eberhard Karls University, Tübingen.

Hill A 1938 The heat of shortening and the dynamic constants of muscle. *Proceedings of the Royal Society of London B* **126**, 136–195.

Jäger R and Haslwanter T 2004 Otolith responses to dynamical stimuli: results of a numerical investigation. *Biological Cybernetics* **90**, 165–175. 10.1007/s00422-003-0456-0.

Keppler V 2003 *Biomechanische Modellbildung zur Simulation zweier Mensch-Maschinen-Schnittstellen* PhD thesis Tübingen.

Keppler V 2010 Analysis of the biomechanical interaction between rider and motorcycle by means of an active rider model *Proceedings, Bicycle and Motorcycle Dynamics 2010 Symposium on the Dynamics and Control of Single Track Vehicles, 20–22 October 2010, Delft, The Netherlands.*

Keppler V 2012 Active Human Body Models for Vehicle Dynamics Simulation. Technical Report 05-18, Biomotion Solutions, Germany. Slides for a presentation given at Open Tech Forum, Vehicle Dynamics Expo, Stuttgart, June 2012.

Keppler V and Günther M 2006 Visualization and quantification of wobbling mass motion–a direct non-invasive method. *Journal of Biomechanics* **39**, S 53.

Limebeer D and Sharp R 2006 Bicycles, motorcycles, and models. *IEEE Control Systems Magazine* (vol. **26**, iss. 5) 34–61.

Maurer C, Mergner T and Peterka R 2006 Multisensory control of human upright stance. *Experimental Brain Research* **171**, 231–250. 10.1007/s00221-005-0256-y.

Mergner T, Maurer C and Peterka R 2003 A multisensory posture control model of human upright stance *Neural Control of Space Coding and Action Production* vol. 142 of *Progress in Brain Research* Elsevier, Prablanc C, Pelisson D and Rossetti Y pp. 189–201.

Mutschler H, Hermle M and Keppler V 2004 Digitaler komfort-dummy. *Humanschwingungen, VDI* **1821**, 349–362.

NASA 1978 *NASA Reference Publication 1024: The internal properties of the body and its segments.* NASA.

Pellionisz A and Llinas R 1985 Tensor network theory of the meta organization of functional geometries in the central nervous system. *Neuroscience* **16**, 245–273.

Pick A and Cole D 2006 Neuromuscular dynamics in the driver-vehicle system. *Vehicle System Dynamics* **44**(sup1), 624–631.

Schmitt, S. and Günther M 2011 Human leg impact: energy dissipation of wobbling masses. *Archive of Applied Mechanics* **81**, 887–897.

Sharp R 2007 Motorcycle steering control by road preview. *Journal of Dynamic Systems, Measurement and Control* **127**, 373.

SIM 2011 SIMPACK. Documentation: VI-CE:168 Automotive Track Sensor.

Smith J and Montgomery T 1996 Biomechanically induced and controller coupled oscillations experienced on the f-16xl aircraft during rolling maneuvers. *NASA Technical Memorandum 4752*.

van Ingen Schenau G, van Soest A, Gabreëls F and Horstink M 1995 The control of multi-joint movements relies on detailed internal representations. *Human Movement Science* **14**, 511–538.

van Soest A 1992 Jumping from structure to control: a simulation study of explosive movements PhD thesis Vrije Universiteit Amsterdam.

7

A Virtual-Reality Framework for the Hardware-in-the-Loop Motorcycle Simulation

Roberto Lot and Vittore Cossalter
University of Padova, Italy

7.1 Introduction

Simulators have been used in the aviation and automotive industries for many years for hardware-in-the-loop testing, ergonomics, training and other applications. By contrast, the range of motorcycle riding simulators is very limited. Honda started to develop a series of motorcycle simulators in 1988: its first prototype consisted of a five degrees of freedom (DOF) mock-up with lateral, yaw, roll, pitch and steer motion on a swinging system for restitution of longitudinal acceleration and was based on a linear four-DOF motorcycle dynamics model. In 1996, as a consequence of a change to the Japanese Road Traffic Act, which required the use of simulators in riding school lessons, Honda put a mass-produced model on the market. This second simulator had a simplified three-DOF mock-up (roll, pitch and steer motions) and it was based on a suitably tuned empirical motorcycle model (Yamazaki 1996). In 2002, Honda presented a third prototype which consisted of a six-DOF manipulator for the mock-up motion, a head-mounted display for visual projection, a four-DOF model for lateral motorcycle dynamics and single-DOF model for longitudinal dynamics (Chiyoda et al. 2000 and Miyamaru et al. 2002). The University of Padova started the development of a riding simulator in the 2000's and presented the first prototype in 2003 (Cossalter 2003). In 2003, the PERCRO laboratory also presented its riding simulator with a real scooter mock-up, mounted on a Stewart platform (Ferrazzin et al. 2003) and in 2007

Modelling, Simulation and Control of Two-Wheeled Vehicles, First Edition.
Edited by Mara Tanelli, Matteo Corno and Sergio M. Savaresi.
© 2014 John Wiley & Sons, Ltd. Published 2014 by John Wiley & Sons, Ltd.

INRETS presented a riding simulator based on a three-DOF platform and a linear five-DOF motorcycle mathematical model (Nehaoua et al. 2007).

The motorcycle simulator of the University of Padova (Cossalter et al. 2011b) is most likely one of the most advanced and reliable, as it is able to reproduce important motorcycle behaviours such as handlebar counter-steering, capsize, weave and wobble instabilities, wheelies, skidding, kick-back, etc., as illustrated in following sections.

7.2 Architecture of the Motorcycle Simulator

A simulator is a complex system that aims to reproduce a real environment in a restricted and controlled area where it is possible to simulate any actions under totally safe conditions. On the motorcycle simulator of the University of Padova (UNIDP) (Figure 7.1), the rider sits on a motorcycle mock-up, equipped with controls as on a real bike. The handlebar and footpads are sensorized, the rider's control actions are transferred to the realtime multibody model of the motorcycle, and the simulated dynamics are then converted into references for the motion and visual cues.

7.2.1 Motorcycle Mock-up and Sensors

The rider sits on a motorcycle mock-up and operates the same inputs available on a real motorcycle. The rider actions are monitored by measuring the steering torque, the body lean, the throttle position, the front brake lever and the rear brake pedal pressures, the clutch position and the gearshift lever position. In particular, the steering torque is measured by means of a transducer on the handlebar steering axis, the throttle position is measured with a linear position transducer connected to the accelerator cable, the front/rear brake actions and the clutch position are measured by means of pressure sensors, and the gearshift command is detected by two micro-switches connected to the gearshift lever. Body lean is measured by different techniques, one that yields significant improvement of riding feeling is based on the measurement of the foot peg forces and indirect estimation of body lean detects. In more detail, when the rider leans to the right, the right foot-peg sensor has an increase in the force, while when the rider leans to the left, the left foot-peg sensor has an increase in force. The variation in the force magnitude between the right and left foot-pegs is related to the rider's movement on the saddle.

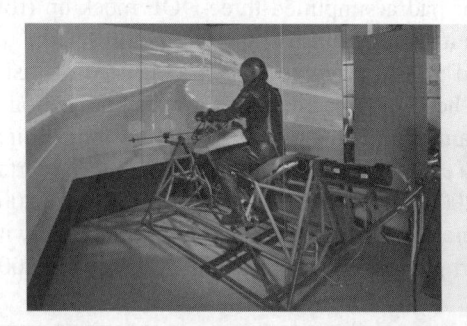

Figure 7.1 The UNIDP motorcycle riding simulator

7.2.2 Realtime Multibody Model

For hardware-in-the-loop applications, it is essential to have a detailed and reliable multibody model of the motorcycle. The simulator is equipped with a realtime version of the validated code (Cossalter et al. 2011a), with which it is able to capture all of the most important vehicle dynamics such as counter-steering, capsize, weave and wobble instabilities, wheelies, skidding and kick-back. The model complexity has been reduced to capture the essential features to give a reliable and fast code suited to realtime simulation, and the simulator mathematical model is nonlinear and has 14 DOFs (Figure 7.2), corresponding to the position and orientation of the chassis, the steering angle, the front and rear suspension travels, the front and rear wheel rotations, the engine spin rate, the front frame bending deflection (modelled with a spring–damper element which restrains the rotation of the front frame with respect to the chassis about the double arrow axis shown in Figure 6.2) and the sprocket absorber deflection. Note that the structural flexibility is important to properly model the wobble mode (Cossalter et al. 2007; Koenen and Pacejka 1982; Sharp 2001; Spierings 1981). Indeed, when neglecting this flexibility it is common to have strong high speed wobble instability thus making the simulator behaviour very different from that of a real bike. The modelling of the sprocket absorber (which accounts for the elastic element between the chain sprocket and the rear wheel rim) is important for proper modelling of the engine-to-slip dynamics which are essential when it comes to traction control design and test (Corno and Savaresi 2010; Massaro et al. 2011). Moreover, this flexibility is related to the *chatter* instability that can appear in a racing motorcycle during braking (Cossalter et al. 2008). As far as the road–tyre model is concerned, the force generation is computed according to a nonlinear dynamic tyre model. In particular, the forces are computed using the well-known magic formula (Pacejka 2006), and the tyre transient behaviour related to the longitudinal and lateral carcass flexibility is managed by means of the relaxation equations. More details of the road–tyre 3D model may be found in (Lot 2006).

The inputs of the mathematical model are those measured by sensors on the mock-up: steering torque, engine torque (which depends on throttle position and engine spin rate), front and rear brake pressures, clutch position, gear ratio and rider lateral position estimated from the differential footrest pressures.

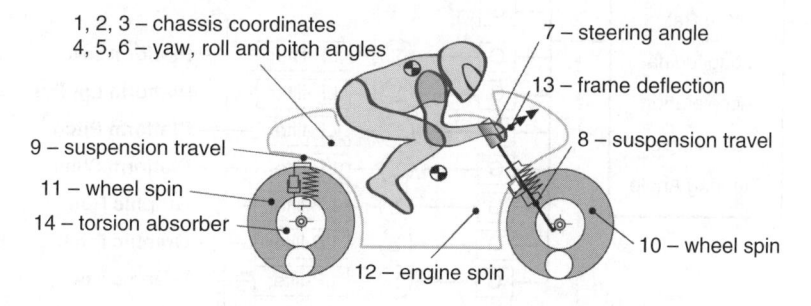

Figure 7.2 The realtime multibody model

7.2.3 Simulator Cues

Once the motion of the virtual bike has been computed, it is important to generate the mock-up motion and the audiovisual cues in order to give the rider a proper riding feeling. In particular, the mock-up should provide motion sensations as close as possible to those of real riding, and the audiovisual system should generate the proper environmental cues. Unfortunately, simulators suffer from the mock-up's limited motion capability, limited resolution and range of the projection system, and limited fidelity of audio systems. Therefore, to obtain a realistic riding feeling, the proper setup of the system cues is essential.

7.2.3.1 Washout Filter

On a simulator, it is physically impossible to precisely replicate real cues in term of motion, visual and acoustic stimuli. The most relevant problem is associated with the replication of accelerations and angular speeds, with conflicts with the workspace constraints of the motion platform. Therefore, the *washout filter* technique has been employed to provide a reduced, yet realistic, riding feeling (Grant 1997). The fundamental idea is to separate accelerations and angular speeds into low and high frequency components. Accelerations at low frequency cannot be reproduced realistically without involving large mockup displacement, so they are scaled and then generated by using the gravitational effect, that is by slowly tilting the simulator. However, medium and high frequency acceleration with zero mean may be realistically reproduced by actually shaking the simulator. As shown in Figure 7.3, the implemented washout is made up of two parts: a first layer includes a matrix that combines the various inputs in a linear combination, and, after that, a second layer of low-pass filtering which provides the output for the platform. As the riding feeling is a holistic experience,

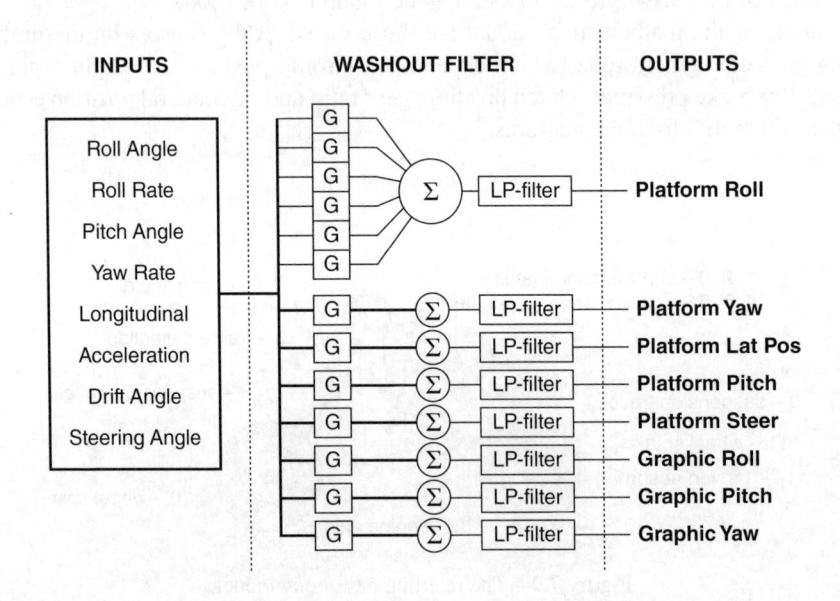

Figure 7.3 The washout filter architecture

the washout filter is used to drive the scenario visualization too. Washout parameters have been tuned using a trial-and-error procedure based on the subjective evaluation of feelings. Appropriate tuning led to a different washout for the visual and motion systems. While cornering, for example, the roll angle is divided into two parts: the biggest one is used to tilt the virtual horizon on the screen, while a smaller part is used to give a motion cue by rotating the mock-up motorcycle.

7.2.3.2 Motion Cues

Figure 7.4 shows a schematic of the motorcycle mock-up whose serial kinematic chain is composed of four mobile frames plus a fixed one to reproduce the motion of the vehicle in terms of lateral displacement, yaw, roll, pitch and steer rotations. The first mobile member is the yaw frame, which is linked to the fixed frame by a pin-in-slot joint and has two degrees of freedom: lateral displacement and yaw rotation. The range of the lateral displacement is ± 0.3m (speed ± 0.3m/s) and that of yaw rotation is $\pm 20°$ (speed $\pm 20°$/s). In order to reduce the load on the pin-in-slot joint the yaw frame is suspended by means of four long steel cables attached to the ceiling. Two servomotors equipped with ball screws A_2 and A_3 drive the yaw rotation and the lateral displacement: when the two servomotors rotate in the same direction lateral motion is generated; when the two servomotors rotate in opposite directions yaw rotation is generated. The second mobile member is the roll frame, which is linked to the yaw frame by means of two revolute joints which define the roll axis of the roll frame with respect to the yaw frame. The roll range is $\pm 20°$ (speed $\pm 60°$/s), and it is driven by a servomotor R_1 equipped with a 50:1 speed reducer. Note that even if the real vehicle reaches roll angles up to $50°$, it is not necessary to reproduce such angles on the simulator. Indeed, due to the combination of the gravity force and the centrifugal force, in the real environment the resulting force acting on the rider lies always in the neighbourhood of the vehicle symmetry

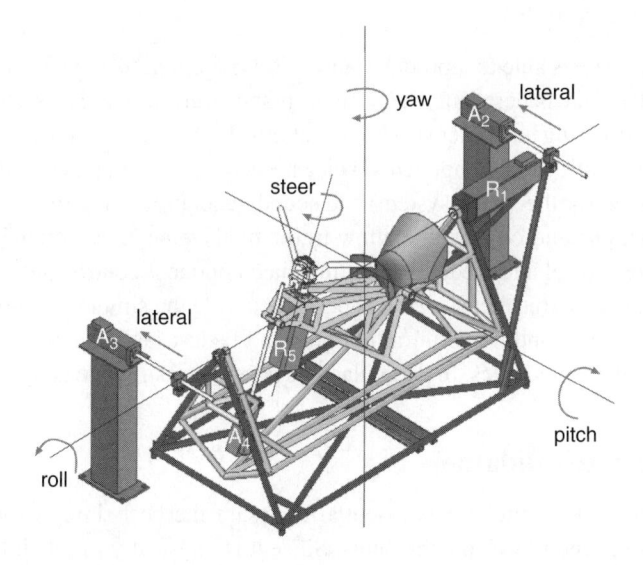

Figure 7.4 Motion cue capabilities of the riding simulator

plane. Therefore only limited roll angles are required to give the same motion cue on the virtual environment of the simulator. The third mobile member is the pitch frame, which is mounted on the roll frame with another revolute joint. It makes rotation about the pitch axis possible and is driven by a servomotor equipped with a ball screw A_4. The pitch range is $\pm 10°$ ($\pm 50°$/s). Finally, the handlebar frame is moved by a servomotor R_5 mounted on the pitch frame and is equipped with a 10:1 speed reducer; the steering range is $\pm 10°$ ($\pm 50°$/s). Each axis of the simulator is equipped with a brush-less servomotor and position loop control.

It has been found by (Chiyoda et al. 2000) that the position of the roll axis is very important for a proper reproduction of the riding cues, in particular when entering a curve. In this simulator the position of the instantaneous roll axis is adjustable by varying the parameters: the ratio between the lateral velocity (motors A_2 and A_3) and the roll rate (motor R_1).

7.2.3.3 Visual and Acoustic Cue

The visual scenario is projected onto a wide screen in front of the rider plus another two positioned at the sides with a 45° angle in order to obtain a 180° horizontal field of view, which remarkably improves the sensation of speed, as captured by the eyes' side portions (Brandt 1973). A 5.1 surround sound system generates the environmental sounds encompassing the rider. Special attention has been devoted to reproduction of the engine sound. The engine sound mainly depends on crankshaft spin rate and throttle position (i.e. engine load), and so the simulator engine sound also depends on these two parameters. In order to have a reliable reproduction, the engine sound of a real motorcycle has been recorded at different engine spin rates, from 1000 rpm (idle) to 12,000 rpm (maximum spin rate). The reproduced sound is obtained by modulating the loudness and pitch according to the engine speed.

7.2.4 Virtual Scenario

The scenario software is able to reproduce various sets of simulation conditions: race tracks, a city with traffic and pedestrians, rural environments and test courses for slalom, lane change, steady cornering and other specific situations. In the virtual scenario, the user interacts with several multimedia components: voice messages, text, graphics (rendered through the 3D engine used for the visual system) and sound generation. Moreover, the core engine has custom functions and objects that allow traffic management, scene environment management, vehicle control and feedback, multimedia component control and so on. From the practical point of view, this means that it is possible to ride the simulator among other vehicles (motorcycles, cars and trucks), traffic lights and pedestrians that behave according to programmer-implemented rules. In particular, it is possible to simulate dangerous situations.

7.3 Tuning and Validation

The proper tuning of a simulator is essential to assure the best riding experience within strict technological and physical constraints. Since it is physically impossible to reproduce accelerations as they are in real life using the simulator, the tuning of the washout filter is crucial and has been carried out by expert riders coping with different riding aspects and in particular:

- the perception of the speed;
- the braking feeling and the feeling while riding on bumpy roads;
- the feeling during transient cornering;
- the vehicle responsiveness during lane changes, overtaking and obstacle avoidance manoeuvres;
- the riding experience at low speed.

The quality of tuning has been assessed by a final validation conducted on a sample of riders having different ages and levels of experience and skill, using both objective and subjective criteria. The aim of the validation process is sketched in Figure 7.5. The interaction between a rider and a motorcycle is represented on the right: the rider controls the vehicle motion by operating the handlebar, throttle and other commands, simultaneously a feedback is given to the rider in terms of motion, acoustic and visual cues. The interaction between a rider and a simulator is represented on the left: the rider actions on the motorcycle mock-up are quite similar to those in the real vehicle, such actions are monitored by means of appropriate sensors installed on the handlebar, brakes, clutch lever and foot pegs, as illustrated in Section 7.2. Rider actions are used as inputs for the motorcycle multibody software, which gives as output the simulated vehicle motion. The latter is then used to feed the washout filter and finally to provide motion, visual and acoustic cues, that is to produce a feedback that closes the loop. Both the simulation of vehicle dynamics and the generation of proper riding cues are essential in providing a realistic riding experience and hence need validation. In particular, the response of the virtual motorcycle to the given rider inputs, may be objectively validated by comparing the virtual and real motion in analogous experimental manoeuvres. Objective validation of the fidelity rider feeling appears not to be feasible, so it has been evaluated by means of subjective rating based on focused questionnaires.

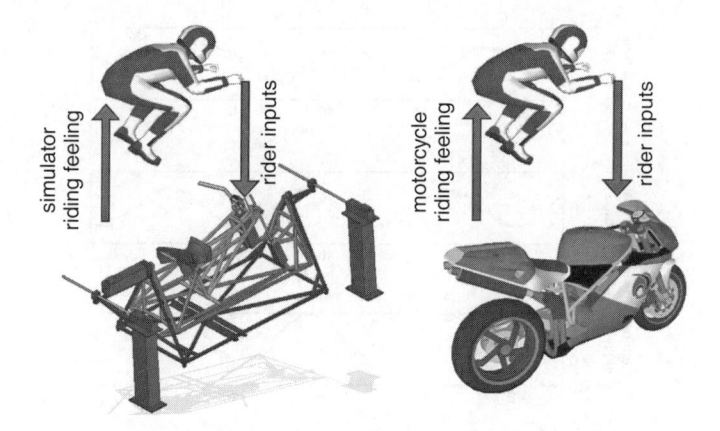

Figure 7.5 Objective and subjective validation of the motorcycle simulator

7.3.1 Objective Validation

The objective evaluation of a simulator may be performed by comparing the behaviour of the real and virtual motorcycles during the same riding actions. Despite the fact that there are many riding conditions with several uncontrolled parameters, the literature concerning the objective evaluation of motorcycle handling characteristics helps to select manoeuvres that are representative of the more general vehicle behaviour (Cossalter 2006; Cossalter 2007; Cossalter and Sadauckas 2006; Ferrazzin et al. 2003; Rice 1978; Weir 1978). In particular, the slalom, lane change and steady turning manoeuvres have been selected for the validation of the University of Padova simulator, as these manoeuvres also belong to the set commonly used by motorcycle manufacturers to develop their own vehicles. Tests were carried out by two skilled riders, and the motorcycle used for the tests was equipped with a special handlebar with steering torque and steering angle sensor, foot pegs with load cells, GPS and an inertial measurement unit with accelerometers and gyrometers. As an example, Figure 7.6 shows the comparison between a slalom manoeuvre performed on the riding simulator and on the motorcycle. Figure 7.6a depicts the rider steering torque, which is the input of the system, and the following plots represent the motorcycle outputs in terms of roll rate (7.6b), yaw rate (7.6c) and steering angle (7.6d), showing a good agreement between simulator and road tests.

7.3.2 Subjective Validation

Since no suitable objective tools are available for the verification of the appropriateness of the simulator's visual, acoustic and motion cues, a subjective evaluation based on the rider's rating has been performed, as motorcycle manufacturers usually do when developing new

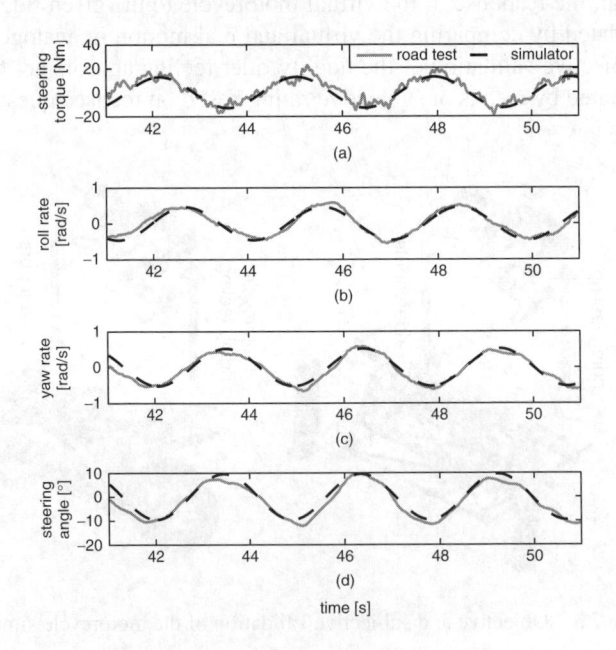

Figure 7.6 Objective validation: slalom

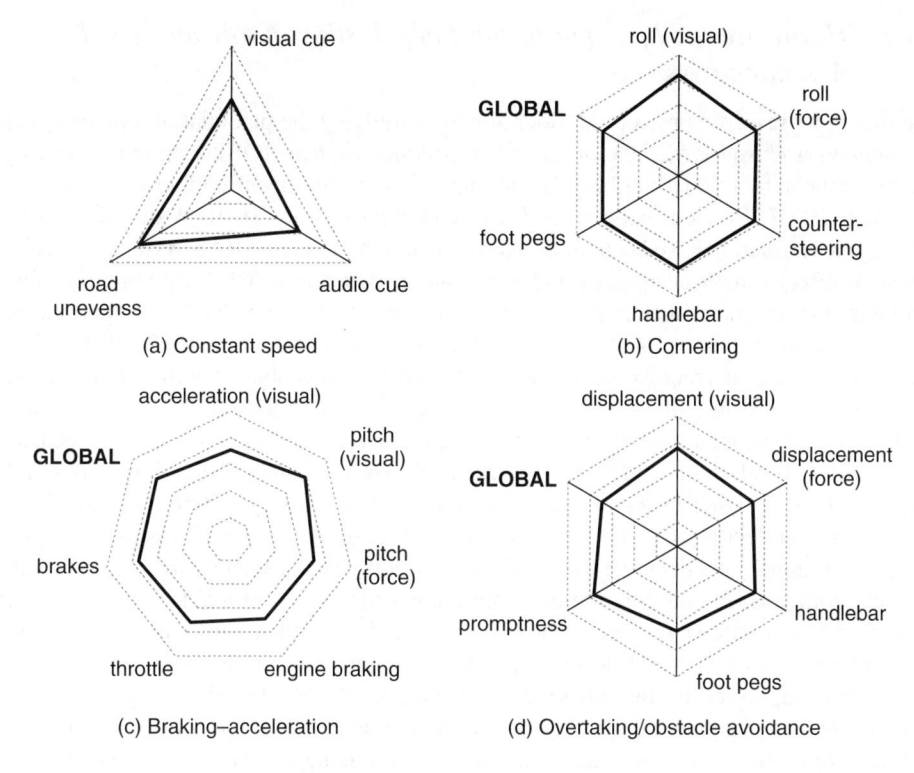

Figure 7.7 Subjective rating

products. Riding sensations have been collected by means of a questionnaire (Cossalter et al. 2011b) for a sample of 20 riders of different age, sex, experience and skill. The questionnaire, which included both technical questions and questions about perception and cognitive processes, was developed with the aid of two skilled riders who are also experts in motorcycle dynamics. The questionnaire focused on different aspects and situations including speed perception, the feeling accompanying braking and acceleration, the feelings of cornering, overtaking and obstacle avoidance. Also, for each situation, they rated the fidelity of the simulator response to rider input, motion cues (in particular roll motion feeling and longitudinal acceleration feeling), and audio/visual cues. Examples of validation results are shown in Figure 7.7 by using radar charts for the different evaluation points. It is worth noting that these results are consistent with the judgements that expert riders made during simulator tuning.

7.4 Application Examples

A motorcycle simulator may be employed in several applications, which includes hardware-in-the-loop testing (ABS, TC, etc.), road education, rider training, ergonomics, etc. Some application examples are given in the following sections.

7.4.1 Hardware- and Human-in-the-Loop Testing of Advanced Rider Assistance Systems

The European research project Saferider aimed at studying the potential of *advanced rider assistance systems* (ARAS) and *on-board information systems* (OBIS) on motorcycles for the most crucial functionalities, and developing efficient and rider-friendly *human–machine interfaces* (HMI) for rider comfort and safety (Bekiaris et al. 2009). Within this project, motorcycle simulators have been used to develop and implement ARAS systems, as well all to test the effectiveness, acceptance and workload of some ARAS/HMI combinations. Three different functions have been implemented: the curve warning, which acts to support a rider to positively and safely negotiate a curve in the road ahead of the vehicle (Biral et al. 2010; Huth et al. 2012), the frontal collision warning, which detects objects in front of the vehicle and provides a warning message in case of danger (Biral and Sartori 2010), and the intersection support, which assists the rider in managing possibly dangerous situations located at intersections (Biral et al. 2012). These different functions share a common modular architecture, which is based on *perception–decision–action* layers paradigm. The *perception* layer aims to collect all the data from the sensors (i.e. linear and angular positions, speeds and accelerations), other vehicles (possible presence of obstacles) and infrastructures (intersection geometry). In addition, the perception layer merges all available data to make the scenario reconstruction and identify the relevant case. Data is then processed by the *decision* layer, which assesses the risk level of the scenario and interacts with the other systems; this layer is supported by the ARAS control module (ACM). Finally, the *action* layer is responsible for managing the warnings to the rider by means of proper HMI elements. This architecture is precisely replicated in the simulator implementation (Figure 7.8), where the perception layer includes emulated GPS, inertial measurement unit (IMU), vehicle interface module (VIF, which gathers vehicle built-in sensors such as wheel spin rates, etc.) and laser scanner. All these (virtual) sensors were connected to a devoted CAN bus, where they are indistinguishable from real sensors, due to common timing and protocol. The decision

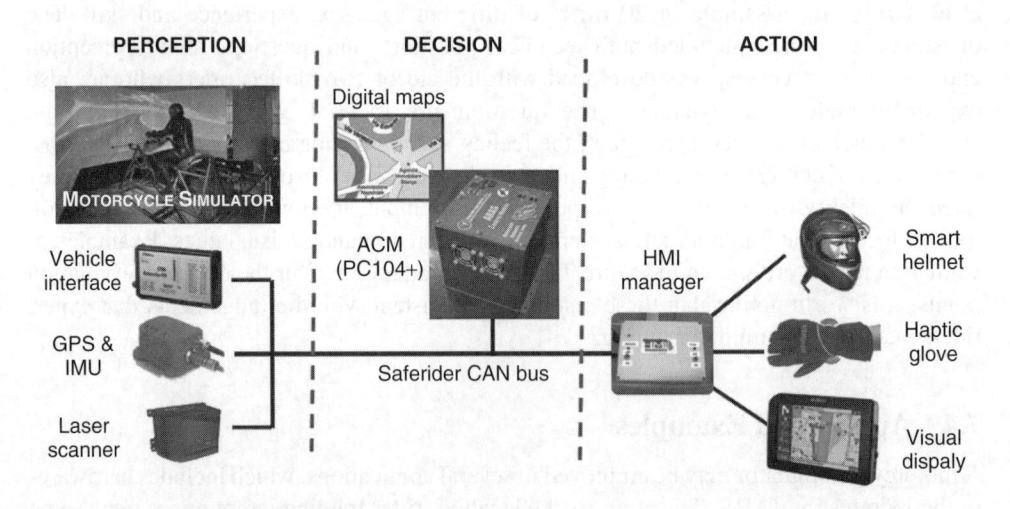

Figure 7.8 The H2IL architecture adopted in the Saferider project

layer consists of the ACM, which manages the ARAS software and interactions with the other systems, hosted by a PC/104+ with a 1.4 3GHz CPU running Linux OS. Finally, the *action* layer includes the HMI manager and a set of HMI elements: a visual display, a force feedback haptic throttle and a vibrating haptic glove (M. et al. 2010), which have been employed with the aim of reducing the visual messages. The HMI manager processes the warning provided by the ACM and properly activates the various HMI elements.

The utilization of a riding simulator in this contest gives the possibility of operating in a flexible environment where both real (hardware-in-the-loop) and emulated devices have been used together. Moreover, testing of such safety-critical functions in a virtual environment avoids any risk to the user. As an example, the *curve warning* (CW) system is now illustrated in more detail. The CW function warns the rider in the case of excessive speed or inadequate behaviour so that the road ahead can be positively and safely negotiated. The system is based on the calculation of a reference *optimal safe* manoeuvre, predicting speed and roll patterns by processing perception layer data at a frequency of 5–10 Hz. Compared with existing systems, the CW function presented is not based on a set of heuristic rules, nor does it relate to the legal or to any assumed speed constraint. To better understand the CW concept let us consider how it works when the motorcycle is approaching a curve. Based on real road geometry and the current vehicle state the CW function uses a receding horizon method to compute a preview of the evolution of vehicle dynamics (i.e., velocity, lateral and longitudinal acceleration, roll angle etc.) at the maximum speed compatible with the fixed safety and comfort requirements. Figure 7.9(a–c) shows the speed and acceleration profiles calculated from their given initial values. Figure 7.9a shows that the preview speed decreases and reaches its minimum in the middle of the curve and finally it increases again at the curve exit. The calculated preview manoeuvre is just one of the possible paths round the curve ahead and in particular it represents the fastest manoeuvre that complies with the given specifications for safety and comfort. Therefore, if the rider is actually riding faster, or accelerating more than the preview manoeuvre, they are potentially in danger and the CW provides a warning. The potentially dangerous behaviour is identified on the basis of

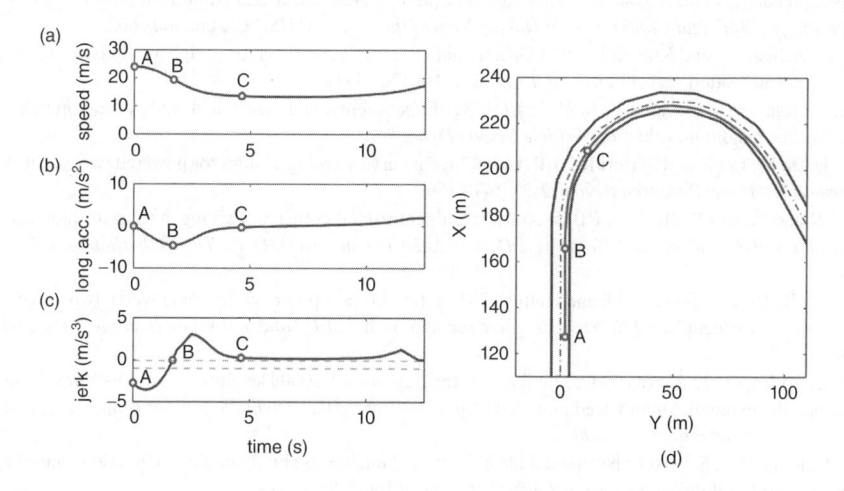

Figure 7.9 Exemplary operation of simulated curve warning

preview jerk (i.e., the time derivative of the acceleration, Figure 7.9c): as the jerk becomes more negative, the urgency of reducing acceleration (or decelerating even more) increases, so two jerk thresholds have been selected for cautionary and imminent warning. As the rider's behaviour differs from the preview manoeuvres, these must be continuously updated to real speed and acceleration, as also to the changing road scenario conditions (e.g., road geometry). Consequently as the vehicle approaches the curve, a sequence of manoeuvres is computed as fast as possible.

7.4.2 Training and Road Education

Simulators are only a representation of the reality, and experiments in educational psychology have shown that there is no transfer if the learning environment clearly deviates from reality (Groeger 2000). However, training with simulators also has some advantages in terms of both didactics and safety-related aspects. From a didactic point of view (Fuller 2008 and SWOV 2010) demonstrated that simulators benefit from a fast exposure to a wide variety of traffic situations, improved possibilities for feedback from different perspectives, unlimited repetition of educational moments, computerized and objective assessment and so on. From a safety-related perspective, the advantages are even more pronounced, as the virtual environment gives the opportunity to safely experience situations which may be very dangerous in reality, for example slippery asphalt, riding in the fog, accidents and other unexpected events. In addition, such situations may be replicated with and without assistance systems such as ABS, TC (or even advanced, forthcoming systems like CW and FCW), to better realize how such systems work. Also, the initial training of totally inexperienced users about vehicle commands and elementary riding presents safety-related aspects and can take advantage of the safe simulator environment.

References

Bekiaris E, Spadoni A and Nikolaou S 2009 Saferider project: New safety and comfort in powered two wheelers *Proceedings - 2009 2nd Conference on Human System Interactions, HSI '09*, pp. 600–602.

Brandt T., Dichgans J. and Koenig E. 1973 Differential effects of central versus peripherical vision on egocentric and exocentric motion perception. *Exp. Brain Res.* **16**, 476–491.

Biral F., Bortoluzzi D., Cossalter V. and Da Lio 2003 Experimental and theoretical study of motorcycle transfer functions for handling evaluation. *Vehicle System Dynamics*.

Biral F, da Lio M, Lot R and Sartori R 2010 An intelligent curve warning system for powered two wheel vehicles. *European Transport Research Review* **2**(3), 147–156.

Biral F LR and Sartori R, Borin A RB 2010 An intelligent frontal collision warning system for motorcycles *In: proceedings of the Bicycle and Motorcycle Dynamics 2010 Symposium. Delft, The Netherlands, 20 22 October 2010*.

Biral F., Lot R., Rota S., Fontana M. and Huth V. 2012 Intersection support system for powered two-wheeled vehicles. threat assessment based on a receding horizon approach. *IEEE Transactions on Intelligent Transportation Systems*.

Chiyoda S, Yoshimoto K, Kawasaki D, Murakami Y and Sugimoto T 2000 Development of a motorcycle simulator using parallel manipulator and head mounted display *in Proc. of the Driving Simulator Conference, DSC 2000, Paris, France, September 6-7, 2000*.

Corno M and Savaresi S 2010 Experimental identification of engine-to-slip dynamics for traction control applications in a sport motorbike. *European Journal of Control* **16**(1), 88–108.

Cossalter V 2006 *Motorcycle Dynamics*. Lulu.

Cossalter V and Sadauckas J 2006 Elaboration and quantitative assessment of manoeuvrability for motorcycle lane change. *Vehicle System Dynamics* **44**(12), 903–920.

Cossalter V., Lot R. and Doria A. 2003 Sviluppo di un simulatore di guida motociclistico *in Proc. of the 16th AIMETA Congress of Theoretical and Applied Mechanics, Ferrara, Italy, September 9-12, 2003.*

Cossalter V, Lot R and Massaro M 2007 The influence of frame compliance and rider mobility on the scooter stability. *Vehicle System Dynamics* **45**(4), 313–326.

Cossalter V., Lot R. and Peretto M.. 2007 Motorcycles steady turning. *Journal of Automobile Engineering.*

Cossalter V, Lot R and Massaro M 2008 The chatter of racing motorcycles. *Vehicle System Dynamics* **46**(4), 339–353.

Cossalter V, Lot R and Massaro M 2011a An advanced multibody code for handling and stability analysis of motorcycles. *Meccanica* **46**(5), 943–958.

Cossalter V, Lot R, Massaro M and Sartori R 2011b Development and validation of an advanced motorcycle riding simulator. *Proceedings of the Institution of Mechanical Engineers, Part D: Journal of Automobile Engineering* **225**(6), 705–720.

Ferrazzin D, Barbagli F, Avizzano C, Di Pietro G and Bergamasco M 2003 Designing new commercial motorcycles through a highly reconfigurable virtual reality-based simulator. *Advanced Robotics* **17**(4), 293–318.

Fontana, M., Diederichs, F., Bencini G., Lot R., Sartori R., Baldanzini N., Spadoni A., and Bergamasco M. 2010 Saferider: Haptic feedback to improve motorcycle safety. *Proceedings of IEEE Haptics Symposium 2010. Waltham, MA, USA, 2010, MARCH 20-26.*

Fuller R 2008 Driver training and assessment: implications of the task-difficulty homeostasis model *Driver behaviour and training, Volume III; Proceedings of the Third International Conference on Driver Behaviour and Training*, pp. 337–348.

Groeger J 2000 *Understanding Driving-Applying cognitive psychology to a complex everyday task*. Psychology Press Ltd., Routledge.

Huth V, Biral F, Martin O and Lot R 2012 Comparison of two warning concepts of an intelligent curve warning system for motorcyclists in a simulator study. *Accident Analysis and Prevention* **44**(1), 118–125.

Koenen C and Pacejka H 1982 Influence of frame elasticity, simple rider body dynamics and tyre moments on free vibrations of motorcycles in curves. *Dynamics of Vehicles on Roads and on Tracks, Proceedings of IAVSD Symposium (International Associa*, pp. 53–65.

Lot R. and Massaro M. 2006 A combined model of tire and road surface for the dynamics analysis of motorcycle handling *in Proc. of the FISITA World Automotive Congress, Yokohama, Japan, 22-27 October, 2006.*

Massaro M, Sartori R and Lot R 2011 Numerical investigation of engine-to-slip dynamics for motorcycle traction control applications. *Vehicle System Dynamics* **49**(3), 419–432.

Miyamaru Y, Yamasaki G and Aoki K 2002 Development of a motorcycle riding simulator. *JSAE Review* **23**(1), 121–126.

Nehaoua L, Hima S, Arioui H, Séguy N and Ãspié S 2007 Design and modeling of a new motorcycle riding simulator *Proceedings of the American Control Conference*, pp. 176–181.

Pacejka HB 2006 *Tyre and Vehicle Dynamics* 2nd ed. edn. Butterworth-Heinemann, Oxford.

P.R. Grant LR 1997 Motion washout filtering tuning: rules and requirements. *Journal of Aircraft.*

Rice RS 1978 Rider skill influences on motorcycle manoeuvring. *SAE paper 780312.*

Sharp R 2001 Stability, control and steering responses of motorcycles. *Vehicle System Dynamics* **35**(4-5), 291–318.

Spierings P 1981 The effect of lateral front fork flexibility on the vibration modes of straight-running single-track vehicles. *Vehicle System Dynamics.*

SWOV 2010 Fact sheet - simulators in driver training.

Weir JWZ D. H. JWZ 1978 Development of handling test procedures for motorcycles. *SAE paper 780313.*

Yamazaki G 1996 The development of motorcycle riding simulator for training. *JSAE Acad. Lecture Art.* **965**, 173–176.

Part Two

Two-Wheeled Vehicles Control and Estimation Problems

8

Traction Control Systems Design: A Systematic Approach

Matteo Corno and Giulio Panzani
Dipartimento di Elettronica, Informazione e Bioingegneria, Politecnico di Milano, Italy

This chapter presents a systematic approach to the design of traction control systems for motorcycles. Several aspects are considered: activation strategy, reference generation and gain-scheduling. Tests on instrumented motorcycles are used to show several critical aspects.

8.1 Introduction

The development of vehicle dynamics control (VDC) systems for powered two-wheelers (PTW) followed a different path from that of systems for passenger cars. For the latter, braking control systems (mainly anti-lock braking systems–ABS–see Savaresi and Tanelli 2010) were the first commercially successful VDC systems. For PTWs, VDC systems that act on the engine initially had better success. There are several reasons for this. First of all the difficulty in actuation: PTWs are sensitive to mass increase, and the first braking pressure controllers were bulky and heavy (Dardanelli et al. 2010). Cultural reasons also played an important role. Braking and traction control systems for PTWs appeared almost simultaneously (the first ABS system in 1988, and the first traction control in 1992), and they were marketed as safety systems. Unfortunately, their poor performance (a non-professional driver could easily outperform these systems) kindled a doubtful attitude among motorcycle enthusiasts. It was the successful employment of traction control systems by racing teams that encouraged a re-evaluation of these systems. Racing teams sent out the message that now traction control systems are something that even professional riders need, in order to reach the limits of a motorcycle. The change of perspective, from safety to performance-oriented systems, accelerated the development of advanced control systems for PTWs (Corno et al. 2008; De Filippi et al. 2011; Seiniger et al. 2012), and give an overview of the potential effects of safety systems in motorcycles.

Modelling, Simulation and Control of Two-Wheeled Vehicles, First Edition.
Edited by Mara Tanelli, Matteo Corno and Sergio M. Savaresi.
© 2014 John Wiley & Sons, Ltd. Published 2014 by John Wiley & Sons, Ltd.

In the PTW context, traction control (TC) systems represent an interesting field where different philosophies have been tested and applied. TC and ABS systems control similar dynamics, but they refer to two completely different scenarios. The ABS market is dominated by two companies, and motorcycle manufacturer's shop for off-the-shelf ABS systems (sometimes with limited customization). On the other hand, TC is perceived as added value by the customers; so motorcycle manufacturers invest in developing their own interpretation of traction control. This led to a proliferation of systems. At the end of 2008, only Ducati and BMW provided TC-equipped motorcycles, but by the end of 2012 the list of brands sporting their own version of TC included BMW, Ducati, Aprilia, MV-Agusta, Kawasaki, Honda and Yamaha among others.

Most TC systems act by reducing the engine torque to prevent excessive slip (or angular acceleration) of the driving wheel. The scientific literature is rich in studies on traction control applied to four-wheeled vehicles (Austin and Morrey 2000; Chun and Sunwoo 2004; Ferrara and Vecchio 2007; Kabganian and Kazemi 2001; Kang et al. 2005; Liu et al. 2011). Despite the growing industrial interest, TC systems for motorcycles have not received much attention by the scientific community. Only a handful of examples can be found: (Cardinale et al. 2008, Hirsch 2009 and Tanelli et al. 2009b). The first paper shows preliminary results on a traction control system for a super-motard; the proposed control law acts on the spark and is quite empirical. No systematic study is presented but rather a trial and error tuning procedure is employed. Tanelli et al. presented a second-order sliding mode approach and its effectiveness is proven with a multibody motorcycle simulator. However, no experimental validation is provided for the sliding mode approach; also Hirsch (2009) only presents simulation results.

Historically, the first two manufacturers to introduce TC on the market were BWM and Ducati. Two systems were available: the ASC (automatic stability control) implemented on the BMW GS 1200, and the DTC (Ducati traction control) available from 2007 on the 1098 and 1198 Ducati super-sport motorcycles. Being commercial systems, only very limited technical details are available. They implement two different philosophies:

- **Safety oriented**. This approach is followed by BMW on their R 1200 GS. Spark advance is used to prevent real wheel spinning on low-friction surfaces (gravel roads and wet surfaces). The system limits the engine torque whenever rear wheel slip reaches high values. The BMW TC is designed to prioritize safety and therefore its behaviour is very conservative. One of its biggest drawbacks is that once the system activates, the intervention is too abrupt and felt by the rider as uneven vibration in the engine torque.
- **Performance oriented**. This approach is the one followed by most manufacturer's in the past few years. The first commercial motorcycle to provide a race-kit with TC was the Ducati 1098 R. This kind of TC is designed to act on high grip surfaces especially during cornering. Its goal is to reach maximum longitudinal acceleration while providing enough lateral force to negotiate the corner, according to a philosophy similar to the one adopted in (Tanelli et al. 2009a). The reception from the market was not always enthusiastic; for example Chris Signlin (test rider from *Motorcycle USA*) said, "I had high hopes for Ducati's system, but it didn't quite perform. The biggest problem is how abrupt it is and it can really disrupt the chassis. It isn't something I'd be comfortably using in a race." Since 2008, performance-oriented TC systems have been continuously improved. At the beginning of 2013, test riders seem to agree that the Aprilia APRC system (Panzani et al. 2010a,b; Savaresi et al. 2010) had solved most issues. In Signlin's words: "I was blown

away by how good the Aprilia's TC was. I wasn't expecting it to be that good. My favorite feature was how consistent and smooth it felt. I got comfortable with it right away, which gave me confidence to be aggressive on the throttle."

These are closed-loop systems, meaning that the rear wheel slip determines the control action. Some after-market companies adopted an open-loop approach. Open-loop systems, not requiring many measurements from the motorcycle, are easily installed. Open-loop systems rely on static maps applying a predetermined engine spark advance based on engine RPM, gear and throttle position. Open-loop systems work only in the very specific conditions for which they have been tuned and they cannot be made robust against varying road and tyre conditions.

This chapter presents a complete and systematic treatment of the issues that practitioners face when designing a TC system for a road motorcycle. We do not aim at presenting a state-of-the-art TC system; rather we propose and advocate a scientific approach to a problem often approached heuristically. The proposed approach enables us to fill the gap between the two aforementioned philosophies, designing a control system that guarantees safety on difficult roads but at the same time does not limit the bike's performance, and leaves the rider a certain amount of control. In the chapter numerous experimental results exemplify the topics. Real-world testing, although more expensive and time-consuming than simulations, offers more dependable results, especially on low-friction surfaces that are notoriously difficult to model and simulate.

All closed-loop traction control systems can be casted into the general functional schematic of Figure 8.1, where θ is the throttle grip position, u is the control variable of choice, T the engine torque, λ the wheel slip defined in Section 8.2 and a_x the longitudinal vehicle acceleration. The control problem in its most general form is composed of four different components. The first concerns torque modulation. The TC system needs to control the wheel torque; different solutions exist: throttle command, spark advance, ignition control or pressure brake modulation (see Chapter 3 for a detailed discussion). Each method has its own features but in the following a general approach will be proposed for the design, whereas experimental results will be presented on an electronic throttle-equipped motorcycle. The second component is the dynamics under control: the transmission, the motorcycle and the wheel slip dynamics itself. The third issue is that of designing the wheel slip controller. This approach requires a target wheel slip. Most safety-oriented systems employ a constant threshold; this approach has some limitations that will be discussed. In the following, a throttle grip position dependent reference wheel slip will be presented.

The chapter is organized as follows: in Section 8.2, the most relevant concepts regarding wheel slip dynamics are recalled; in Section 8.3, the TC system is designed, taking into account many aspects: activation, deactivation, wheel slip reference generation and robustness. The system is then tuned and tested in Section 8.4.

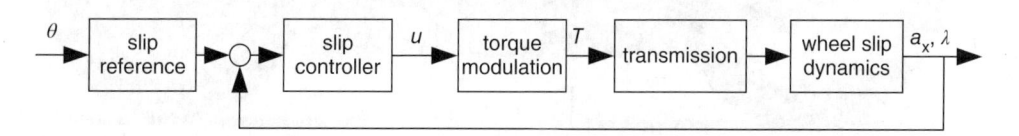

Figure 8.1 A schematic view of the overall TC problem

8.2 Wheel Slip Dynamics

Wheel slip dynamics are often described by the single corner model (e.g., Savaresi and Tanelli 2010). Although (Corno et al. 2009) prove the shortcomings of the single corner model when it comes to motorcycles, it is still very useful for studying the main dynamics that TC systems face. The single corner model is based on the following assumptions:

- load transfer phenomena, which are induced by pitch motion, are neglected;
- the dependence of friction forces from the vertical load is modelled as a proportionality relation;
- the wheel radius is assumed to be constant, even though, during braking, a consequence of the pitch motion is a dynamic change in the wheel radius which is function of the instantaneous vertical load;
- straight-line running;
- tyre relaxation dynamics (Pacejka 2002) are neglected.

The model is given by the following set of equations (Figure 8.2)

$$\left.\begin{array}{l} J\dot{\omega} = -rF_z\mu(\lambda) + T_d \\ m\dot{v} = F_z\mu(\lambda). \end{array}\right\} \quad (8.1)$$

where

- J [kg m^2], m [kg] and r [m] are the moment of inertia of the wheel, the quarter-car mass (m includes both the sprung and unsprung mass of the quarter-car), and the wheel radius, respectively;

Figure 8.2 Single corner model diagram

Traction Control Systems Design: A Systematic Approach

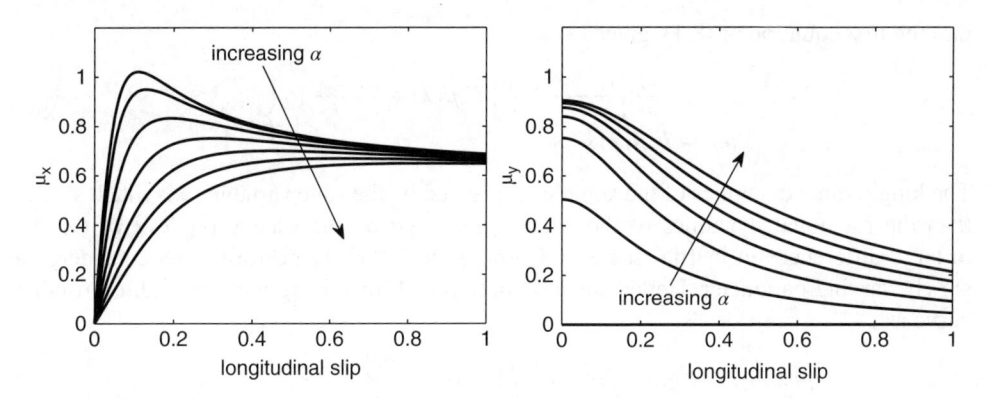

Figure 8.3 Longitudinal and lateral friction coefficient dependency on wheel slip (λ) and sideslip angle (α)

- ω [rad/s] is the angular velocity of the wheel;
- v [m/s] is the longitudinal velocity of the wheel axis (assumed equal to the velocity of the vehicle's centre of gravity);
- T_d [Nm] is the driving torque (which is the input variable);
- F_z [N] is the vertical tyre–road contact force;
- $\mu(\lambda)$ is the longitudinal friction coefficient. This depends on the tyre–road contact characteristic and on the longitudinal wheel slip (among several other parameters). The friction coefficients μ are defined as the ratio of longitudinal (lateral) force and the vertical tyre load.

During traction, the longitudinal wheel slip is defined as

$$\lambda = \frac{\omega r - v}{\omega r} \qquad (8.2)$$

and determines the longitudinal and lateral forces that the tyre exerts, according to nonlinear maps. Figure 8.3 depicts an example of friction coefficients' dependency on longitudinal wheel slip and sideslip–see (Pacejka 2002) for a complete treatise of tyre forces. Inspecting the Figure 8.3 the importance of a traction control system is immediately clear. The longitudinal characteristic has an ascending part, a peak (at λ^*), and a descending part; an excessive longitudinal wheel slip will determine a nonmaximal longitudinal force and a poor lateral stability as, for λ larger than λ^*, the lateral force quickly drops. The goal of an ideal TC system is therefore to make sure that λ does not exceed λ^*.

TC systems are also important from a dynamical point of view; when the wheel slip exceeds λ^* the dynamics become open-loop unstable and the rider cannot control it. To see this, consider system (8.1); the state variables are v and ω. As λ, v and ω are linked by an algebraic relationship, it is possible to replace the state variable ω with the state variable λ. Plugging

$$\dot{\lambda} = \frac{vr}{(\omega r)^2}\dot{\omega} - \frac{1}{r\omega}\dot{v}$$

and

$$\omega r = \frac{v}{(1 - \lambda)}$$

into the first equation of (8.1), gives

$$\begin{aligned}\dot{\lambda} &= -\frac{1-\lambda}{v}\left[\frac{(1-\lambda)r^2}{J}+\frac{1}{m}\right]F_z\mu(\lambda)+\frac{r}{v}\frac{(1-\lambda)^2}{J}T_d\\ m\dot{v} &= F_z\mu(\lambda)\end{aligned}\Biggr\}.\qquad(8.3)$$

The longitudinal dynamics of the vehicle (expressed by the state variable v) is much slower than the rotational dynamics of the wheel (expressed by the state variable λ or ω) due to large differences in inertia (see e.g., Corno et al. 2011). Henceforth, v is considered a slowly varying parameter. Under this assumption and linearizing the longitudinal friction curve as

$$\mu(\lambda(\omega,v)) = \mu(\lambda(\bar{\omega},\bar{v})) + \left.\frac{\partial\mu}{\partial v}\right|_{\bar{\omega},\bar{v}}\delta v + \left.\frac{\partial\mu}{\partial\omega}\right|_{\bar{\omega},\bar{v}}\delta\omega = \qquad(8.4)$$

$$= \mu(\bar{\lambda}) + \left.\frac{\partial\mu}{\partial\lambda}\right|_{\bar{\lambda}}\left.\frac{\partial\lambda}{\partial v}\right|_{\bar{\omega},\bar{v}}\delta v + \left.\frac{\partial\mu}{\partial\lambda}\right|_{\bar{\lambda}}\left.\frac{\partial\lambda}{\partial\omega}\right|_{\bar{\omega},\bar{v}}\delta\omega = \qquad(8.5)$$

$$= \mu_0\bar{\lambda} + \mu_1(\bar{\lambda})\frac{1}{\bar{\omega}r}\delta v + \mu_1(\bar{\lambda})\frac{\bar{v}r}{(\bar{\omega}r)^2}\delta\omega \qquad(8.6)$$

the single corner model can be approximated by the following transfer function

$$G_\lambda(s) = \frac{\left[\frac{(1-\bar{\lambda})^2}{vJ}\right]}{s+\frac{\mu_1(\bar{\lambda})F_z}{M\bar{v}}(1-\bar{\lambda})\left[(1-\bar{\lambda})\frac{r^2M}{J}+1\right]}.\qquad(8.7)$$

The transfer function (8.7) is a simple model which represents only the bulk of the wheel-slip dynamics. Nevertheless, some important features can be observed:

- The stability of the dynamics is uniquely determined by the slope of the friction characteristic at the linearization point μ_1. If the slope is positive the dynamics is open-loop asymptotically stable, whereas if the slope is negative the dynamics is unstable. The change of slope happens at λ^*, which thus separates the longitudinal wheel domain into a stable region and an unstable region. This observation confirms the necessity of traction control systems: not only the longitudinal and lateral forces drop if $\lambda > \lambda^*$, but once that value is crossed the wheel slip will quickly diverge to 1, unless the wheel torque is reduced.
- The linearization velocity affects the position of the pole, but not the gain. In particular, as the velocity decreases, the pole is pushed toward high frequency. Note that as the velocity tends to 0, the dynamics gets infinitely fast. See the upper plots in Figure 8.4.
- The linearization wheel slip affects both the position of the pole and the gain. As shown in the bottom plot of Figure 8.4 the closer the wheel slip is to λ^* the lower the pole is and the higher the DC gain.
- The vertical load has an effect primarily on the gain. In particular, as the vertical load increases the gain decreases.

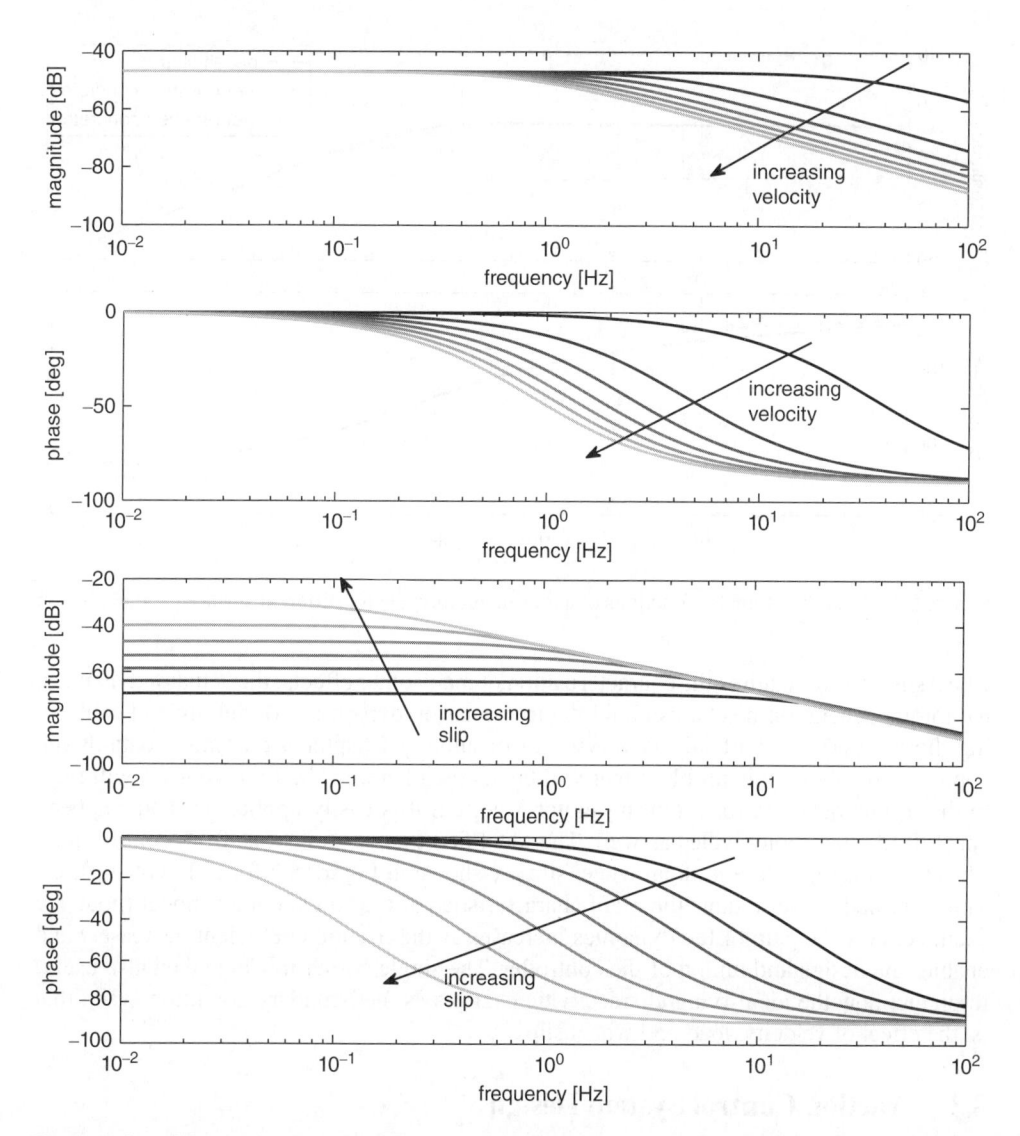

Figure 8.4 Bode plots of the wheel torque to wheel slip transfer functions as modelled by the single corner model. Dependency on longitudinal velocity (upper plot) and linearization wheel slip (lower plot)

As clearly shown by transfer function (8.7), wheel slip control has several advances over the competing approach of wheel acceleration control. The most compelling advantage is that, given a minimum velocity, it is always possible to stabilize wheel slip dynamics with a fixed structure controller (refer to Savaresi and Tanelli 2010 for a detailed discussion on the topic). This, of course, greatly simplifies the control system design.

The single corner only models the wheel slip dynamics; it provides useful indications for the design of the controller, but it is not sufficient. As shown in Figure 8.1, the design of the

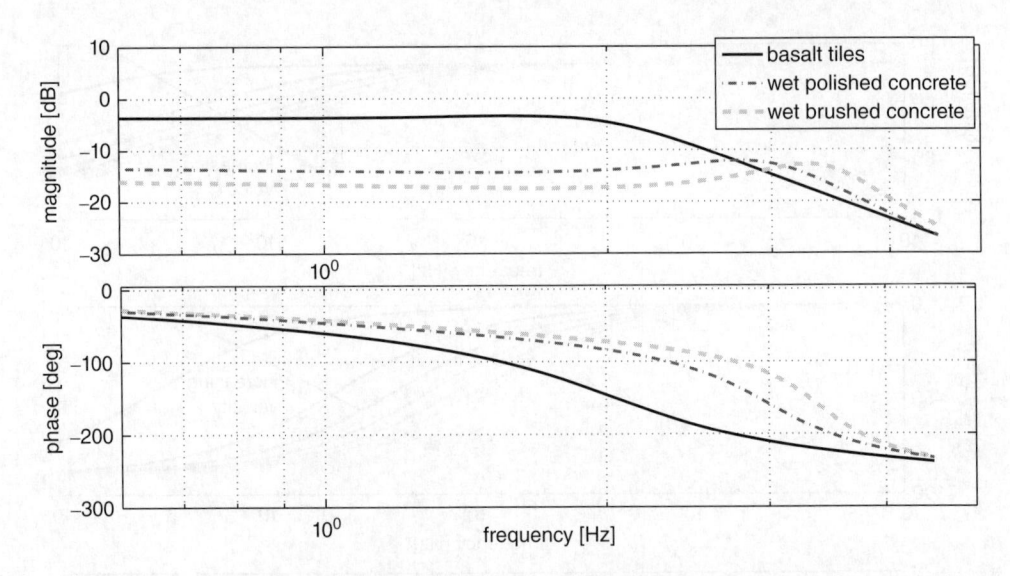

Figure 8.5 Identified transfer functions of a touring motorcycle for different surfaces at $v = 20$ mph

wheel slip control requires the characterization of many other effects: the actuator dynamics, the torque generation mechanism and the transmission dynamics. Modelling and calibrating those aspects individually is very time-consuming; Chapter 3 presents a data-driven approach to solving this problem that will be adopted here. In the following we will refer to the results obtained for a touring motorcycle, but it is easily applicable (and has been applied) to other motorcycles as well. The identified transfer functions from control input (throttle demand) to longitudinal wheel slip are shown in Figure 8.5 for different surfaces. The identified model retains the main characteristics of the single corner model (note, for example, how the gain of the dynamics increases as the friction coefficient decreases) and enables the design and tuning of the controller. The single corner model is then also useful to predict how the identifying characteristic changes as the boundary conditions vary, that is, the effect of velocity, load and wheel slip.

8.3 Traction Control System Design

This section describes to the design of the TC system. Figure 8.6 represents the detailed block diagram of the proposed traction control. The TC system is composed of four subsystems: 1) a supervisor that manages the activation and deactivation of the system; 2) the wheel slip controller which modulates the control variable (in this case the throttle demand) in order to track a desired wheel slip; 3) the slip reference generator; 4) a surface transition recognition that is necessary to deal with sudden changes of friction and to perform the controller gain scheduling. In the following the four subsystems will be described in detail.

8.3.1 Supervisor

The TC system activation must be accurately managed in order to avoid negatively affecting the riding experience. A finite state machine (Figure 8.7) governs the possible states of the

Traction Control Systems Design: A Systematic Approach

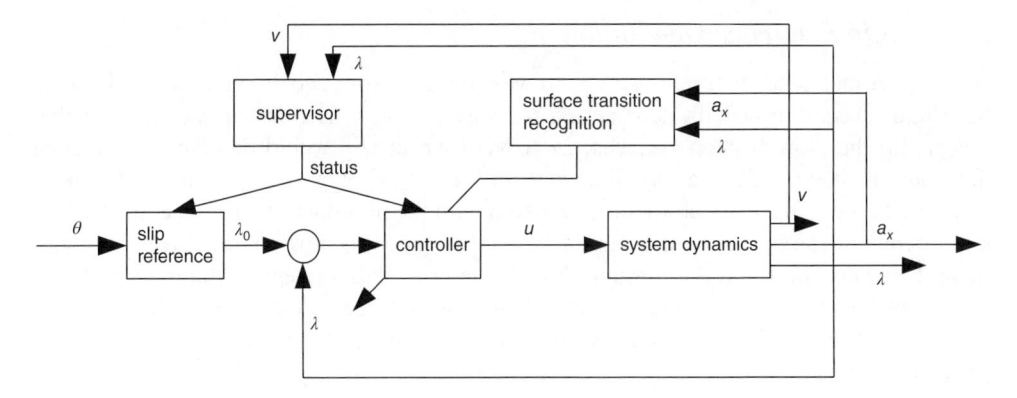

Figure 8.6 Complete traction control system block diagram

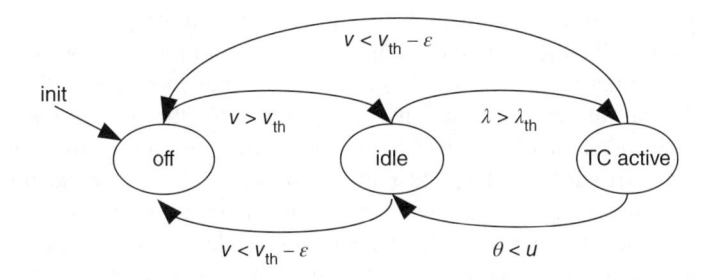

Figure 8.7 Traction system finite state machine

TC system. Depending on the value of velocity and wheel slip, the machine is in one of three states:

- **off**–the traction control system is enabled only above a threshold velocity v_{th}. This is necessary because, as the velocity tends to 0, the wheel dynamics becomes infinitely fast and the wheel slip computation is affected by numerical errors. The use of the deadband ε avoids chattering.
- **idle**–in this state the control system is ready to intervene if the activation rule is triggered.
- **TC active**–in this state the control system is active and modulates the engine torque.

The activation is wheel slip based ($\lambda > \lambda_{th}$); on the other hand the deactivation is based on the value of the driver's requested throttle (measured at the throttle grip) and the TC system requested throttle. If the driver is requesting a lower throttle than currently applied by the TC system, the TC system is shut off. This behaviour is imposed by a motorcycle proprietary ECU, and serves as an additional safety system; the TC system cannot deliver more torque than commanded by the driver. The combination of the deactivation and activation logics does not guarantee the absence of chattering, that is continuous transition between the master 'idle' and 'TC active' state. The TC system must therefore be designed so that in case of chattering of the finite state machine (FSM), there is no chattering in the requested torque.

8.3.2 Slip Reference Generation

A fixed threshold and slip reference system would suffer from two drawbacks: 1) The ideal threshold wheel slip heavily depends on the surface. High friction surfaces need higher wheel slip than low friction surfaces; 2) A fixed wheel slip would deprive the driver of any control when the TC is active. In fact, this is one of the biggest complaints with safety-oriented TC systems. Both problems are solved with the introduction of a scheduled wheel slip reference. The activation threshold is fixed, but once it is crossed, the driver can modulate the wheel slip reference through the throttle grip. This strategy is implemented by a two-threshold mechanism. When the activation logic is triggered the current wheel slip and throttle position are saved and referred to as λ_{hold} and θ_{hold}; these values are then used to generate the map:

$$[\theta_{hold}, 100\%] \rightarrow [\lambda_{threshold}, \lambda_{max}]. \tag{8.8}$$

The rider controls the slip by opening or closing the throttle; when the throttle is fully open the rider is requesting the maximum allowed slip. This mechanism guarantees better safety, robustness and controllability than the usual single-threshold logic. In the single-threshold logic an error in tuning can result in a limited or dangerous motorcycle. By allowing the rider to directly control the reference slip, the system helps the rider to control the bike, without limiting its performance. Map (8.8) can implement different shapes, and the practitioners can use it to give different feelings to the motorcycle. Figure 8.8 plots three examples. In the first case a linear map is employed, which guarantees uniform sensitivity; a convex shape gives a better controllability at low wheel slips, while a concave map is better suited for precisely modulating high wheel slip. Of course, the practitioners have the freedom to implement other ideas and strategies. Note that the maps should also be defined for values lower than θ_{hold} and do not need to intercept the origin. The deactivation logic is not based on wheel-slip, but on θ, so it is possible that during the modulation the throttle grip is closed with respect to the initial position. This strategy also avoids deadlocks in the 'TC active' state. In the remainder of the analysis, a simple linear map between throttle position and slip reference is employed.

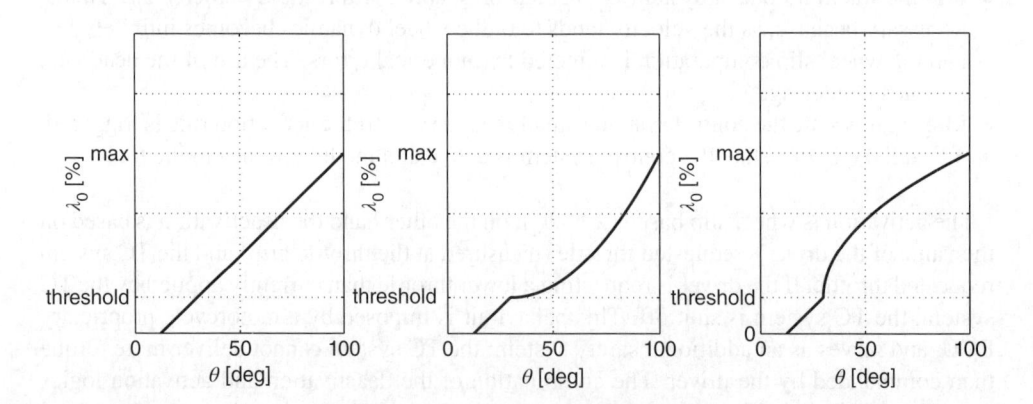

Figure 8.8 Three possible wheel slip reference–throttle grip maps: linear, convex, concave

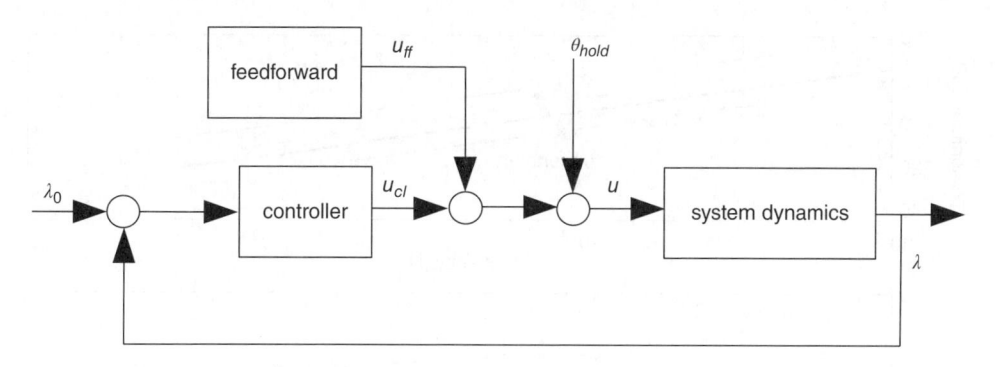

Figure 8.9 Wheel slip control law block diagram

8.3.3 Control Law Design

The wheel slip control law is the heart of the TC system. In order to avoid chattering and the resulting discontinuity in wheel slip, a linearized control scheme is devised (Figure 8.9). The throttle at the activation instant, θ_{hold}, is used as the linearization throttle around which the controller modulates the control variable.

The control action is determined by a closed loop term that provides robustness and tracking and a feedforward term to improve reactiveness. The dynamics shown in Figure 8.5 allows for a model-based control law design, thus simplifying the tuning phase. The dynamics are described by a transfer function with the following structure:

$$G(s) = \mu \frac{\frac{s^2}{\omega_z^2} + 2\frac{\xi_z}{\omega_z} + 1}{\left(\frac{s^2}{\omega_p^2} + 2\frac{\xi_p}{\omega_p} + 1\right)\left(\frac{s}{p_1} + 1\right)} e^{-s\tau} \qquad (8.9)$$

The system's dynamics are subject to considerable variation. The plots show how they are predicted to vary on different surfaces, and the methodological analysis showed that the dynamics depend on the level of wheel slip and velocity. Two strategies are possible: schedule the controller depending on the velocity and wheel slip (as shown in Corno et al. 2009); or approach the problem with a fixed structure controller. The first approach can give good performance, but the scheduling may become difficult to tune and manage. The second approach is simpler to implement but the achievable bandwidth is limited. A third, hybrid approach is possible: a *weakly* gain scheduled controller proves to be robust, reasonably performing and cost-efficient to implement. The controller is based on a PID with lead–lag filter robustly designed for all conditions whose parameters are only marginally adapted as the friction changes.

A 5× multiplicative uncertainty is added to the wheel slip dynamics to account for varying velocity, surface and wheel slip levels. The resulting loop transfer functions are shown in Figure 8.10. The resulting bandwidth varies from 0.06 Hz to 1 Hz, a considerable variation that can be avoided only by the precise knowledge of the friction characteristics or by using more advanced design techniques such as the one described by (Tanelli et al. 2009b).

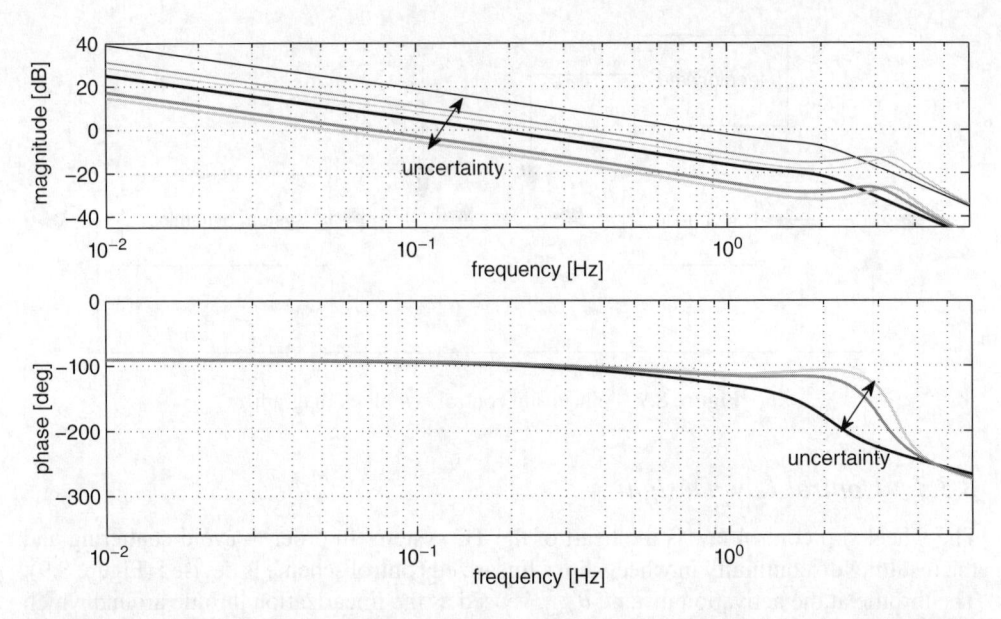

Figure 8.10 Loop transfer function for different surfaces and accounting for a 5× multiplicative uncertainty

The closed-loop term has been tested by using step responses. Figure 8.11 shows the results. Despite the conservative design approach, the control guarantees repeatable responses on different surfaces and a good tracking of the reference.

The controller accurately tracks the desired wheel slip, when requested through the throttle grip, but the system does not react quickly enough to reject a sudden loss of

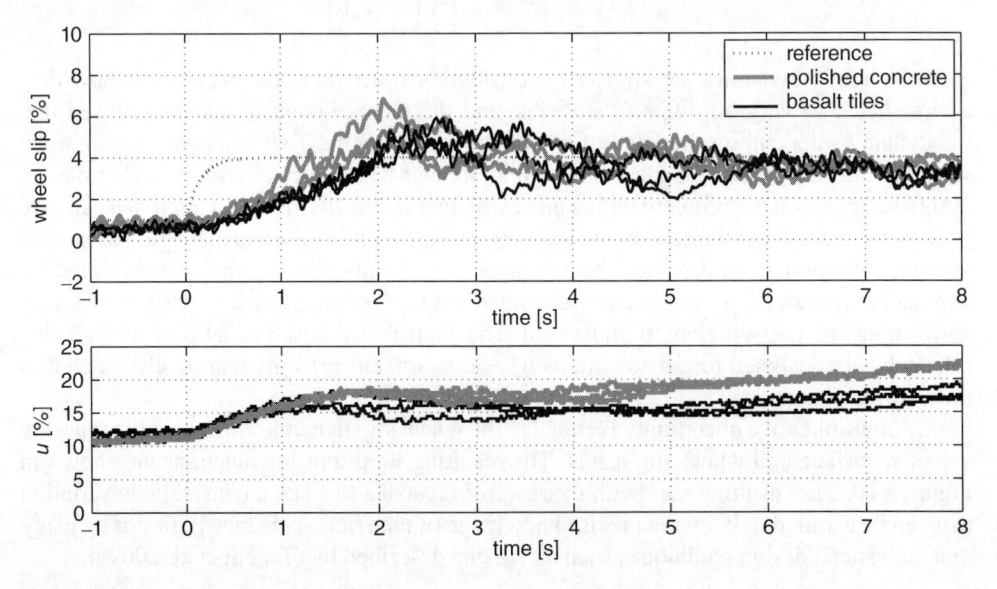

Figure 8.11 Wheel slip reference step responses for two different surfaces. Wheel slip (top plot) and requested throttle (bottom plot)

friction, such as when driving through a puddle. A feedforward action guarantees safety in these cases. The feedforward action is based on a threshold on the time derivative of λ: if, despite the TC being active, an increase of wheel slip is detected, the feedforward action applies a reduction of the throttle proportional to θ_{hold} (with gain K_{ff}). Figure 8.12 shows the feedforward gain tuning effects. At 0 s, the motorcycle crosses from a dry asphalt road to a wet concrete road. Without the feedforward term, the wheel slip reaches 30% before converging to a steady state. This response is suboptimal for two reasons: 1) the high wheel slip may compromise the vehicle stability; 2) the controller overcompensates the initial spike causing an undershoot that limits the longitudinal acceleration. If tuned correctly, the feedforward term solves both issues. The figure helps in understanding the role of K_{ff}. If too high a value is chosen ($K_{ff} = 0.95$) the first problem is solved and thus the manoeuvre is safe, but the motorcycle is not accelerating as much as the rider intends. A less aggressive tuning ($K_{ff} = 0.50$) provides a better response: the initial spike is absent as well as the undershoot. Being an open-loop term, the effect of the feedforward action will depend on the motorcycle conditions; however, the choice of K_{ff} is considerably simplified by the fact that the closed-loop term provides most of the control effort.

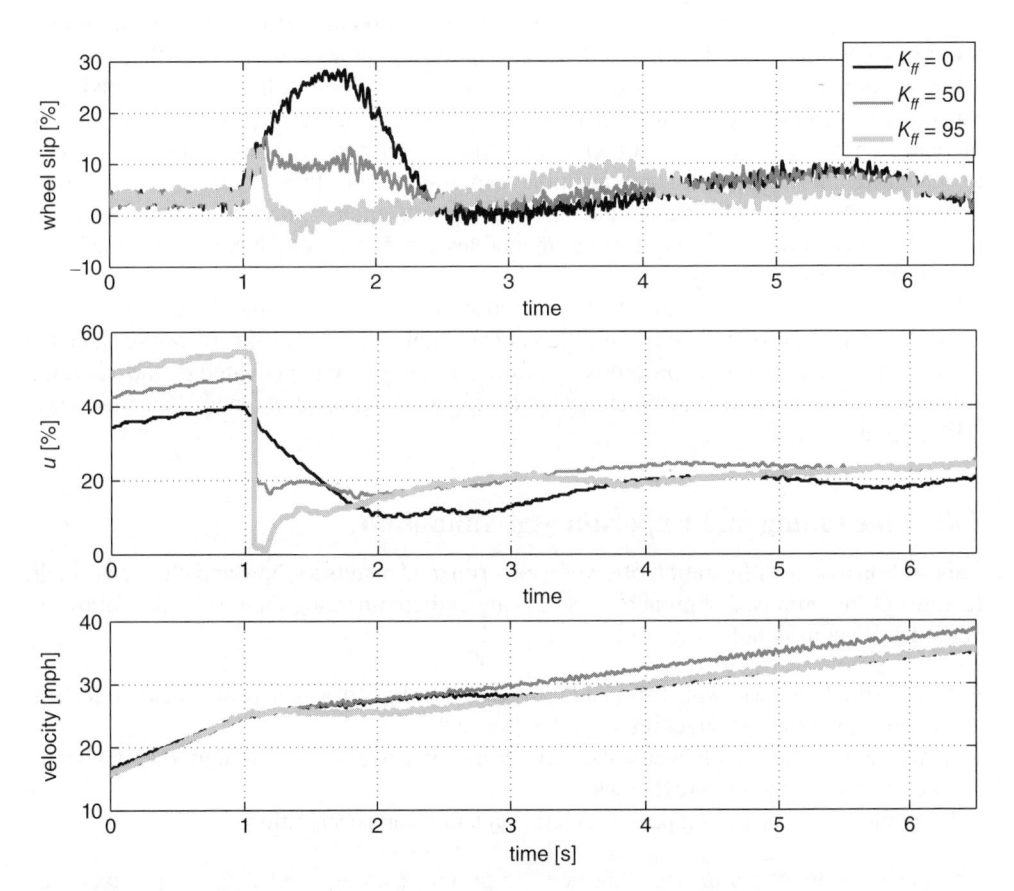

Figure 8.12 Tuning of the feedforward component. Effect on wheel slip (top plot), requested throttle (centre plot) and longitudinal velocity (bottom plot)

8.3.4 Transition Recognition

Although not critical from the safety point of view, the transition from low to high friction also needs to be addressed. When, with the TC activated, the motorcycle goes from a slippery to a high grip surface, the TC may be perceived as sluggish. This is another consequence of the robust tuning of the controller. The only way to improve the response is to introduce gain scheduling, based on the road surface. The main issue with such an approach is the recognition of the road friction coefficient. Many results are available (Rajamani et al. 2012 and works cited therein), showing that robust online road surface estimation is far from being solved. The difficulty is partially solved by scheduling, based on the detection of sudden changes of friction. The sluggish behaviour is perceived only as a consequence of sudden increases of friction. If the road condition changes slowly, the bandwidth of the controller is high enough to track the desired wheel slip. Recognizing sudden friction changes can be easily and robustly done.

The system recognizes the transition using the information on the longitudinal acceleration and wheel slip and schedules the closed loop controller accordingly. In particular, the integral time of the controller is adjusted in order to speed up the transient: the controller starts from the most conservative setting and adjusts itself.

Figure 8.13 presents the idea of the transition recognition algorithm. The plot shows the throttle, longitudinal acceleration and wheel slip during an acceleration test that starts on a low friction surface and then transitions (at around 6 s) onto a high friction road. The transition is characterized by a sudden drop of wheel slip and a longitudinal acceleration increase. These two conditions are characteristic of such a phenomenon and can be used to recognize the transition. The transition recognition algorithm is based on two equal derivative filters that estimate $\dot{a}_x$ and $\dot{\lambda}$: by comparing their sign (using thresholds) it is possible to detect the event. From experimental tests, a threshold that recognizes transition in less than 100 ms is found.

Figure 8.14 plots the comparison of the fixed structure controller and the gain-scheduled controller. The gain-scheduled controller considerably speeds up the response, and the motorcycle accelerates more promptly, a difference that is well appreciated by the test rider. Every time the controller is activated or the feedforward action intervenes, the controller is reinitialized.

8.4 Fine tuning and Experimental Validation

This section uses the illustration of several experimental results to show and clarify the main features of the proposed approach. In analysing and commenting the results the following variables are considered:

- Wheel slip: both peak wheel slip and reference tracking error. The peak weakly relates to stability, and reference tracking to performance.
- Yaw rate: correlates with lateral stability. When the wheel slip is too high, the rear lateral force drops and the motorcycle wags.
- Longitudinal acceleration: peak acceleration and mean acceleration.

The most important parameters that need to be tuned are λ_{th} and λ_{max}, respectively the wheel slip that activates the traction control and the maximum allowed wheel slip. Their

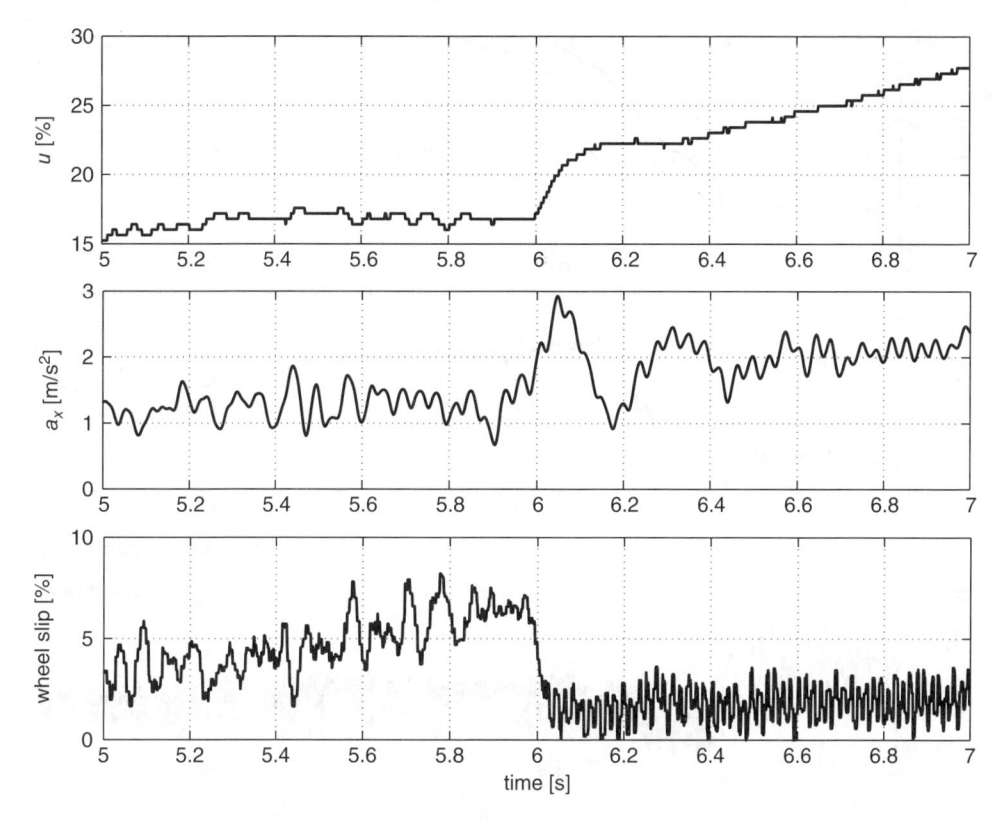

Figure 8.13 Transition from a low to a high friction surface with a fixed structure controller. The transition happens at around 6 s. Requested throttle (top plot), longitudinal acceleration (centre plot) and wheel slip (bottom plot)

tuning is critical because they directly influence the *character* of the motorcycle. Under the hypotheses of perfect wheel slip reference tracking, the choice of λ_{th} only depends on the road surface. As pointed out in the design phase, the wheel slip controller is far from having infinite bandwidth, so the choice of λ_{th} affects the peak wheel slip and the tuning of the feedforward action. Figure 8.15 graphically represents the trade-offs (without the feedforward action). As λ_{th} is increased, the TC intervention is delayed, and the wheel builds up momentum, leading to a higher initial overshoot. On the other hand, choosing too low a threshold heavily affects the acceleration. The plot shows that $\lambda_{th} = 3\%$ leads to a very smooth tracking of the wheel slip reference but considerably limits the acceleration in the first 2 seconds. The final choice does not depend only on quantitative analysis; subjective feedback plays a critical role. Some riders prefers the initial spike as it sends a clear message that he/she has done something wrong.

Also, the choice of λ_{max} mainly affects the riding feeling. The TC stabilizes the wheel slip dynamics: the rider's task is therefore considerably simplified, but they set the wheel slip reference. Depending on the level of expertise, they may be able to drive the motorcycle safely with high wheel slip or not. Some riders may prefer a thrilling motorcycle that is still capable of drifting and thus requires some expertise; others prefer a perfectly safe but thrill-less

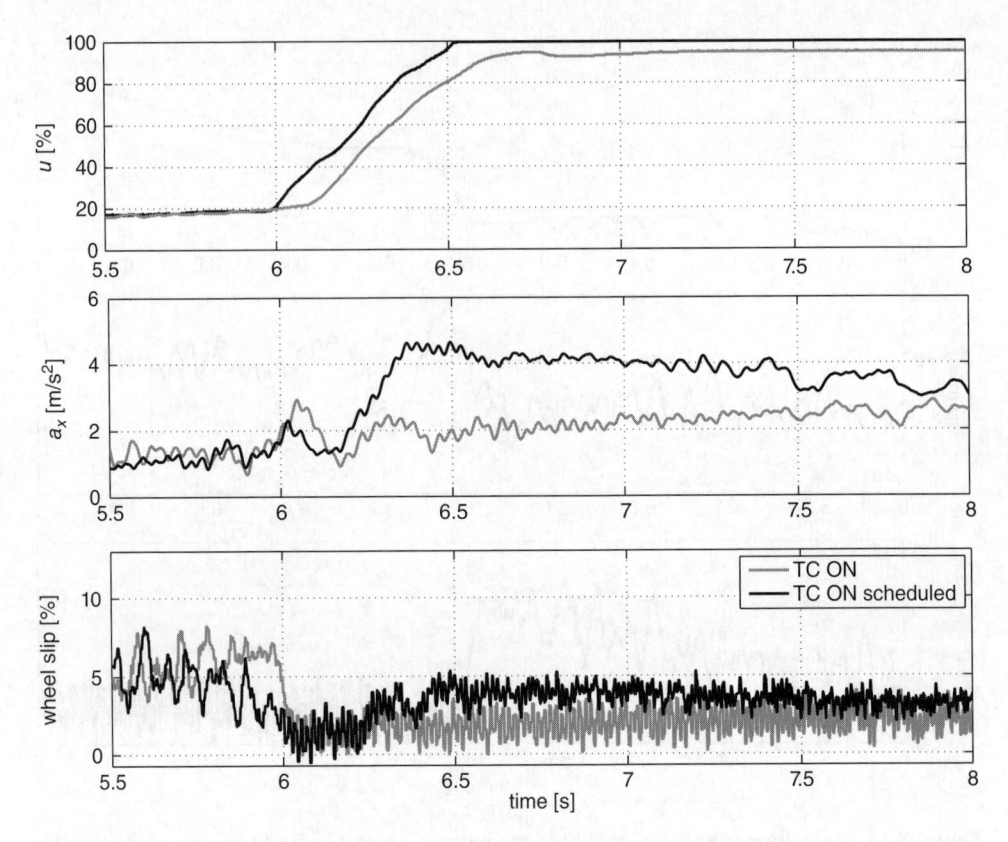

Figure 8.14 Transition from a low to a high friction surface. Comparison of the fixed structure control and the gain-scheduled controller. Requested throttle (top plot), longitudinal acceleration (centre plot) and wheel slip (bottom plot)

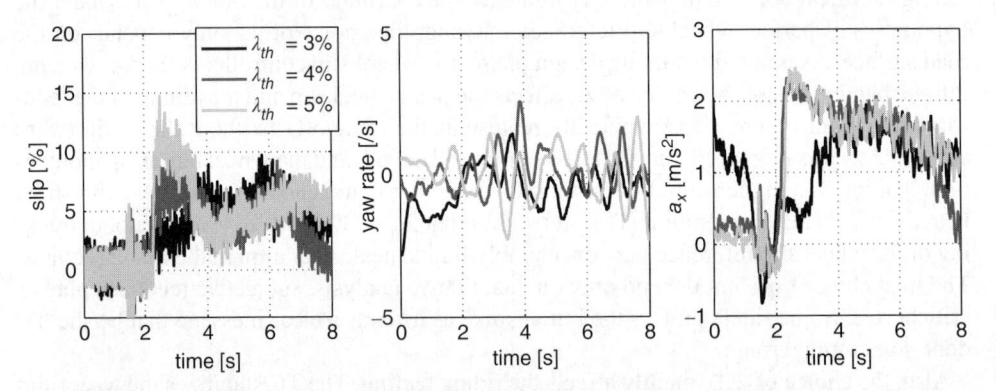

Figure 8.15 Acceleration on wet polished concrete for different levels of λ_{th}, wheel slip, yaw rate and longitudinal acceleration

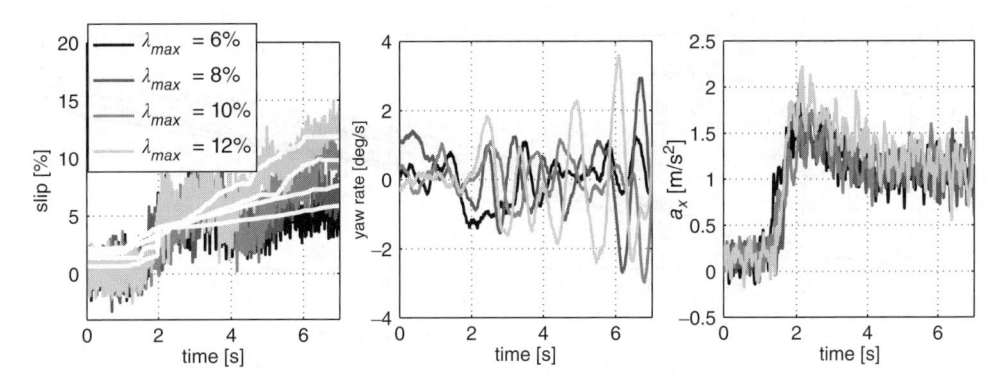

Figure 8.16 Acceleration on wet polished concrete for different levels of λ_{max}, wheel slip, yaw rate and longitudinal acceleration

motorcycle. Figure 8.16 better explains the quantitative advantages and disadvantages of having a high λ_{max}. Note that the wheel slip accurately tracks the reference. The longitudinal acceleration in the second part of the manoeuvre is not affected by the tuning because the friction curve is rather flat around λ^*. On the other hand λ_{max} has a clear effect on the yaw rate. The higher the wheel slip the greater the yaw rate oscillations, indicating an incipient lateral instability. Were an estimation of the vehicle roll angle available (Ryu et al. 2002 and Boniolo et al. 2009), λ_{max} could be scheduled as a reasonable choice between 5% to 6%.

In order to assess the overall effect of the TC system, four different scenarios are investigated: acceleration on basalt tiles, acceleration on polished concrete, transition from high to low friction and transition from low to high friction. The following tests are performed by professional test riders: they are instructed to accelerate as fast as possible, with the TC system assist (TC on) and without (TC off). For safety reasons, the tests could not be properly blinded: the rider was always aware of what kind of system they were testing.

Figure 8.17 plots the results of a series of accelerations on wet basalt tiles, from an initial velocity of 12 mph. The following conclusions can be drawn:

- The traction control system successfully controls the wheel slip in all cases. The TC system clearly outperforms the manual acceleration in terms of wheel slip peak. When the TC is active, the average wheel slip peak is less than 12%, an improvement of 43% with respect to the manual acceleration.
- The TC yields a more stable motorcycle: yaw rate oscillations peaks are reduced of 61% with respect to the manual acceleration.
- The TC does not limit acceleration performance. No differences are observed either in the maximum or the mean acceleration.
- From all standpoints, the manoeuvres performed with TC are more consistent than the manual ones.

The basalt tiles are the most challenging conditions (lowest friction) tested. In these challenging conditions, it can be concluded that TC control improves safety without impeding performance. Tests performed starting from higher velocity ($v = 25$ mph) confirm the conclusions.

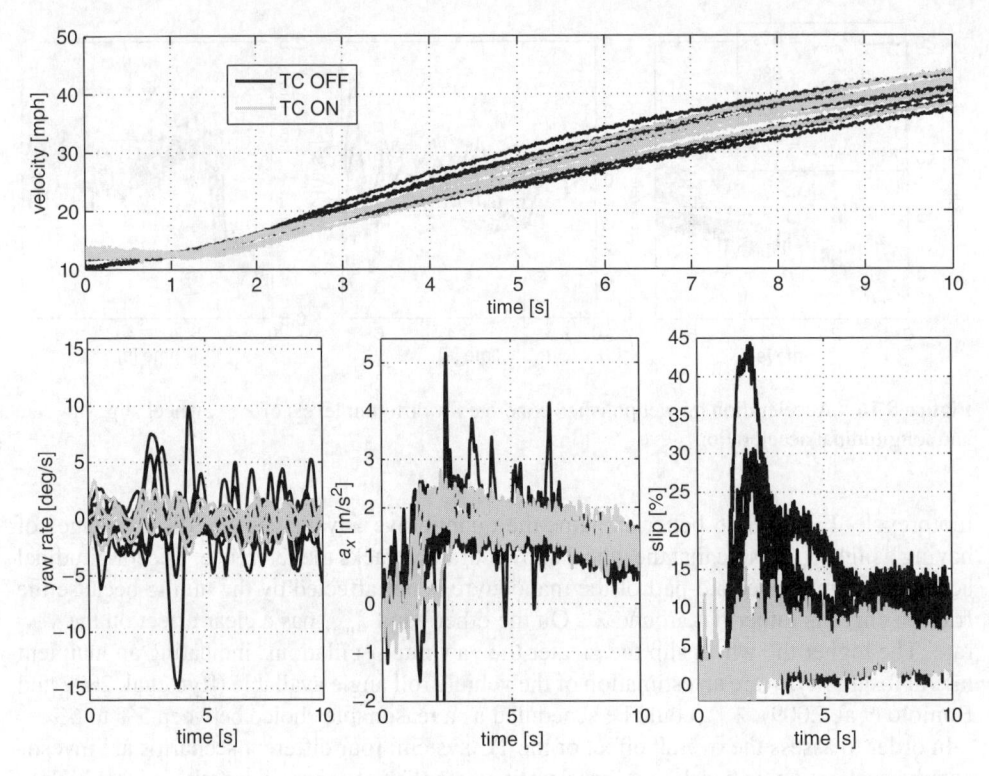

Figure 8.17 Acceleration on wet basal tiles from 12 mph. Velocity (top plot), yaw rate (bottom left plot), longitudinal acceleration (bottom centre plot) and wheel slip (bottom right)

A series on analogous tests has been performed on wet polished concrete. Wet polished concrete has a higher friction coefficient than basalt tiles and is a more common condition for road motorcycles. The tests performed starting at 25 mph are shown in Figure 8.18. The following conclusions are due:

- The traction control system successfully controls the wheel slip. When the TC is active the wheel slip peak is 9%, which represents an improvement of 50% with respect to the manual manoeuvre.
- The TC system makes the manoeuvres more stable. The peak yaw rate during the test is reduced of 70% with respect to the manual acceleration.
- In this case the TC affects the acceleration profile. During the first phase of the acceleration, the rider reaches higher acceleration without the traction control; however, as the velocity increases, the motorcycle loses stability and the rider is forced to close the throttle. With the TC on, the motorcycle reaches a lower maximum acceleration during the first phase, but then, by precisely modulating the wheel slip, the rider is better able to accelerate at higher velocity. As a consequence the peak acceleration is reduced by 15% on average, while the average acceleration over the entire manoeuvre improves by about 7.5%. The difference with respect to the previous scenario is due to the fact that the two surfaces have different optimal wheel slip. The chosen λ_{th} is closer to λ^* of the basalt tiles than of the polished concrete.
- The TC helps the rider accelerate more consistently.

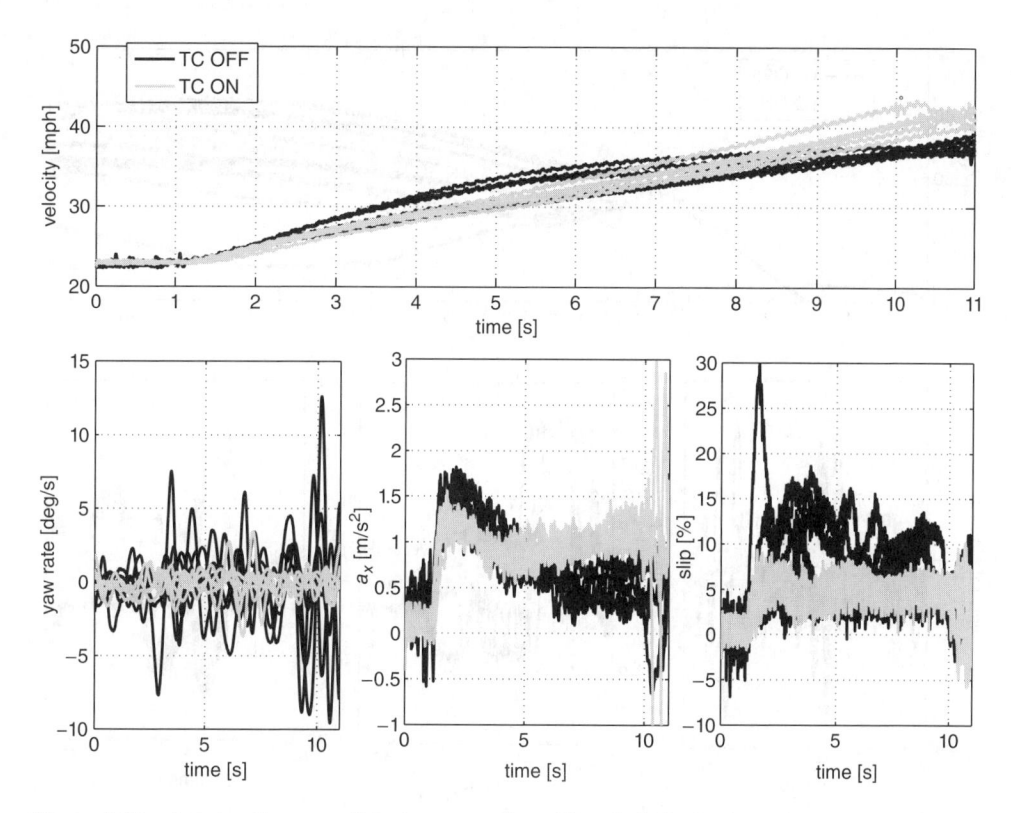

Figure 8.18 Acceleration on polished concrete from 25 mph. Velocity (top plot), yaw rate (bottom left plot), longitudinal acceleration (bottom centre plot) and wheel slip (bottom right)

As pointed out in the design section, sudden transitions can strain the TC system. In the remainder of the section two μ-jump, that is sudden changes of friction coefficient, are tested, Figure 8.19 plots the results of a series of tests where wet asphalt suddenly changes to wet basalt tiles. The rider is instructed to get to at the friction jump at around 25 mph. The following conclusions can be drawn:

- This is a very challenging manoeuvre even for professional riders. The rider is not able to manually deliver a consistent acceleration. In some cases, they are forced to brake the rear wheel to stabilize the motorcycle (negative rear wheel slip).
- The post jump maximum wheel slip with and without the TC system are comparable, but note that the TC is able to recover from the initial peak faster than the rider.
- The better control of the wheel slip is also seen in terms of yaw rate. The maximum yaw rate peaks are lower with the TC activated and are rejected more quickly. The oscillations without the TC ends at 6 s, while the TC rejects them at around 4.2 s.
- In terms of acceleration, the TC is not able to perform as well as the best manual trial, but on average it delivers more consistent performance and a better overall (across the tests) acceleration.

In conclusion, the TC system considerably improves stability and repeatability, even if it limits the best achievable performance.

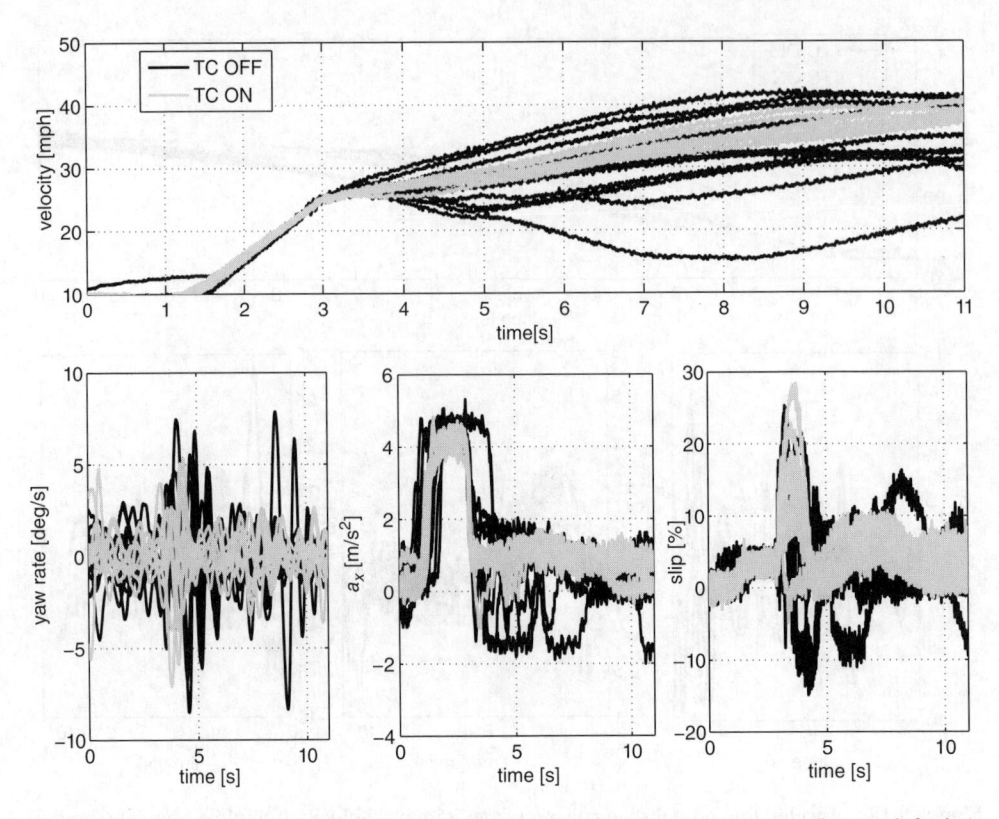

Figure 8.19 μ jump: high-to-low friction transition. Velocity (top plot), yaw rate (bottom left plot), longitudinal acceleration (bottom centre plot) and wheel slip (bottom right)

The reverse μ-jump (from low to high friction) validates the scheduling approach. The rider accelerates on wet basalt tiles that suddenly change into a wet asphalt road. This test is non-safety-critical, and the TC system is not expected to improve either safety or performance. A good TC system should not interfere with or impede the manoeuvre. Figure 8.20 summarizes the results. It can be seen that the acceleration performance is marginally affected. The TC reduces the mean acceleration of 4% and the peak acceleration of 20%. In analysing these results, one should also keep in mind that, for safety reasons, it is not possible to properly *blind* the tests. The rider is always aware of what system is being testing and the transition is clearly visible; it is therefore impossible to exclude some level of anticipation of the opening of the throttle when the TC was inactive.

8.5 Conclusions

This chapter presents a systematic approach to the design of TC systems for motorcycles. The design of a TC system for a touring motorcycle is taken as an example to quantitatively describe different aspects. After a brief introduction to the problem and the state of the art, the controller architecture has been detailed. Particular attention has been given to the three main aspects of the design: the activation logic, the set point generation and the control law tuning.

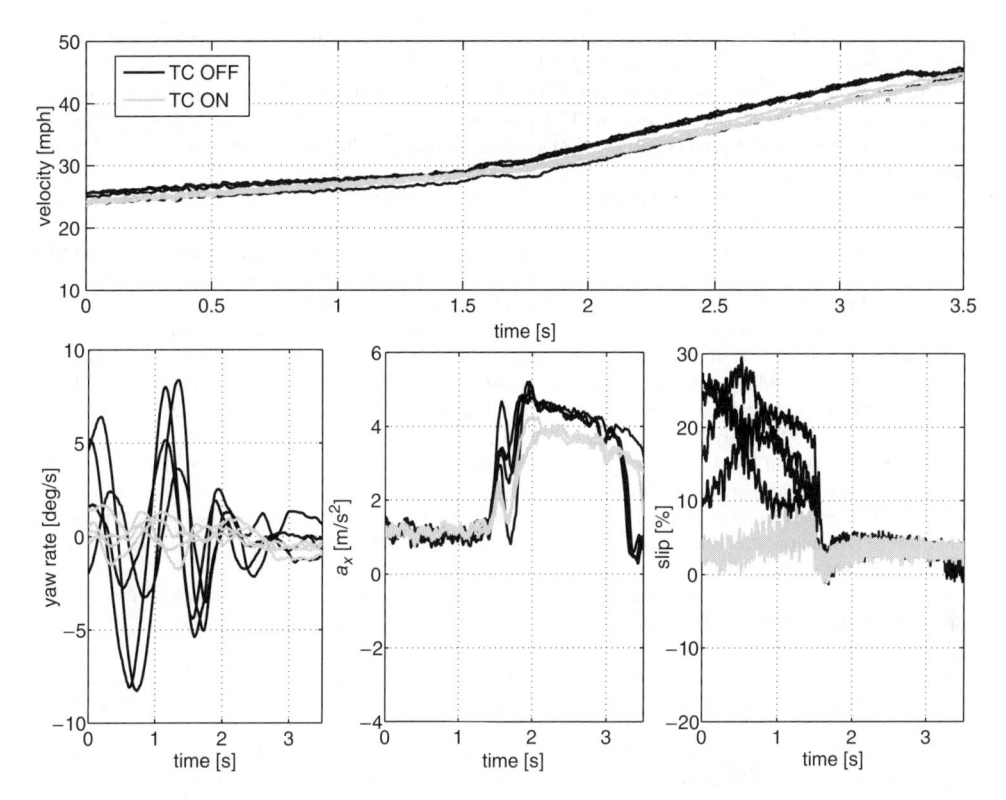

Figure 8.20 μ jump: low-to-high friction transition. Velocity (top plot), yaw rate (bottom left plot), longitudinal acceleration (bottom centre plot) and wheel slip (bottom right)

The proposed control architecture offers a compromise between the two approaches currently followed by the industry: performance- and safety-oriented solutions. This is obtained by letting the rider use the throttle to specify a reference slip. By setting a limit to the maximum slip, safety is increased without impeding performance. This aspect is believed to be crucial for the success of a traction control system for motorcycles.

The slip dynamics model identified in the previous chapter was employed to design a robust controller that was then refined using experimental results. The experimental results confirmed the thesis presented in (Corno et al. 2009) that a fixed structure slip controller cannot guarantee both robustness and performance. These limitations have been solved by introducing a feedforward action and a scheduling of the controller based on the recognition of changes in friction.

References

Austin L and Morrey D 2000 Recent advances in antilock braking systems and traction control systems. *Proceedings of the Institution of Mechanical Engineers, Part D: Journal of Automobile Engineering* **214**(6), 625–638.

Boniolo I, Savaresi S and Tanelli M 2009 Roll angle estimation in two-wheeled vehicles. *IET Control Theory and Applications* **3**(1), 20–32.

Cardinale P, D'Angelo C and Conti M 2008 A traction control system for motocross and supermotard *Intelligent Solutions in Embedded Systems, 2008 International Workshop on*, pp. 1–15 IEEE.

Chun K and Sunwoo M 2004 Wheel slip control with moving sliding surface for traction control system. *International Journal of Automotive Technology (S1229-9138)* **5**(2), 123–133.

Corno M, Gerard M, Verhaegen M and Holweg E 2011 Hybrid ABS control using force measurement. *Control Systems Technology, IEEE Transactions on* **PP**(99), 1–13.

Corno M, Savaresi S, Tanelli M and Fabbri L 2008 On optimal motorcycle braking. *Control Engineering Practice* **16**(6), 644–657.

Corno M, Savaresi S and Balas G 2009 On linear parameter varying (LPV) slip-controller design for two-wheeled vehicles. *International Journal of Robust and Nonlinear Control* **19**(12), 1313–1336.

Dardanelli A, Alli G and Savaresi S 2010 Modeling and control of an electro-mechanical brake-by-wire actuator for a sport motorbike *Proceedings of the 5th IFAC Symposium on Mechatronic Systems, Cambridge, MA, USA*.

De Filippi P, Tanelli M, Corno M and Savaresi SM 2011 Enhancing active safety of two-wheeled vehicles via electronic stability control *Proceedings of the 18th IFAC world congress on automatic control*, pp. 638–643.

Ferrara A and Vecchio C 2007 Low vibration vehicle traction control to solve fastest acceleration/deceleration problems via second order sliding modes *American Control Conference, 2007. ACC '07*, pp. 5236–5241 IEEE.

Hirsch V 2009 Curve safe traction control for racing motorcycles *Proceedings of the SAE World Congress 2009*. Publikationsnotiz: SAE World Congress, 20–23 April 2009, Detroit, Michigan, USA.

Kabganian M and Kazemi R 2001 A new strategy for traction control in turning via engine modeling. *Vehicular Technology, IEEE Transactions on* **50**(6), 1540–1548.

Kang S, Yoon M and Sunwoo M 2005 Traction control using a throttle valve based on sliding mode control and load torque estimation. *Proceedings of the Institution of Mechanical Engineers, Part D: Journal of Automobile Engineering* **219**(5), 645–653.

Liu Z, Wan R, Shi Y and Chen H 2011 Simulation analysis of traction control system for four-wheel-drive vehicle using fuzzy-pid control method. *Advances in Information Technology and Education* pp. 250–257.

Pacejka H 2002 *Tyre and Vehicle Dynamics*. Buttherworth Heinemann, Oxford.

Panzani G, Corno M, Savaresi S, Fabbri L, Ricci A, Fioravanzo F and Lisanti P 2010a Metodo di controllo della trazione per partenze da fermo in un motoveicolo ed apparato implementante lo stesso Italian Patent MI2010A000877.

Panzani G, Corno M, Savaresi S, Fabbri L, Ricci A, Fioravanzo F and Lisanti P 2010b Metodo per il riconoscimento dell'impennata e per la gestione della trazione in un motoveicolo Italian Patent MI2010A000878.

Rajamani R, Phanomcheoeng G, Piyabongkarn D and Lew JY 2012 Algorithms for real-time estimation of individual wheel tire-road friction coefficients. *Mechatronics, IEEE/ASME Transactions on* **17**(6), 1183–1195.

Ryu J, Rossetter E and Gerdes J 2002 Vehicle sideslip and roll parameter estimation using gps *Proceedings of the International Symposium on Advanced Vehicle Control (AVEC), Hiroshima, Japan*, pp. 373–380.

Savaresi S and Tanelli M 2010 *Active Braking Control Systems Design for Vehicles*. Springer Verlag.

Savaresi S, Corno M, Formentin S and Fabbri L 2010 System and method for controlling traction in a two-wheeled vehicle US Patent 20100312449.

Seiniger P, Schroter K and Gail J 2012 Perspectives for motorcycle stability control systems. *Accident Analysis and Prevention* **44**(1), 74–81.

Tanelli M, Corno M, Boniolo I and Savaresi SM 2009a Active braking control of two-wheeled vehicles on curves. *International Journal of Vehicle Autonomous Systems* **7**(3), 243–269.

Tanelli M, Vecchio C, Corno M, Ferrara A and Savaresi SM 2009b Traction control for ride-by-wire sport motorcycles: A second-order sliding mode approach. *IEEE Transactions on Industrial Electronics* **56**(9), 3347–3356.

9

Motorcycle Dynamic Modes and Passive Steering Compensation

Simos A. Evangelou[a] and Maria Tomas-Rodriguez[b]
[a]*Department of Electrical and Electronic Engineering, Imperial College, London, UK*
[b]*School of Engineering and Mathematical Sciences, City University, London, UK*

9.1 Introduction

In this chapter we introduce the main characteristics of motorcycle oscillatory modes and then discuss how their damping can be increased by employing steering compensation. Two of the modes, weave and wobble, have a major influence on vehicle lateral stability. If they become lightly damped or undamped, under certain operating conditions, they can have a detrimental effect on rider safety, and also they significantly reduce the riding comfort and manoeuvrability of the vehicle.

Steering damping, which is the most common solution to alleviate the instability problems of wobble mode, has a negative effect on weave stability. Mechanical steering compensators comprising springs, dampers and inerters, when employed as replacements for conventional steering dampers are thought to provide simultaneous stabilization of wobble and weave. Here, we analyse the influence of such passive mechanical networks on vehicle stability, and describe how they can be designed and implemented, the focus being on high-performance sports motorcycles.

Our approach is influenced by classical passivity ideas from circuit theory as well as analogies between electrical and mechanical networks. In the standard electrical-mechanical current-force analogy, an inductor corresponds to a spring, while a damper can be represented by a resistor. In order to complete this analogy, another mechanical component is needed to represent a capacitor. Although a mass is analogous to a capacitor with one terminal grounded, a new device is required that does not have any restriction on its terminal connections. The inerter, as proposed by (Smith 2002), is a suitable mass-like element, and it can be of either translational or rotational nature.

In this chapter we also introduce the phenomenon of burst oscillations in racing motorcycles that are accelerating at high speed. Such steering oscillations reduce rider confidence and are therefore harmful to lap times, and potentially dangerous for riders. We study and explain the bursting phenomenon using a high-fidelity computer model, and then propose a design methodology using mechanical steering compensators to damp these oscillations. The methodology is analogous to that used to improve the constant-speed dynamic performance of road-going sports machines already mentioned. An important difference in the study of bursting oscillations is the need to deal with acceleration, since it is under these conditions that the bursting occurs.

The chapter is divided in the following sections. Section 9.2 introduces the main motorcycle modes and discusses qualitatively their characteristics. Section 9.3 describes a high-fidelity motorcycle model that is then used in Section 9.4 to illustrate quantitatively the characteristics of the oscillatory modes and the influence of steering damping on machine stability. Section 9.5 describes a design methodology for steering compensation that improves modal damping. Section 9.6 provides an analysis of burst oscillations followed by a design of steering compensation to suppress these oscillations. Finally, conclusions are given in Section 9.7.

9.2 Motorcycle Main Oscillatory Modes and Dynamic Behaviour

The free steering system of motorcycles and the corresponding self-steering action is the cause of their oscillatory nature. For obvious safety reasons there is a need to control the oscillatory tendency of motorcycles throughout their operating range. Also, it is required that motorcycles respond to the rider's commands. Therefore stability is pursued, taking into account handling qualities. The dynamic response of motorcycles is a consequence of these requirements, which are the currently adopted design principles for modern machines.

Broadly speaking, motorcycle natural modes can be divided in two groups, *in-plane* and, *out-of-plane* modes (Cossalter et al. 2004; Limebeer et al. 2002; Sharp et al. 2004):

- *In-plane modes*

These are modes whose shape involves freedoms in the motorcycle's plane of symmetry, such as pitch, bounce and front/rear wheel hop. They usually only affect rider comfort and do not represent a threat to the machine's stability.

- *Out-of-plane modes*

These involve lateral motion, and, depending on their frequency and damping properties, they can seriously affect motorcycle stability, so they have the potential to compromise rider safety. These modes are *capsize*, *weave* and *wobble*.

Capsize mode is a slightly unstable slow roll mode that disappears when speed is increased above a certain limit. A speed of approximately 7 m/s will usually stabilize this mode, with its natural frequency being approximately 0.8 Hz at that speed. This mode is usually easily dealt with by the rider's action and therefore it does not represent a concern when discussing stability of motorcycles.

Weave mode is a fishtailing-type motion involving yaw and roll of the motorcycle body and steering system oscillations. It is well damped at low to moderate speeds but becomes less damped as the speed is increased. The natural frequency of this mode is zero at very low speed and increases to somewhere in the range of 2–4 Hz, depending on the mass and size of the machine, with the lowest frequencies corresponding to the heaviest motorcycles.

Wobble mode is a steering oscillation that is reminiscent of the supermarket cart front-wheel caster shimmy. The few existing and properly documented wobble oscillations involve moderate speeds, although there are many anecdotal accounts of wobble at high speeds. Theoretical results indicate that the torsional stiffness of the motorcycle frame at the steering head determines whether it will be prone to wobbling at medium speeds or at high speeds, corresponding to compliant and stiffer framed machines respectively (Foale 2002; Sharp and Alstead 1980; Spierings 1981). The natural frequency of this mode does not change much with speed. It is normally in the range 6–9 Hz and it is governed primarily by the mechanical trail, the front frame steer inertia and the front tyre cornering stiffness (Evangelou and Limebeer 2000a). The stiff framed machines, being prone to wobbling at high speed, are often fitted with a steering damper in order to achieve wobble stability, but a steering damper will normally have a destabilizing effect on high speed weave.

The *in-plane* and *out-of-plane* modes in *straight running* conditions are uncoupled for small perturbations and they can be studied separately. On the other hand, when the motorcycle is *cornering*, the out-of-plane (lateral) modes and the in-plane modes associated with tyre deflections and suspension motions become coupled (Koenen 1983; Sharp et al. 2004). Linearization for small perturbations from cornering trim states allows modes to be calculated and analysed, but the mode shapes become complex. Weave couples strongly with bounce and pitch, since their natural frequencies are similar; similarly, wobble interacts with front suspension motions (Limebeer et al. 2002). Under cornering, the motorcycle becomes prone to resonate in response to regular road forcing when the combination of road undulation wavelength and vehicle speed produces a forcing frequency that is matched to the natural vibration frequencies of the machine. Moderate lean angles are found to represent worst case conditions (Limebeer et al. 2002).

In addition to small oscillations, motorcycles can obviously undergo more general motions. The full motorcycle/rider equations that describe the general motion are mathematically complex and contain general nonlinear terms. The virtual machine and, by implication, real machines, can be expected to demonstrate the behavioural properties of general nonlinear dynamical systems (Nayfeh and Mook 1979). General motions will naturally be associated with larger amplitudes of vibration, in which the nonlinear terms increase in relative importance in the equations of motion. The experience of the rider will arise mostly from operation near trim states, and as this experience does not consider nonlinear phenomena, suddenly encountering new situations of different character can be a source of difficulty for the rider (Shaeri et al. 2004).

There are several works that deal with unusual operating circumstances. Some of them have shown by accurate simulations the important role that road undulations, road camber and acceleration play in the weave and wobble mode stability and damping (Evangelou et al. 2008, 2012; Limebeer et al. 2002). An improved understanding of the small perturbation and nonlinear behaviour of motorcycles brings implications for motorcycle design, rider training, road maintenance and accident investigation.

9.3 Motorcycle Standard Model

We make use of a mathematical model to analyse and illustrate the behaviour of motorcycles. The model comprises multiple rigid bodies introduced in a tree structure, and it is an evolution of prior computer models (Evangelou and Limebeer 2000a,b; Sharp and Limebeer 2001; Sharp et al. 2003, 2004, 2005). It represents the class of modern road-going sports machines, and has a parameter set based on the Suzuki GSX-R1000. The machine geometry is shown in Figure 9.1, in which the motorcycle's seven constituent masses are represented by circles that have a diameter that is proportional to the mass of the associated body. All the critical points, such as mass centres, body joints and so on, are individually marked.

The model involves a main frame that is allowed unrestricted motion with six degrees of freedom. The swinging arm, upper body of the rider and front frame are attached to it with pitching, leaning and steering freedoms respectively. Twisting of the front frame is also allowed, and it involves small angular displacements at the steering head about an axis that is perpendicular to the (inclined) steering axis. The model also has spinning road wheels, standard rear monoshock suspension, and telescopic front forks that allow linear in-line displacements. The mathematical model employed here also takes into account the aerodynamic forces and moments, which are proportional to the square of the speed. Wide road tyre models with vertical compliance are used to track dynamically the migration of both of the ground-contact points as the machine rolls, pitches and steers. The forces and moments due to the interaction between the ground and the tyre are generated from the

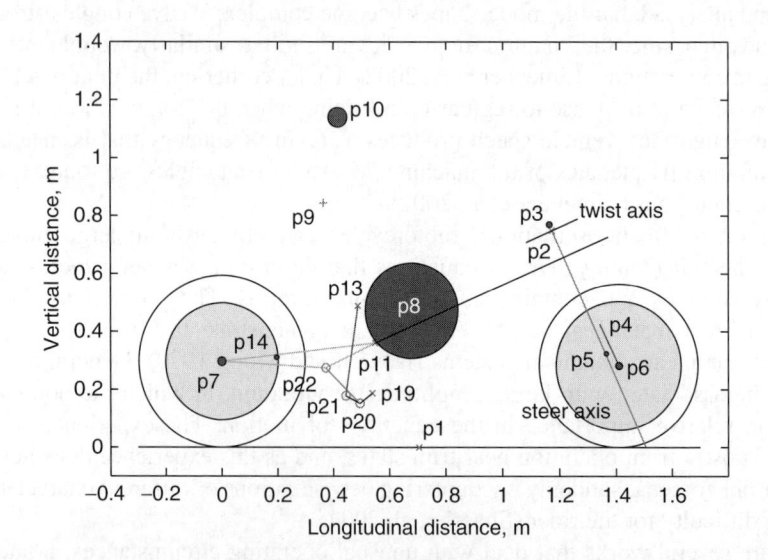

Figure 9.1 Motorcycle model in its nominal configuration. The constituent bodies are shown as circles, with their radii proportional to their mass. All the critical points of the model are individually marked

combined slip, normal load and camber angle relative to the road using 'magic formulae'. The lateral tyre carcass compliance is modelled using standard linear first-order relaxation type tyre models (de Vries and Pacejka 1997, 1998; Pacejka 2002; Tezuka et al. 2001).

The overall model is built with the multibody modelling code VehicleSim® (formerly called AutoSim, Anon. 1998) and it can be obtained from the website.* The VehicleSim model can be configured to generate C++ code that numerically integrates the nonlinear equations of motion, or it can be used to generate a MATLAB script with a symbolic state-space representation that describes small perturbations around a prescribed trim condition. Here, the trim states are found by nonlinear simulation in which the machine is controlled at constant forward speed and roll angle by speed-error to rear-wheel-drive-moment and roll-angle-error to steering-torque PI and PID feedback loops respectively. Once obtained, the trim states are used to set up, in numerical state-space form, the open-loop linearized equations of motion.

One of our purposes is to describe a phase-compensation-based design for better influencing the self-steering action as compared to using conventional steering dampers. Such compensation is achieved with the use of passive mechanical networks consisting of springs, dampers and inerters. The effects of a steering damper (or a more general steering compensator) can be introduced in the motorcycle model through the differential equations that describe it. For the particular purposes of our study, the generalized regulator feedback structure (Green and Limebeer 1995), shown in Figure 9.2, is used to represent the integration of the steering compensation system with the rest of the model. The steering compensator appears as the feedback element, $K(s)$. The generalized plant, $P(s)$, is a linear time-invariant model of the subject motorcycle. It has vertical road displacement disturbances $d(s)$ and steering torque $T_s(s)$ as inputs, and the steering angle $\delta(s)$ and the steering velocity as outputs. The vertical road displacement disturbances are introduced to take into account the effect of road profiling on steering oscillations.

Although Figure 9.2 shows a frequency-domain model of the linearized system, it is representative of the physical interconnection of the steering compensator and machine in general, and therefore it is also applicable to nonlinear time-domain studies.

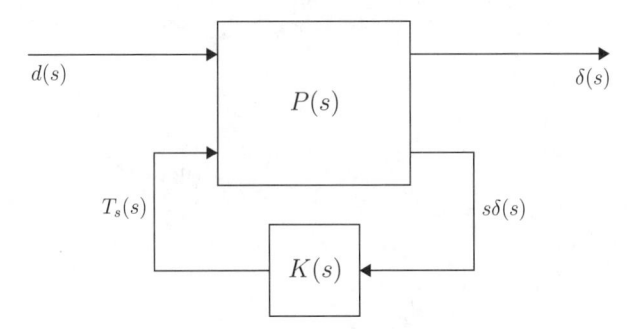

Figure 9.2 Diagram representing the steering compensator function as a feedback arrangement. $P(s)$ is the linearized motorcycle model, $K(s)$ is the steering compensator, signal $d(s)$ represents vertical road displacement disturbances, $T_s(s)$ is the steering torque and $\delta(s)$ is the steering angle. s is the standard Laplace transform variable

---

* http://www.imperial.ac.uk/controlandpower/motorcycles/

If $P(s)$ is partitioned as:
$$P(s) = \begin{bmatrix} P_{11}(s) & P_{12}(s) \\ P_{21}(s) & P_{22}(s) \end{bmatrix},$$
then the generalized regulator configuration in this case is defined by
$$\begin{bmatrix} \delta(s) \\ s\delta(s) \end{bmatrix} = \begin{bmatrix} P_{11}(s) & P_{12}(s) \\ sP_{11}(s) & sP_{12}(s) \end{bmatrix} \begin{bmatrix} d(s) \\ T_s(s) \end{bmatrix},$$
and
$$T_s(s) = K(s)s\delta(s),$$
which gives
$$\delta(s) = (I - sP_{12}(s)K(s))^{-1} P_{11}(s) d(s).$$

Repeated reference will be made to the Nyquist criterion of the open-loop system $sK(s)P_{12}(s)$, in which $P_{12}(s)$ maps the steering torque $T_s(s)$ into the steering angle $\delta(s)$.

9.4 Characteristics of the Standard Machine Oscillatory Modes and the Influence of Steering Damping

Many modern road motorcycles are equipped with a conventional steering damper. The mechanical function of a damper is to create a moment that opposes the angular velocity of the steering assembly with respect to the main frame. In respect of stability, its purpose is to alleviate high-speed wobble oscillations in the case of stiff-framed motorcycles, and medium-speed oscillations in the case of older and more flexible-framed vehicles.

The characteristics of the main motorcycle oscillatory modes, wobble and weave, are now illustrated in the root-locus diagrams of Figures 9.3 and 9.4. The diagrams are generated

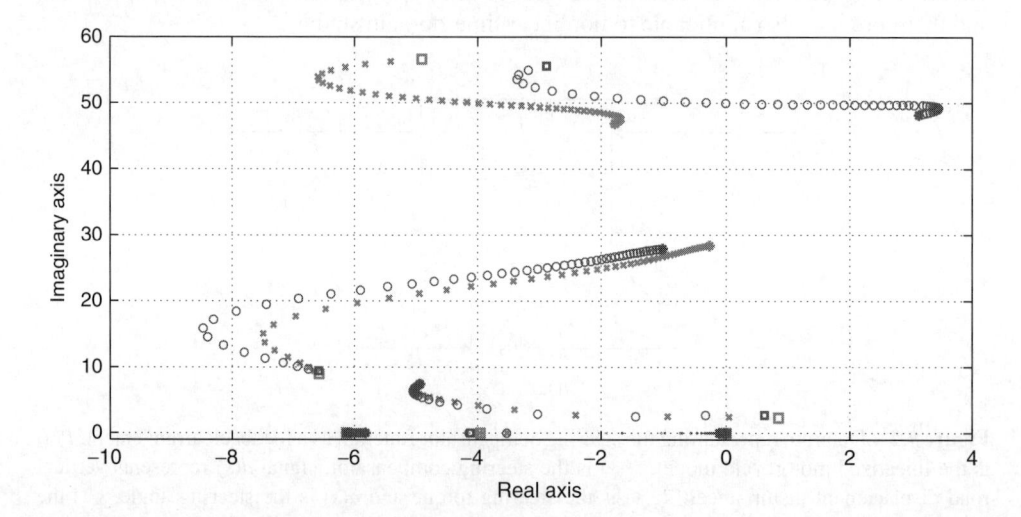

Figure 9.3 Straight running root-loci with forward speed the varied parameter, for the machine with no steering damping (∘) and with 6.944 Nms/rad (nominal) steering damping (×). The speed is increased from 5 m/s (□) to 75 m/s (★)

using linearizations of the standard motorcycle model described in Section 9.3. Figure 9.3 loci correspond to the case that the machine is operating under straight-running conditions, and both free-steering (open loop) and damper-fitted (closed loop) cases are shown. The forward speed of the vehicle ranges from 5 to 75 m/s in steps of 2 m/s. It can be seen from this diagram that the wobble-mode frequency varies between 47 and 57 rad/s while the weave mode frequency varies between 10 and 28 rad/s. It can also be seen clearly that when a steering damper is not fitted, the damping of the wobble mode decreases with increasing speed, and the mode becomes unstable at approximately 25 m/s, while the weave mode is stable for all speeds. On the other hand, once the nominal steering damper with damping constant of 6.944 Nms/rad is fitted, satisfactory wobble mode stability is achieved, but at the expense of weave mode damping. This clearly represents a conflict in the choice of suitable steering damper coefficients that can provide simultaneous wobble and weave damping.

Figure 9.4 shows root-loci for the machine fitted with the nominal steering damper for three values of lean angle. It can be seen that increased values of roll angle tend to increase the high-speed weave-mode damping. Since the coupling between the in-plane and out-of-plane dynamics increases with roll angle, the weave mode vulnerability to road displacement forcing is expected to maximize at an intermediate value of roll angle ($\sim 15°$) (Limebeer et al. 2002). It can also be seen from Figure 9.4 that for roll angles up to 30°, the high-speed wobble-mode damping increases with roll angle. At low speeds, the wobble-mode damping decreases monotonically with roll angle, and the vulnerability of this mode is worst at low speed and high roll angles.

The effect of changing the steering damper coefficient can be studied further by generating the Nyquist diagram shown in Figure 9.5. The frequency response for straight running at 75 m/s is considered. In the case of a steering damper as the fitted compensator in the feedback loop of Figure 9.2, $K(s)$ becomes a constant, c, say. According to the well-known Nyquist criterion, closed-loop stability is achieved when there are N counter-clockwise

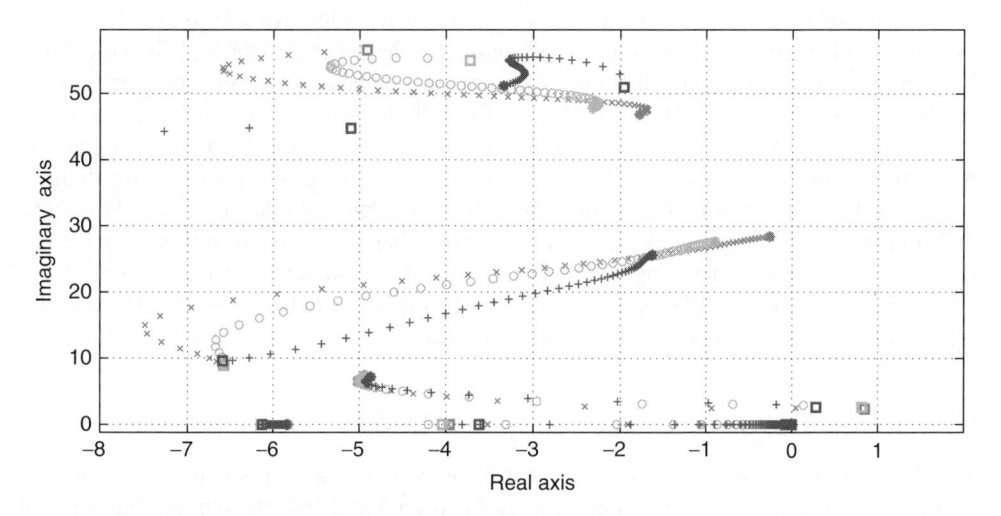

Figure 9.4 Root-locus plots with forward speed the varied parameter, for three cases of roll angle: straight running (×), 15° (∘) and 30° (+). The nominal steering damper is fitted. The speed is increased from 7 m/s (□) to 75 m/s (⋆).

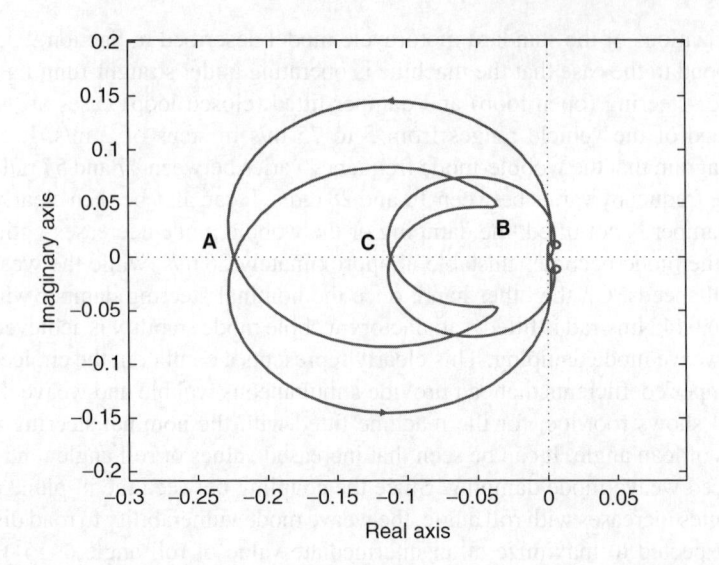

Figure 9.5 Straight-running Nyquist diagram for the open-loop motorcycle model for a forward speed of 75 m/s. The frequency at A is 47.6 rad/s, at B it is 33.8 rad/s and at C it decreases to 28.4 rad/s

encirclements of the $-1/c$ point, where N is the number of unstable poles of the open-loop system and c is the value of the steering damping. As can be seen from Figure 9.3, at 75 m/s the wobble mode of the nominal motorcycle, without its steering damper, is unstable with a corresponding complex conjugate pair of poles in the right-half plane. Therefore, under these operating conditions, two counter-clockwise encirclements of the $-1/c$ point are required. This is achieved for a (small) range of damper values, as is obvious from the Nyquist diagram. Indeed, if the amount of steering damping is too low, the wobble mode becomes unstable, and if it is too high, the weave mode becomes unstable. With reference to Figure 9.5, this aspect of the motorcycle behaviour can be better illustrated. When the steering damping value is low, such that the $-1/c$ point is located at A, the system is marginally stable and will oscillate at 47.6 rad/s, which is the wobble-mode frequency. When the steering damping is now increased, the $-1/c$ point lies between A and C where it is encircled twice in a counter-clockwise sense and the machine will be stable. When the steering damper is increased further, to cause the $-1/c$ point to be coincident with the point C, the machine will oscillate at 28.4 rad/s, with the weave mode now being marginally stable. Any further increases in the steering damping will render the machine unstable due to the absence of any encirclements of the $-1/c$. The nominal steering damper value of 6.944 Nms/ rad locates the $-1/c$ point at -0.144, which is approximately midway between points A and C in Figure 9.5.

9.5 Compensator Frequency Response Design

The first step in the design of passive steering compensator networks is to understand the influences of the spring and the inerter, as isolated components – the damper has already been studied in the previous section. Their effect on the wobble- and weave-mode damping and stability is particularly of interest.

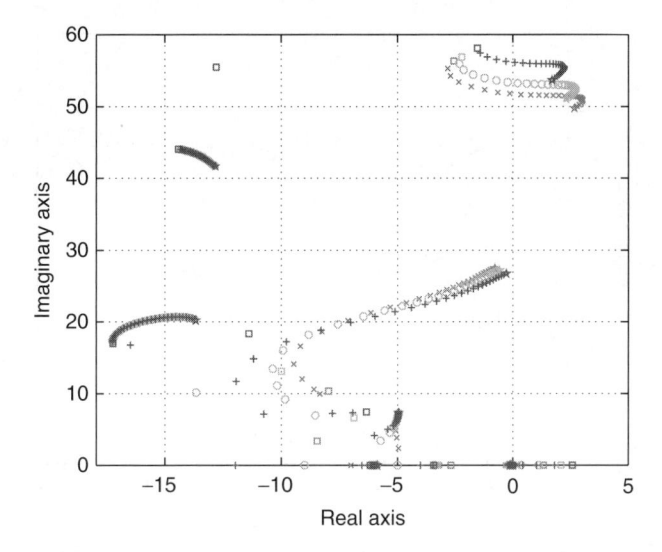

Figure 9.6 Straight-running root-loci with forward speed as the varied parameter for three cases of torsional spring: 100 Nm/rad (×), 200 Nm/rad (∘) and 400 Nm/rad (+). The speed is increased from 5 m/s (□) to 75 m/s (⋆). All cases correspond to the nominal machine without a steering damper

The root-loci in Figure 9.6 show the effect of a simple torsional spring on the modal damping of the machine. As the spring stiffness is increased, the wobble-mode natural frequency increases. At intermediate and high speeds, the wobble mode is unstable for all spring stiffness values but it becomes less unstable as the stiffness increases. The spring has a negligible effect on the weave mode, for example at high speed, where the weave-mode damping is low. When the spring is considered as the feedback element in Figure 9.2, $K(s) = k/s$, where k is the spring stiffness.

The inerter is a two-terminal mechanical device with the property that an equal and opposite force applied at its terminals is proportional to the relative acceleration between them (Smith 2002). There are linear and rotational versions of the inerter, and their inertance is measured in kilograms and kgm^2 respectively. In its rotational form, the inerter provides a resisting moment M that is proportional to the relative angular acceleration between its terminals. In mathematical form,

$$M = b(\dot{\omega}_1 - \dot{\omega}_2),$$

where ω_1 and ω_2 are the angular velocities at the two terminals, and b is the inertance. Figure 9.7 illustrates the effect on the machine modal damping characteristics of introducing an inerter, for which now $K(s) = bs$. It can be seen that the wobble mode natural frequency reduces as the value of inertance is increased. This is not surprising as the change is similar to increasing the moment of inertia of the front frame assembly, whose rotation is the main component of the wobble mode shape. Increased inertance also makes the wobble mode more unstable and increases the damping of the high-speed weave mode.

When comparing Figures 9.3 and 9.7, it is suggested that an effective steering compensator should behave like an inerter at low frequencies (2–3 Hz) in order to improve the damping of

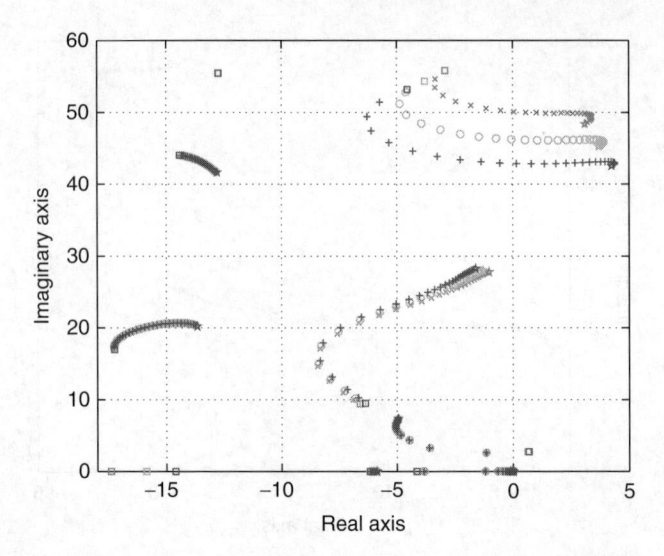

Figure 9.7 Straight-running root-loci with forward speed as the varied parameter for three cases of steering inertance: no inertance (×), 0.1 kgm² (∘) and 0.2 kgm² (+). The speed is increased from 5 m/s (□) to 75 m/s (⋆). All cases correspond to the nominal machine without a steering damper

the weave mode, while acting as a damper at higher frequencies (5–9 Hz) in order to stabilize the wobble mode. This can be interpreted as a form of lead compensation. Over the still lower frequency range of 0–0.5 Hz used by the rider (Aoki 1979; Sugizaki and Hasegawa 1988), for balancing and path-following control, the compensator network must apply minimal torque to the steering system to allow unhindered rider steering action. Therefore high values of constant gain (damper-like) or spring-like properties are undesirable in that frequency range (Evangelou et al. 2006).

The three simple passive elements already considered can be combined into networks, the simplest containing only two components, to offer more possibilities. It is easy to show that a damper in series with an inerter provides the compensator function $K(s) = scb/(sb + c)$, where b and c are the inertance and damping coefficient respectively. The frequency response of this function is very similar to that of an inerter at frequencies below the break frequency c/b, and to that of a damper at frequency above c/b. This frequency response function can therefore be beneficial for simultaneous control of weave and wobble. Extending the above type of reasoning to networks with three components allows further networks to be utilized with potentially more promising properties. A related network is one comprising the series connection of a damper, an inerter and a spring (Figure 9.8).

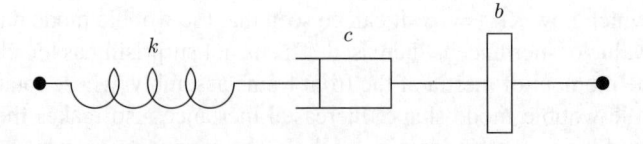

Figure 9.8 Simple steering compensation network comprising the series connection of a damper (c Nms/rad), inerter (b Kgm²) and spring (k Nm/rad)

Although being mostly similar, this network offers one important generalization in comparison to the two-component device $scb/(sb+c)$ identified above: it allows a rapid phase change in the neighbourhood of the resonant frequency $\omega_n = \sqrt{k/b}$. As with the series combination of damper–inerter, it has inerter-like characteristics at low frequencies. In the neighbourhood of ω_n it is damper-like and at high frequencies it has the characteristics of a spring. In our application, ω_n will be tuned to the frequency of the wobble mode so as to introduce damping there. The compensator function provided by this network is:

$$K(s) = k \frac{s}{s^2 + s\frac{k}{c} + \frac{k}{b}}$$

It may be observed that this is a second-order rational function and it is a special case of a positive-real biquadratic function. It is possible to re-parametrize the network in terms of its undamped natural frequency and damping ratio as

$$K(s) = k \frac{s}{s^2 + 2\zeta\omega_n s + \omega_n^2}$$

in which

$$\omega_n = \sqrt{\frac{k}{b}}$$

and

$$\zeta = \frac{\sqrt{bk}}{2c}.$$

The frequency response characteristics of the network are shown in Figure 9.9, where it can be verified that this network introduces inerter-like behaviour at low frequencies and

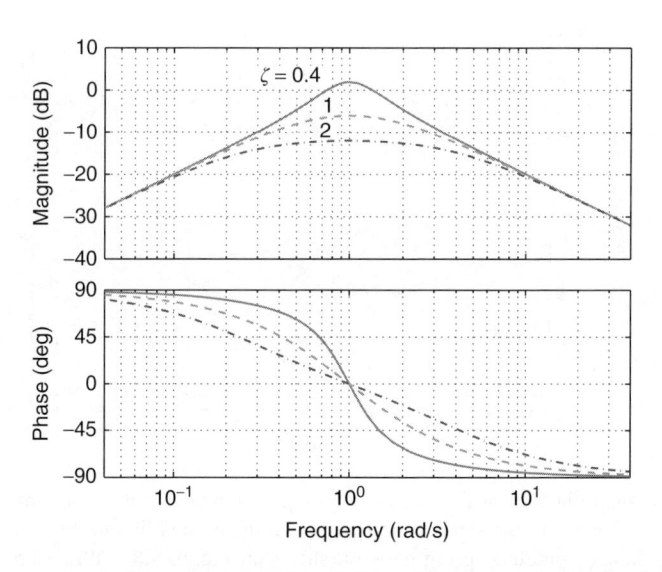

Figure 9.9 Frequency response characteristics of the series damper–inerter–spring network with the resonant frequency normalized to $\omega_n = 1$. Three values of damping ratio ζ are illustrated

damping in the vicinity of ω_n, which can be tuned to the frequency of the wobble mode. The frequency response magnitude peak occurs exactly at the frequency ω_n and takes the value of the damper constant c. The damping ratio ζ is another parameter that can be used to tune the frequency response. It determines the sharpness of the magnitude peak and the rate of change of phase with frequency. Smaller values of ζ make the transition from inerter-like to damper-like faster. The larger ζ is, the wider the frequency range, where the damper-like behaviour persists. The parameters ω_n and ζ fully define the phase characteristics of the network.

The steering compensator is designed by selecting appropriate values for ω_n, ζ and k. Trial parameters $\omega_n = 50$ rad/s, closely matching wobble-mode frequency, and $\zeta = 0.4$ are selected. With reference to the Nyquist diagram for the 75 m/s straight-running condition in Figure 9.5, parameter k is chosen to position the point -1 at an appropriate location in the stable k-value range. It is observed that $k = 320$ Nm/rad places the point -1 in the middle of the range; however, since it does not produce adequate wobble-mode damping performance at high lean angles, the spring stiffness is increased to $k = 500$ Nm/rad. This change globally improves the wobble-mode damping but some weave-mode stability is sacrificed. After back substitution, inerter and damper parameter values of $b = 0.2$ kgm^2 and $c = 10$ Nms/rad, respectively, are found. The influence of this particular choice of parameters on the Nyquist diagram is illustrated in Figure 9.10. It may be observed that the network opens up the interval over which two counter-clockwise encirclements can be achieved, by moving the real-axis crossing point associated with weave-mode instability towards the origin, and the wobble-mode crossing point to the left of the diagram. The resulting root locus plot, when this mechanical network is applied, is shown in Figure 9.11. Although the design is based on a single high-speed straight-running linearized model, substantial improvements in the modal damping of both wobble and weave under all operating conditions are achieved.

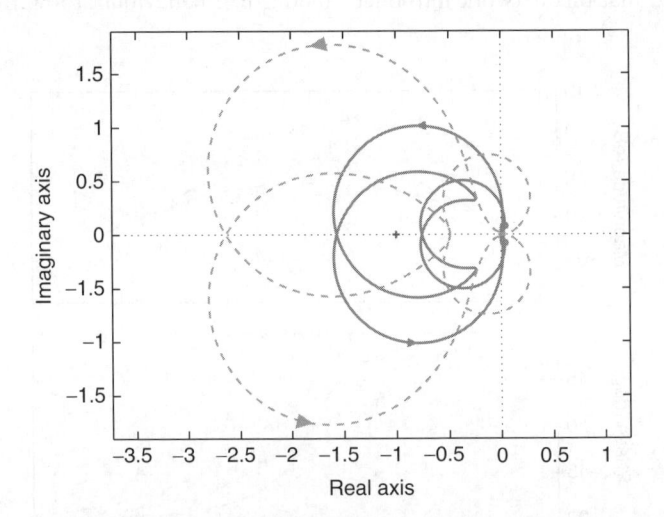

Figure 9.10 Nyquist diagram of the straight-running motorcycle with a forward speed of 75 m/s. The solid line corresponds to the standard machine, and the dashed line to the compensated system using the series damper–inerter–spring network shown in Figure 9.8, with design values $\omega_n = 50$ rad/s, $\zeta = 0.4$ and $k = 500$ Nm/rad

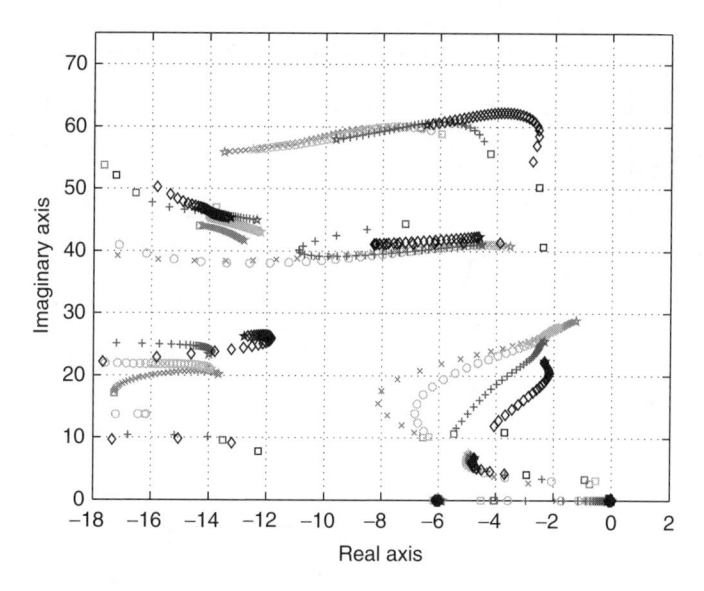

Figure 9.11 Root loci for the compensated motorcycle with forward speed the varied parameter. Four values of roll angle are illustrated: straight running (×), 15° (∘), 30° (+), and 45° (◊). The speed is varied from 7 m/s (□) to 75 m/s (⋆). The machine is fitted with the series damper–inerter–spring network shown in Figure 9.8, with the parameter values $b = 0.2$, $c = 10$ and $k = 500$

9.6 Suppression of Burst Oscillations

9.6.1 Simulated Bursting

Race-track measurement data presented in (Evangelou et al. 2012) reveals the presence of steering oscillations under high-speed acceleration conditions. Once the oscillations begin, they persist for a few seconds with a frequency of the order of 28 rad/s, which is consistent with weave-type behaviour. We attempted to replicate with a simulation model the measured bursting behaviour. The model of the subject motorcycle used is described in (Evangelou et al. 2012) and it is very similar to the one already described in Section 9.3. A simulation is set up in which the vehicle is operating under straight-running conditions with a speed reference of $v = v_0 + A \sin(0.2\,t)$, where $v_0 = 70$ m/s and $A = 25$ m/s. The output from this simulation is shown Figure 9.12. The motorcycle forward speed varies sinusoidally between 45 and 95 m/s, while its acceleration varies between ± 5 m/s^2. In order to represent the influence of road roughness, a low-amplitude steering torque disturbance is introduced into the simulation. The discontinuities in the acceleration signal originate from the 'snap' operation in the chain drive. Burst oscillation in the steering angular rate signal with amplitude of almost 2 rad/s can be observed when the machine is accelerating. The bursting appears to begin soon after the motorcycle reaches the peak acceleration of 5 m/s^2. This oscillation is also visible in the forward acceleration signal of the machine. The spectral content of the steering velocity bursting signal is shown in Figure 9.13. The data between 10 s and 20 s in Figure 9.12 was used to produce this result. It can be seen that the bursting frequency increases from approximately 33 rad/s to approximately 38 rad/s, indicating a

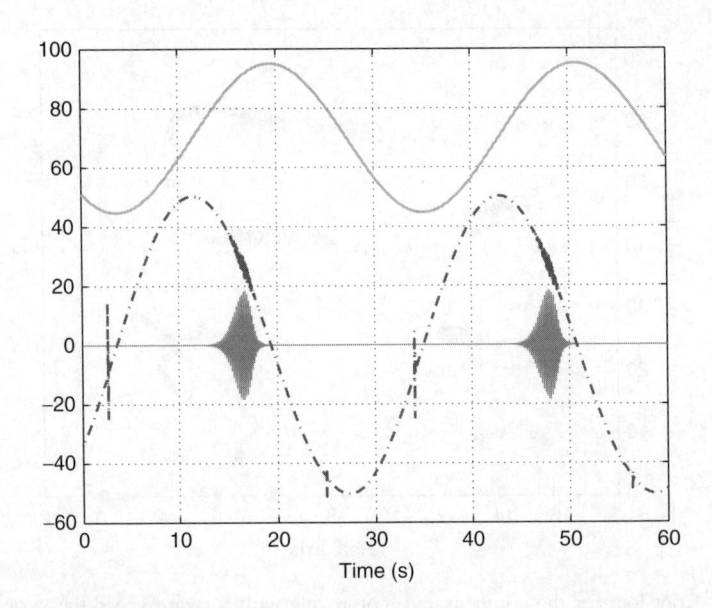

Figure 9.12 Simulated weave-mode type burst oscillations under straight-running conditions in the subject motorcycle. In the simulation the machine speed is varied according to $v = v_0 + A\sin(0.2\ t)$ m/s, while a small steering torque $t_p = \varepsilon \cos(2\pi 4.5\ t)$ Nm is applied simultaneously. The solid (green) line is the motorcycle speed, the dot-dashed (red) line is $10\times$ the acceleration, while the solid (blue) bursting characteristic is $10\times$ the steering velocity in rad/s. The parameter values used in the simulation are $v_0 = 70$ m/s, $A = 25$ m/s, and $\varepsilon = 5.0 \times 10^{-5}$ N-m

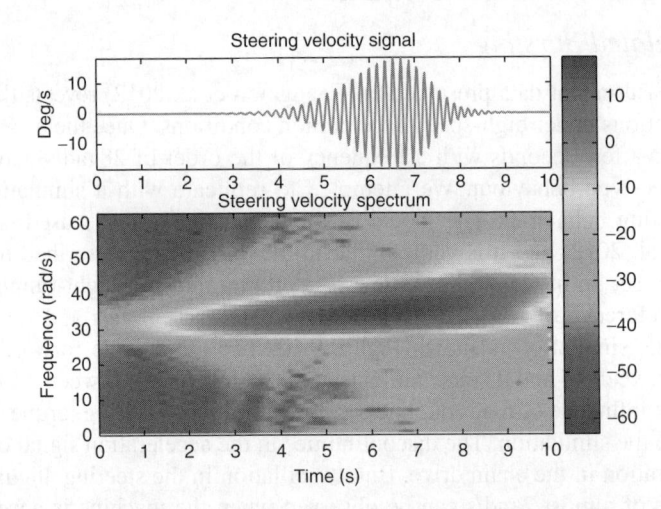

Figure 9.13 Short-period Fourier transform of the steering velocity signal given in Figure 9.12. Data between 10 s and 20 s was used – bursting is apparent between 14 s and 18 s (in Figure 9.12)

correspondence with the machine's weave mode (Figure 9.3). The slight increase in frequency is in agreement with increases in the machine speed.

9.6.2 Acceleration Analysis

The influence of acceleration on machine stability is now investigated on the basis of a linear time-invariant small perturbation analysis. It is observed that bursting-type instabilities occur on time scales over which vehicle speed variations caused by acceleration are relatively unimportant and, therefore, for the analysis the machine speed can be assumed constant. This situation is very similar to the analysis of the constant-speed machine on an inclined road surface where gravitational forces produce an analogous effect to longitudinal acceleration-related forces of inertia. The analysis framework is based on modelling ideas that make use of d'Alembert acceleration-related forces (Evangelou et al. 2012; Limebeer and Sharma 2010; Meijaard and Schwab 2006). When the motorcycle is accelerating at a m/s^2, then inertial forces of magnitude $F_i = m_i a$ are applied at the mass centres of each of the motorcycle's constituent masses. These forces act in the rear-wheel longitudinal direction in the case of the main frame, the rear swing arm and the rear wheel, while the inertial forces act in the front-wheel longitudinal direction in the case of the front steered body, the front suspension body and the front wheel.

Figure 9.14 illustrates the effect of accelerating and braking on the machine straight-running modal characteristics, and provides an insight into the causes of the bursting shown in Figure 9.12. Taking the constant-speed plots as reference, it can be seen that braking

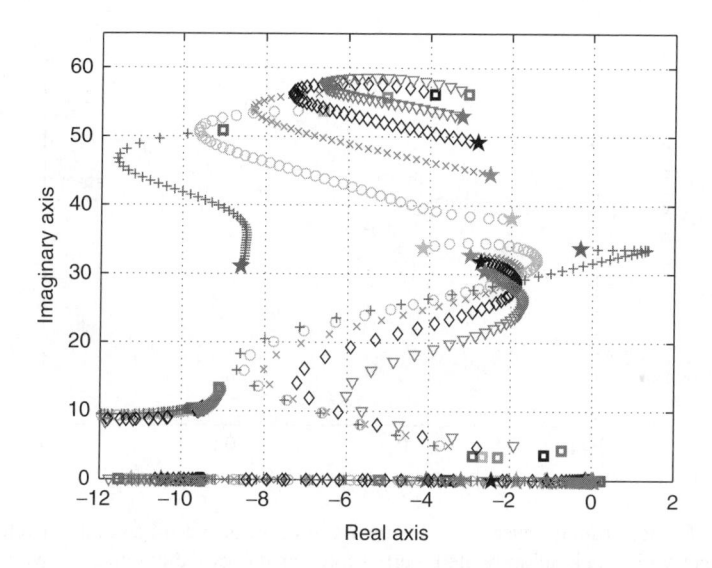

Figure 9.14 Straight-running wobble- and weave-mode eigenvalue root-loci of the subject motorcycle fitted with a steering damper, showing the effect of acceleration-related inertial forces. The speed is varied between 9 m/s (□) and 95 m/s (★) for five cases of acceleration-related inertial force: −4 m/s^2 (▽), −2 m/s^2 (◊), 0 m/s^2 (×), 2 m/s^2 (∘) and 4 m/s^2 (+)

causes the wobble-mode frequency to increase, while acceleration causes it to reduce. It is also clear that braking causes the weave-mode frequency to decrease, while acceleration causes it to increase. Of importance to the present research is the destabilizing interaction that occurs between the wobble and weave modes under steady acceleration at elevated speeds. This interaction increases the wobble-mode damping substantially, while at the same time causes the weave-mode damping to reduce and even become unstable over the range of speeds from 60 to 93 m/s. The design challenge is to avoid these destabilizing interactions.

The nature of the machine oscillatory behaviour is further understood from Figure 9.15, which shows the shape of the straight-running weave-mode eigenvector at 70 and 80 m/s, for a range of d'Alembert acceleration and braking forces. It is evident that firm acceleration reduces the relative magnitudes of the roll, yaw and lateral translation components of the weave-mode eigenvector, thereby making this mode more 'wobble-like'.

In Figure 9.16 the bursting phenomenon is investigated from a frequency-response perspective using open-loop Nyquist diagrams for the steering damper loop illustrated in Figure 9.2. The frequency responses are for the open-loop transfer function mapping $T_s(s)$ to $s\delta(s)$ at 80 m/s with the damper gain included, for various acceleration and braking cases. The open loop stability of the subject motorcycle is as in the (o) plot in Figure 9.3, indicating that two counter-clockwise encirclements of the -1 point are required for closed-loop stability. The fitted damper will stabilize the machine at each illustrated value of acceleration and braking in Figure 9.16, except for the 4 m/s² acceleration case where

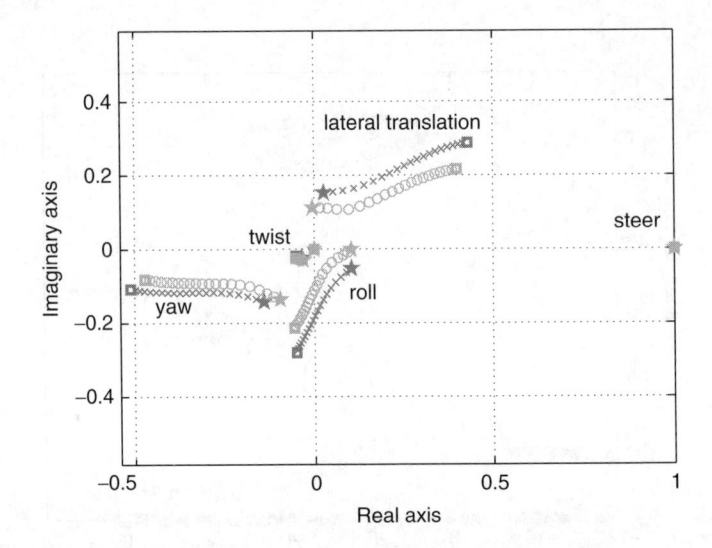

Figure 9.15 Straight-running weave-mode eigenvector loci of the subject motorcycle fitted with a steering damper with acceleration-related inertial force as the varied parameter. Two cases of speed are shown, 70 m/s (×) and 80 m/s (o), while the acceleration-related inertial forces are varied between -4 m/s² (□) and 4 m/s² (★). The 13 eigenvector components corresponding to the generalized coordinates are shown and are normalized so that the steer angle component is $+1$. As with the classical weave mode, the five dominant components are: the machine's yaw, roll, frame twist and steer angles, and the lateral translation

there is no point on the negative real axis that is encircled twice, and therefore no damper will work. What can be concluded from this observation is that some form of phase compensation is required.

9.6.3 Compensator Design and Performance

A steering compensator is synthesized using the approach taken in Section 9.5, to address bursting instabilities. It is clear from Figure 9.16 that phase-lag compensation is required in the 4 m/s² acceleration case. The network illustrated in Figure 9.17 offers the desired phase characteristics and it is therefore chosen for further study.

The compensator function for the spring–damper network is $K(s) = k/(s + \frac{k}{c})$, defined completely by selecting values for c and k. Trial parameter values are chosen and subsequently optimized using the robust frequency-domain optimization approach employed in (Evangelou et al. 2012). These component values are constrained to be positive, in order to maintain network realizability. After optimization calculations are performed, the optimal damper value is found to be $c = 7.30$ Nms/rad, while the optimal spring stiffness is

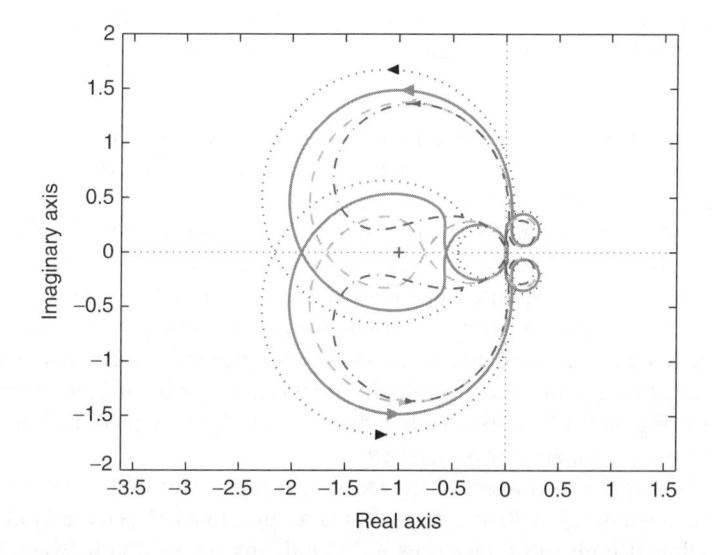

Figure 9.16 Nyquist diagrams of the steering compensation loop for the straight-running motorcycle at 80 m/s with the damper gain included. Four acceleration-related inertial forces are illustrated: -2 m/s² (dotted line), 0 m/s² (solid line), 2 m/s² (dashed line) and 4 m/s² (dash-dot line)

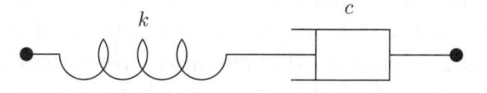

Figure 9.17 Series spring (k Nm/rad) damper (c Nms/rad) steering compensation network with phase-lag characteristics. The optimized spring–damper values are $k = 921.6$ Nm/rad and $c = 7.30$ Nms/rad

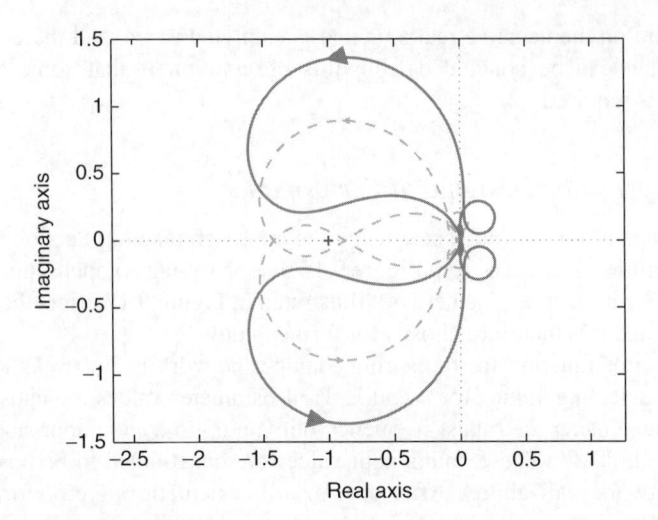

Figure 9.18 Nyquist diagram for the straight-running subject motorcycle at 80 m/s with a steering damper fitted (solid line), and with the steering phase-lag compensator fitted (dashed line). The acceleration-related inertial force is 4 m/s^2. This diagram illustrates the stability margin improvement at high speed and high acceleration, brought about by a steering compensator

$k = 921.6$ Nm/rad. These figures correspond to a compensation network that is a slightly smaller damper than the nominal one on the subject motorcycle, in series with a stiff spring. Therefore, large changes in behaviour are not expected.

Figure 9.18 shows the result of steering compensation on the Nyquist diagram of the machine under high-speed straight-running conditions, where acceleration-related burst instability is the problem identified in Figure 9.16. It is clear that bursting under steady acceleration can be prevented using a spring–damper phase-lag compensating network, since it is able to maintain two counter-clockwise encirclements, in contrast to the steering damper. The amplitude of the steering velocity burst is of the order of 2 rad/s when a damper is installed (see Figure 9.12). With no other changes, this figure reduces to 1.58×10^{-4} rad/s when the optimized compensator is installed.

Figure 9.19 illustrates the impact of the steering compensator on the wobble and weave modes for straight-running at 80 m/s. It shows from a different perspective from Figure 9.14 how acceleration at high speed produces a destabilizing interaction between wobble and weave modes. The acceleration-related inertial forces are swept between ± 5 m/s^2. The wobble mode, which is the higher frequency mode seen in the figure, is well damped both with the steering damper and the optimized spring–damper compensator, so steering compensation has no significant impact on this mode under the present operating conditions. As a result of the modal interaction the weave mode goes unstable under steady acceleration. In contrast to the wobble mode case, the compensator has a strong influence on the weave mode stability at high levels of acceleration. Although the weave mode still appears to be 'unstable', we are not considering conventional stability but fast growth rate (bursting) behaviours.

Figure 9.20 is the counterpart to Figure 9.14 when the steering compensator is fitted. In comparison to the steering damper fitted motorcycle, the machine's wobble-mode frequency increases when the steering compensator is fitted, which reduces the interaction between this

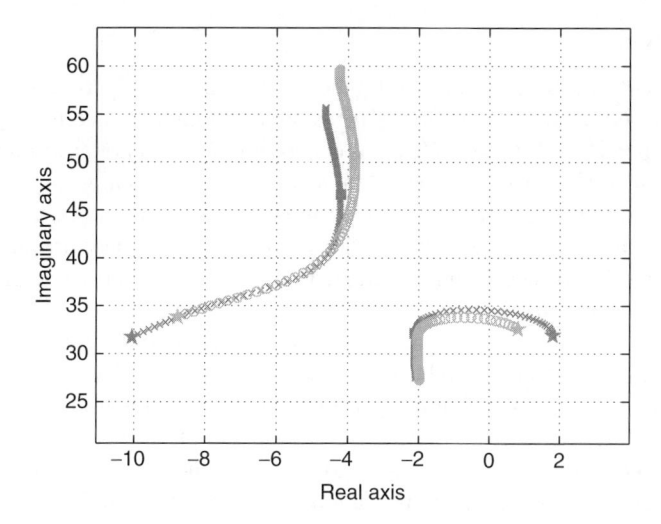

Figure 9.19 Root-loci for the straight-running subject motorcycle at 80 m/s showing the wobble- and weave-mode eigenvalues, with the acceleration-related inertial force varied from -5 m/s^2 (□) to 5 m/s^2 (★). The (×) plot is the case when the steering damper is fitted and the (∘) corresponds to the case when the phase-lag compensator is fitted (damper 7.3 Nms/rad; spring 921.6 Nm/rad)

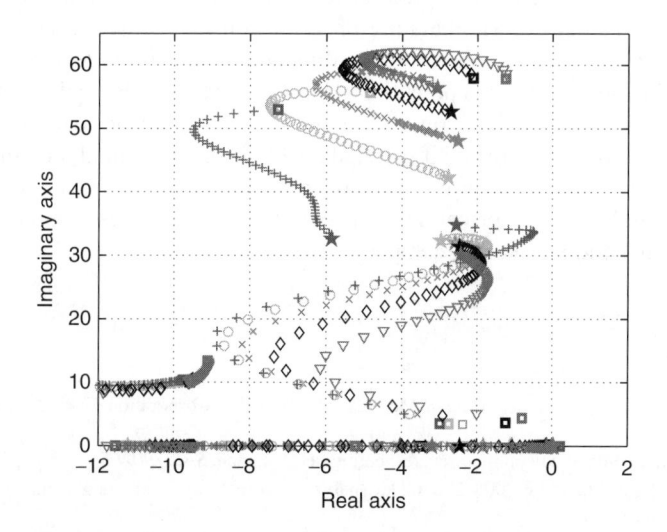

Figure 9.20 Straight-running root-loci of the phase-lag compensator fitted motorcycle with speed as the varied parameter, for acceleration-related inertial force of -4 m/s^2 (▽), -2 m/s^2 (◊), 0 m/s^2 (×), 2 m/s^2 (∘) and 4 m/s^2 (+). The speed is varied from 9 m/s (□) to 95 m/s (★) in steps of 2 m/s

mode and the weave mode. Indeed, at 4 m/s^2, the compensator is successful in stabilizing the high-speed weave mode in the sense of moving into the left half-plane the eigenvalues of the constant-speed model augmented with acceleration-related forces. Although under low-speed braking conditions the steering compensator has a destabilizing effect on the wobble mode, it has no other negative effects.

9.7 Conclusions

The design of motorcycles is influenced by the need for them to operate stably while being responsive to the commands of the rider. The existence of the free-steering system is fundamental to the operation of motorcycles, but depending on the amount of impedance in the steering freedom, the balance between machine stability and handling can be influenced.

The primary motorcycle modes that have the potential to compromise rider safety are oscillatory in nature and involve lateral motions of the vehicle. These are known as wobble and weave modes. Their nature and characteristics have been described with reference to a high-fidelity motorcycle mathematical model.

It has been explained how steering damping that is commonly used to reduce wobble mode oscillations, has a detrimental effect on weave mode stability. Mechanical networks consisting of the basic passive elements, springs, dampers and inerters, have been used to influence the motorcycle steering impedance so that stability of both wobble and weave is improved, without compromising the manoeuvrability of the vehicle. A simple but effective design methodology for steering compensation has been presented. Further improvements in the design are possible by optimizing the network parameters using robust frequency-domain optimization. The optimization criterion can take into account the need for robust stability, and also the role played by road displacement forcing in triggering unstable weave and wobble phenomena.

Burst oscillations arising in racing motorcycles under steady acceleration at high speeds have also been investigated. The bursting phenomenon is a consequence of destabilizing interactions between the classical wobble and weave modes in the region of weave-mode frequencies. This phenomenon is potentially dangerous to the rider and can be detrimental to lap times by undermining rider confidence. A simple mechanical steering compensator is proposed as a curative measure and it is tested on a high-fidelity simulation model. The stabilizing influence of the compensator arises from the phase lag it introduces in the steering-damper loop. The compensator can be constructed from the series combination of a stiff spring and a conventional steering damper.

References

Anon. 1998 *Autosim 2.5+ Reference Manual* Mechanical Simulation Corporation 709 West Huron, Ann Arbor MI.
Aoki A 1979 Experimental study on motorcycle steering performance. *SAE 790265*.
Cossalter V, Lot R and Maggio F 2004 The modal analysis of a motorcycle in straight running and on a curve. *Meccanica* **39**(1), 1–16.
de Vries EJH and Pacejka HB 1997 Motorcycle tyre measurements and models In *Proc. 15th IAVSD Symposium on the Dynamics of Vehicles on Roads and on Tracks* (ed. Palkovics L), Budapest Hungary. Suppl. Vehicle System Dynamics, **29**, 1998, 280–298.
de Vries EJH and Pacejka HB 1998 The effect of tyre modeling on the stability analysis of a motorcycle *Proc. AVEC'98*, pp. 355–360 SAE of Japan, Nagoya.
Evangelou S and Limebeer DJN 2000a Lisp programming of the "sharp 1971" motorcycle model Unpublished report, http://www3.imperial.ac.uk/controlandpower/research/motorcycles/reports.
Evangelou S and Limebeer DJN 2000b Lisp programming of the "sharp 1994" motorcycle model Unpublished report, http://www3.imperial.ac.uk/controlandpower/research/motorcycles/reports.
Evangelou S, Limebeer DJN and Tomas Rodriguez M 2008 Influence of road camber on motorcycle stability. *ASME J. Applied Mechanics* **75**(6), 061020 (12 pages).

Evangelou S, Limebeer DJN, Sharp RS and Smith MC 2006 Control of motorcycle steering instabilities–passive mechanical compensators incorporating inerters. *IEEE Control Systems Magazine* **26**, 78–88. ISSN: 1066-033X.

Evangelou SA, Limebeer DJN and Tomas Rodriguez M 2012 Suppression of burst oscillations in racing motorcycles. *ASME J. Applied Mechanics* **80**(1), 011003 (14 pages).

Foale T 2002 Motorcycle handling and chassis design – the art and science.

Green M and Limebeer DJN 1995 *Linear Robust Control*. Prentice Hall, Englewood Cliffs, New Jersey 07632.

Koenen C 1983 *The dynamic behaviour of motorcycles when running straight ahead and when cornering* PhD thesis Delft University of Technology.

Limebeer DJN and Sharma A 2010 Burst oscillations in the accelerating bicycle. *ASME Journal of Applied Mechanics* **77**(6), 0610126.

Limebeer DJN, Sharp RS and Evangelou S 2002 Motorcycle steering oscillations due to road profiling. *Transactions of the ASME, Journal of Applied Mechanics* **69**(6), 724–739.

Meijaard JP and Schwab AL 2006 Linearized equations for an extended bicycle model In *3rd European Conference on Computational Mechanics, Structures and Coupled Problems in Engineering* (ed. Mota Soares et al. CA), pp. 1–18, Lisbon, Portugal.

Nayfeh AH and Mook DT 1979 *Nonlinear Oscillations*. Wiley-Interscience, NY.

Pacejka HB 2002 *Tyre and Vehicle Dynamics* Butterworth Heinemann Oxford. ISBN 0-7506-5141-5.

Shaeri A, Limebeer DJN and Sharp RS 2004 Nonlinear steering oscillations of motorcycles *Proceeding of 43rd CDC*, Atlantis, Paradise Island, Bahamas.

Sharp RS and Alstead CJ 1980 The influence of structural flexibilities on the straight running stability of motorcycles. *Vehicle System Dynamics* **9**(6), 327–357.

Sharp RS and Limebeer DJN 2001 A motorcycle model for stability and control analysis. *Multibody System Dynamics* **6**(2), 123–142.

Sharp RS, Evangelou S and Limebeer DJN 2003 Improved modelling of motorcycle dynamics In *ECCOMAS Thematic Conference on Advances in Computational Multibody Dynamics* (ed. Ambrósio J), Lisbon. MB2003-029 (CD-ROM).

Sharp RS, Evangelou S and Limebeer DJN 2004 Advances in the modelling of motorcycle dynamics. *Multibody System Dynamics* **12**(3), 251–283.

Sharp RS, Evangelou S and Limebeer DJN 2005 Multibody aspects of motorcycle modelling with special reference to Autosim In *Advances in Computational Multibody Systems* (ed. Ambrsio JG), pp. 45–68. Springer-Verlag, Dordrecht, The Netherlands.

Smith MC 2002 Synthesis of mechanical networks: The inerter. *IEEE Trans. Automatic Control* **47**(10), 1648–1662.

Spierings PTJ 1981 The effects of lateral front fork flexibility on the vibrational modes of straight-running single-track vehicles. *Vehicle System Dynamics* **10**(1), 21–35.

Sugizaki M and Hasegawa A 1988 Experimental analysis of transient response of motorcycle rider systems. *SAE 881783*.

Tezuka Y, Ishii H and Kiyota S 2001 Application of the Magic Formula tire model to motorcycle maneuverability analysis. *JSAE Review* **22**, 305–310.

10

Semi-Active Steering Damper Control for Two-Wheeled Vehicles

Pierpaolo De Filippi, Mara Tanelli, and Matteo Corno
Dipartimento di Elettronica, Informazione e Bioingegneria, Politecnico di Milano, Italy

This chapter is devoted to the analysis and control of steering dynamics of powered two-wheelers (PTWs). It is shown that under certain conditions the steering dynamics can become unstable. Steering instabilities can be controlled via semi-active steering dampers. Two different control strategies, focused on the control of low- and high-frequency oscillations, are presented. The control strategies are validated by simulation and on an instrumented motorcycle.

10.1 Introduction and Motivation

Powered two-wheelers are prone to instabilities of some of their vibrational modes. There are two principal modes of vibration in straight-running and steady-state cornering: weave and wobble (see Cossalter 2002; Limebeer et al. 2002). Weave is a low-frequency oscillation of the vehicle chassis. Wobble is a higher frequency oscillation of the steering handle around its axis. Some operating conditions can trigger instabilities or under-damped behaviour of these modes. When this happens, most riders are incapable of controlling the oscillations. Steering instability is therefore a safety-critical issue for PTWs (e.g. Duke 1997).

Several authors have analysed these dynamics (e.g., Limebeer et al. 2002; Sharp and Limebeer 2004; Cossalter et al. 2002). These analyses are carried out using multibody models. They enable an accurate description of the phenomenon and detailed sensitivity analyses showing that wobble and weave depend on many static and dynamic parameters: steering axle caster angle, wheelbase and wheel inertia, roll angle and forward velocity being the most important. Simulations show that passive steering dampers, generating a

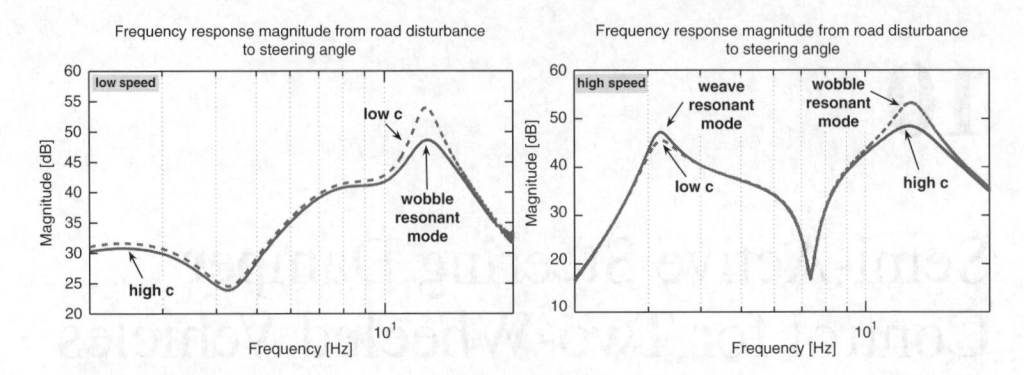

Figure 10.1 Magnitude of the frequency response $G_{d\delta}(j\omega)$ with roll angle $\varphi = 30°$ and (a) speed $v = 50$ km/h and (b) speed $v = 140$ km/h; maximum (dashed line) and minimum (solid line) steer damper value. See also (De Filippi et al. 2011), reproduced with permission from IEEE

moment opposite to the steering handle–chassis relative angular velocity, can improve vehicle stability. The tuning of these devices is however not straightforward.

Figures 10.1a,b and 10.2a,b summarize the issue. They show the frequency responses $G_{d\delta}(j\omega)$ and $G_{T_s\delta}(j\omega)$, the dynamics between a road disturbance d and, respectively, the steering torque T_s and the steering angle δ for a sports motorcycle at low and high speed as a function of the steering damping coefficient.

The frequency responses are obtained from simulations performed using Bikesim®, tuned to fit a sports motorcycle. The figure agrees with the results in (Evangelou et al. 2006), showing that:

- at low speed the wobble mode is under-damped and the weave mode is well-damped, while at high speed both modes' resonances are present;
- increasing the steering damping coefficient damps the wobble mode, but excites the weave mode, while decreasing the damping has the opposite effect.

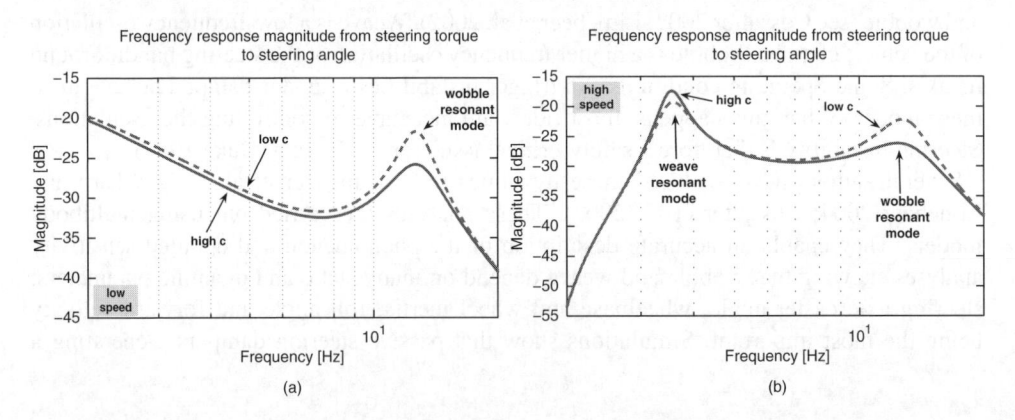

Figure 10.2 Plot of the magnitude of the frequency response $G_{T_s\delta}(j\omega)$ with roll angle $\varphi = 0°$ and (a) speed $v = 50$ km/h and (b) speed $v = 140$ km/h; maximum (solid line) and minimum (dashed line) steer damper value. See also (De Filippi et al. 2011), reproduced with permission from IEEE

Given the safety relevance and the complex dynamic relationship, devising suitable industrial control strategies for steering instabilities is non-trivial. This work presents genuine semi-active control laws to tackle this problem.

Other researchers have looked at the issue. Scientifically, the interest has mainly been on the analysis of the steering dynamic behaviour (Limebeer et al. 2002; Sharp and Limebeer 2004 and many others). To the best of the authors' knowledge, the only attempt at studying the control problem is presented in (Evangelou et al. 2006; Evangelou et al. 2007; Limebeer et al. 2006). Specifically, (Evangelou et al. 2006 and Evangelou et al. 2007) present the design of a mechanical compensator. The device is composed of a spring, a damper and an inerter (Smith 2002). The approach has been tested in simulation, obtaining good results. However, the device is complex and still at the prototype phase. Another approach is shown in (Limebeer et al. 2006), where active compensators, powered by electric motors, are considered.

Industrially, the most common solution is based on a tunable passive steering damper, which the rider can adjust to their liking. Recently, some high-end motorcycles have been being equipped with semi-active steering dampers (e.g. Wakabayashi and Sakai 2004). From what is known of these systems, they are adaptive systems; the damping coefficient is varied according to static maps (usually a function of the longitudinal velocity and/or acceleration). The damping coefficient is varied quasi-statically, thus yielding a limited performance improvement with respect to passive solutions.

The approach presented in this chapter is appealing from both the scientific and industrial standpoint: it proposes novel semi-active control algorithms that improve the trade-off between weave and wobble via feedback control and it is also based on available off-the-shelf semi-active dampers.

The chapter is structured as follows. An analytical model is derived in Section 10.2, and this is validated against both simulation and experimental data. The model reveals a parallelism between vertical and steering dynamics. Based on this, in Section 10.3 the concept of sky-hook and ground-hook (e.g. Fischer and Isermann 2004; Savaresi et al. 2003; Williams 1997) are adapted to the steering dynamics domain. Two control laws are proposed: the rotational ground-hook, aimed at minimizing the absolute vibration of the steering axis, and the rotational sky-hook, focused on the chassis oscillations. In Section 10.4, the proposed strategies are validated on challenging manoeuvres performed in simulation. Finally, Section 10.5 presents experimental results obtained on an instrumented bike.

10.2 Steering Dynamics Analysis

In this section a control-oriented model of the motorcycle steering dynamics is derived. Two bodies are considered to constitute the vehicle: the main frame and the steering assembly. These can rotate around the steering axis (see Figures 10.3a,b). The symbols c_θ and s_θ stand for $\cos(\theta)$ and $\sin(\theta)$, respectively; the subscripts f and r indicate the front and rear wheel, respectively.

The model is derived based on the following assumptions:

- The forward speed is constant in magnitude v, i.e, only steady-state cornering or straight-running is considered.
- The longitudinal force exerted by the tyre is equal to zero, as the forward speed is constant and tyre–road friction and aerodynamic forces are neglected.

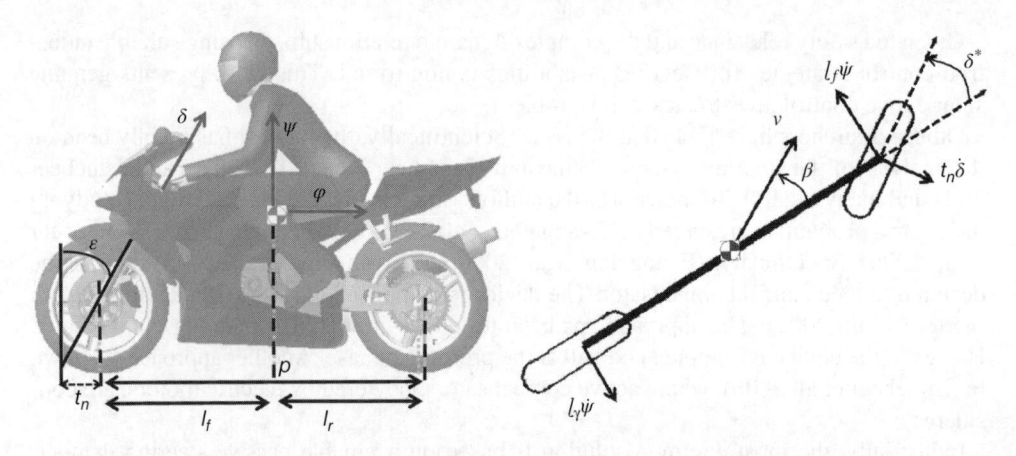

Figure 10.3 Side view (left) and view from above (right) of the motorcycle with reference frames and notation. Adapted from (De Filippi et al. 2011), reproduced with permission from IEEE

- The tyre moments are neglected.
- The tyre sideslip angles α_i, the tyre roll angles φ_i, $i = \{f, r\}$ variations around the nominal condition and the steering angle δ are small.
- The product of inertia between the steering assembly and the main frame is neglected.
- The gyroscopic effects are neglected.
- The pitch dynamic is neglected.

The following reference frames (Figure 10.4) need to be defined: O_{x_r, y_r, z_r} is a moving reference frame, its origin fixed at the intersection, denoted by p_r, between the motorcycle vertical axis and the road plane. The vertical axis $\mathbf{z_r}$ can rotate of an angle ψ_r with respect to the inertial reference frame; the vector $\mathbf{x_r}$ is directed as the longitudinal axis of the motorcycle; $\mathbf{y_r}$ completes the frame. The origin of the second moving reference frame O_{x_b, y_b, y_b} coincides with the one of the first and the second reference rolls (with the bike) at an angle φ.

Thanks to the above assumptions, the acceleration a_r of the point p_r is written as (see also Figure 10.3a,b)

$$a_r = -v(\dot{\beta} + \dot{\psi}_r)s_\beta \mathbf{x_r} + v(\dot{\beta} + \dot{\psi}_r)c_\beta \mathbf{y_r}, \qquad (10.1)$$

where β is the chassis sideslip angle, i.e. the angle between the centre of mass velocity and the vehicle longitudinal axis (Figure 10.3b). If the motorcycle is cornering (i.e. $\varphi \neq 0$), the relation between $\dot{\psi}_r$ and the angular velocity $\dot{\psi}$, as measured in the reference frame O_{x_b, y_b, z_b}, is given by

$$\dot{\psi}_r = c_\varphi \dot{\psi} + s_\varphi \dot{\varphi}.$$

Finally, if a small vehicle sideslip angle β and steady-state cornering ($\dot{\varphi} = 0$) are assumed, the absolute value of the acceleration is

$$|a_r| \approx |a_y| = (\dot{\beta} + \dot{\psi}c_\varphi)v, \qquad (10.2)$$

where a_y is the vehicle lateral acceleration in the reference frame O_{x_r, y_r, z_r}.

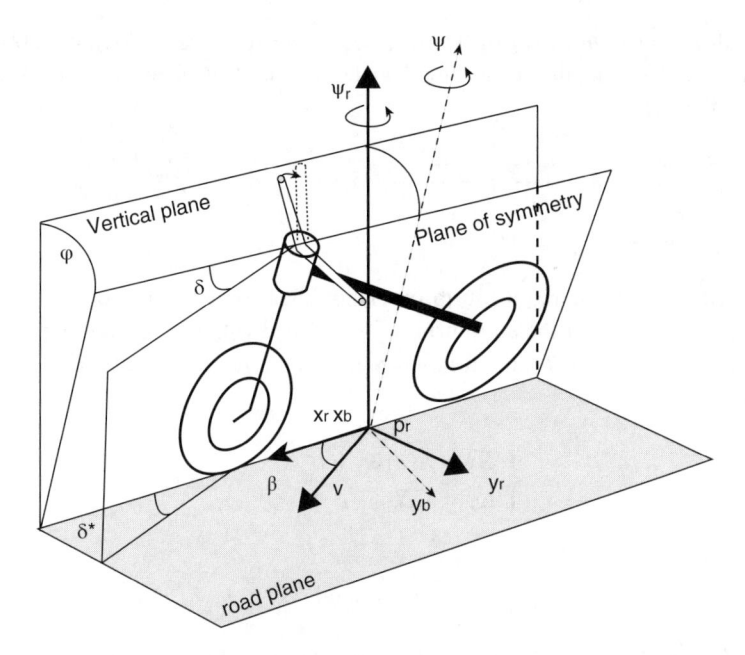

Figure 10.4 Schematic view of the main reference frames and kinematic coordinates. Adapted from (De Filippi et al. 2011), reproduced with permission from IEEE

As the product of inertia between the steering assembly and the main frame is neglected, the lateral force balance in the body reference frame (see also (10.2)) yields

$$Mv(\dot{\beta} + \dot{\psi}c_\varphi) - F_{yf}c_{\delta^*} - F_{yr} = 0, \tag{10.3}$$

where F_{yf} and F_{yr} are the tyre lateral forces. Now $\delta^* = \delta c_\delta c_\varphi$ where ε indicates the caster angle (Figures 10.3a,b) is the angle between the vehicle and the tyre longitudinal axes due to a steering manoeuvre possibly while cornering. The torque balance computed around z_b yields

$$J_z\ddot{\psi} - F_{yf}c_\varphi l_f c_{\delta^*} + F_{yr}l_r c_\varphi = 0, \tag{10.4}$$

where J_z is the yaw inertia. The torque balance around the steering axis takes the form

$$J_s\ddot{\delta} + c\dot{\delta} + F_{yf}t_n c_\varepsilon c_{\varphi_f} = T_s, \tag{10.5}$$

where J_s is the steer inertia, T_s is the steering torque, t_n is the normal trail and c is the steering damping coefficient. Assuming small tyre sideslip angles α_i, the tyre lateral forces result

$$F_{yi} = F_{zi}(k_{\alpha_i}\alpha_i + k_{\varphi_i}\varphi_i), \quad i = \{f, r\}, \tag{10.6}$$

where

$$k_{\alpha_i} := \frac{1}{F_{zi}}\frac{\partial |F_{y_i}|}{\partial \alpha_i}\bigg|_{\alpha_i=\bar{\alpha}_i, \varphi_i=\bar{\varphi}_i}, \quad k_{\varphi_i} := \frac{1}{F_{zi}}\frac{\partial |F_{y_i}|}{\partial \varphi_i}\bigg|_{\alpha_i=\bar{\alpha}_i, \varphi_i=\bar{\varphi}_i}, \quad i = \{f, r\}, \tag{10.7}$$

are the *cornering* and *camber* (or *roll*) stiffnesses, respectively (e.g. Cossalter 2002; Pacejka 2002), and F_{zi} in (10.7) are the vertical load on the wheels. The wheel sideslip and roll angles are (Figure 10.3(b))

$$\alpha_f = -\beta - \frac{l_f c_\varphi}{v}\dot\psi + \frac{t_n c_\varepsilon c_\varphi}{v}\dot\delta + \delta^*, \quad \alpha_r = -\beta + \frac{l_r c_\varphi}{v}\dot\psi \qquad (10.8)$$

$$\varphi_f = \varphi - \delta s_\varepsilon, \quad \varphi_r = \varphi.$$

Equations (10.3)-(10.8) define a fourth-order nonlinear dynamical model with states $x = [\beta\ \dot\psi\ \dot\delta\ \delta]^T$. Let us consider the equilibrium $\bar x$ associated with the constant input $\bar T_s$. Letting $\Delta x = x - \bar x$ and $\Delta T_s = T_s - \bar T_s$, the following linear parametrically varying (LPV) model is obtained:

$$\begin{cases} \Delta\dot x = A(\rho)\Delta x + B\,\Delta T_s \\ \Delta y = C\Delta x \end{cases}, \qquad (10.9)$$

where

$$\rho = [v,\ \varphi] \qquad (10.10)$$

is the vector collecting the time-varying parameters.

The roll dynamics are not modelled; the model is, however, capable of capturing the effect of a roll angle with a time-varying parameter.

The dynamic matrix $A(\rho)$ of model (10.9) has some interesting features, defining a_{ij} the ij-th element of matrix $A(\rho)$:

- It is strongly dependent on the linearization point. Wobble and weave strongly depend on both velocity and roll angle.
- The *direct* terms ($a_{11}, a_{12}, a_{21}, a_{22}, a_{33}, a_{34}$) are predominantly negative and thus have a stabilizing effect; the damping coefficient c is part of them (see a_{33}).
- The *cross* terms ($a_{13}, a_{14}, a_{23}, a_{24}, a_{31}, a_{32}$) are predominantly positive and have a destabilizing effect.

10.2.1 Model Parameters Estimation

The control-oriented model depends on several parameters, some of them (lengths, wheelbases and masses) are known, others (such as tyre stiffnesses) are not. The latter are identified from data; in this first phase simulation data is employed.

The simulation experiment consists of a double-impulse steering torque input, performed at different constant velocities. This type of experiment can be also carried out on a real bike and hence simplifies the direct comparison between the simulator and the experimental data.

The parameter vector θ to be identified is composed of: the normal trail t_n, the yaw inertia J_z, the steer inertia J_s and the tyre stiffnesses k_α and k_{φ_i}, $i = \{f, r\}$. The parameter estimation procedure consists of two steps.

Table 10.1 Identified parameters: simulation and experimental results

Parameter	Simulation	Experimental
normal trail t_n [m]	0.12	0.085
Front wheel cornering stiffness k_{α_f} [rad^{-1}]	16	20
Rear wheel cornering stiffness k_{α_r} [rad^{-1}]	15.8	18
Front wheel roll stiffness k_{φ_f} [rad^{-1}]	0.3	0.7
Rear wheel roll stiffness k_{φ_r} [rad^{-1}]	1.8	0.7
Yaw inertia J_z [kg m^2]	30	22
Steer inertia J_s [kg m^2]	0.6	0.45

1. The parameter space is gridded, and the optimal point θ^o selected as the minimizer of the steering angle simulation error, i.e.

$$\theta^o = \mathrm{argmin}_\theta J(\theta) = \left(\frac{1}{N}\sum_{k=1}^{N}(\delta_{\mathrm{sim}}(k) - \delta_{\mathrm{mod}}(k;\theta))^2\right)^{1/2}, \quad (10.11)$$

where N is the number of data points and $\delta_{\mathrm{sim}}(k)$ and $\delta_{\mathrm{mod}}(k)$ are the simulator and model steering angles, respectively.
2. The optimal point θ^o provided by the previous step is employed as initialization for a gradient-based algorithm which refines the optimization based on the same cost function (10.11), yielding a final parameter vector denoted as θ^*. Table 10.1 summarizes the identified parameters.

Figure 10.5a plots the comparison between the steering angle of the control-oriented model and of the multibody simulator at $v = 140$ km/h. The optimal cost function is $J(\theta^*) = 0.023°$ at $v = 140$ km/h, computed over the time span $t \in [0, 3]$ s.

The model is further validated via the yaw rate comparison shown in Figure 10.5b. The simulation error for the yaw rate is 0.355°/s, computed over the time span $t \in [10, 13]$ s. Note that this validation compares a variable that does not appear in the optimization cost function.

The frequency analysis of the proposed model yields interesting considerations. Figure 10.6 plots the model eigenvalues as a function of the speed (from 50 to 170 km/h) and the magnitude Bode diagram of $G_{T_s\delta}(j\omega)$ at high speed and for different values of the damping coefficient.

The model accurately describes the weave and wobble modes characteristics. Inspecting Figure 10.6, one can conclude that:

- the weave frequency moves, as a function of the speed, within a frequency range of [2.8, 4] Hz, whereas the wobble frequency is within [8, 10] Hz.

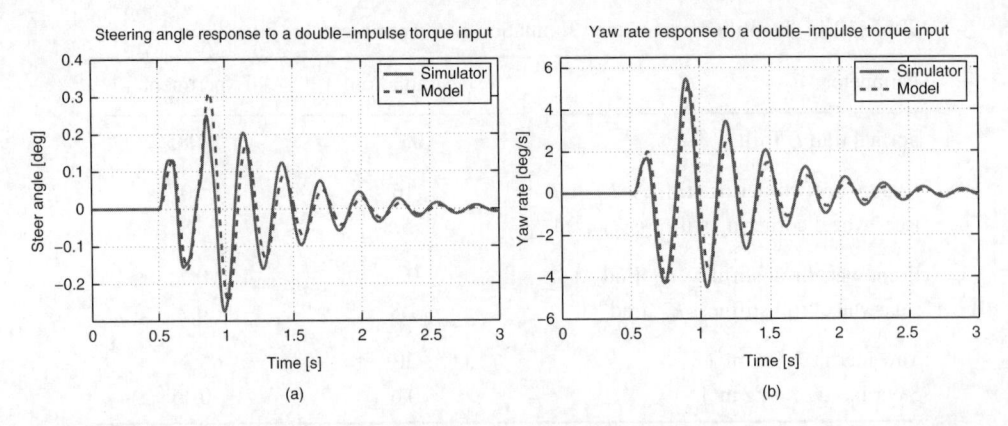

Figure 10.5 Time history of (a) the steering angle response and (b) the yaw rate response to a double-impulse torque input at speed $v = 140$ km/h; model (dashed line) and simulator (solid line). Adapted from (De Filippi et al. 2011), reproduced with permission from IEEE

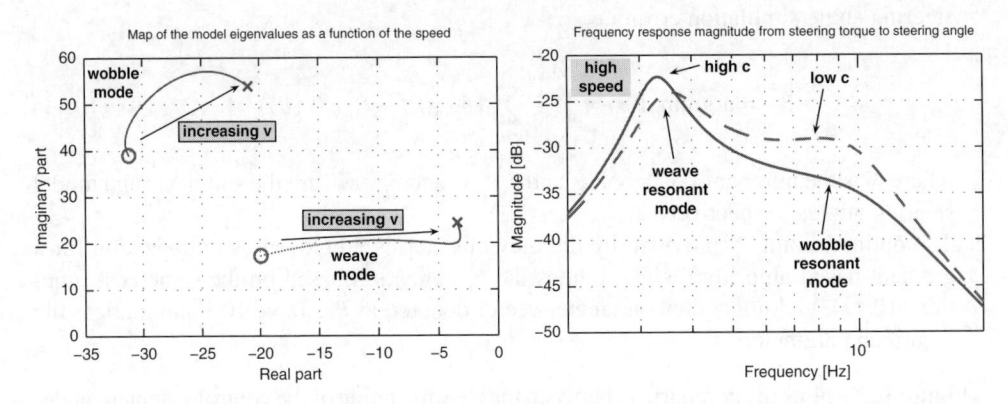

Figure 10.6 Map of the model eigenvalues as a function of speed (left): speed is increased from 50 km/h (circle) to 170 km/h (cross) and plot of the magnitude of the transfer function $G_{T_s\delta}(s)$ derived from the analytical model (right) with roll angle $\varphi = 0°$ and speed $v = 140$ km/h: maximum (solid line) and minimum (dashed line) steering damper coefficient value. Adapted from (De Filippi et al. 2011), reproduced with permission from IEEE

- The damping of the weave mode monotonically decreases as speed increases; the same is true for the wobble mode, although its damping variation is more limited.
- The magnitude plot of $G_{T_s\delta}(j\omega)$ is in general accordance with Figure 10.2b. The wobble mode is more damped; this is probably due to the absence of tyre relaxation dynamics in the model.
- The dependence of weave and wobble modes on the steering damper coefficient is correctly reproduced.

It can be concluded that, under the adopted simplifying assumptions, the model correctly describes the resonant modes of interest.

10.2.2 Comparison between Vertical and Steering Dynamics

The control-oriented model helps to highlight the similarities between the steering and vertical dynamics of a motorcycle. To do this, the control-oriented model is further simplified, considering only the steering rotational dynamics. If $\varphi = 0$ and $a_y = 0$, then the weave and wobble cause only a perturbation of the chassis sideslip angle and $\psi = -\beta$ (Figure 10.3a,b). Under these assumptions, the lateral force balance (10.3) can be neglected, obtaining

$$M_\phi \ddot{\phi} + R_\phi(\rho)\dot{\phi} + K_\phi \phi = Q_{T_s} T_s, \quad \phi = \begin{bmatrix} \psi & \delta \end{bmatrix}^T, \quad (10.12)$$

where

$$M_\phi = \begin{bmatrix} J_z & 0 \\ 0 & J_s \end{bmatrix}, \quad K_\phi = \begin{bmatrix} F_{zr}k_{\alpha_r}l_r - F_{zf}k_{\alpha_f}l_f & -F_{zf}k_{\alpha_f}l_f c_\varepsilon \\ F_{zf}k_{\alpha_f}t_n & F_{zf}k_{\alpha_f}t_n c_\varepsilon \end{bmatrix} \quad (10.13)$$

$$R_\phi(v) = \frac{1}{v}\begin{bmatrix} F_{zr}k_{\alpha_r}l_r^2 + F_{zf}k_{\alpha_f}l_f^2 & -F_{zf}k_{\alpha_f}l_f t_n \\ -F_{zf}k_{\alpha_f}l_f t_n & cv + F_{zf}k_{\alpha_f}t_n^2 \end{bmatrix} = \frac{1}{v}\tilde{R}_\phi + \begin{bmatrix} 0 & 0 \\ 0 & c \end{bmatrix}$$

are mass, stiffness and damping matrices, respectively, and $Q_{T_s} = [0, 1]^T$. This model has some interesting features:

- The mass matrix is positive definite and diagonal; the elements on the diagonal are inertias. The fact that the chassis inertia is greater than the steering inertia is reflected in the numerical value of the matrix elements.
- For all speed values $v > 0$ the damping matrix $R_\phi(v)$ is symmetric, positive definite and inversely proportional to v.
- The stiffness matrix is not symmetric, as the force field is not conservative. Further, the yaw stiffness (see the element $k_{\phi_{11}}$) depends on the difference between front and rear tyre stiffness, while the steer stiffness (see $k_{\phi_{22}}$) depends only on the front tyre stiffnesses and normal trail.

The modal matrix analysis (Meirovitch 1975) reveals further considerations. The modal matrices Φ_v for speed $v = 50$ and $v = 140$ km/h (where each column is normalized with respect to the element of largest magnitude) are

$$\Phi_{50} = \begin{bmatrix} 0.82 & -1 \\ -1 & 0.2 \end{bmatrix}, \Phi_{140} = \begin{bmatrix} 0.8 & -1 \\ -1 & 0.2 \end{bmatrix}. \quad (10.14)$$

The natural frequencies associated with the weave mode are 3.9 Hz at 50 km/h and 3.5 Hz at 140 km/h, while the ones associated with the wobble mode are 8.4 Hz at 50 km/h and 9.5 Hz at 140 km/h. Moreover, the weave-mode column exhibits opposite yaw and steer components, while the wobble-mode column presents a dominant value of the steer component. This is consistent with the observation that, when the weave is excited, the steering assembly and the chassis oscillate in counter-phase, whereas the wobble mainly affects the steering axle. It can be concluded that the model retains the main characteristics of the considered modes.

Model (10.12) points to a parallelism between the vertical and steering dynamics of PTWs. The vertical dynamics of a single-corner system (see (SAE 1992) for details) is

$$M\ddot{x} + R\dot{x} + Kx = Q_d F, \quad x = \begin{bmatrix} z & z_t \end{bmatrix}^T, \quad Qd = \begin{bmatrix} 0 & 1 \end{bmatrix}^T, \quad (10.15)$$

Table 10.2 Variables and parameters mapping

Physical meaning	Vertical Symbol	Steering Symbol	Physical meaning
Sprung mass	M	J_z	Yaw inertia
Unsprung mass	m	J_s	Steer inertia
Sprung mass vertical velocity	$\dot{z}$	$\dot{\psi}$	Chassis yaw rate
Unsprung mass vertical velocity	$\dot{z}_t$	$\dot{\delta} + \dot{\psi}$	Steering assembly yaw rate
Suspension elongation	$\Delta z = z - z_t$	δ	Steering angle

where F is the vertical force and z and z_t are the vertical position of the body and of the unsprung mass, respectively, and

$$M = \begin{bmatrix} M & 0 \\ 0 & m \end{bmatrix}, R = \begin{bmatrix} c & -c \\ -c & c \end{bmatrix}, K = \begin{bmatrix} k & -k \\ -k & k + k_t \end{bmatrix} \qquad (10.16)$$

are the mass, damping and stiffness matrix, with c, k and k_t being the damping coefficient and spring elastic constant of the suspension and the elastic constant of the tyre. These matrices have the same structure as those in (10.13); in particular, the mass matrix is diagonal, with elements of different magnitude. The body mass M is associated with the resonance of the vehicle chassis, while the unsprung mass m is associated with the wheel resonance. Thus, in the proposed parallelism, the body resonance can be linked to the weave resonance, while the wheel resonance is associated with wobble. Table 10.2 summarizes the parallelism between the two domains.

The problem of controlling the steering dynamics is dynamically similar to that of controlling the vertical dynamics. In the next section, the concepts of sky-hook (SH) and ground-hook (GH) will be extended to the out-of-plane modes of interest.

10.3 Control Strategies for Semi-Active Steering Dampers

In vertical dynamics control, a classical strategy used when a comfort objective is pursued is the sky-hook damping (e.g. Fischer and Isermann 2004; Savaresi et al. 2003; Williams 1997). The SH suspension control rationale is to isolate the sprung mass from the road: this idea can be envisioned as an ideal shock absorber that links the chassis and a fixed reference frame (the 'sky'). According to this configuration, in the ideal formulation of SH control, there is a shock absorber delivering a force proportional to the chassis vertical velocity:

$$F_{SH}(t) = c_{SH}\dot{z}(t). \qquad (10.17)$$

Conversely, when handling is the main objective, the GH algorithm is employed, in which the ideal shock absorber is linked to the tyre of the vehicle and to a fixed reference frame ('the ground') and delivers a force proportional to the tyre vertical velocity:

$$F_{GH}(t) = -c_{GH}\dot{z}_t(t), \qquad (10.18)$$

providing a good isolation of the tyre from road disturbances.

These two control strategies pursue different objectives: SH yields the minimization of the body vertical acceleration (comfort objective), at the cost of worsening the wheel resonance; on the other hand, GH minimizes the wheel vertical acceleration (handling objective) while achieving a suboptimal damping of the chassis resonance. A trade-off arises between the two objectives and the two control strategies alone cannot provide simultaneous isolation of both the body and the wheel.

According to the previously mentioned similarities, the SH and GH concepts can be extended to motorcycle steering dynamics.

10.3.1 Rotational Sky-Hook and Ground-Hook

The rotational sky-hook (RSH) strategy (Figure 10.7) is ideally characterized by a shock absorber linked to the chassis and to an appropriate frame. This frame translates with the motorcycle and rotates only around the longitudinal and transverse axes, following the pitch and roll rotations of the motorcycle chassis (Table 10.2). This configuration focuses on the isolation of the chassis from road or steering torque disturbances.

The ideal shock absorber delivers a torque proportional to the yaw rate:

$$\tau_{sh}(t) = c_{sh}\dot{\psi}(t), \qquad (10.19)$$

while the real steering damper can only deliver a torque proportional to the steering angular velocity:

$$\tau(t) = c(t)\dot{\delta}(t). \qquad (10.20)$$

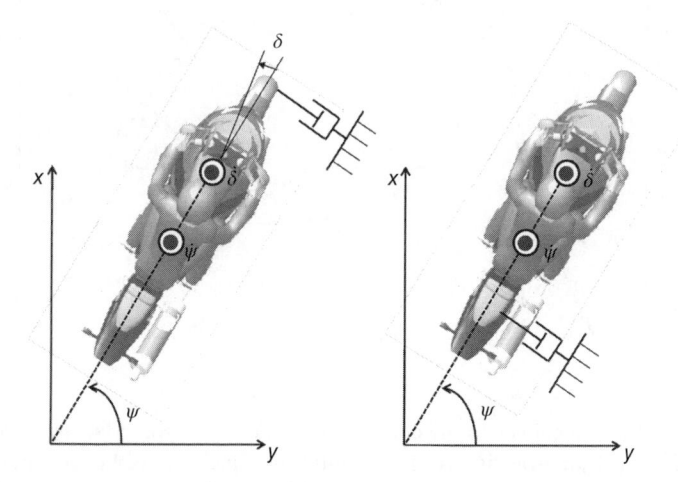

Figure 10.7 Schematic view of the ideal ground-hook (left) and sky-hook (right) damping schemes. Adapted from (De Filippi et al. 2011), reproduced with permission from IEEE

The ideal RSH strategy is emulated by a time-varying $c(t)$; the desired $c(t)$ is determined equalling the ideal and the actual torques in (10.19) and (10.20). If the industrial standard of a two-state damper is assumed, $c(t)$ can take only two values, which will be referred to as c_{min} and c_{max}. This choice leads to a two-state approximation of the RSH control algorithm (Rajamani 2006; Savaresi and Spelta 2007; Savaresi et al. 2005), given by

$$c(t) = \begin{cases} c_{max} & \text{if } \dot{\psi}(t)\dot{\delta}(t) \geq 0 \\ c_{min} & \text{if } \dot{\psi}(t)\dot{\delta}(t) < 0 \end{cases}. \tag{10.21}$$

The rotational ground-hook (RGH) control strategy (Figure 10.7) is based on an ideal shock absorber connected to the steering assembly and the same reference frame described for the RSH case. This configuration yields a better rejection of both road and steering torque disturbances. In the RGH approach, the ideal shock absorber delivers a torque proportional to the sum of the yaw rate and steering angular velocities:

$$\tau_{gh}(t) = c_{gh}(\dot{\delta}(t) + \dot{\psi}(t)), \tag{10.22}$$

while a real steering damper always delivers the torque computed in (10.20). Following the same rationale as before, the controlled damping coefficient $c(t)$ is selected as

$$c(t) = \begin{cases} c_{max} & \text{if } \dot{\delta}(t)\left(\dot{\delta}(t) + \dot{\psi}(t)\right) \geq 0 \\ c_{min} & \text{if } \dot{\delta}(t)(\dot{\delta}(t) + \dot{\psi}(t)) < 0 \end{cases}. \tag{10.23}$$

Both control strategies require yaw rate and steering velocity measurements, for which sensors are readily available.

The RSH and RGH control strategies are designed to provide high attenuation of the motorcycle chassis oscillations and the front-assembly oscillations respectively. Figure 10.8

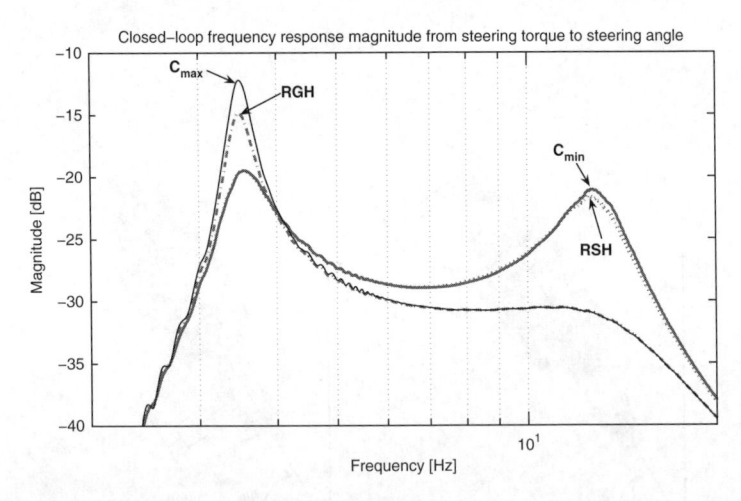

Figure 10.8 Plot of the magnitude of the closed-loop frequency response $G_{T_s\delta}(j\omega)$ with roll angle $\varphi = 0°$ and speed $v = 140$ km/h obtained with c_{min} (solid thick line), c_{max} (solid thin line), RSH (dash-dotted line) and RGH (dotted line). Adapted from (De Filippi et al. 2011), reproduced with permission from IEEE

depicts the frequency response $G_{T_s\delta}(j\omega)$, relating steering torque to steering angle at a forward velocity of 140 km/h. To estimate $G_{T_s\delta}(j\omega)$, constant-speed straight running has been simulated using Bikesim, and a frequency sweep torque disturbance is applied to the steer, ranging from 1 to 20 Hz (Wellstead 1981). Two linear passive steering dampers ($c_{min} = 0.016$ Nm/s and $c_{max} = 0.044$ Ns/m) and the RSH and RGH control algorithms have been tested.

From Figure 10.8, the following remarks are in order:

- The RSH algorithm provides an attenuation comparable to that achieved by c_{min} around the weave resonance frequency.
- The RGH algorithm provides a damping comparable to that achieved by c_{max} around the wobble resonance.
- The RSH algorithm yields an intermediate attenuation at high frequency, slightly better than c_{min}, but worse than c_{max}.
- The RGH algorithm provides an intermediate attenuation at low frequency, better than c_{max}, but worse then c_{min}.
- The two approaches are not symmetric. The damping difference between RSH and c_{min} at high frequency is smaller than that between RGH and c_{max} at low frequency. This phenomenon is understood by recalling that Figure 10.8 considers a steering torque disturbance and by analysing the ideal RSH and RGH strategies in (10.21) and (10.23). The RSH, with its torque proportional to the chassis yaw rate, tends not to intervene when a high frequency torque is applied to the steering assembly. Conversely, the RGH strategy, by exerting a torque proportional to $\dot{\delta} + \dot{\psi}$, acts also when the excitation is at low frequency. The opposite is true in case of road disturbances, as will be shown later.

10.3.2 Closed-Loop Performance Analysis

To better appreciate the differences between the control algorithms, the cost functions J_s and J_{sN} are defined:

$$J_s = \frac{1}{N} \sum_{k=1}^{N} (\delta(k) - \bar{\delta})^2, \qquad (10.24)$$

where $\delta(k)$ is the steering angle, $\bar{\delta}$ is the steady-state steering angle and N is the number of samples in the test. The second cost function is obtained by normalizing (10.24) by the worst case of all the considered situations (control architectures and velocity).

$$J_{sN} = \frac{J_s}{J_{sMax}}, \qquad (10.25)$$

This cost function is better suited to comparing different strategies. In Table 10.3 the values of (10.25) resulting from steering torque disturbance tests performed at 50 km/h and 140 km/h are collected. From Table 10.3, we can say:

- At low speed the RGH provides the best results, close to the c_{max} passive setting.
- RSH outperforms c_{min}. It yields roughly the same damping for the weave mode, but RSH delivers better performance than c_{min} around the wobble mode. An improvement of 8% at low speed and 4% at high speed is obtained.

Table 10.3 Values of the normalized cost function J_{sN} in (10.25) for $v = 50$ km/h and $v = 140$ km/h, obtained with the different passive and semi-active control strategies

Control strategy	Low speed	High speed	Mean value
c_{min}	1	1	1
c_{max}	0.58	0.79	0.55
RSH	0.91	0.95	0.94
RGH	0.56	0.60	0.48

- RGH outperforms c_{max}: it reaches similar damping for the wobble mode, and RGH damps the weave better than c_{min}. An improvement of 4% at low speed and 23% at high speed is observed.
- Considering the mean value of the cost function, RSH yields an improvement of 6% with respect to c_{min} and RGH an improvement of 7% against c_{max}.

The closed-loop performance in case of road disturbance is also evaluated. Figure 10.9 shows the frequency response $G_{d\delta}(j\omega)$ with roll angle $\varphi = 30°$ and speed $v = 140$ km/h for the different control strategies. To identify $G_{d\delta}(j\omega)$, the motorcycle has been simulated in steady-state cornering with the road height profile defined by a frequency sweep (exciting the 1 to 20 Hz frequency range).

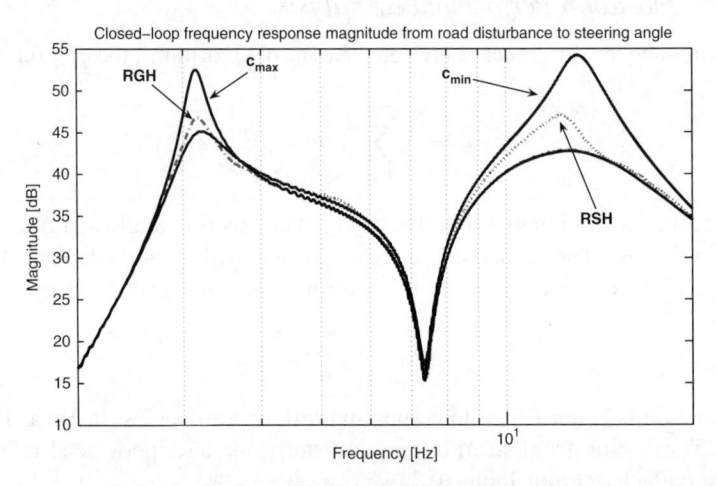

Figure 10.9 Plot of the magnitude of the closed-loop frequency response $G_{d\delta}(j\omega)$ with roll angle $\varphi = 30°$ and speed $v = 140$ km/h obtained with RSH (dotted line), RGH (dashed-dotted line), c_{max} and c_{min} (solid lines) control strategies. Adapted from (De Filippi et al. 2011), reproduced with permission from IEEE

10.4 Validation on Challenging Manoeuvres

The initial validation has been performed using frequency sweeps; to further validate the proposed control strategies, some realistic but challenging manoeuvres have been considered in simulation.

10.4.1 Performance Evaluation Method

In order to assess and compare the control strategies, a mixed frequency/time-domain evaluation is proposed. A stability control system for two-wheeled vehicles must consider:

- the attenuation of the oscillation of the steering assembly at the weave and wobble natural frequencies;
- the minimization of the effects of external disturbances on the steering assembly and on the chassis of the vehicle.

Thus, in order to evaluate the closed-loop performance of a controlled semi-active steering damper, both the steer rate $\dot{\delta}$ and the roll rate $\dot{\varphi}$ of the vehicle are considered.

The magnitude spectrum $\dot{\Delta}(j\omega)$ is employed to define the following cost functions:

$$J_{weave,i} = \frac{\dot{\Delta}_i(j\omega_{weave})}{\dot{\Delta}_{c_{max}}(j\omega_{weave})} \tag{10.26}$$

$$J_{wobble,i} = \frac{\dot{\Delta}_i(j\omega_{wobble})}{\dot{\Delta}_{c_{min}}(j\omega_{wobble})}, \tag{10.27}$$

where $\dot{\Delta}_i(j\omega_{weave})$ and $\dot{\Delta}_i(j\omega_{wobble})$ are the amplitude spectra of the steer rate evaluated at the weave and wobble natural frequencies when algorithm i is implemented. These cost functions are normalized with respect to the worst case, namely, the value obtained by a passive steering damper with a high and low damping coefficient for the weave- and wobble-oriented cost function, respectively.

Cost functions (10.26) and (10.27) evaluate the performance of the control strategies around the weave- and wobble-mode natural frequencies. Another interesting aspect is the roll dynamics is that if the roll rate of the bike is minimized in response to external disturbances, the overall stability of the vehicle is improved. Given these two control objectives, the H_2-norm of the steer and roll rate has been considered. The H_2-norm is appropriate for signals that decay to zero as time progresses and is related to the energy of the signal. The following cost functions are defined:

$$J_{steer,i} = \frac{\|\dot{\delta}_i\|_2}{\max_i \|\dot{\delta}_i\|_2} \tag{10.28}$$

$$J_{chassis,i} = \frac{\|\dot{\varphi}_i\|_2}{\max_i \|\dot{\varphi}_i\|_2} \tag{10.29}$$

where the cost functions are normalized with respect to the worst case.

10.4.2 Validation of the Control Algorithms

In the following simulations, the rider is simulated as a basic roll stabilizing controller with a bandwidth of 0.5 Hz; the rider's body does not move with respect to the vehicle frame. The steering damper actuator bandwidth is 30 Hz, comparable with that of a commercial semi-active steering damper. The front and rear torque actuators (brakes and engine) have a bandwidth of 10 Hz (Dardanelli et al. 2010; Panzani et al. 2011). The manoeuvres are designed to excite the bike, accounting for the main source of disturbances: wheel torque, steering torque and road unevenness.

10.4.2.1 Panic Braking Manoeuvre

The first test simulates a perturbation on the front and rear wheel torque. This is obtained by simulating a braking during high-speed cornering: the rider suddenly pulls the front brake lever and releases the throttle. In this situation, the longitudinal tyre force is increased while the lateral force is decreased, yielding a disturbance that acts on both the steering system and the rear frame. Figures 10.10a,b plot steer rate and roll rate. The following conclusions can be drawn:

- As soon as the torque is applied, the steering assembly oscillates while the motorcycle tends to steer out of the corner.
- Analysing the steer rate, weave and wobble resonances are clearly visible: during the first part of the transient response, a passive steering damper with a low value of the damping coefficient (c_{min}) yields the worst performance, while during the second part it guarantees the best attenuation of the resonance and vice versa for the passive steering damper with c_{max}. This test confirms that the RSH and RGH algorithms achieve intermediate performance: the RSH performs better than c_{min} during the first part of the transient response and as c_{min} during the second part. The same consideration is valid comparing the RGH algorithm to c_{max}.

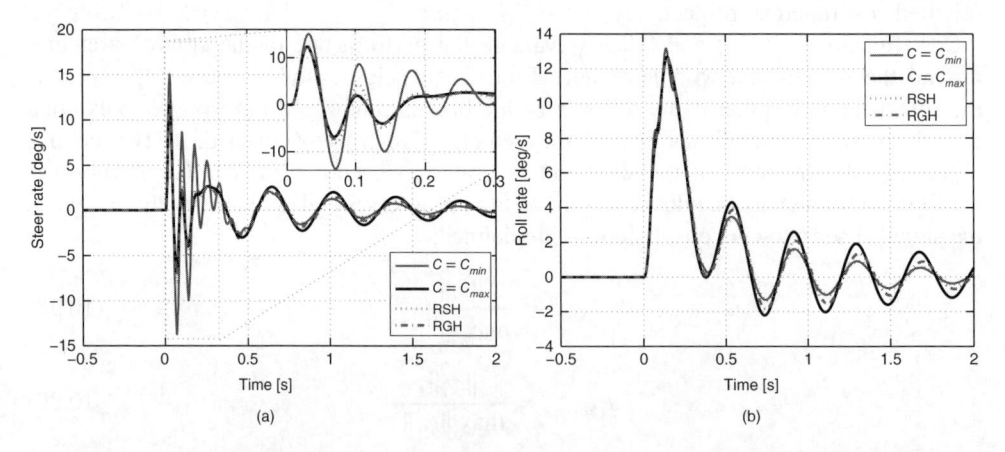

Figure 10.10 Time histories of the steer rate (a) and roll rate (b) in response to a step disturbance on the front braking torque (100 Nm) with a sudden closing of the throttle at $v = 130$ km/h and $\varphi = 30°$

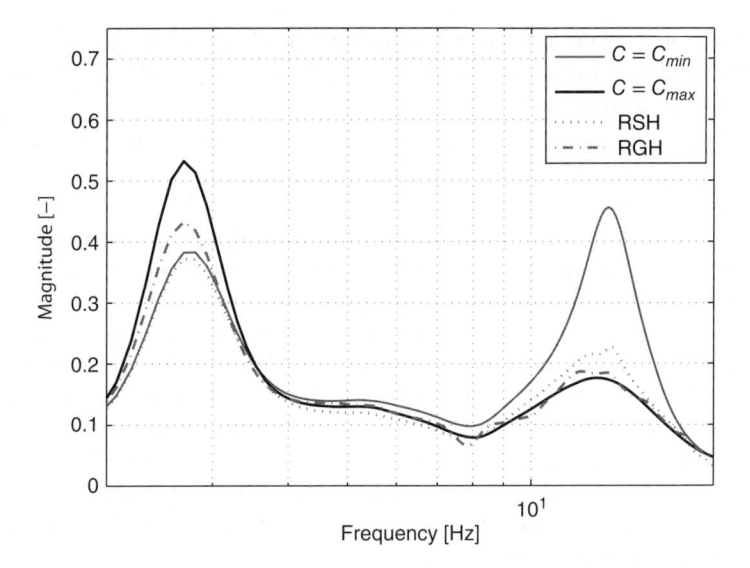

Figure 10.11 Amplitude spectra of the steer rate calculated in response to a step disturbance on the front braking torque (100 Nm) with a sudden closing of the throttle at $v = 130$ km/h and $\varphi = 30°$

Figure 10.11 plots the frequency-based analysis. The magnitude spectrum of the steer rate is depicted. The weave and wobble resonances are clearly visible.

To better appreciate the performance of the control strategies, Figures 10.12a,b depict the values of the cost functions (10.26)–(10.29) for different values of the passive damping coefficient and for different semi-active control strategies. A large value of the damping coefficient increases the attenuation of the wobble mode but has a detrimental effect on the weave. The RSH and RGH algorithms achieve very good results around the weave and wobble mode, respectively, and guarantee better overall performance than passive

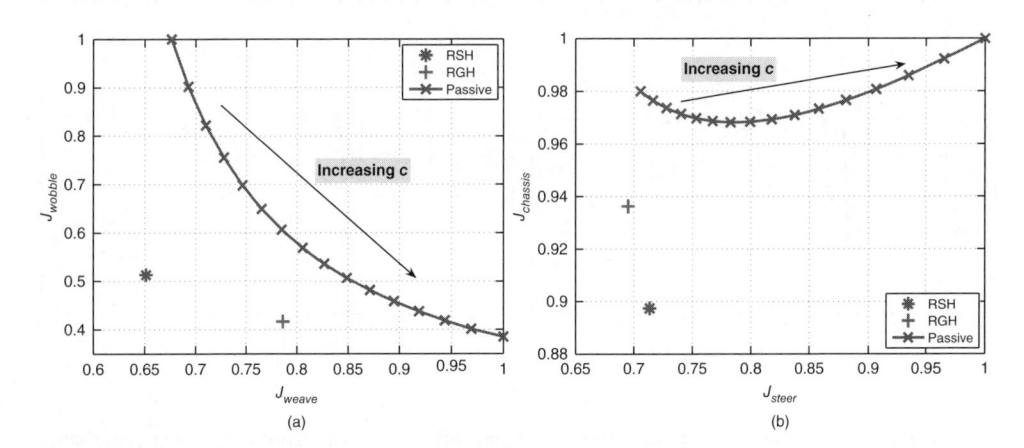

Figure 10.12 Values of the cost functions (10.26), (10.27) (a) and (10.28), (10.29) (b) calculated during a braking cornering manoeuvre at $v = 130$ km/h and $\varphi = 30°$

configurations. Once again, it is clear that tuning a passive steering damper is non-trivial; a semi-active steering damper can damp both the weave and wobble mode, increasing the overall stability of the motorcycle.

10.4.2.2 Lowside

The second test evaluates the performance of the control algorithms when the rider sharply accelerates while exiting a corner. To simulate this situation, the so-called *lowside* phenomenon, a step of the rear wheel torque is applied. The disturbance acts on the rear frame and, to a lesser extent, on the steering assembly. Figures 10.13a,b show the time histories of the steer rate and roll rate during the manoeuvre. Wobble oscillations of the steering system are naturally well damped, while weave oscillations are visible in the second part of the transient. The RGH outperforms the RSH during the high frequency oscillations, while the RSH algorithm attenuates the weave oscillations better than the RGH. These control strategies achieve intermediate performance with respect to the passive configurations c_{min} and c_{max}: the RGH algorithm outperforms c_{max} at low frequency, while the RSH algorithm outperforms c_{min} at high frequency.

Figures 10.14a,b show the values of the cost functions (10.26)–(10.29). As expected, c_{min} and the RSH better damp the weave mode, while c_{max} and the RGH algorithm improve the attenuation of the wobble mode. Also, the best passive configuration is c_{min}, because the lowside manoeuvre mainly excites the weave mode.

10.4.2.3 μ-Jump

Another challenging and interesting manoeuvre is the so-called μ-jump test. In this case, a sudden change of the tyre–road friction coefficient, from $\mu = 0.7$ to $\mu = 1$ and then back to $\mu = 0.7$ is simulated. A 10-metre wide friction patch is simulated. Figures 10.15a,b plot

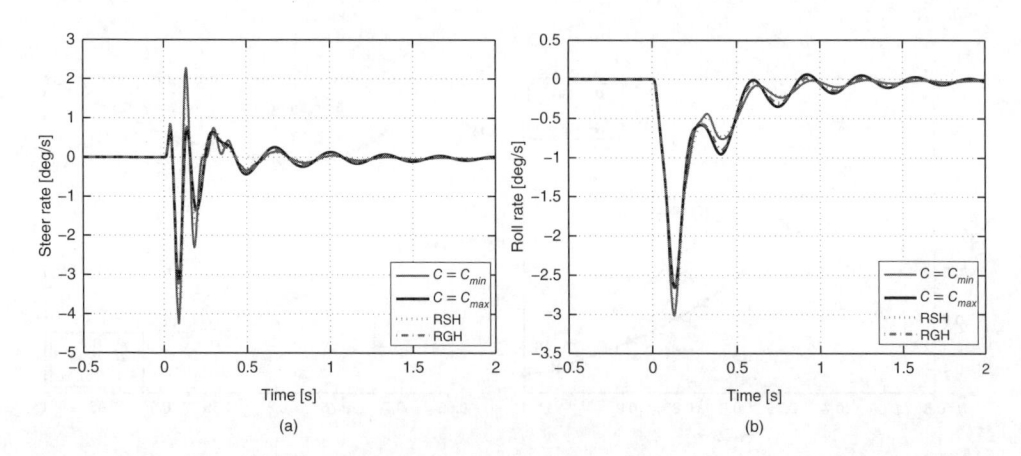

Figure 10.13 Time histories of the steer rate (a) and roll rate (b), in response to a step disturbance on the rear traction torque (200 Nm) at $v = 130$ km/h and $\varphi = 30°$

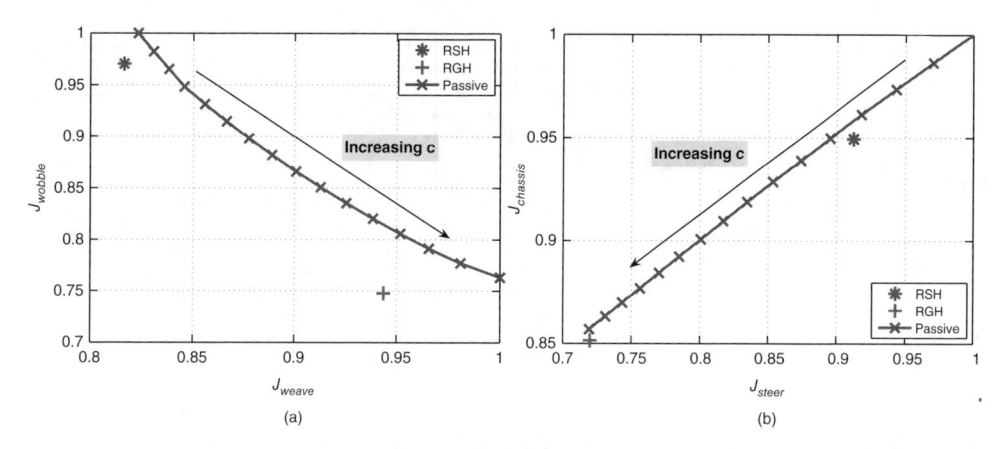

Figure 10.14 Values of the cost functions (10.26), (10.27) (a) and (10.28), (10.29) (b), calculated during a lowside manoeuvre at $v = 130$ km/h and $\varphi = 30°$

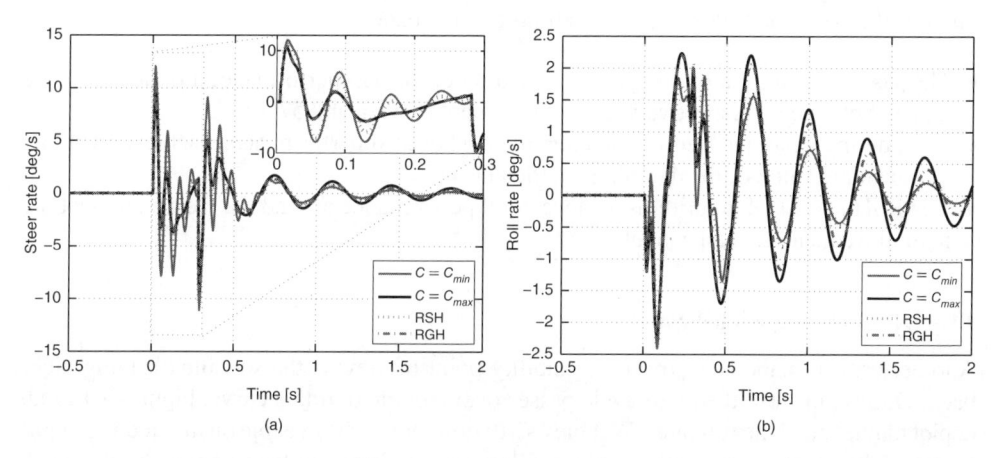

Figure 10.15 Time histories of the steer rate (a) and roll rate (b) to a double step variation of the tyre–road friction coefficient at $v = 130$ km/h and $\varphi = 30°$

the resulting steer and roll rates. At approximately $t = 0.3$ s, when the front wheel exits the high friction patch, the motorcycle is subject to a second disturbance.

Both weave and wobble modes are excited. As soon as the front tyre passes over the first transition, the motorcycle tends to steer into the corner, as the friction coefficient increases, while the steering assembly starts to oscillate. The considerations outlined in the previous section still hold, showing that the semi-active control strategies consistently achieve the stabilization goal. Namely, c_{min} and the RSH algorithm better damp the weave mode, while c_{max} and the RGH algorithm guarantee good performance around the wobble natural frequency.

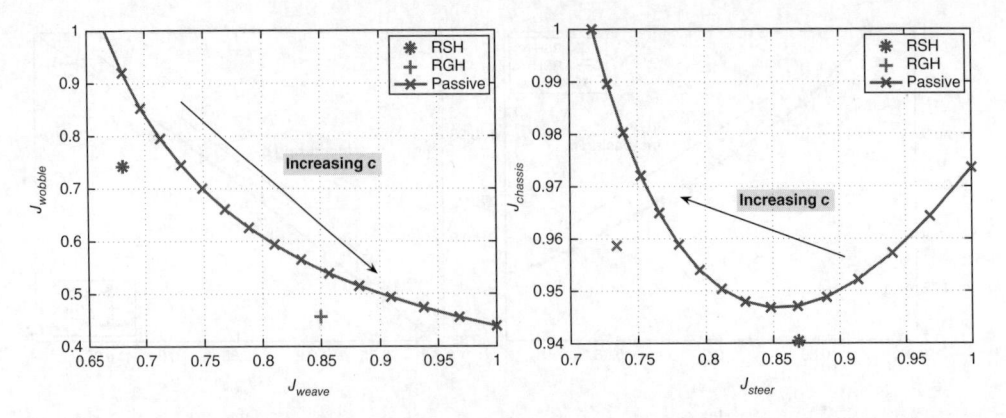

Figure 10.16 Values of the cost functions (10.26), (10.27) (a) (10.28), (10.29) (b), calculated during a μ-jump test at $v = 130$ km/h and $\varphi = 30°$

Moreover, by inspecting the values of the cost functions (10.26)–(10.29) depicted in Figures 10.16a,b, the following observations can be made:

- The passive configurations c_{min} and c_{max} guarantee the best performance around the weave and wobble mode, respectively: an improvement of 34% and 57%.
- Considering a passive steering damper, the best overall performance is achieved with an intermediate value of the damping coefficient.
- The RSH and RGH algorithms achieve good performance around the weave and wobble natural frequencies, respectively.

10.4.2.4 Leaning Kick-back

Another critical maneouvre from the stability point of view is the so-called leaning kick-back. During this test, the motorcycle passes over a bump during a curve. Figures 10.17a,b depict the steer and the roll rates. Wobble oscillations are clearly visible on the steering angular rate, while both weave and wobble oscillations are visible on the roll rate. As expected, the passive configuration with the maximum value of the damping coefficient and the RGH algorithm better damp the steering assembly and chassis wobble oscillations. However, the passive configuration with the minimum value of the damping coefficient and the RSH algorithm increase the damping of the chassis weave oscillations.

The cost functions (10.26)–(10.29) shown in Figures 10.18a,b confirm the above considerations. RGH is the best overall solution, reducing both steering assembly and chassis oscillations. Among the passive configurations, c_{max} behaves the best.

10.4.2.5 Kick-back

Another challenging manoeuvre is the kick-back test. The motorcycle passes over a bump while riding straight, then while the front wheel is lifted, the steer is rotated so that, when the front wheel touches the ground, its plane is out of the driving direction. It has been experimentally shown (Lot and Massaro 2007) that such a test excites both weave and wobble

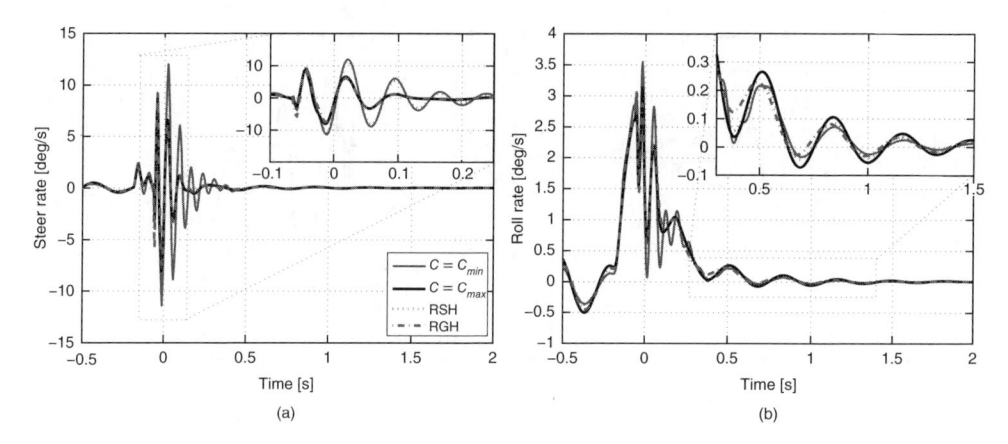

Figure 10.17 Time histories of the steer rate (a) and roll rate (b) to a leaning kick-back at $v = 130$ km/h and $\varphi = 30°$

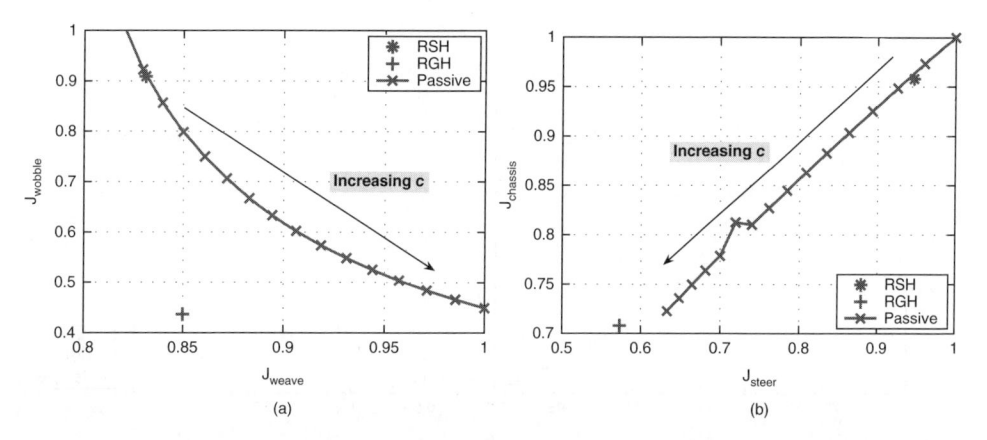

Figure 10.18 Values of the cost functions (10.26), (10.27) (a) (10.28), (10.29) (b), calculated during a leaning kick-back test at $v = 130$ km/h and $\varphi = 30°$

modes. Figures 10.19a,b plot the time histories of the steer and roll rates, confirming the observations outlined so far.

Figures 10.20a,b depict the values of the cost functions (10.26)–(10.29) calculated during the kick-back test. By inspecting the figure, the following conclusions are drawn:

- During such a critical manoeuvre, a passive steering damper with c_{max} improves the damping of the wobble mode; however, the minimum value of the damping coefficient c_{min} cannot guarantee a good attenuation of the weave resonance.
- The RSH algorithm achieves the best attenuation of the weave mode: an improvement of 27% with respect to c_{max} and 6% with respect to the best passive configuration (c_{min}).
- The RSH algorithm achieves the best overall performance: a reduction of 33% of the steering assembly oscillations and of 20% of the chassis oscillations.

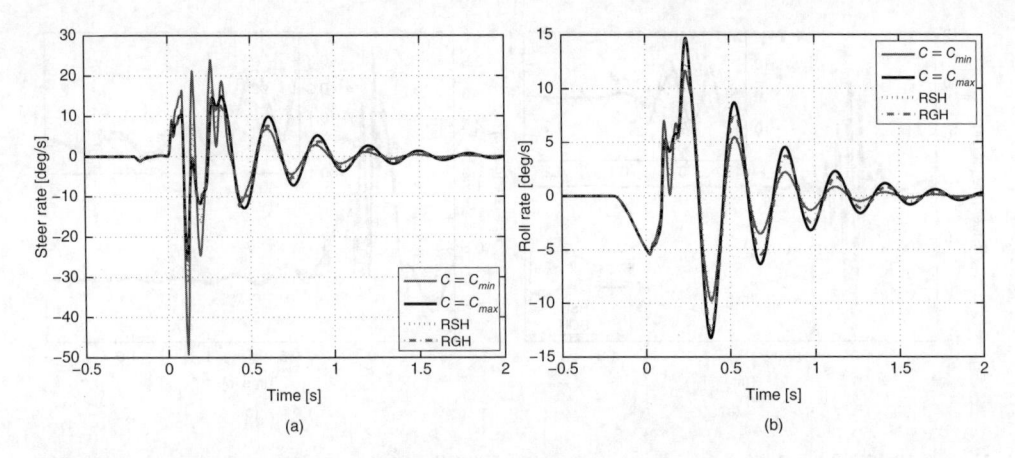

Figure 10.19 Time histories of the steer rate (a) and roll rate (b) to a kick-back at $v = 130$ km/h on straight running

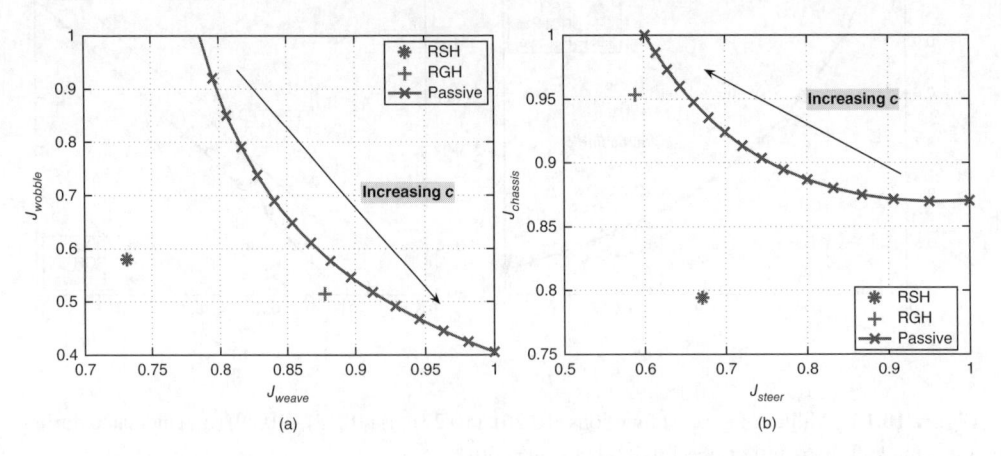

Figure 10.20 Values of the cost functions (10.26), (10.27) (a) and (10.28), (10.29) (b), calculated during a kick-back test at $v = 130$ km/h in straight running

10.4.2.6 Highside

The most challenging test is the *high-side* manoeuvre. The high-side is a dangerous phenomenon that happens while entering or exiting a curve (Cossalter 2002). The manoeuvre is simulated as an acceleration during a curve. The total friction force exerted by the tyre increases as the vertical load at the rear wheel increases. When the total friction force reaches its limit value, the rear wheel loses grip and the lateral force drops. As a consequence, the motorcycle slip outwards; the driver tries to control the vehicle by counter-steering and suddenly closing the throttle. By doing so, the rear wheel regains adherence and a step-like lateral force is generated that twists and pushes the motorcycle upwards.

Semi-Active Steering Damper Control for Two-Wheeled Vehicles

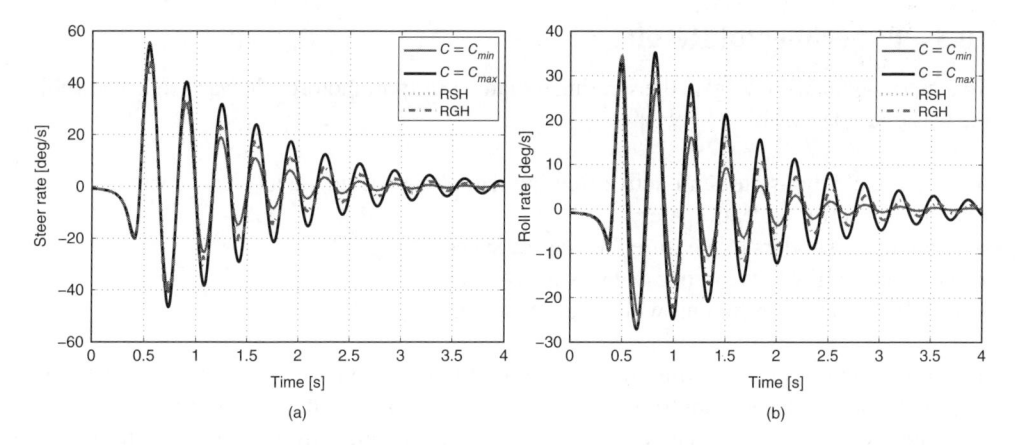

Figure 10.21 Time histories of the steer rate (a) and roll rate (b) obtained during the highside manoeuvre

Figures 10.21a,b plot the time histories of the steer rate and roll rate. Only the weave mode is excited, and thus the passive configuration c_{min} and the RSH control strategy achieve the best performance. RGH outperforms the passive configuration c_{max}, achieving intermediate performance.

Figure 10.22 depicts functions (10.28)–(10.29) evaluated for the highside test. The best overall performance is obtained with the passive configuration c_{min}. The RSH algorithm guarantees an attenuation level of the chassis oscillation similar to that obtained with c_{min} and a slight loss of performance is registered regarding the steering assembly oscillations.

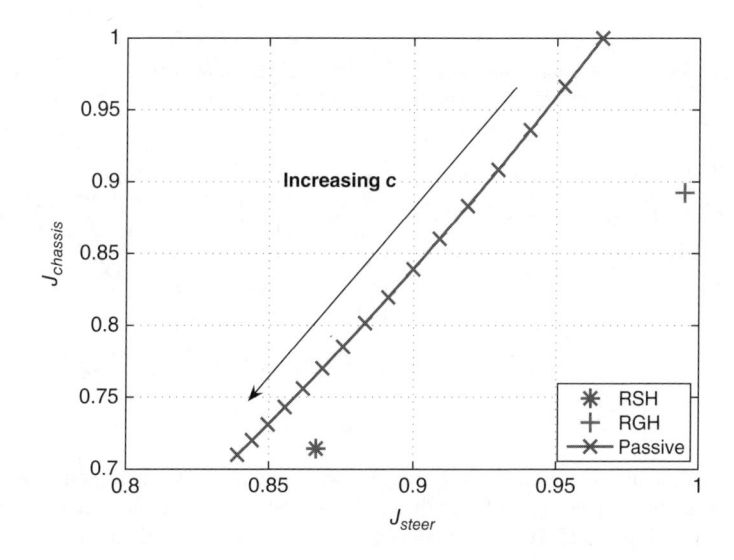

Figure 10.22 Values of the cost functions (10.28), (10.29) obtained during the highside manoeuvre

10.5 Experimental Results

To experimentally investigate the steering dynamics, a sport motorcycle was used. The vehicle was equipped with

- a removable two-state semi-active steering damper, actuated via a solenoid valve;
- a linear steering potentiometer;
- two wheel encoders to measure the vehicle speed;
- a one-axis MEMS gyroscope to measure yaw rate.
- a 1 kHz vehicle control unit with logging capabilities.

The following test protocol was adopted: while riding straight at a constant speed, the rider applies a double torque impulse to the steering handle, letting the handle go after the perturbation. The test was repeated at different speeds, ranging from 50 to 140 km/h. This test – which can be safely performed by a professional rider – excites only the weave mode, because of the limited power of the input signal.

First, the control-oriented model is identified and validated. The unknown parameters are identified using the modified gradient-based algorithm described above and initialized with the parameters identified from the simulator. Note that, because of the lack of steering torque sensor, the actual input is unknown. Thus also the input parameters were fed to the optimization and identified.

Figures 10.23a,b report the validation results at $v = 140$ km/h. The plots show the simulated and measured steering angle and yaw rate. The fit is satisfactory: the steering angle average error is $0.028°$, while for yaw rate it is $0.514°/s$, both computed over a time span of 1.5 s. The sensitivity to the damping coefficient of the steering damper is also investigated. Three damping levels have been tested: c_{min}, c_{max} and c_0. The first two obtained with minimum and maximum current applied to the semi-active steering damper, and the last one

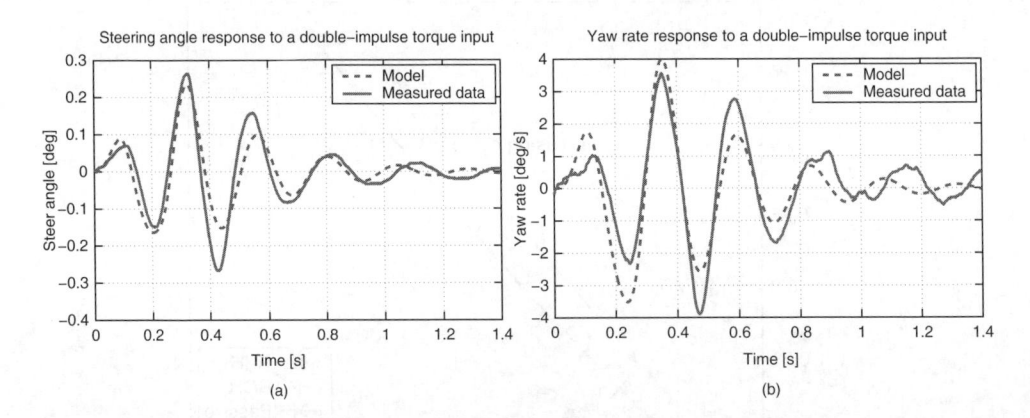

Figure 10.23 Time history of the steering angle response (a) and yaw rate response (b) to a double-impulse torque input at speed $v = 140$ km/h; model (dashed line) and measured data (solid line). Adapted from (De Filippi et al. 2011), reproduced with permission from IEEE

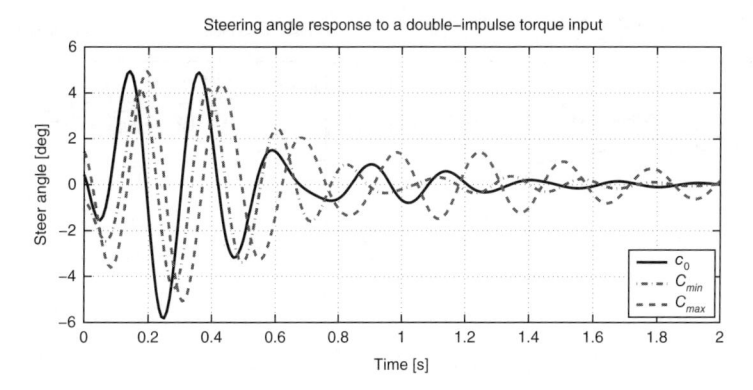

Figure 10.24 Time history of the steering angle response to a double-impulse torque input at $v = 140$ km/h: c_0 (solid line), c_{min} (dashed dotted line) and c_{max} (dashed line). Adapted from (De Filippi et al. 2011), reproduced with permission from IEEE

without any added steering damping (a certain level of damping is always present, due to friction).

Figure 10.24 shows the expected dependency of the weave mode damping on the damping coefficient (see also Figure 10.2(b)). In fact, the oscillations settle more quickly as c decreases. Also, it is evident that the actual rider input torque varies from one test to another, making it impossible to compare the results by simply comparing the time response. To overcome the limitation, the damping of the modes for each experiments is evaluated via Kung's subspace identification algorithm (Kung 1978; Lovera and Previdi 2000). Specifically, a fourth-order linear time-invariant dynamical model was identified from the measured impulse responses. The left panel of Figure 10.25 shows the map of weave eigenvalues as a function of the steering damper coefficient at a forward speed of 110 km/h. As expected, the lowest available damping coefficient yields the best damping. The right panel of Figure 10.25 maps the identified weave eigenvalues as a function of speed. The weave mode is well-damped at low speed, while it is resonant at high speed. The experimental characteristics are in agreement with what was discussed in Section 10.2.

Figure 10.26 shows the final validation. The RSH and RGH are tested using the same protocol at $v = 110$ km/h. The normalized cost function J_{sN} in (10.25) is plotted for the different damping solutions

10.6 Conclusions

In this chapter, the steering dynamics of PTWs have been discussed from a control-oriented point of view. The analysis underlined a parallelism between vertical and steering dynamics that has then been exploited to propose two innovative control strategies for semi-active steering damper control. A rotational sky-hook and a rotational ground-hook algorithm have been worked out to effectively damp the weave and the wobble resonance, respectively.

The performance of the control strategies has been evaluated both via a full-fledged motorcycle simulator and with an instrumented vehicle.

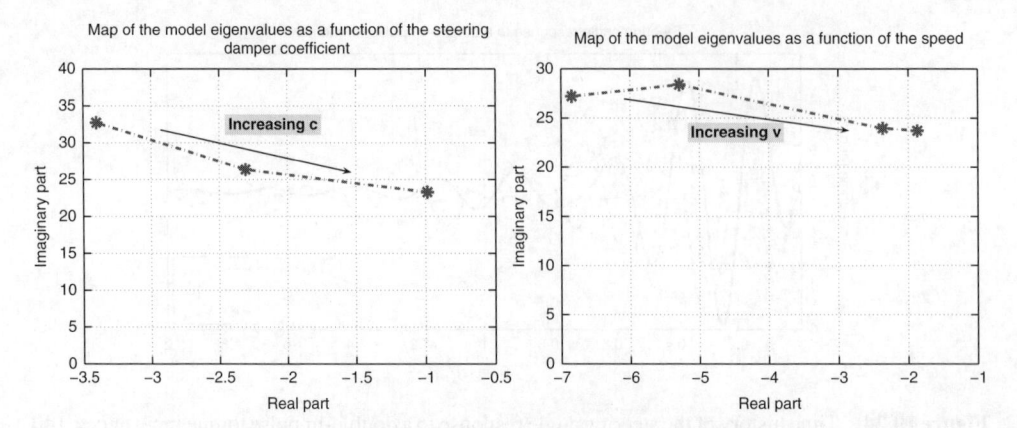

Figure 10.25 Map of the model eigenvalues as a function of the steering damper coefficient values c_0, c_{min} and c_{max} at speed $v = 110$ km/h (left) and of the vehicle speed values 50, 80, 110 and 140 km/h for minimum steer damper value c_{min}. Adapted from (De Filippi et al. 2011), reproduced with permission from IEEE

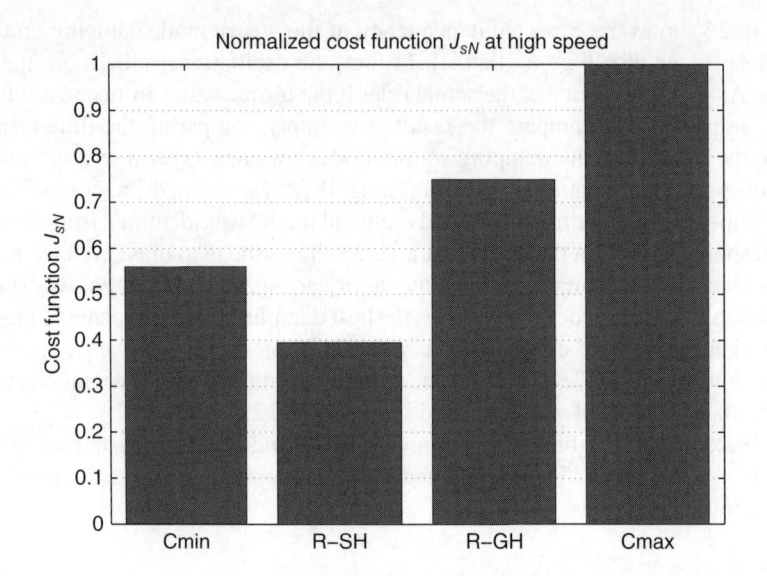

Figure 10.26 Values of the normalized cost function J_{sN} in (10.25) for $v = 110$ km/h, obtained from experimental data. Adapted from (De Filippi et al. 2011), reproduced with permission from IEEE

References

Cossalter V 2002 *Motorcycle Dynamics*. Race Dynamics, Milwaukee, USA.

Cossalter V, Lot R and Maggio F 2002 The influence of tire properties on the stability of a motorcycle in straight running and curves *SAE Automotive Dynamics & Stability Conference and Exhibition (ADSC), Detroit, Michigan, USA*.

Dardanelli A, Alli G and Savaresi S 2010 Modeling and control of an electro-mechanical brake-by-wire actuator for a sport motorbike *Proceedings of the 5th IFAC Symposium on Mechatronic Systems, Cambridge, MA, USA*.

De Filippi P, Tanelli M, Corno M, Savaresi SM and Fabbri L 2011 Semi-active steering damper control in two-wheeled vehicles. *IEEE Transactions on Control Systems Technology* **19**(5), 1003–1020.

Duke O 1997 Planet Bike – Radical Thriller or Flawed Killer. *Bike* **39**, 14–17.

Evangelou S, Limebeer D, Sharp R and Smith M 2006 Control of motorcycle steering instabilities. *IEEE Control Systems Magazine* **26**(5), 78–88.

Evangelou S, Limebeer D, Sharp R and Smith M 2007 Mechanical steering compensators for high-performance motorcycles. *Journal of Applied Mechanics* **74**(2), 332–346.

Fischer D and Isermann R 2004 Mechatronic semi-active and active vehicle suspensions. *Control Engineering Practice* **12**(11), 1353–1367.

Kung S 1978 A new identification and model reduction algorithm via singular value decomposition *Proceedings of 12th Asilomar conference on circuits systems and computers*.

Limebeer D, Sharp R and Evangelou S 2002 Motorcycle steering oscillations due to road profiling. *Journal of Applied Mechanics* **69**(6), 724–739.

Limebeer D, Sharp R, Evangelou S and Smith M 2006 *An H_∞ Loop-Shaping Approach to Steering Control for High-Performance Motorcycles* vol. 329 of Lecture Notes in Control and Information Sciences Springer pp. 257–275. In: Control of Uncertain Systems: Modeling, Approximation, and Design.

Lot R and Massaro M 2007 The Kick-Back of Motorcycles: Experimental and Numerical Analysis *Proceedings of the ECCOMAS Thematic Conference, Multibody Dynamics, Milan, Italy*.

Lovera M and Previdi F 2000 Identification of linear models for the dynamics of a photodetector. *Control Engineering Practice* **8**(10), 1149–1158.

Meirovitch L 1975 *Elements of vibration analysis*. Mc-Graw Hill, New York, USA.

Pacejka HB 2002 *Tyre and Vehicle Dynamics*. Buttherworth Heinemann, Oxford.

Panzani G, Corno M and Savaresi S 2011 Design of an adaptive throttle-by-wire control system for a sport motorcycle *Proceedings of the 18th IFAC World Congress on Automatic Control, Milan, Italy*.

Rajamani R 2006 *Vehicle Dynamics and Control*. Springer, Mechanical Engineering Series, New York.

SAE 1992 Fundamentals of Vehicle Dynamics.

Savaresi S and Spelta C 2007 Mixed Sky-Hook and ADD: approaching the filtering limits of a semi-active suspension. *Journal of dynamic systems, measurement, and control* **129**(4), 382–392.

Savaresi S, Silani E and Bittanti S 2005 Acceleration-driven-damper (ADD): an optimal control algorithm for comfort-oriented semi-active suspensions. *Journal of Dynamic Systems, Measurement and Control* **127**, 218–229.

Savaresi S, Silani E, Bittanti S and Porciani N 2003 On performance evaluation methods and control strategies for semi-active suspension systems *Proceedings of the 42nd IEEE Conference on Decision and Control (CDC), Maui, Hawaii, USA*, pp. 2264–2269.

Sharp R and Limebeer D 2004 On steering wobble oscillations of motorcycles. *Proceedings of the Institution of Mechanical Engineers, Part C: Journal of Mechanical Engineering Science* **218**(12), 1449–1456.

Smith M 2002 Synthesis of mechanical networks: the inerter. *IEEE Transactions on Automatic Control* **47**(10), 1648–1662.

Wakabayashi T and Sakai K 2004 Development of electronically controlled hydraulic rotary steering damper for motorcycles *Proceedings of the 5th International Motorcycle Safety Conference*, pp. 1–22.

Wellstead P 1981 Non-parametric methods of system identification. *Automatica* **17**, 55–69.

Williams R 1997 Automotive active suspensions part 1: basic principles. *Proceedings of the Institution of Mechanical Engineers, Part D: Journal of Automobile Engineering* **211**, 415–426.

11

Semi-Active Suspension Control in Two-Wheeled Vehicles: a Case Study

Diego Delvecchio[a] and Cristiano Spelta[b]
[a]Dipartimento di Elettronica, Informazione e Bioingegneria, Politecnico di Milano, Italy
[b]Dipartimento di Ingegneria, Università degli Studi di Bergamo, Italy

The topic of this chapter is the experimental analysis and development of a control system for a semi-active suspension in a two-wheeled vehicle. The control system is implemented via a semi-active electro-hydraulic damper, located at the rear suspension of a motorcycle. The entire design and analysis procedure is carried out: the semi-active damper is characterized; a model to effectively describe the suspension system is introduced and analysed; a wide range of control strategies are considered. The strategies are then implemented in the electronic control unit of the motorbike. Tests at bench have been carried out in an innovative layout to measure the tyre deflection and to have a comprehensive picture of both the so-called road-holding performance and comfort performance. The chapter presents a detailed evaluation of the experiments in both the time and the frequency domain.

11.1 Introduction and Problem Statement

The topic of this chapter is the introduction and the experimental analysis of a semi-active suspension system for motorcycles (Ahmadian et al. 2004; Bosch 2000; Canale et al. 2006; Choi et al. 2000; Giua et al. 2004; Hong et al. 2003; Hrovat 1997; Kawabe et al. 1998; Kiencke and Nielsen 2000; Poussot-Vassal et al. 2006; Savaresi et al. 2005; Savaresi et al. 2010; Tseng and Hedrick 1994). Among the many different types of electronically controlled suspension systems (Canale et al. 2007; Balas et al. 2003; Filardi 2003; Williams

---

Modelling, Simulation and Control of Two-Wheeled Vehicles, First Edition.
Edited by Mara Tanelli, Matteo Corno and Sergio M. Savaresi.
© 2014 John Wiley & Sons, Ltd. Published 2014 by John Wiley & Sons, Ltd.

1997; Spelta et al. 2009), semi-active suspensions seem to provide an attractive compromise between cost (energy consumption and actuators/sensors hardware) and performance. In the field of road and off-road vehicles, semi-active suspensions have been recently introduced in production cars, heavy trucks, trains and agricultural tractors, but they are still under preproduction testing on motorcycles. For this reason, the topic of controllable suspensions in vehicle applications has received considerable interest among both academic and industrial researchers. This topic is characterized by two main streams: the research and development of the technology suitable for controllable suspension and the research, design and testing of new algorithms for controllable suspension systems (Poussot-Vassal et al. 2012 and Savaresi et al. 2010 give a recent survey of the existing literature).

This chapter presents a complete case study of the design, implementation and testing of an electronic control system for a semi-active rear suspension on a high-performance motorcycle. Semi-active systems are traditionally based on a layout comprising two sensors (Silani 2004), such as one accelerometer and one potentiometer. The accelerometer is placed either on the body side or the wheel side. The potentiometer is used to monitor the suspension deflection. An alternative layout is made up of two accelerometers one on the body side and one on the wheel side: in this layout the deflection velocity of the suspension can be obtained by integrating the difference of the two accelerometers. In this research area a major improvement would be a semi-active suspension system equipped with only one sensor, without significant degradation of performance, and with an evident advantage in terms of reduction of cost and complexity (Poussot-Vassal et al. 2006). Recently, a control strategy has been theoretically developed in this direction: the Mix-1-Sensor. This rationale uses only one body accelerometer and it has been shown to ensure quasi-optimal performance (Savaresi and Spelta 2009). The goal of this work is the practical implementation and the illustration of results from real tests, moving from an idealized environment to a real vehicle. Part of this chapter is rooted in the work by (Savaresi et al. 2008).

The chapter outline is as follows: in Section 11.2 the semi-active damper is described and modelled, and its main features and control-relevant characteristics are illustrated. In Section 11.3 the 'quarter-car' model describing the suspension system is considered. In Section 11.4 the performance evaluation methods for controllable suspensions are briefly introduced. Section 11.5 focuses on the presentation of three different semi-active control algorithms. Section 11.6 is devoted to the illustration of the test-rig and the experimental protocol used for the analysis. The control algorithms are implemented on a motorcycle and their performances are experimentally evaluated and compared in Section 11.7. Some concluding remarks end the chapter.

11.2 The Semi-Active Actuator

The basic concept of electro-hydraulic semi-active damper is depicted in Figure 11.1. Compared to the classical passive element, the electro-hydraulic device comprises electronic valves instead of passive valves, so that by the means of an electric signal, it is possible to vary the valve characteristic and thus the resulting damping (for details of the state-of-the-art technologies, see Savaresi et al. 2010).

The semi-active shock absorber presented here is an electro-hydraulic semi-active actuator that has been developed specifically for motorcycle applications. This component is equipped with one current-driven solenoid electro-hydraulic valve, which can continuously vary the damping level within its controllability range. The electro-hydraulic valve has

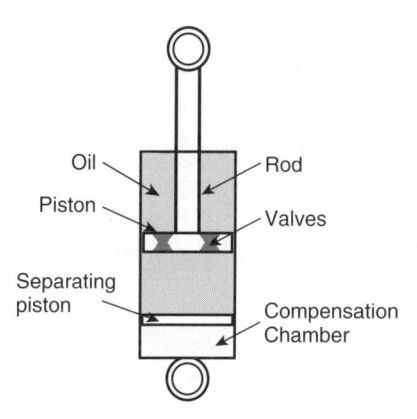

Figure 11.1 Description of an electro-hydraulic semi-active shock absorber (This picture appeared originally in: Savaresi et al. *Semi-Active Suspension Control for Vehicle*. Butterworth Heinemann, Elsevier. August 2010). Reproduced with permission from Elsevier

no embedded electronics, so it must be commanded by an external electronic control unit (ECU), which implements a fast servo-loop to control the desired level of current. This shock absorber can be addressed as an electronically controlled device, but not as a 'smart' device (Savaresi et al. 2006).

The following equations provide a model of the adjustable shock absorber:

$$F_d = c(t)\,(\dot{z}(t) - \dot{z}_t(t))$$
$$\dot{c}(t) = -\alpha c(t) + \alpha c_{in}(t-\tau)$$
$$c_{in}(t) \in [c_{min}, c_{max}] \tag{11.1}$$

where F_d represents the damping force delivered by the shock absorber; $\dot{z}(t) - \dot{z}_t(t)$ is the stroke speed defined as the difference between the vertical velocities of the vehicle body and the wheel, respectively (Figure 11.5); $c_{in}(t)$ and $c(t)$ stand for the requested and actual damping ratio, respectively; τ is the physical delay between the request and the actuation of the damping coefficient; c_{min} and c_{max} represent the minimum and maximum level of damping that can be requested to the actuator; α is the closed-loop actuation bandwidth: semi-active electrohydraulic shock absorbers usually feature a bandwidth of 20–30 Hz. Note that in (11.1) $\dot{z}(t) - \dot{z}_t(t)$ and $c_{in}(t)$ are the inputs, and the output is the damping force F_d. The dynamical behaviour of the damping coefficient is described by a first-order differential equation with delayed input. The force delivered by the shock absorber is proportional to the stroke speed, scaled by the actual damping. Notice that the model of a passive, i.e., non-controllable, shock absorber can be obtained from (11.1) by simply setting $c_{in}(t)$ as constant.

To characterize the passive-like behaviour of the shock absorber, this has been tested with a 10 Hz sinusoidal excitation. Figure 11.2 depicts the classical speed–force domain response of the shock absorber in two extreme configurations: $c_{in} = c_{min}$ and $c_{in} = c_{max}$. The ratio between the minimum and maximum damping is the so-called controllability range of the shock absorber. Notice that this ratio is about 1:3, and it is large enough to obtain good results with semi-active algorithms (Figure 11.9). Consider that a different mechanical layout or the use of other technologies based on magneto-rheological fluids or electro-rheological fluids might provide a wider controllability range (up to a 1:10 ratio) (Savaresi et al. 2005).

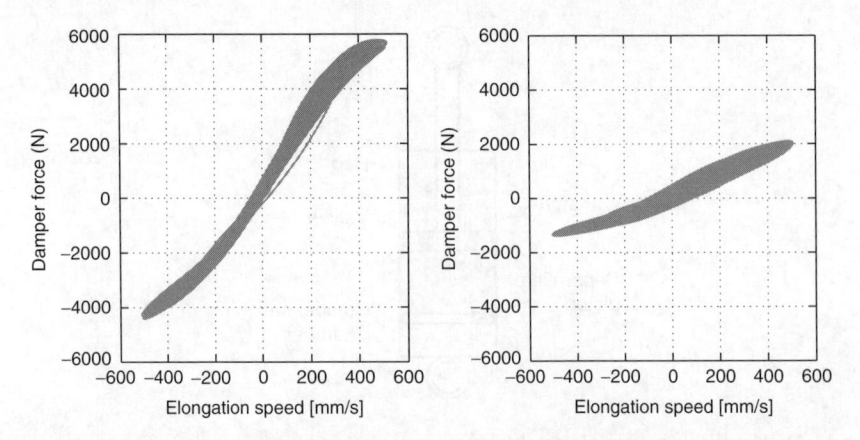

Figure 11.2 Damper characteristics in the speed–force domain. Left: minimum damping (c_{min}); right: maximum damping (c_{max}) (This picture appeared originally in: Spelta et al. *Experimental analysis and development of a motorcycle semi-active 1-sensor rear suspension*. Control Engineering Practice. 2010. vol. 18. pp. 1239–1250). Reproduced with permission from Elsevier

In Figure 11.2, the behaviour of the maximum and minimum fixed damping can be observed and compared to the linear approximation (obtained by the linear regression of the experimental data), as proposed by (11.1). Notice that the speed–force trajectories show a slight hysteresis and a little regressive behaviour for very high elongation speed around ±400 mm/sec. This is mainly due to the asymmetric characteristics of the hydraulics circuits that manage the bound and rebound phase.

To highlight the damping actuation dynamic the following test can be performed: the shock absorber is tested by a constant elongation speed excitation; during the excitation, a switch from the minimum to the maximum damping (and vice versa) can be requested. An

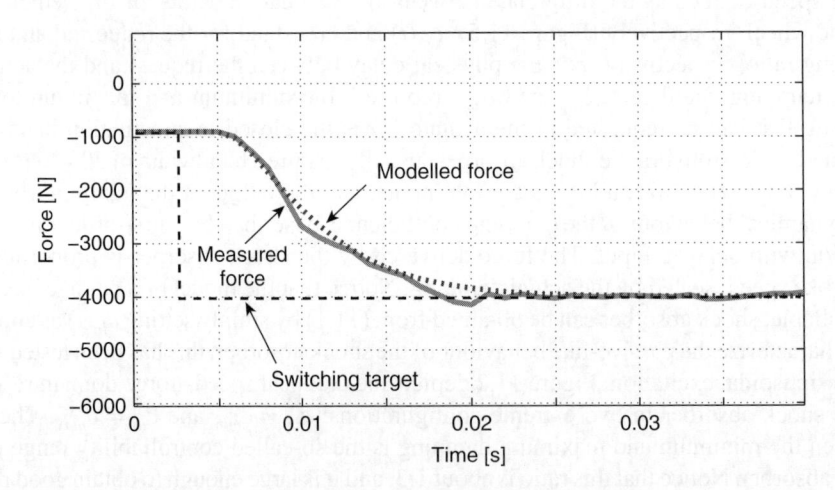

Figure 11.3 Details of the transient behaviour of the damper, subject to a step-like variation of the damping request

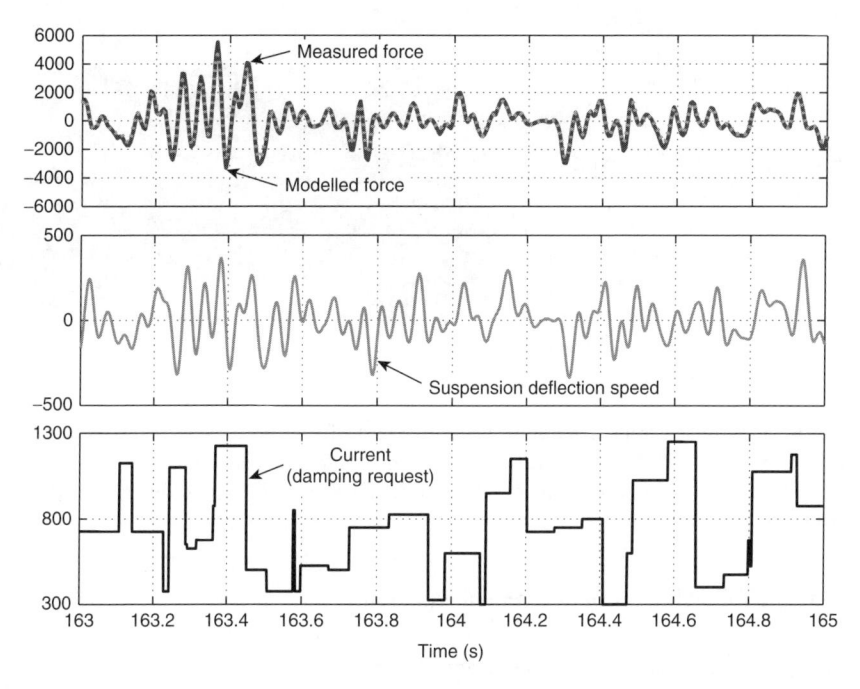

Figure 11.4 Time history of the shock absorber response during a test reproducing realistic on-road conditions. From top to bottom: damping force (measured and simulated); elongation speed $z(t) - z_t(t)$; damping request (expressed by the valve current set-point)

example of this test is displayed in Figure 11.3 where, during the compression phase (negative forces), the damping request switches from its minimum value c_{min} to its maximum value c_{max}. Notice that the force response to a step on the damping request shows approximately a linear first-order dynamic behaviour, as assumed by (11.1), although it looks slightly perturbed in the middle of the rise time: the force reaches the target in less than 20 ms, with a delay of 5 ms. Note that the ECU servo-loop controlling the damping actuation runs at a frequency of 1 kHz.

A final validating test of (11.1) can be performed at bench as follows: The shock absorber is excited with an elongation speed signal resembling the response of the suspension to a realistic road profile (Robson and Doddas 1970). The damping request signal is designed as a pseudo-random signal with steps of variable duration and amplitude as it can measures in common semi-active suspension control (Savaresi et al. 2005). The results of this analysis are depicted in Figure 11.4, which shows the adherence of the experimental data to the signal obtained by simulation according to (11.1).

11.3 The Quarter-Car Model: a Description of a Semi-Active Suspension System

The dynamic model of a semi-active suspension can be described effectively by the quarter-car model (Figure 11.5 – Gillespie 1992; Isermann 2003). Taking into account the dynamical description of the electronic shock absorber given by (11.1), the quarter-car model can be

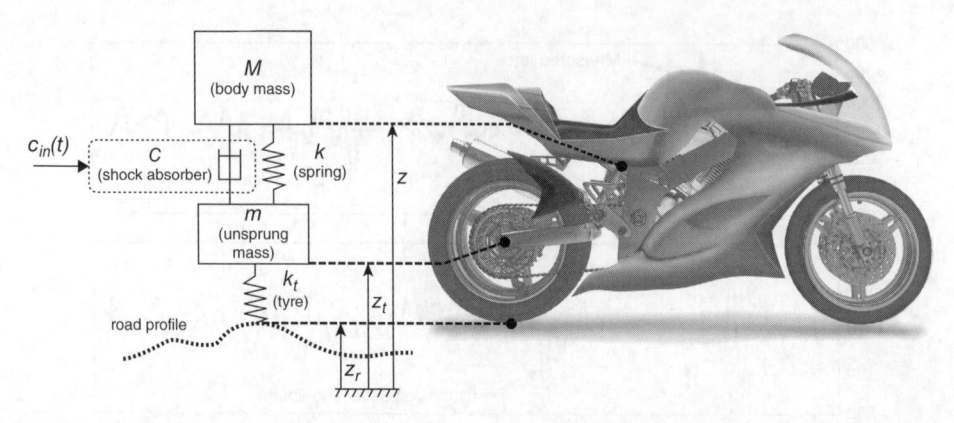

Figure 11.5 Quarter-car representation of the rear part of the motorcycle (This picture appeared originally in: Spelta et al. *Experimental analysis and development of a motorcycle semi-active 1-sensor rear suspension*. Control Engineering Practice. 2010. vol. 18. pp. 1239–1250). Reproduced with permission from Elsevier

defined by the following set of differential equations:

$$\begin{cases} M\ddot{z}(t) = -k(z(t) - z_t(t) - \Delta_k)c(t)(\dot{z}(t) - \dot{z}_t(t)) - Mg \\ m\ddot{z}_t(t) = k(z(t) - z_t(t) - \Delta_k) + c(t)(\dot{z}(t) - \dot{z}_t(t)) - k_t(z_t(t) - z_r(t) - \Delta_t) - mg \\ \dot{c}(t) = -\beta c(t) + \beta c_{in}(t) \\ z_t(t) - z_r(t) > \Delta_t \end{cases} \quad (11.2)$$

where $z(t)$, $z_t(t)$ and $z_r(t)$ are the vertical positions of the body vehicle, the unsprung mass and the road profile, respectively; M is the quarter-car body mass; m is the unsprung mass (tyre, wheel, brake calliper, suspension links etc.); k and k_t are the stiffness of the suspension spring and the tyre, respectively; $c(t)$ and $c_{in}(t)$ are the actual and the requested damping respectively, as described by (11.1); and Δ_k and Δ_t are elongation of the unloaded spring and the elongation of tyre, respectively. The model of (11.2) is a fifth-order nonlinear dynamic model. It is nonlinear as the damping coefficient $c(t)$ is a state variable. The constraint $z_t(t) - z_r(t) > \Delta_t$ ensures the avoidance of the tyre–road contact-loss phenomenon.

Remark: Stability properties of the semi-active quarter-car model *The controllable device in (11.2) is a shock absorber, which has no influence on the equilibrium point. The steady state derives from the stiffness values of the springs (and their unloaded lengths) and from the comprised masses. Due to the presence of only variable damping, (11.2) can be regarded as strictly passive, so the equilibrium is stable with respect to any control law of damping and with any values of the system parameters. (Kahlil 2002) gives the definition of passive systems and studies the stability property of such systems. To give an example for the stability analysis of (11.2), assume that the damping control represents by a switching rule. Note that this is a realistic assumption in a digital semi-active suspension system (Savaresi et al. 2005b). In this case, the quarter-car model can be regarded as a system commutating between two different stable linear modes defined by the damping c_{min} and c_{max}. It can be shown, by algebraic manipulation, that each mode of the switching system* is associated with a state matrix: $A_i = A_{min}, A_{max}$. *It is easy to see (by solving the related LMI, as described in Boyd et al. 1994) that there is a positive definite matrix P so that $A_i^T P + A_i P < 0$ for any i. According to the results on stability of switching systems reported by Liberzon, 2003, the model of (11.2) is thus stable with respect to any possible switching control (global uniform asymptotic stability).*

The general high-level structure of a semi-active control architecture is as follows:

- The **control variable** is the requested damping coefficient $c_{in}(t)$.
- The **measured output signals** are usually two: the vertical acceleration $\ddot{z}(t)$ of the motorcycle body, and the suspension displacement $z(t) - z_t(t)$ (alternatively: the vertical acceleration $\ddot{z}(t)$ and the vertical acceleration of the unsprung mass $\ddot{z}_t(t)$).
- The **controlled variables** are the chassis vertical acceleration $\ddot{z}(t)$ for a comfort-oriented strategy design and assessment, and the tire deflection $z_t(t) - z_r(t)$ for a road-holding-oriented strategy design and assessment.
- The **input disturbance** is the road profile $z_r(t)$; it is assumed to be a non-measurable and unpredictable signal (no road preview by sonar, laser or video camera is available).

The design of a control algorithm delivering the best possible performance is a non-trivial issue, since (11.2) is nonlinear with respect to the control variable c_{in} which is limited by strongly asymmetric thresholds (Guzzella and Isidori 1993; Koo et al. 2004; Savaresi et al. 2005b; Savaresi and Spelta 2007; Savaresi and Spelta 2009; Spelta et al. 2011).

11.4 Evaluation Methods for Semi-Active Suspension Systems

Regarding the suspension systems (either passive, semi-active or active), it is of great importance to have specific evaluation tools to measure the efficiency of a new structure, a novel control algorithm in order to evaluate the improvements brought with respect to a nominal reference system. More specifically, when dealing with suspension systems, the two main aspects of interest (for both academic and industrial applications) are:

- comfort characteristics;
- road-holding (or handling) characteristics.

When the **comfort objective** is considered, it mainly concerns the passenger's comfort. Generally speaking, the passenger's comfort is determined by several elements such as the human state (feeling, age, health, general abilities etc.) and chassis vibrations as a result of the road disturbance transmission. It is worth noting that only the latter element can be controlled by the suspension. Although this is addressed as a comfort goal, it has been of interest for the automotive community (both for industrial and academic research) since the comfort feeling due to vibrations and road unevenness may be of impact on the driving capabilities and on the decision abilities. Therefore the comfort objective also has an effect on the general safety level of the vehicle. The literature provides the following index to have a measure of the comfort level perceived by the driver (Savaresi et al. 2010b):

$$\int_0^T (\ddot{z}(t))^2 dt \qquad (11.3)$$

where T is an observation period for the metric commutation. Index (11.3) represents a concise index to measure the level of the vibrations transmitted to the chassis. Note that in the case $T \to \infty$ then (11.3) is the L_2 norm of the signal $\ddot{z}(t)$. The following normalized index can also be adopted for suspension system evaluation, in order to provide a measure of how the road unevenness is transmitted to the body.

$$J_c = \frac{\int_0^T (\ddot{z}(t))^2 dt}{\int_0^T (\ddot{z}_r(t))^2 dt} \tag{11.4}$$

Note that to maximize the comfort level, J_c must be minimized.

The road-holding is the vehicle capability to keep the contact with the road surface, aiming at maximizing the wheel tracking. Generally speaking, the interaction between the tyre and the road surface represents the actuation of the longitudinal and lateral forces that govern the vehicle dynamics. Concisely, the tyre provides longitudinal and lateral forces according to the following relations:

$$F_x = F_x(F_n, \lambda, \alpha, \gamma, v) \tag{11.5}$$
$$F_y = F_y(F_n, \lambda, \alpha, \gamma, v)$$

where F_x, F_y are the longitudinal and lateral forces as a nonlinear function of F_n, α and λ, γ; F_n is the vertical load acting on the tyre; λ is the tyre slip; α is the sideslip angle; γ is the Camber angle; and v is the vehicle speed. F_n is defined as:

$$F_n = (M + m)g - k_t(z_t(t) - z_r(t) - \Delta_t) \tag{11.6}$$

The symbols used in Equation (11.6) are consistent with the symbols used in (11.2). It is assumed that the nonlinear functions F_x and F_y increase monotonically with respect to the vertical load F_n (Tanelli and Savaresi 2006). Therefore, the literature usually assesses the road-holding objective with the following index (Savaresi et al. 2010):

$$\int_0^T (z_t(t) - z_r(t))^2 dt \tag{11.7}$$

Equation (11.7) represents a concise index to measure the variations of the vertical forces acting on the tyre, so of a loss of lateral and longitudinal traction capability. Note that if $T \to \infty$ then (11.7) stands for the L_2 norm of the signal $z_t(t) - z_r(t)$. The following normalized index can also be adopted for suspension system evaluation, in order to provide a measure of how the road unevenness reflects on the road-holding loss:

$$J_{rh} = \frac{\int_0^T (z_t(t) - z_r(t))^2 dt}{\int_0^T (z_r(t))^2 dt} \tag{11.8}$$

To maximize the comfort level, J_{rh} must be minimized. Note that the knowledge of the road vertical profile $z_r(t)$ is necessary for computing index J_{rh}. Since $z_r(t)$ is a non-measurable signal on road, the road-holding can be assessed only either in a simulation environment or on the test bench (as described in the next section).

Indexes J_c and J_{rh} give a simple concise measure of suspension system performance in terms of comfort and road-holding, respectively. In order to provide a more detailed analysis of the suspension behaviour it is interesting to have a frequency domain perspective. Note that, since the quarter-car-based semi-active model is highly nonlinear, the classical frequency response (FR) cannot be used. So it has been substituted with another similar frequency-domain analysis tool, called approximate frequency response (Gelb and der Velde 1968; Williams 1997). This tool provides the approximate FR from the vertical acceleration of the road disturbance $\ddot{z}_r(t)$ to the vertical body acceleration $\ddot{z}(t)$, and the approximate FR from the road disturbance $z_r(t)$ to the tyre deflection $z_t(t) - z_r(t)$.

11.5 Semi-Active Control Strategies

The algorithms are presented using the classical notation of a quarter-car model (11.2). Some of these algorithms belong to the class of comfort-oriented algorithms, in the sense that their main goal is to provide a high-quality filtering of the road disturbances, without deteriorating road-contact performance. Despite this comfort-oriented flavour, these algorithms are typically also used on high-performance sports vehicles. One of the presented algorithms is the so-called ground-hook that aims at addressing explicitly the road-holding performance.

11.5.1 Sky-hook Control

One of the most widely used semi-active control strategies is based on the heuristic approximation of the ideal concept of sky-hook (SH) damping (Karnopp et al. 1974; Karnopp et al. 1983). The two-state approximation of the sky-hook requires an ON-OFF controllable shock absorber; the control law is given by:

$$\begin{cases} c_{in} = c_{max} & \text{if } \dot{z}\left(\dot{z} - \dot{z}_t\right) \geq 0 \\ c_{in} = c_{min} & \text{if } \dot{z}(\dot{z} - \dot{z}_t) < 0. \end{cases} \tag{11.9}$$

If a continuous modulation of the damping coefficient is available, a slightly more sophisticated expression of the SH algorithm can be implemented:

$$c_{in} = \underset{[c_{min}, c_{max}]}{\text{sat}} \left[\frac{c_{SH}\dot{z}}{\dot{z} - \dot{z}_t} \right] \tag{11.10}$$

where c_{SH} is a tuning parameter, and represents the desired ideal SH damping ratio. The classical choice for this parameter is simply $c_{SH} = c_{max}$. The drawback of rule (11.10) is that all the damping force is spent for the body, thus leaving the unsprung mass badly damped; this behaviour can be corrected by evenly distributing the damping force between the body-mass and the wheel-mass, obtaining:

$$c_{in} = \underset{[c_{min}, c_{max}]}{\text{sat}} \left[\frac{\frac{1}{2}c_{SH}\dot{z} + \frac{1}{2}c_{SH}\left(\dot{z} - \dot{z}_t\right)}{\dot{z} - \dot{z}_t} \right] \tag{11.11}$$

Algorithm (11.11) is called the continuous SH. Note that algorithms (11.9)–(11.11) require knowledge of $\dot{z}$ and $(\dot{z} - \dot{z}_t)$, so in principle two sensors are required. The usual sensor layout for semi-active applications consists of the use of two vertical accelerometers, one body-side and one wheel-side. Alternatively, a stroke sensor can be adopted (such as a potentiometer) instead of one of the two accelerometers. The use of a single sensor is possible if complemented by an observer system to estimate the non-measurable signals—see an example provided by (Delvecchio et al. 2011).

11.5.2 Mix-1-Sensor Control

Similarly to SH, this strategy requires a two-level damper, the control law being given by:

$$\begin{cases} c_{in} = c_{max} & \text{if } (\ddot{z}^2 - \alpha^2 \dot{z}^2) \leq 0 \\ c_{in} = c_{min} & \text{if } (\ddot{z}^2 - \alpha^2 \dot{z}^2) > 0 \end{cases} \quad (11.12)$$

This control law is extremely simple since, like SH, it is based on a static rule which makes use of $\dot{z}, \ddot{z}$ only. The Mix-1-Sensor algorithm is demonstrated to be an approximation of the optimal damping control for a semi-active suspension system (Savaresi and Spelta 2009). The key idea of (11.12) is condensed to $(\ddot{z}^2 - \alpha^2 \dot{z}^2)$. In fact, (11.12) selects, at the end of every sampling interval, the maximum damping c_{max} or the minimum damping c_{min}, according to the current value of $(\ddot{z}^2 - \alpha^2 \dot{z}^2)$: if $(\ddot{z}^2 - \alpha^2 \dot{z}^2) > 0$, c_{min} is requested, otherwise c_{max} is selected. The amount $(\ddot{z}^2 - \alpha^2 \dot{z}^2)$ can therefore be considered as a frequency-range selector working in the time domain. Parameter α is the only tuning knob that characterizes algorithm (11.12). The value of α represents the desired cross over frequency of the trade-off between under-damped (c_{min}) and over-damped (c_{max}) suspension. For further details on the design of this control strategy and on the effectiveness of the frequency-range selector $(\ddot{z}^2 - \alpha^2 \dot{z}^2)$, the reader is referred to (Savaresi and Spelta 2007). Since (11.12) computes only the amount $(\ddot{z}^2 - \alpha^2 \dot{z}^2)$ it requires the use of one sensor only (typically an accelerometer) to monitor $\ddot{z}$, then by integration $\dot{z}$ can be derived. This feature makes the Mix-1-Sensor extremely appealing for real implementation.

11.5.3 The Ground-Hook Control

The ground-hook control (GH) is the semi-active heuristic approximation of the ideal concept of ground-hook damping (Hrovat 1997); it is the most widely used control strategy in semi-active suspension systems for addressing the road-holding objective. The two-state approximation of the ground-hook requires an ON-OFF controllable shock absorber; the control law is given by:

$$\begin{cases} c_{in} = c_{max} & \text{if } \dot{z}_t (\dot{z} - \dot{z}_t) < 0 \\ c_{in} = c_{min} & \text{if } \dot{z}_t (\dot{z} - \dot{z}_t) \geq 0 \end{cases} \quad (11.13)$$

Similarly to the SH, the GH control also requires the use of two sensors.

To conclude this section, note that SH and GH algorithms have no tuning knobs, even though evolutions of the basic SH algorithm, equipped with tuning parameters, have been recently presented in (Poussot-Vassal et al. 2006 and Sammiar et al. 2000). Instead, the Mix algorithms have a key tuning parameter, which is the so-called cross-over frequency. This parameter is typically set at the classical c-invariant frequency of the suspension, but it can be moved from that position for fine-tuning purposes.

11.6 Experimental Set-up

The experimental facility presented here is a four-post test-rig, adapted for a two-wheeled vehicle. For motorcycle testing, since a two-wheeled vehicle has no an intrinsic equilibrium condition at stand still, the test-rig has been equipped with an additional structure to keep the vehicle in the vertical position, while leaving all the in-plane movements (pitch and heave) completely free. The test-rig displacement is measured by the test-rig acquisition system, and then synchronized with the vehicle signals. The control algorithms are implemented in the vehicle ECU, according to the following experimental layout:

- The basic suspension sensor set includes a body-side vertical MEMS accelerometer, having a range of 10 g; a wheel-side vertical MEMS accelerometer, range of 25 g; a stroke sensor, consisting of a potentiometer, with range of 0–60 mm.
- To measure the tyre deflection a distance sensor was installed at the wheel hub to acquire the distance between the wheel and the bench plate. The distance sensor is from Baumer (OADM 13I6475/S35A) featuring a response time of less than 0.9 ms, appropriate for this kind of application.
- The shock absorber, introduced in the previous section, is driven by a feedback control of the current. The current is measured by a Hall-effect based transducer by LEM, having the range $\pm 5A$;
- The electronic control unit is based on a Freescale microcontroller desinged for automotive applications. The suspension code runs at a frequency of 1 kHz, so the algorithms, the filters and the observers are digitalized accordingly, by using a Tustin transformation.

The testing protocol used for evaluating the performance of the semi-active algorithms consists of three types of experiments. The first type of test-rig experiment is a time-varying sinusoidal excitation (usually called a frequency sweep). The explored frequency range is 0.1–30 Hz. The whole experiment lasts about 2 minutes, and the amplitude of the sinusoidal excitation decreases as the frequency increases. The test-rig displacement for this excitation is illustrated in Figure 11.6. This amplitude profile (as a function of the frequency) is designed in order to achieve the maximum amount of excitation, while avoiding loss of contact of the wheel. Figure 11.6 displays also an example of the signals measured on the vehicle (the two accelerations and the suspension stroke) when the damping coefficient is set at its minimum value (c_{min}). Notice that the occurrence of the two classical main resonances (the body and the wheel resonance) is clearly shown by the behaviour of the stroke signal.

The filtering effect of the suspension is clearly visible by the direct comparison of the body and wheel accelerations. The second type of excitation is a broad band signal, which reproduces a typical road profile. This excitation signal is displayed in Figure 11.7, both in the time domain and in the frequency domain. Notice that the amplitude (peak-to-peak) of this excitation is about 6 cm. The spectral content of this signal is similar to a band-limited integrated white noise, which is commonly used to resemble a realistic road profile (Robson and Dodds 1970). The last type of experiment consists in a series of impulse excitation as described in Figure 11.8.

11.7 Experimental Evaluation

In this section the test-rig and the on-road results are presented and discussed. This study was carried out according to the testing protocol introduced in the previous section. For conciseness, the section is mainly focused on the response to the 'sweep' test, as the driven

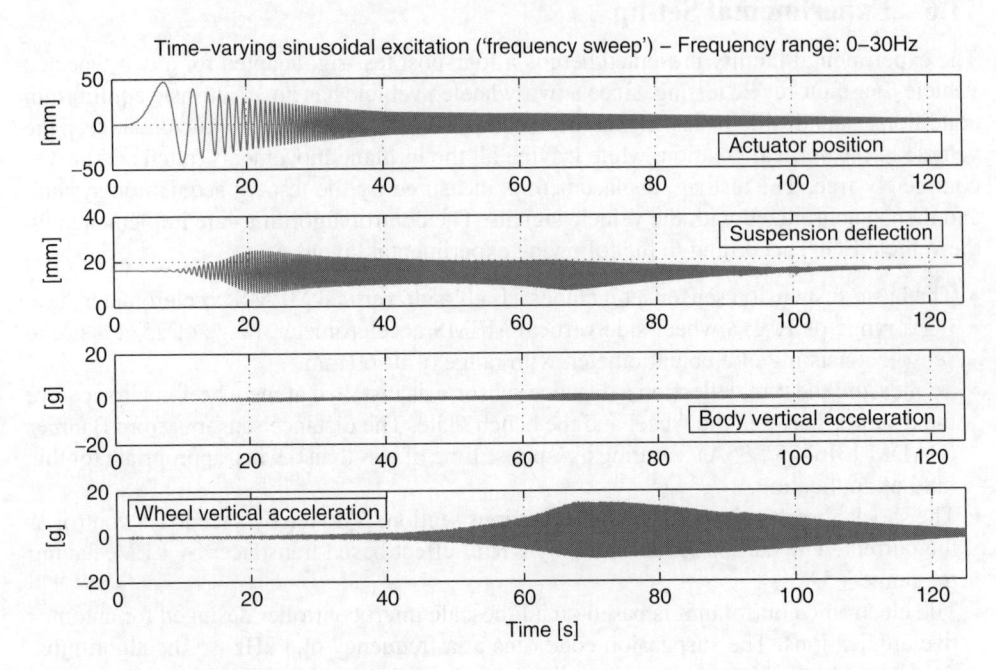

Figure 11.6 Example of the time-history of the bench acquisition. From top to bottom: test rig displacement, suspension stroke, body acceleration, wheel acceleration

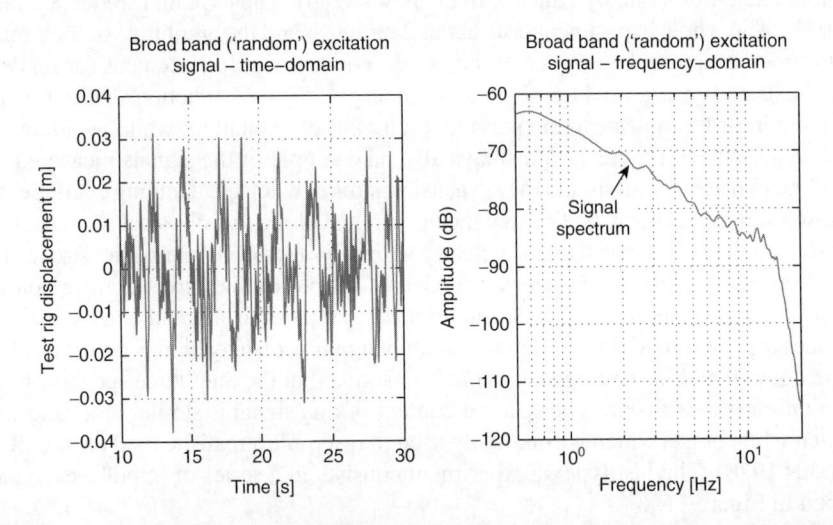

Figure 11.7 Broadband excitation in the time domain (left) and in the frequency domain (right) (This picture appeared originally in: Spelta et al. *Experimental analysis and development of a motorcycle semi-active 1-sensor rear suspension*. Control Engineering Practice. 2010. vol. 18. pp. 1239–1250). Reproduced with permission from Elsevier

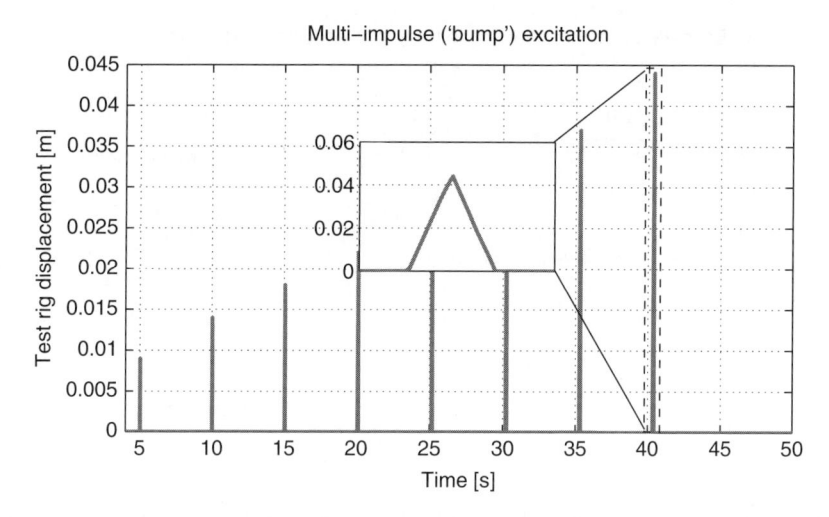

Figure 11.8 Series of impulsive road excitations with increasing amplitude (This picture appeared originally in: Spelta et al. *Experimental analysis and development of a motorcycle semi-active 1-sensor rear suspension*. Control Engineering Practice. 2010. vol. 18. pp. 1239–1250). Reproduced with permission from Elsevier

conclusions are consistent also with the 'random' tests. First, the analysis is made in the frequency domain, by analysing the estimated frequency response from the road profile to the body displacement. Time-domain performances are then considered, using a condensed performance index and a direct time-domain signal analysis. For a comprehensive description of suspension performance evaluation, both in time and frequency domain see (Savaresi et al. 2010a), and the reference cited therein.

In order to understand the basic behaviour of the rear suspension of the vehicle, first it is interesting to analyse the performance of the vehicle without control algorithms (namely in a passive-like configuration). For this purpose, Figures 11.9 and 11.10 depict the suspension system performances when the damping coefficient is kept fixed at c_{min} and c_{max}. In particular, Figure 11.9 shows the approximate FR from the road vertical acceleration $\ddot{z}_r(t)$ to the body vertical acceleration $\ddot{z}(t)$ (comfort-oriented evaluation in the frequency domain), while Figure 11.10 depicts the approximate FR from the road vertical displacement $z_r(t)$ to the tyre deflection $z_t(t) - z_r(t)$ (road-holding-oriented evaluation in the frequency domain). The obtained results are very clear and show the classical behaviour of an under-damped and an over-damped suspension and its trade-off: when c_{min} is used, the two main resonances (the body resonance at 2.5 Hz and the wheel resonance at 12 Hz) are clearly visible since they are poorly damped. Inspecting Figures 11.9 and 11.10 some considerations can be drawn:

- From the comfort point of view, the use of c_{max} setting ensures a better damping of the body resonance, but it suffers from a bad filtering at mid to high frequencies, where the low damping setting c_{min} clearly out performs. This is known as comfort trade-off, and any passive setting represents an acceptable compromise between damping and filtering.
- From the road-holding point of view the filtering performances highlight that the use of high damping (c_{max}) improves the road-holding quality at both the wheel and body

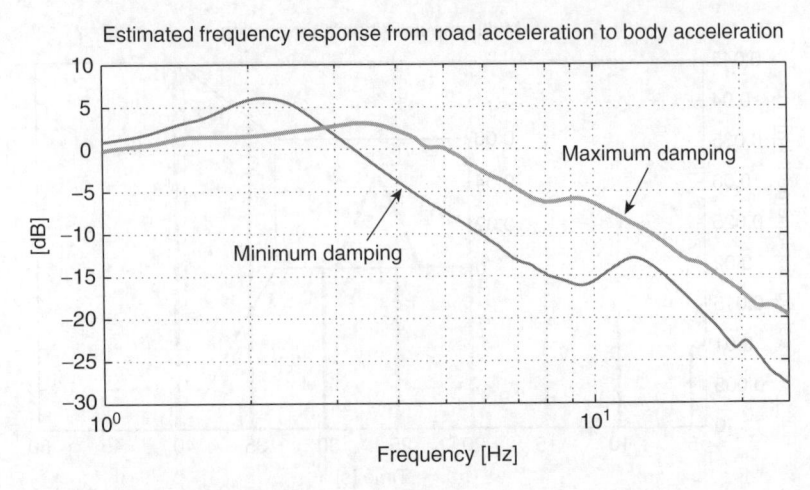

Figure 11.9 Comfort-oriented filtering performance of the passive-like suspension configuration, depicted in the frequency domain (This picture appeared originally in: Spelta et al. *Experimental analysis and development of a motorcycle semi-active 1-sensor rear suspension*. Control Engineering Practice. 2010. vol. 18. pp. 1239–1250). Reproduced with permission from Elsevier

resonances, but it shows a bad filtering behaviour for the mid frequencies, where the low damping setting (c_{min}) provides a better filtering. This recalls the road-holding trade-off, and any passive setting is a compromise between the best filtering at mid frequencies of a low-damped suspension and the best damping of the high-frequency wheel resonance.

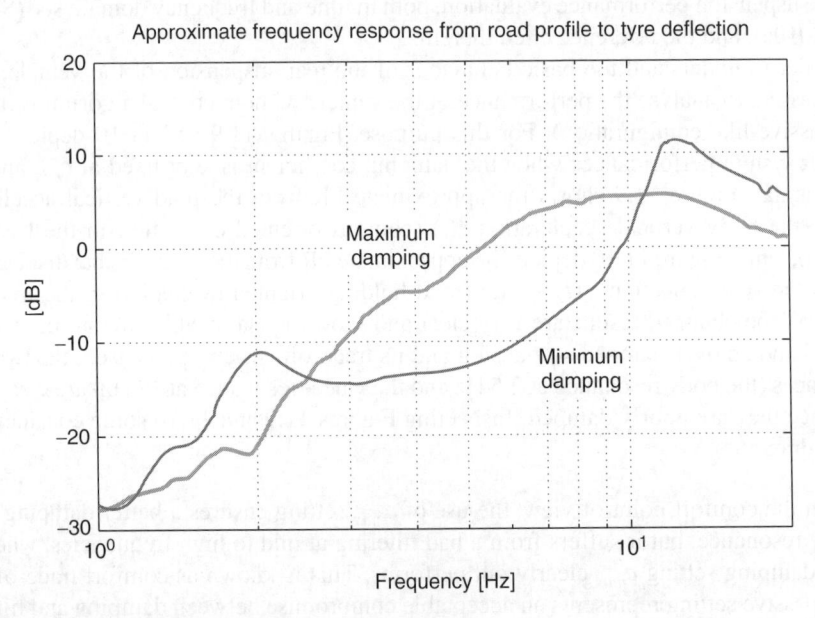

Figure 11.10 Road-holding-oriented filtering performance of the passive-like suspension configuration, depicted in the frequency domain

- In the frequency range up to about 10 Hz the passive-like settings behave similarly in terms of both comfort performance and road-holding performance. The use of c_{max} setting provides the best results around the body resonance, while the use of c_{min} setting ensures good results in terms of filtering beyond the body resonance and up to 10 Hz. This is not surprising since controlling the body movements of the motorcycle increases the general level of driveability of the vehicle. Beyond 10 Hz the performance shows a clear trade-off between comfort and road-holding: the best setting in terms of comfort is the worst in terms of road-holding, and vice versa. From this point, in the frequency range up to 10 Hz an improvement in terms of comfort also corresponds to an improvement in terms of road-holding.
- Any semi-active control of damping aims to remove the comfort trade-off, without the road-holding trade-off deteriorating.

Figures 11.11 and 11.12 show the frequency domain performance of the semi-active algorithms introduced in Section 11.5. To help the analysis, the performances of the passive settings c_{min} and c_{max} are also reported. The results may be easily interpreted and the following conclusions can be drawn:

- In terms of comfort (Figure 11.11) both the SH and the Mix-1-Sensor provide optimal performance over the entire range of frequency (Savaresi and Spelta 2009 gives the optimality analysis of the Mix-1-Sensor). They show a good damping of the body resonance as with the c_{max} setting and, at the same time, a good filtering as provided by the c_{min} setting. This behaviour of both SH and Mix-1-Sensor clearly shows that the comfort trade-off is fully removed. The performance of GH is extremely poor both in terms of resonance damping

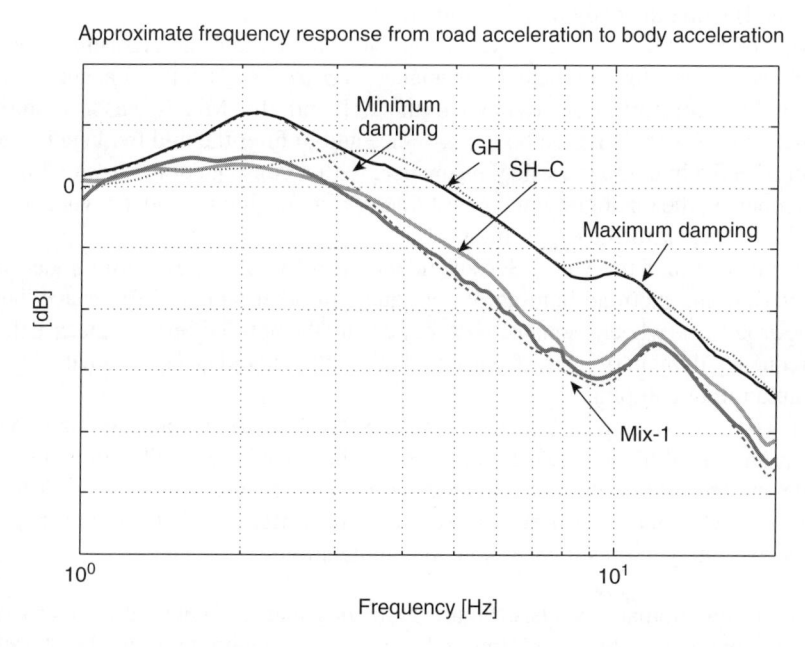

Figure 11.11 Comfort-oriented filtering performance of the suspension, depicted in the frequency domain: comparison between the proposed control strategies

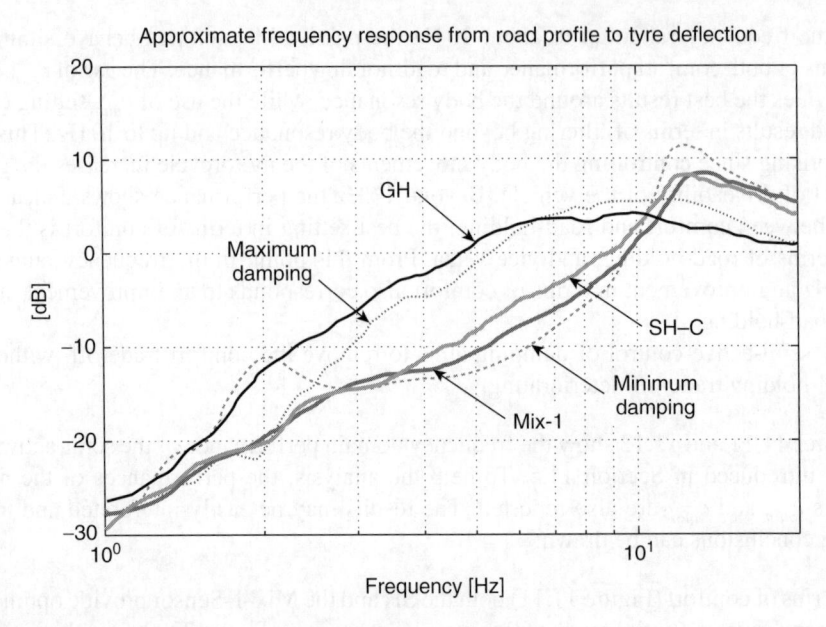

Figure 11.12 Road-holding-oriented filtering performance of the suspension, depicted in the frequency domain: comparison between the proposed control strategies

and high-frequency filtering. This is not surprising as the ideal concept of the ground-hook is based on the assumption that the wheel is ideally linked to a fixed reference (instead of the ground) while the body remains undamped.

- In terms of handling (Figure 11.12), as already mentioned, the algorithm's behaviour resembles the comfort-oriented performances up to the wheel resonance. Therefore, despite their comfort-oriented flavour, the SH and the Mix-1-Sensor represent the optimal strategies to damp the body resonance and to filter the mid frequencies, but they cannot effectively damp the wheel resonance, with a consequent loss of road-holding at this frequency; the GH represent the best trade-off algorithm around the wheel resonance as it also outperforms the c_{max} setting.
- Interestingly enough the Mix-1-Sensor and the SH fully solve the comfort trade-off, without deteriorating the road-holding performance. In other words, they behave like a c_{min} setting around the wheel resonance, but they ensure the optimal performances in the rest of the frequency domain of interest. This kind of result cannot be achieved with any passive configuration of damping.
- In terms of results, the Mix-1-Sensor and the SH differences are almost negligible. They are also similarly characterized by a low demanding rule in terms of computational complexity, but the former algorithm requires only a single sensor (an accelerometer placed body-side), while the latter needs two sensors. This difference, in terms of sensor layout, may be extremely interesting for industrial applications.

Moving to time-domain analysis, Figure 11.13 provides a condensed comparison of the performance of all the tested algorithms and fixed-damping configurations. The performance index used in Figure 11.13 is defined by (11.4) and (11.8). For comparison purposes, the

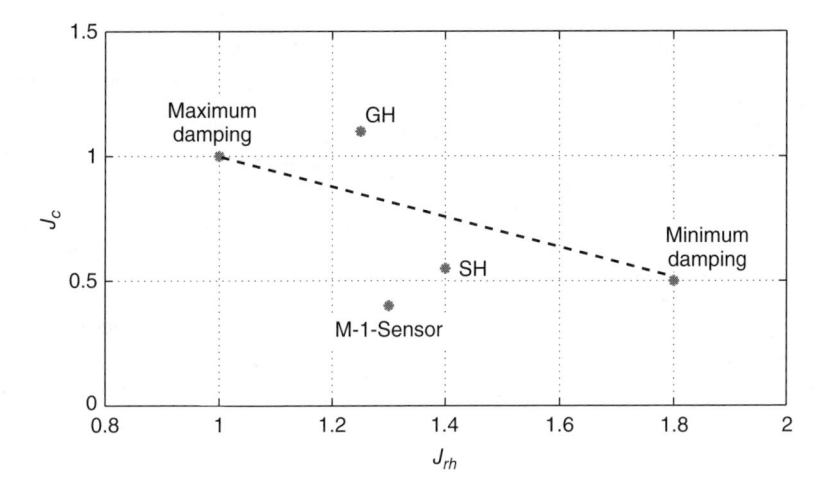

Figure 11.13 Passive settings performance and semi-active performance mapped in terms of indexes J_c and J_{rh}

indexes are normalized with respect to the value achieved by the c_{max} setting. These results were obtained from the test-rig experiments with the random-walk excitation (see Figure 11.7). According to the definition of these indexes, the smaller J_c and J_{rh} are, the better the filtering performance is. Figure 11.13 confirms most of the results drawn from the frequency-domain analysis. It is clear also how the semi-active control is able to overcame the compromises of a standard (non-controllable) suspension. Mix-1-Sensor exceeds the comfort performances of both the c_{max} and c_{min} settings without a deterioration of the road-holding index, as it provides a value of J_c similar to that of an under-damped suspension and a value of J_{rh} similar to that of an over-damped suspension. It is also interesting to observe that, from Figure 11.13, we could conclude that the SH and the fixed damping provide almost the same performance; however, this condensed performance index hides the fact that SH provides the same filtering performance but with the advantage of a much better resonance damping; this fact can only be appreciated by the direct frequency-domain analysis.

To conclude the analysis, the performance of the system has been evaluated for the bump test, as illustrated in Figure 11.14. As the results depict, when a c_{min} fixed damping is used, the suspension reacts to the bump with a large stroke movement (almost 30 mm peak-to-peak); the consequence is a good filtering (the acceleration peaks body-side are small). The drawback is that, when the bump is passed, the settling time is relatively long and characterized by serious undamped oscillations. On the other hand, when a c_{max} fixed damping is used, the suspension reacts to the bump with a small stroke movement (less than 15 mm peak-to-peak); the consequence of this is poor filtering (the acceleration peaks body-side are large). The benefit of this fixed tuning is clear in the second part of the transient: when the bump is passed, the settling time is relatively short and very well damped. The semi-active Mix-1-Sensor algorithm inherits the best of the two fixed settings: in the first part of the transients it keeps the damping low, to get a good filtering (low acceleration peaks), and in the second part of the transient it sets the damping to the maximum value, in order to provide a short settling time.

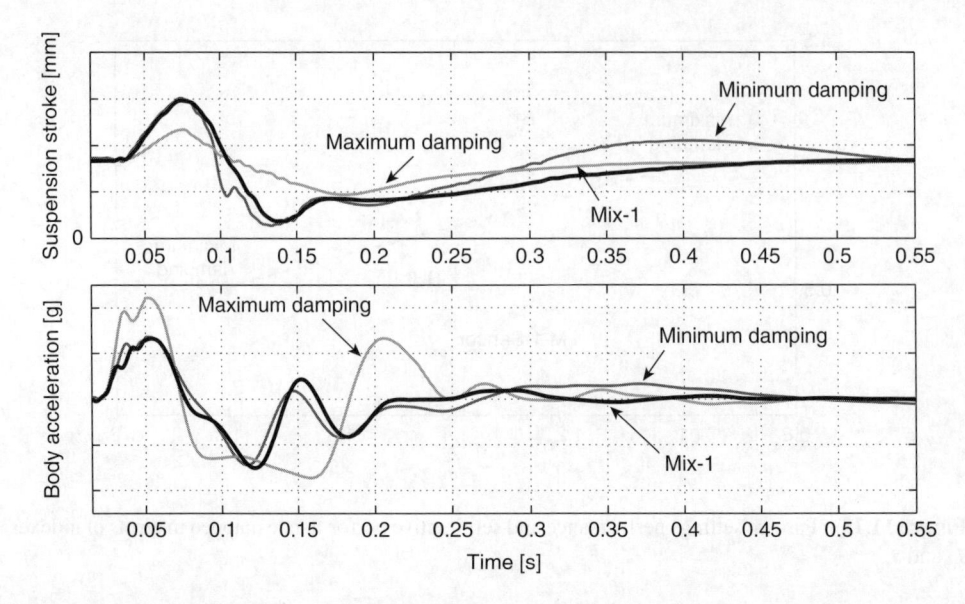

Figure 11.14 Example of the suspension response to a 45 mm bump. Suspension stroke (top); body acceleration (bottom)

Focusing on the bump experiment, it is interesting to inspect the handling behaviour of the algorithm. In Figure 11.15 the tyre deflection is displayed for the 45 mm bump experiment (where the zero stands for the steady-state condition of the tyre deflection). Notice that after encountering the bump (dashed line), the tyre deflection decreases since the tyre is compressed. This phase obviously is not critical from the road-holding point of view, since the vertical load increases. The critical part is when the tyre deflection gets larger

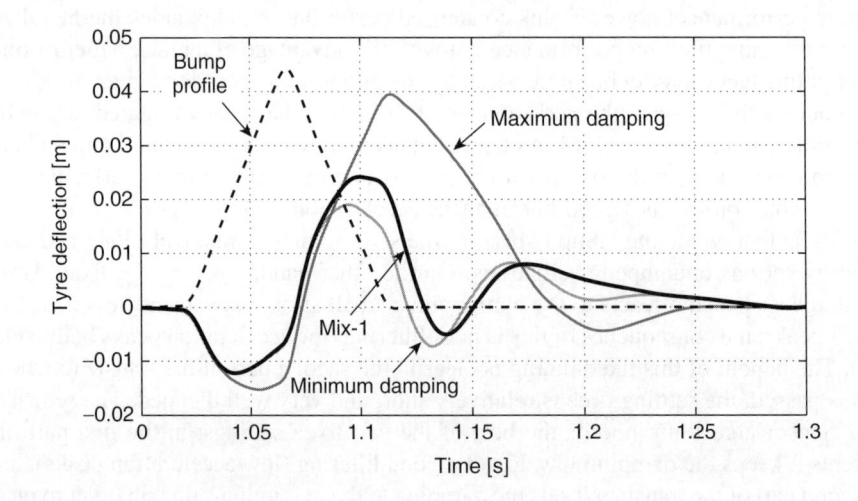

Figure 11.15 Example of the suspension response to a 45 mm bump. Tyre deflection

than its steady-state value, and in this case the vertical load decreases and the available lateral and longitudinal forces are smaller. The road-holding of the vehicle therefore deteriorates. The fixed-damping c_{max} provides the worst performance, because the high hydraulic friction makes the recovery from the compression phase slower. On the other hand, the fixed-damping c_{min} provides the best performance, since the suspension rapidly returns to the original position; its problem (as usual) is the undamped behaviour. Notice, however, that this behaviour has little impact of the handling performance. Finally, semi-active algorithms (like Mix-1-Sensor) provide a good compromise: the tyre deflection rapidly returns to the initial condition, and the damping after the initial phase is very good. As already observed, even if these algorithms can be classified as comfort-oriented, it is interesting to observe that their handling performance is good on large road disturbances that are able to excite all the range of frequencies.

11.8 Conclusions

In this chapter, the semi-active suspension system design has been proposed and experimentally studied, in terms of both the classical comfort objective and the road-holding objective. The semi-active actuator specifically developed for motorcycle applications has been presented. Some state-of-the-art semi-active algorithms have been considered, then implemented on the vehicle control unit and finally tested on a two-post test-bench. The testing layout provided a comprehensive analysis in terms of both comfort and road-holding performance. The experiments confirm the effectiveness of the use of semi-active technology in this class of vehicles, highlighting how the appropriate control algorithm is able to remove the compromises that characterize the traditional (non-controllable) system. In particular, the experimental work has shown that so-called comfort-oriented control strategies are able to also provide good road-holding performance.

Current research is focusing on the study of new road-holding-oriented control design, capable of optimizing the wheel dynamics while also guaranteeing the best comfort behaviour.

References

Ahmadian M, Song X and Southward C 2004 No-jerk skyhook control methods for semiactive suspensions. *Journal of Vibration and Acoustics* **126**, 580–585.

Balas GJ, Bokor J and Szabo Z 2003 Invariant subspaces for LPV systems and their application. *IEEE Transaction on Automatic Control* **48**, 2065–2069.

Bosch 2000 *Automotive Handbook, 5th Edition*. Bosch GmbH.

Boyd S, El-Ghaoui L, Feron E and Balakrishnan V 1994 *Linear Matrix Inequalities in System and Control Theory*. SIAM Studies in Applied Mathematics.

Canale M, Fagiano L, Milanese M and Borodani P 2007 Robust vehicle yaw control using an active differential and IMC techniques. *Control Engineering Practice* **15**(8), 923–941.

Canale M, Milanese M and Novara C 2006 Semi-active suspension control using fast model-predictive techniques. *IEEE Transaction on Control System Technology* **14**(6), 1034–1046.

Choi S, Nam M and Lee B 2000 Vibration control of a mr seat damper for commercial vehicles. *Journal of Intelligent Material Systems and Structures* **11**, 936–944.

Delvecchio D, Spelta C and Savaresi S 2011 Estimation of the tire vertical deflection in a motorcycle suspension via Kalman-filtering techniques *Proceedings of the 2011 IEEE Multi-conference on Systems and Control*, Denver, Colorado.

Filardi G 2003 Robust control design strategies applied to a DVD-video player PhD thesis Joseph Fourier University, Laboratoire d'Automatique de Grenoble Grenoble.

Gelb A and der Velde WV 1968 *Multiple-Input Describing Functions and Nonlinear System Design*. McGraw-Hill, New York.

Gillespie T 1992 *Fundamental of vehicle dynamics*. Society of Automotive Engineers.

Giua A, Melas M, Seatzu C and Usai G 2004 Design of a predictive semiactive suspension system. *Vehicle System Dynamics* **41**(4), 277–300.

Guzzella L and Isidori A 1993 On approximate linearization of nonlinear control systems. *International Journal of Robust and Non Linear Control* **3**, 261–276.

Hong K, Sohn H and Hedrick J 2003 Modified skyhook control of semi-active suspensions: A new model, gain scheduling, and hardware-in-the-loop tuning. *ASME Journal of Dynamic Systems, Measurement, and Control* **124**, 158–167.

Hrovat D 1997 Survey of advanced suspension developments and related optimal control application. *Automatica* **33**(10), 1781–1817.

Isermann R 2003 *Mechatronic Systems: Fundamentals*. Springer-Verlag.

Kahlil K 2002 *Nonlinear Systems. Third Edition*. Prentice Hall: Upple Saddle River, NJ.

Karnopp D 1983 Active damping in road vehicle suspension systems. *Vehicle System Dynamics* **12**(6), 296–316.

Karnopp D, Crosby M and Harwood R 1974 Vibration control using semi-active force generators. *Journal of Engineering for Industry* **96**, 619–626.

Kawabe T, Isobe O, Watanabe Y, Hanba S and Miyasato Y 1998 New semi-active suspension controller design using quasi-linearization and frequency shaping. *Control Engineering Practice* **6**, 1183–1191.

Kiencke U and Nielsen L 2000 *Automotive Control Systems*. Springer-Verlag.

Koo JH, Goncalves F and Ahmadian M 2004 Investigation of the response time of magnetorheological fluid dampers. *SPIE* **5386**, 63–71.

Liberzon D 2003 *Switching in Systems and Control*. Birkhauser. Boston, MA.

Poussot-Vassal C, Sename O, Dugard L, Ramirez-Mendoza R and Flores L 2006 Optimal Skyhook control for semi-active suspensions *Proceedings of the 4th IFAC Symposium on Mechatronics Systems*, pp. 608–613, Heidelberg, Germany.

Poussot-Vassal C, Spelta C, Sename O, Savaresi S and Dugard L 2012 Survey and performance evaluation on some automotive semi-active suspension control methods: A comparative study on a single-corner model. *Annual Reviews in Control* **36**, 148–160.

Robson J and Dodds C 1970 The response of vehicle component to random road-surface undulations *Proceedings of the 13th FISITA congress*, Brussels, Belgium.

Sammier D, Sename O and Dugard L 2000 $\mathcal{H}_\infty$ control of active vehicle suspensions *Proceedings of the IEEE International Conference on Control Applications (CCA)*, pp. 976–981, Anchorage, Alaska.

Savaresi S and Spelta C 2007 Mixed sky-hook and ADD: Approaching the filtering limits of a semi-active suspension. *ASME Transactions: Journal of Dynamic Systems, Measurement and Control* **129**(4), 382–392.

Savaresi S and Spelta C 2009 A single-sensor control strategy for semi-active suspensions. *IEEE Transaction on Control System Technology* **17**(1), 143–152.

Savaresi S, Bittanti S and Montiglio M 2005a Identification of semi-physical and black-box models: the case of MR-dampers for vehicles control. *Automatica* **41**, 113–117.

Savaresi S, Poussot-Vassal C, Spelta C, Sename O and Dugard L 2010a *Semi-Active Suspension Control Design for Vehicles*. Butterworth-Heinemann: Oxford.

Savaresi S, Silani E and Bittanti S 2005b Acceleration driven-damper (add): an optimal control algorithm for comfort-oriented semi-active suspensions. *ASME Transactions: Journal of Dynamic Systems, Measurement and Contro* **127**, 218–229.

Savaresi S, Spelta C, Moneta A, Tosi F, Fabbri L and Nardo L 2008 Semi-active control strategies for high-performance motorcycles *Proceedings of the 2008 IFAC World Congress*, pp. 4689–4694, Seoul, South Korea.

Savaresi S, Tanelli M, Langthaler P and Re LD 2006 Identification of tire-road contact forces by in-tire accelerometers *Proceedings of the 14th IFAC Symposium on System Identification (SYSID)*, Newcastle, Australia.

Silani E, Savaresi S, Bittanti S, Fischer D and Isermann R 2004 Managing information redundancy for the design of fault-tolerant slow-active controlled suspension. *Tire Technology International* **2004**, 128–133.

Spelta C, Previdi F, Savaresi S, Bolzern P, Cutini M, Bisaglia C and Bertinotti S 2011 Performance analysis of semi-active suspension with control of variable damping and stiffness. *Vehicle System Dynamics* **42**(1-2), 237–256.

Spelta C, Previdi F, Savaresi S, Fraternale G and Gaudiano N 2009 Control of magnetorheological dampers for vibration reduction in a washing machine. *Mechatronics* **19**, 410–421.

Spelta et al. Experimental analysis and development of a motorcycle semi-active 1-sensor rear suspension. *Control Engineering Practice*. 2010. vol. **18**. pp. 1239–1250).

Tanelli M and Savaresi S 2006 Friction curve peak detection by wheel deceleration measurement *Proceedings of the IEEE Conference on Intelligent Transportation Systems (ITSC)*, Toronto, Canada.

Tseng H and Hedrick J 1994 Semi-active control laws – optimal and sub-optimal. *Vehicle System Dynamics* **23**(1), 545–569.

Williams R 1997 Automotive active suspensions part 1 and part 2. *IMechE* **211**, 415–426.

12

Autonomous Control of Riderless Motorcycles

Yizhai Zhang[a], Jingang Yi[a], and Dezhen Song[b]
[a]Rutgers University, USA
[b]Texas A&M University, USA

12.1 Introduction

In Chapter 2, we presented a new dynamic model of autonomous motorcycles for agile manoeuvres. The new features of the dynamic model in Chapter 2 are the relaxation of the zero lateral velocity non-holonomic constraint of the wheel contact points and the integration of the tyre–road friction models. The control inputs to the motorcycle dynamic model are the steering angle and the wheel angular velocities. The objective of this chapter is to develop a simultaneously trajectory-tracking and path-following control system using the model developed in Chapter 2.

Control of an autonomous motorcycle only using the steering and speed changes as inputs is challenging due to the platform's non-minimum phase and under-actuation properties.* For such systems, there does not exist an analytical causal compensator for *exactly* output (trajectory) tracking while keeping the internal (balancing) stability (Grizzle et al. 1994). With an additional rider lean (weight shifting) as a control input, it has been shown that manoeuvring a bicycle becomes easier because adding the extra control input essentially eliminates the right half-plane zeros (Åström et al. 2005). In (Beznos et al. 1998), an autonomous bicycle is designed and balanced using gyroscopic actuators. The controller in (Beznos et al. 1998) is based on a linearized bicycle model. In (Getz 1995), a nonlinear control method is designed for trajectory tracking and balancing. In (Lee and Ham 2002),

---

* An under-actuated mechanical system refers to a mechanical dynamic system in which the number of control inputs is less than the number of the generalized coordinates (Bullo and Lewis 2004). Readers can also refer to (Sastry 1999) for an overview of the control of nonlinear non-minimum phase systems.

a balancing and tracking control mechanism is designed by on-board shifting weights. In (Tanaka and Murakami 2004, 2009), a simplified inverted pendulum model is utilized for bicycle balancing. A proportional derivative (PD) controller with a disturbance observer is employed to design a controller to balance the bicycle. The authors, however, focus on balancing the bicycle in straight-line motion.

In this chapter, we employ and extend the control design in (Getz 1995 and Yi et al. 2006). In (Getz 1995), an external/internal convertible (EIC) dynamical system is presented, and the motorcycle dynamics are an example of an EIC system. A nonlinear tracking control design is also discussed for the non-minimum phase bicycle dynamic systems. In our previous work (Yi et al. 2006), we extended the dynamic models to consider motorcycle geometric and steering mechanism properties. In both (Getz 1995 and Yi et al. 2006), non-holonomic constraints of zero lateral velocity at the rear wheel contact point are enforced, and only rear wheel friction force is considered for traction/braking forces. In Chapter 2, we relaxed the non-holonomic constraints assumptions and considered that both wheels can produce braking actuation though traction is only from the rear wheel. With the new model, the EIC model based control design is presented in this chapter. The control systems design takes advantage of the control actuation flexibility and reduces the design complexity compared to those in (Getz 1995 and Yi et al. 2006). Two simulation examples demonstrate the effectiveness and efficacy of the control systems design.

We also present a path-following design to overcome the large errors shown in trajectory tracking. For autonomous vehicles, particularly under-actuated mechanical systems, manoeuvre regulation or path following control has demonstrated a superior performance comparing with the trajectory tracking design in the time domain (Aguiar and Hespanda 2007; Aguiar et al. 2005; Al-Hiddabi and McClamroch 2002; Hauser and Hindman 1995; Skjetne et al. 2004). In the path-following control design, the desired velocity profile along the trajectory is obtained either using a Lyapunov-based approach or requiring online solving of an optimization problem, which is non-causal for non-minimum phase dynamical systems such as motorcycle dynamics. We use a velocity field concept to generate the desired velocity profile for motorcycle systems. Our approach is inspired by the work in (Li and Horowitz 2001) of passivity-based control of fully actuated robot manipulators. We integrate the velocity field concept with the EIC control design of under-actuated non-minimum phase motorcycle dynamics. The presented trajectory-tracking and path-following control design is an extension of the work presented in the conference papers (Zhang and Yi 2010; Yi et al. 2009).

The remainder of the chapter is organized as follows. In Section 12.2, we present the EIC-based tracking and balancing control design. We then present a tracking error improvement by path-following control design in Section 12.3. Both trajectory-tracking and path-following control designs are validated through simulation results. Finally, we conclude the chapter and discuss future research directions in Section 12.4.

12.2 Trajectory Tracking Control Systems Design

12.2.1 External/Internal Convertible Dynamical Systems

The motorcycle dynamics with tyre models is presented by (2.35) in Chapter 2. We now consider how to put such a system into the form of an external/internal convertible (EIC) dynamical system. The EIC form of a nonlinear dynamical system can be viewed as a special case of the normal form.

Definition 1 (Getz 1995) *A single-input, single-output, n $(= m + p)$-dimensional time-invariant nonlinear control system is called in an* external/internal convertible form *if the system is of the form*

$$\Sigma(u) \begin{cases} \dot{x}_i = x_{i+1}, & i = 1, \cdots, m-1, \\ \dot{x}_m = u, \\ \dot{\alpha}_i = \alpha_{i+1}, & i = 1, \cdots, p-1, \\ \dot{\alpha}_p = f(\mathbf{x}, \boldsymbol{\alpha}) + g(\mathbf{x}, \boldsymbol{\alpha})u, \\ y = x_1, \end{cases} \qquad (12.1)$$

with input $u \in \mathbb{R}$, output $y \in \mathbb{R}$, state variables $(\mathbf{x}, \boldsymbol{\alpha})$, with $\mathbf{x} := (x_1, \cdots, x_m) \in \mathbb{R}^m$ and $\boldsymbol{\alpha} := (\alpha_1, \cdots \alpha_p) \in \mathbb{R}^p$. The coordinates $(\mathbf{x}, \boldsymbol{\alpha})$ are assumed to be defined on the open ball $\mathbf{B}_r \subset \mathbb{R}^n$ around the origin. The origin is assumed to be an equilibrium of the system, namely, $f(\mathbf{0}, \mathbf{0}) = 0$. The functions $f(\mathbf{x}, \boldsymbol{\alpha})$ and $g(\mathbf{x}, \boldsymbol{\alpha})$ are C^n in their arguments, and $g(\mathbf{x}, \boldsymbol{\alpha}) \neq 0$ for all $(\mathbf{x}, \boldsymbol{\alpha}) \in \mathbf{B}_r$. Moreover, we refer to the external subsystem *of $\Sigma(u)$ as*

$$\Sigma_{\text{ext}}(u) \begin{cases} \dot{x}_i = x_{i+1}, & i = 1, \cdots, m-1, \\ \dot{x}_m = u \end{cases} \qquad (12.2)$$

and the internal subsystem *of $\Sigma(u)$ as*

$$\Sigma_{\text{int}}(u) \begin{cases} \dot{\alpha}_i = \alpha_{i+1}, & i = 1, \cdots, p-1, \\ \dot{\alpha}_p = f(\mathbf{x}, \boldsymbol{\alpha}) + g(\mathbf{x}, \boldsymbol{\alpha})u. \end{cases} \qquad (12.3)$$

Figure 12.1 shows the structure of an EIC system. An EIC system is *convertible* because under a simple state-dependent input and an output transformation, the internal system is converted to an external system, and the external system is converted to an internal system (dual structure). To see such a property, let

$$u = g(\mathbf{x}, \boldsymbol{\alpha})^{-1}[v - f(\mathbf{x}, \boldsymbol{\alpha})] \qquad (12.4)$$

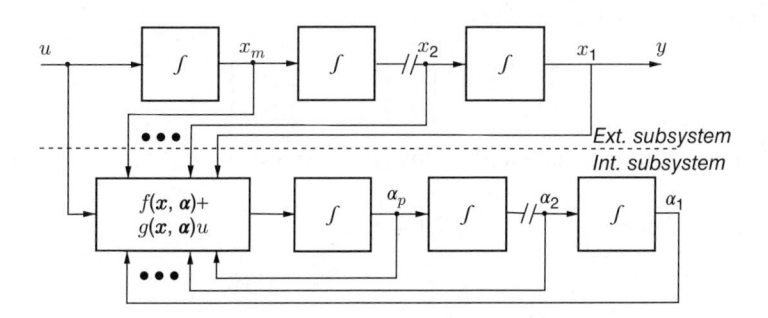

Figure 12.1 An external/internal convertible system

define a state-dependent input transformation, $u \mapsto v$. Define $\xi = \alpha^1$ as the *dual output*. Apply transformation (12.4) to the EIC system (12.1) and the resulting system is referred to as the dual of $\Sigma(u)$.

$$\Sigma_d(v) \begin{cases} \dot{x}_i = x_{i+1}, & i = 1, \cdots, m-1, \\ \dot{x}_m = -g(\mathbf{x},\boldsymbol{\alpha})^{-1}f(\mathbf{x},\boldsymbol{\alpha}) + g(\mathbf{x},\boldsymbol{\alpha})^{-1}v, \\ \dot{\alpha}_i = \alpha_{i+1}, & i = 1, \cdots, p-1, \\ \dot{\alpha}_p = v, \\ \xi = \alpha_1. \end{cases} \quad (12.5)$$

Thus the use of input transformation (12.4) and the output assignment $\xi = \alpha_1$ converts the internal dynamics of $\Sigma(u)$ to the external dynamics of $\Sigma_d(v)$, and the external dynamics of $\Sigma(u)$ to the internal dynamics of $\Sigma_d(v)$.

Since the EIC form is a special normal form of nonlinear dynamical systems, we can apply the input–output linearization method (Isidori 1995; Sastry, 1999) to convert (2.35) into an EIC form. Let $M_{22} \in \mathbb{R}^{2 \times 2}, B_{22} \in \mathbb{R}^{2 \times 2}$ and $K_2 \in \mathbb{R}^2$ denote the block elements of matrices **M**, **B** and **K**, given by (2.22), (2.37) and (2.36), respectively. Using the input transformation

$$\mathbf{u}_\lambda = B_{22}^{-1}M_{22}[M_{22}^{-1}(M_{21}\ddot{\varphi} - K_2 - B_{21}\omega_\sigma) + \mathbf{u}_a], \quad (12.6)$$

Equation (2.35) becomes

$$\begin{cases} M_{11}\ddot{\varphi} = K_1 - M_{12}\mathbf{u}_a + B_{11}\omega_\sigma, \\ \begin{bmatrix} \dot{v}_{rx} \\ \dot{v}_{ry} \end{bmatrix} = \begin{bmatrix} a_{rx} \\ a_{ry} \end{bmatrix} =: \mathbf{u}_a, \end{cases} \quad (12.7)$$

where $\mathbf{u}_a$ is the controlled acceleration of point C_2 in the xyz coordinate system. We also define the controlled jerk of point C_2 and yaw acceleration as

$$\mathbf{u}_j := \begin{bmatrix} u_{rx} \\ u_{ry} \\ u_\psi \end{bmatrix} = \begin{bmatrix} \dot{a}_{rx} \\ \dot{a}_{ry} \\ \ddot{\psi} \end{bmatrix} = \begin{bmatrix} \dot{\mathbf{u}}_a \\ \frac{v_{rx}\omega_\sigma + \sigma a_{rx}}{l} \end{bmatrix}, \quad (12.8)$$

where we use kinematics $l\dot{\psi} = \sigma v_{rx}$ in the calculation. Let (X, Y) denote the coordinates of the contact point C_2 and then we have

$$\begin{bmatrix} v_X \\ v_Y \end{bmatrix} = \begin{bmatrix} \dot{X} \\ \dot{Y} \end{bmatrix} = \begin{bmatrix} c_\psi & -s_\psi \\ s_\psi & c_\psi \end{bmatrix} \begin{bmatrix} v_{rx} \\ v_{ry} \end{bmatrix}.$$

Differentiating the above equation twice (dynamic extension), we obtain

$$\begin{bmatrix} \dddot{v}_X \\ \dddot{v}_Y \end{bmatrix} = \mathbf{U} + \mathbf{u}_J, \quad (12.9)$$

where

$$\mathbf{U} = \begin{bmatrix} -2\dot{v}_{rx}s_\psi - 2\dot{v}_{ry}c_\psi - v_{rx}\dot{\psi}c_\psi + v_{ry}\dot{\psi}s_\psi \\ 2\dot{v}_{rx}c_\psi - 2\dot{v}_{ry}s_\psi - v_{rx}\dot{\psi}s_\psi - v_{ry}\dot{\psi}c_\psi \end{bmatrix} \dot{\psi}$$

and
$$\mathbf{u}_J := \begin{bmatrix} c_\psi & -s_\psi \\ s_\psi & c_\psi \end{bmatrix} \begin{bmatrix} u_{rx} \\ u_{ry} \end{bmatrix} + \begin{bmatrix} -v_{rx}s_\psi - v_{ry}c_\psi \\ v_{rx}c_\psi - v_{ry}s_\psi \end{bmatrix} u_\psi. \tag{12.10}$$

We define the new inputs u_X and u_Y such that
$$\mathbf{u}_J = -\mathbf{U} + \begin{bmatrix} u_X \\ u_Y \end{bmatrix} \tag{12.11}$$

and then the motorcycle dynamics (12.7) are in the EIC form as

$$\Sigma_{\text{ext}} : \left\{ \begin{bmatrix} \ddot{v}_X \\ \ddot{v}_Y \end{bmatrix} = \begin{bmatrix} u_X \\ u_Y \end{bmatrix}, \right. \tag{12.12a}$$

$$\Sigma_{\text{int}} : \ddot{\varphi} = \frac{g}{h}\left(s_\varphi + \frac{bl_t c_\xi \dot{\psi}}{hv_{rx}}c_\varphi\right) - \frac{1}{h}\left(1 - \frac{h\dot{\psi}}{v_{rx}}s_\varphi\right)\dot{\psi}v_{rx}c_\varphi - \frac{1}{h}c_\varphi u_{\psi y}, \tag{12.12b}$$

where
$$u_{\psi y} := bu_\psi + a_{ry}. \tag{12.13}$$

Remark 1 *When the motorcycle runs along a straight line, $\sigma = 0$ and matrix B_{22} becomes singular and we cannot use input transformation (12.6). In this case, we calculate the total braking force from the second equation of the motion and split the front and rear wheels in a way not producing any net moments around mass centre G. A similar approach is discussed in (Gerdes and Rossetter 2001). If the resultant total force is traction, then it must be produced by the rear wheel.*

12.2.2 Trajectory Tracking Control

12.2.2.1 Control System Overview

The trajectory control system then guides the motorcycle to follow the desired trajectory $\mathcal{T}$: $(X_d(t), Y_d(t))$ while keeping the platform balanced and stable. We here employ and extend the control design approach in (Getz 1995). Figure 12.2 illustrates such a control scheme. The trajectory control design consists of two steps. The first step is to design a tracking control $\mathbf{u}_{\text{ext}}$ of the external subsystem Σ_{ext} for the desired trajectory $\mathcal{T}$. The second step is to design a balancing controller for the internal subsystem Σ_{int} around the internal equilibrium manifold, denoted as $\mathcal{E}(t)$. The internal equilibrium manifold $\mathcal{E}(t)$ is an embedded sub-manifold in the state space and is dependent on the external control $\mathbf{u}_{\text{ext}}$ and the external subsystem. Estimations of internal equilibrium $\mathcal{E}(t)$ and its derivatives are obtained by a dynamic inversion technique (Getz 1995). The final causal control system is a combination of external and internal design.

12.2.2.2 Approximate Tracking Control

We assume that the desired trajectory $\mathcal{T}$: $(X_d(t), Y_d(t))$ is differentiable at least up to fourth order, that is, C^4.[†] This is feasible since the motion planning algorithm can usually generate a set of piecewise circular curves (C^∞) for $\mathcal{T}$ (Song et al. 2007).

[†] For the external subsystem control, we only need $\mathcal{T}$ to be C^3. The requirement for C^4 is due to the estimation of the internal (roll angle) equilibrium and its derivatives by a dynamic inversion technique.

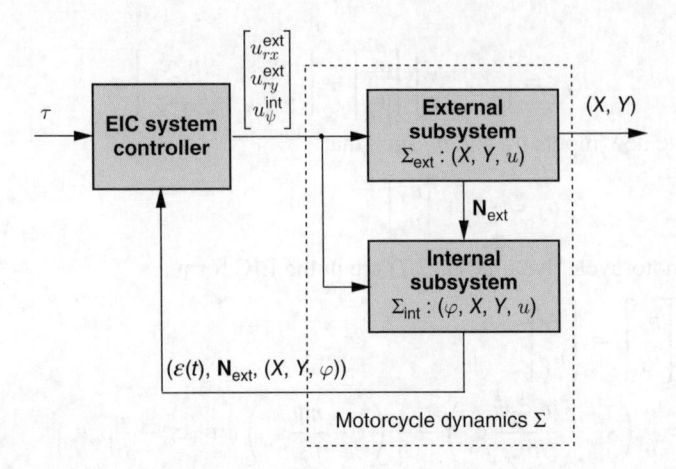

Figure 12.2 EIC-based approximate output tracking control of the autonomous motorcycle dynamics

We design a controller $\boldsymbol{u}_{\text{ext}}$ to track the desired trajectory $(X_d(t), Y_d(t))$ for the external subsystem Σ_{ext} (12.12a) disregarding, for the moment, the evolution of the internal subsystem Σ_{int} (12.12b).

$$\boldsymbol{u}^{\text{ext}} := \begin{bmatrix} u_X^{\text{ext}} \\ u_Y^{\text{ext}} \end{bmatrix} = \begin{bmatrix} X_d^{(3)} \\ Y_d^{(3)} \end{bmatrix} - \sum_{i=1}^{3} b_i \begin{bmatrix} X^{(i-1)} - X_d^{(i-1)} \\ Y^{(i-1)} - Y_d^{(i-1)} \end{bmatrix}, \quad (12.14)$$

where constants b_i, $i = 1, 2, 3$, are chosen such that the polynomial equation $s^3 + b_3 s^2 + b_2 s + b_1 = 0$ is Hurwitz. Under such a control, we define a nominal external vector field $\mathbf{N}_{\text{ext}}$ as

$$\mathbf{N}_{\text{ext}} := \begin{bmatrix} \dot{X}(t) \\ \ddot{X}(t) \\ X_d^{(3)} - \sum_{i=1}^{3} b_i \left(X^{(i-1)} - X_d^{(i-1)} \right) \\ \dot{y}(t) \\ \ddot{Y}(t) \\ Y_d^{(3)} - \sum_{i=1}^{3} b_i (Y^{(i-1)} - Y_d^{(i-1)}) \end{bmatrix}. \quad (12.15)$$

By external control (12.14) and the input transformation (12.11), we find the input $u_J^{\text{ext}} = -\mathbf{U} + \boldsymbol{u}^{\text{ext}}$. From (12.10), we obtain u_j^{ext} as

$$\begin{bmatrix} u_{rx} \\ u_{ry} \end{bmatrix} + \begin{bmatrix} -v_{ry} \\ v_{rx} \end{bmatrix} u_\psi = \begin{bmatrix} c_\psi & s_\psi \\ -s_\psi & c_\psi \end{bmatrix} \boldsymbol{u}_J. \quad (12.16)$$

Note that $\boldsymbol{u}_J \in \mathbb{R}^2$ and $\boldsymbol{u}_j \in \mathbb{R}^3$ and the above equation is under-determined. There are many options to determine $\boldsymbol{u}_j$ by (12.16). Here we propose to choose $u_\psi = \ddot{\psi} = 0$ because such a choice significantly reduces the complexity of the control design as shown in the following:

$$\boldsymbol{u}_j^{\text{ext}} = \begin{bmatrix} u_{rx}^{\text{ext}} \\ u_{ry}^{\text{ext}} \\ u_{\psi}^{\text{ext}} \end{bmatrix} = \begin{bmatrix} \boldsymbol{R}(\psi)\boldsymbol{u}_J^{\text{ext}} \\ 0 \end{bmatrix} = \begin{bmatrix} \boldsymbol{R}(\psi)\left(-\boldsymbol{U} + \boldsymbol{u}^{\text{ext}}\right) \\ 0 \end{bmatrix}. \quad (12.17)$$

Next, we consider the internal (roll angle) equilibrium, denoted as φ_e, by substituting u_{ψ}^{ext} and u_{ry}^{ext} above into the internal subsystem dynamics (12.12b). We define the implicit function F_φ of φ as

$$F_\varphi := g\left(\tan\varphi + \frac{bl_t\dot\psi c_\xi}{hv_{rx}}\right) - \left(1 - \frac{h\dot\psi s_\varphi}{v_{rx}}\right)\dot\psi v_{rx} - u_{\psi y}^{\text{ext}}, \quad (12.18)$$

$u_{\psi y}^{\text{ext}} = bu_\psi^{\text{ext}} + a_{ry} = a_{ry}$, and the roll angle equilibrium $\varphi_e := \varphi_e(\psi, v_{rx}, \boldsymbol{u}_j^{\text{ext}})$ is a solution of the algebraic equation $F_{\varphi_e} = 0$. We define an internal (roll angle) equilibrium manifold $\mathcal{E}(t)$ as

$$\mathcal{E}(t) = \left\{\left(X^{(0,2)}, Y^{(0,2)}, \varphi^{(0,1)}\right) \mid \varphi = \varphi_e, \ \dot\varphi = 0\right\}. \quad (12.19)$$

The internal equilibrium manifold $\mathcal{E}(t)$ can be viewed as a time-dependent graph over the six-dimensional (X, Y)-subspace in $\mathbb{R}^6$ of the external subsystem (12.12a) that is evolved with the external nominal vector field $\boldsymbol{N}_{\text{ext}}$ (12.15) under the external subsystem control $\boldsymbol{u}^{\text{ext}}$.

For motorcycle balance systems, we like to control the roll angle φ around $\mathcal{E}(t)$ while the external subsystem is tracking $\mathcal{T}$ under the control of $\boldsymbol{u}^{\text{ext}}$. Note that $\dot\varphi_e \neq 0$ and $\ddot\varphi_e \neq 0$ in general, and here we approximate the derivatives $\dot\varphi_e$ and $\ddot\varphi_e$ by using directional derivatives (Isidori 1995; Sastry 1999) along the vector field $\boldsymbol{N}_{\text{ext}}$. We define the directional derivative (or Lie derivative) as $\overline{L}_{\boldsymbol{N}_{\text{ext}}}\varphi_e := L_{\boldsymbol{N}_{\text{ext}}}\varphi_e + \frac{\partial\varphi_e}{\partial t}$ and $\overline{L}^2_{\boldsymbol{N}_{\text{ext}}}\varphi_e := \overline{L}_{\boldsymbol{N}_{\text{ext}}}\overline{L}_{\boldsymbol{N}_{\text{ext}}}\varphi_e$. With the above approximations for $\dot\varphi_e$ and $\ddot\varphi_e$, the stabilizing control of the internal subsystem Σ_{int} (12.12b) around $\mathcal{E}(t)$ is then given by the following feedback linearization:

$$u_{\psi y}^{\text{int}} = \left(\frac{c_\varphi}{h}\right)^{-1}\left[\frac{g}{h}\left(s_\varphi + \frac{bl_t c_\xi \dot\psi}{hv_{rx}}c_\varphi\right) - \frac{1}{h}\left(1 - \frac{h\dot\psi}{v_{rx}}s_\varphi\right)\dot\psi v_{rx}c_\varphi - v_{\psi y}\right], \quad (12.20a)$$

$$v_{\psi y} = \overline{L}^2_{\boldsymbol{N}_{\text{ext}}}\varphi_e - \sum_{i=1}^{2} a_i\left(\varphi^{(i-1)} - \overline{L}^{i-1}_{\boldsymbol{N}_{\text{ext}}}\varphi_e\right), \quad (12.20b)$$

where constants a_1 and a_2 are chosen such that the polynomial equation $s^2 + a_2 s + a_1 = 0$ is Hurwitz. Therefore, the internal control is obtained from (12.13) as

$$u_\psi^{\text{int}} = \frac{1}{b}(u_{\psi y}^{\text{int}} - a_{ry}). \quad (12.21)$$

The *final* control system design of the motorcycle balance system (12.9) combines the above development in (12.21) and (12.17) as

$$\boldsymbol{u}_j = \begin{bmatrix} u_{rx}^{\text{ext}} \\ u_{ry}^{\text{ext}} \\ u_{\psi}^{\text{int}} \end{bmatrix}. \quad (12.22)$$

It is noted that the coupling between the external- and internal-subsystem control designs is through the introduction of the internal equilibrium manifold $\mathcal{E}(t)$. By defining $\mathcal{E}(t)$, we approximately decouple the external and internal subsystems using the EIC dual structural properties of the motorcycle system.

We define $\vartheta(t) = [X(t)\ v_X(t)\ \dot{v}_X(t)\ Y(t)\ v_Y(t)\ \dot{v}_Y(t)]^T$ as the state variables of the external subsystem and $\varrho(t) = [\varphi(t)\ \dot{\varphi}(t)]^T$ as the state variables of the internal subsystem. We also define the output $\zeta(t) = [X(t)\ Y(t)]^T$ and the desired output $\zeta_d(t) = [X_d(t)\ Y_d(t)]^T$. We assume that the desired trajectory $\zeta_d(t)$ and its derivatives (up to the fourth order) are bounded by a positive number $\varepsilon > 0$, namely, $\zeta_d(t) \in \mathbf{B}_\varepsilon^{(4)} := \{x(t)\ |\ \|x^{(0,4)}(t)\|_\infty < \varepsilon\}$, where $\|x^{(0,n)}(t)\|_\infty := \sup_{t \geq 0} \|x^{(0,n)}(t)\|_\infty$. We also define the tracking errors $e_i^\vartheta = \vartheta_i - X_d^{(i-1)}$, $e_{i+3}^\vartheta = \vartheta_{i+3} - Y_d^{(i-1)}$, $i = 1, 2, 3$, $e_j^\varphi = \varphi^{(j)} - \varphi_e^{(j)}$, $j = 0, 1$ and $\mathbf{e} := [e_1^\vartheta, \cdots, e_6^\vartheta, e_1^\varphi, e_2^\varphi]^T$. We also define the perturbation error $p_\varphi (= O(\|\zeta_d^{(0,4)}(t)\|\,\|\mathbf{e}\|))$ as the approximation errors by using the directional derivatives for $\dot{\varphi}_e$ and $\ddot{\varphi}_e$ in the internal subsystem control design (12.20b), namely,

$$p_\varphi = \bar{L}_{\mathbf{N}_{\text{ext}}}^2 \varphi_e - \ddot{\varphi}_e + \sum_{i=1}^{2} a_i(\varphi_e^{(i-1)} - \bar{L}_{\mathbf{N}_{\text{ext}}}^{i-1} \varphi_e).$$

We similarly define another two perturbation errors $p_X (= O(\|\zeta_d^{(0,4)}(t)\|\,\|\mathbf{e}\|))$ and $p_Y (= O(\|\zeta_d^{(0,4)}(t)\|\,\|\mathbf{e}\|))$ due to the resulting errors in the external subsystem state $\vartheta(t)$ using the internal subsystem control $u_{\psi y}^{\text{int}}$ in the external subsystem (12.22). An explicit formulation for p_X and p_Y can similarly be found by the dual structure of the EIC system (Getz 1995). We consider the perturbation vector for the error dynamics of $\Sigma(u)$ (12.11) under control (12.22) as

$$p(\zeta_d^{(0,4)}(t), \mathbf{e}) = [0,\ 0,\ p_X,\ 0,\ 0,\ p_Y,\ 0,\ p_\varphi]^T.$$

We assume an affine perturbation for $p(y_d^{(0,4)}(t), e)$, namely, there exist constants $k_1 > 0$ and $k_2 > 0$, such that $\|p(\zeta_d^{(0,5)}(t), \mathbf{e})\|_\infty \leq k_1 \varepsilon + k_2 \|\mathbf{e}\|_\infty$.

We only state the convergence properties of the approximate tracking control design. The proof of these properties follows directly from Proposition 6.7.4 and Theorem 6.7.6 in Getz (1995) and we omit them here.

Theorem 1 *For the balance system (12.11), assuming that the desired trajectory $\zeta_d(t) \in \mathbf{B}_\varepsilon^{(4)}$ for some $\varepsilon > 0$, and if the affine perturbation constant $k_2 > 0$ is a sufficiently small real number, then there exists a class-$\mathcal{K}$ function $r(\varepsilon)$ such that for all $(\mathbf{e}^\vartheta(0), \mathbf{e}^\varphi(0)) \in \mathbf{B}_{r(\varepsilon)}$, $(\mathbf{e}^\vartheta(t), \mathbf{e}^\varphi(t))$ converges to zero exponentially until $(\mathbf{e}^\vartheta(t), \mathbf{e}^\varphi(t))$ enters $\mathbf{B}_{r(\varepsilon)}$. Once $(\mathbf{e}^\vartheta(t), \mathbf{e}^\varphi(t))$ enters $\mathbf{B}_{r(\varepsilon)}$, it will stay in $\mathbf{B}_{r(\varepsilon)}$ thereafter.*

12.2.2.3 Estimation of the Internal Equilibrium Manifold

A dynamic inversion technique approach in (Getz 1995) is used to estimate the internal equilibrium state φ_e in (12.20b). To illustrate the dynamic inversion technique, we differentiate

$F_\varphi = 0$ with time, and using the fact that $u_\psi^{\text{ext}} = \ddot{\psi} = 0$ we obtain

$$\dot{\varphi}_e = \frac{1}{g\sec^2\varphi_e + h\dot{\psi}c_{\varphi_e}} \left(\frac{gbl_t c_\xi \dot{\psi} \dot{v}_{rx}}{h v_{rx}^2} + \dot{\psi} \dot{v}_{rx} + u_{ry}^{\text{ext}} \right)$$

$$=: E(\varphi_e, \dot{\psi}, v_{rx}, \dot{v}_{rx}, u_{ry}^{\text{ext}}). \tag{12.23}$$

A dynamic inverter for an estimate $\hat{\varphi}_e$ of the internal equilibrium φ_e is designed as

$$\dot{\hat{\varphi}}_e = -\beta F_{\hat{\varphi}} + E(\hat{\varphi}_e, \dot{\psi}, v_{rx}, \dot{v}_{rx}, u_{ry}^{\text{ext}}), \tag{12.24}$$

where $F_{\hat{\varphi}_e}$ is given by (12.18), and $\beta > 0$ is the inverter gain. The proof of the exponential convergence of the estimation (12.24) follows directly from the development of the dynamic inversion technique in (Getz 1995).

The estimate of the directional derivative $\overline{L}_{\mathbf{N}_{\text{ext}}} \varphi_e$ in (12.20b) is obtained by (12.24), namely, $\overline{L}_{\mathbf{N}_{\text{ext}}} \varphi_e = E(\varphi_e, \dot{\psi}, v_{rx}, \dot{v}_{rx}, u_{ry}^{\text{ext}})$. The estimate of $\overline{L}_{\mathbf{N}_{\text{ext}}}^2 \varphi_e$ is obtained by directly taking one more directional derivative of $\overline{L}_{\mathbf{N}_{\text{ext}}} \varphi_e$ along $\mathbf{N}_{\text{ext}}$. For brevity, we give the derivation in Appendix A. We also list the calculation of $\overline{L}_{\mathbf{N}_{\text{ext}}} u_{rx}^{\text{ext}}$ and $\overline{L}_{\mathbf{N}_{\text{ext}}} u_{ry}^{\text{ext}}$ in Appendix A. Such calculations are needed for computing $\overline{L}_{\mathbf{N}_{\text{ext}}}^2 \varphi_e$. The approximation errors in estimating φ_e (by $\hat{\varphi}_e$) and its directional derivatives $\overline{L}_{\mathbf{N}_{\text{ext}}} \varphi_e$ and $\overline{L}_{\mathbf{N}_{\text{ext}}}^2 \varphi_e$ (by $\overline{L}_{\mathbf{N}_{\text{ext}}} \hat{\varphi}_e$ and $\overline{L}_{\mathbf{N}_{\text{ext}}}^2 \hat{\varphi}_e$, respectively) are considered as additional terms in the perturbation $p(\zeta_d^{(0,4)}(t), \mathbf{e})$. Therefore, the stability results of the approximate control design in the previous section are still held.

Remark 2 *Although the above control system design is similar to those in (Getz, 1995), the final form is much simpler because we have chosen $u_\psi^{\text{ext}} = 0$ in (12.17). We have such a flexibility by (12.16) to determine $\mathbf{u}_j$ because we now have three control input variables, while in (Getz 1995) only the rear wheel driving torque and the steering angle are controlled. Because of this difference, we only require that the trajectory $\mathcal{T}$ is at least C^4 rather than C^5 as required by the controller in (Getz 1995). Using optimization techniques by considering the input constraints for determining $\mathbf{u}_j$ by (12.16) is an extension of the control design and is currently ongoing research.*

12.2.3 Simulation Results

In this section, we demonstrate the control systems design through two numerical examples. The first example is taken from (Yi et al. 2006) for showing a general motorcycle trajectory, and the second example illustrates aggressive manoeuvres with large sideslip angles.

We use a racing motorcycle prototype in (Sharp et al. 2004 and Corno et al. 2008) in our simulation. The motorcycle parameters are listed in Table 12.1. We use the tyre 160/70 in (Sharp et al. 2004) for the racing motorcycle since the test data is available. The tyre stiffness coefficients listed in Table 12.1 are calculated under the nominal load $F_z = 1600$ N.

Figure 12.3 shows the tracking performance of a general trajectory. The position errors under the control system in Figure 12.3b are within 1 metre with the centre line of the

Table 12.1 Motorcycle model parameters

m(kg)	b(m)	l(m)	l_t(m)	h(m)	ξ(deg)	r(m)	λ_{sm}	$\lambda_{\gamma m}$(deg)	μ_m	k_λ(N)	k_φ(N/rad)	k_γ(N)
274.2	0.81	1.37	0.15	0.62	26.1	0.3	0.1	6	3	41504	23968	1227

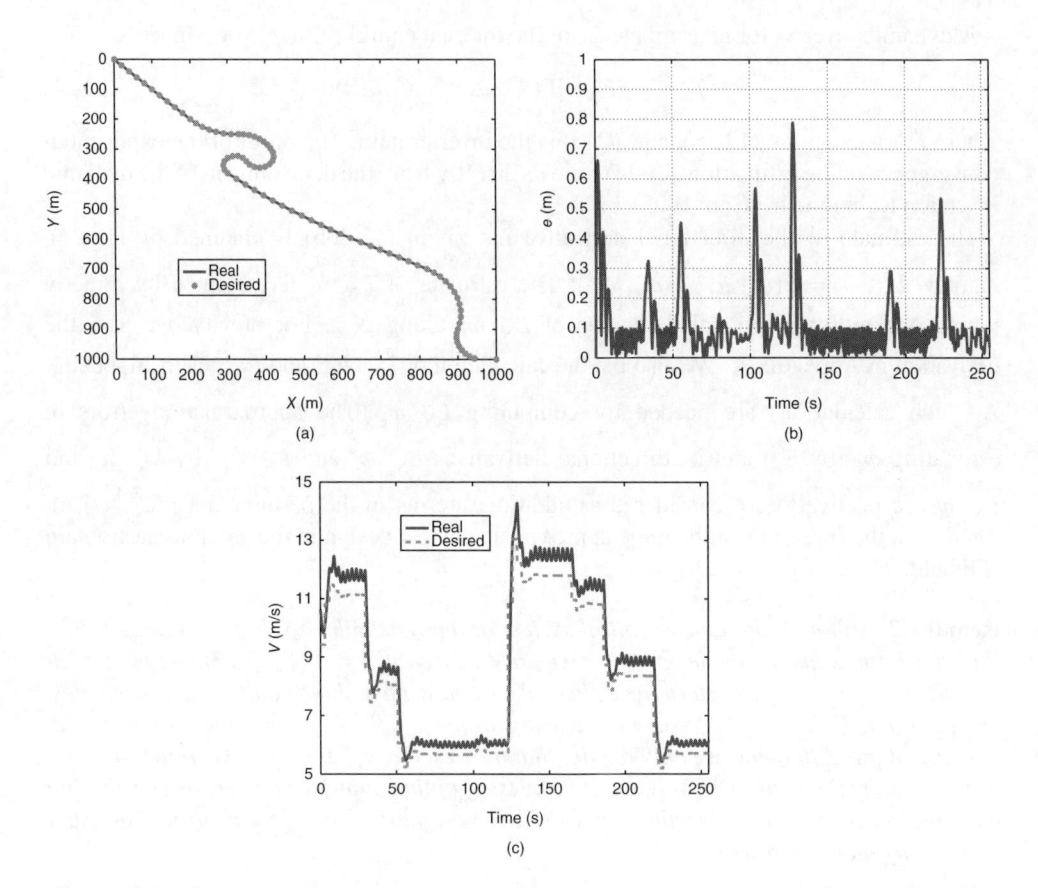

Figure 12.3 Tracking performance of a general trajectory. (a) Trajectory positions. (b) Tracking position error. (c) Rear wheel contact point velocity magnitude

track throughout the entire course. The desired velocity in Figure 12.3c is determined by the curvature of the trajectory. Figure 12.4 shows the roll angle φ, the body-frame velocities v_{rx} and v_{ry} of rear wheel contact point C_2, and steering angle ϕ. From Figure 12.4a we clearly see that the lateral velocity v_{ry} is quite small most of the time because the motorcycle is generally running along a straight line. At turning locations, the longitudinal velocity is reduced and the lateral velocity increases. The roll angle and steering angle are small for such a small-curvature trajectory.

Figure 12.5 shows the longitudinal slips and sideslip angles of the front and rear wheels. Again, it is clear that the slip values at both wheels are small. The front wheel only brakes

Autonomous Control of Riderless Motorcycles

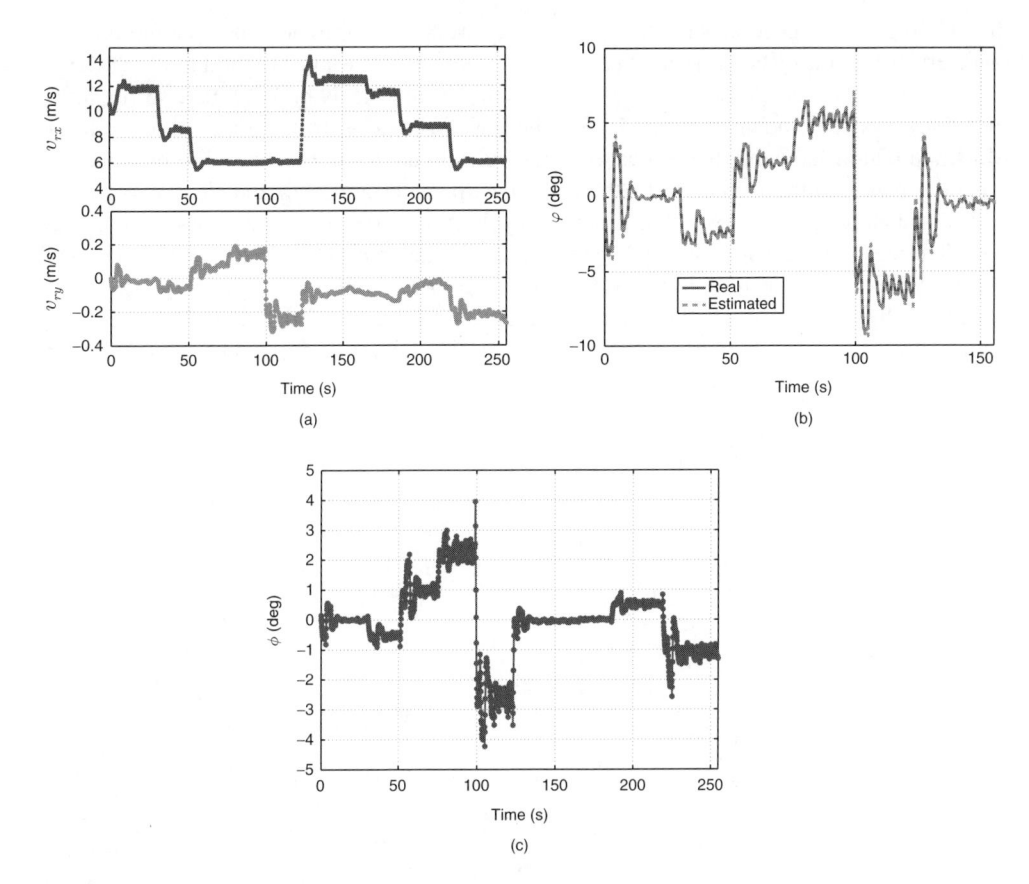

Figure 12.4 Roll angle and steering angle of the general trajectory tracking. (a) Rear wheel contact point body-frame velocities v_{rx} and v_{ry}. (b) Roll angle φ. (c) Steering angle ϕ

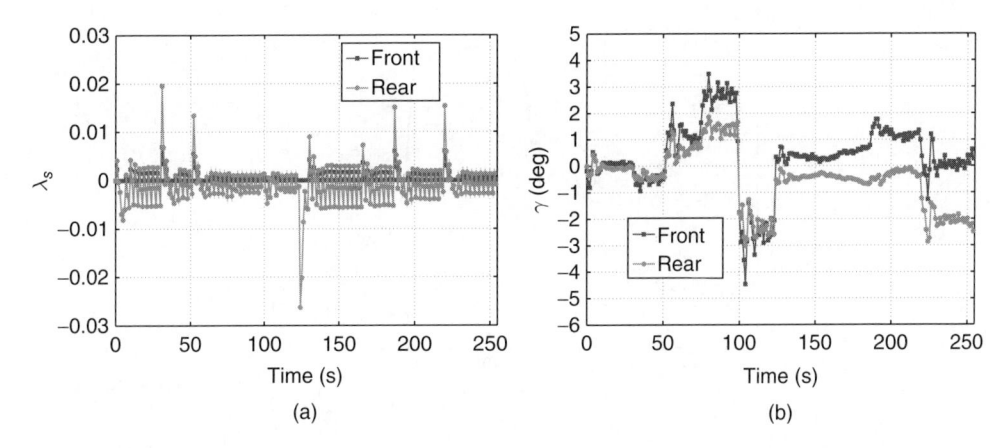

Figure 12.5 Longitudinal slips and slip angles at the front and rear wheels. (a) Slip ratio λ_{fs} and λ_{rs}. (b) Slip angles γ_f and γ_r

and the rear wheel generates traction or braking forces. For example, when the motorcycle accelerates around 120 s, the rear wheel slip has a large negative spike to produce the traction force. When the vehicle needs to reduce velocity, both wheels brake with a set of large positive slip spikes shown in Figure 12.5a. The sideslip angles shown in Figure 12.5b clearly illustrate that at large-curvature locations, the sideslip angles are increased to produce the lateral forces for turning. Typically, the rear sideslip angles are small and close to zero.

The second example shows that the motorcycle runs under a more aggressive manoeuvre. The desired trajectory is ("8"-shape) with circular radius of 25 metres; see Figure 12.6a. In a figure, the motorcycle starts from the origin and moves along the direction indicated by the arrows in the figure. The desired velocity of the motorcycle moving along the "8"-shape trajectory is designed to be varying significantly as shown in

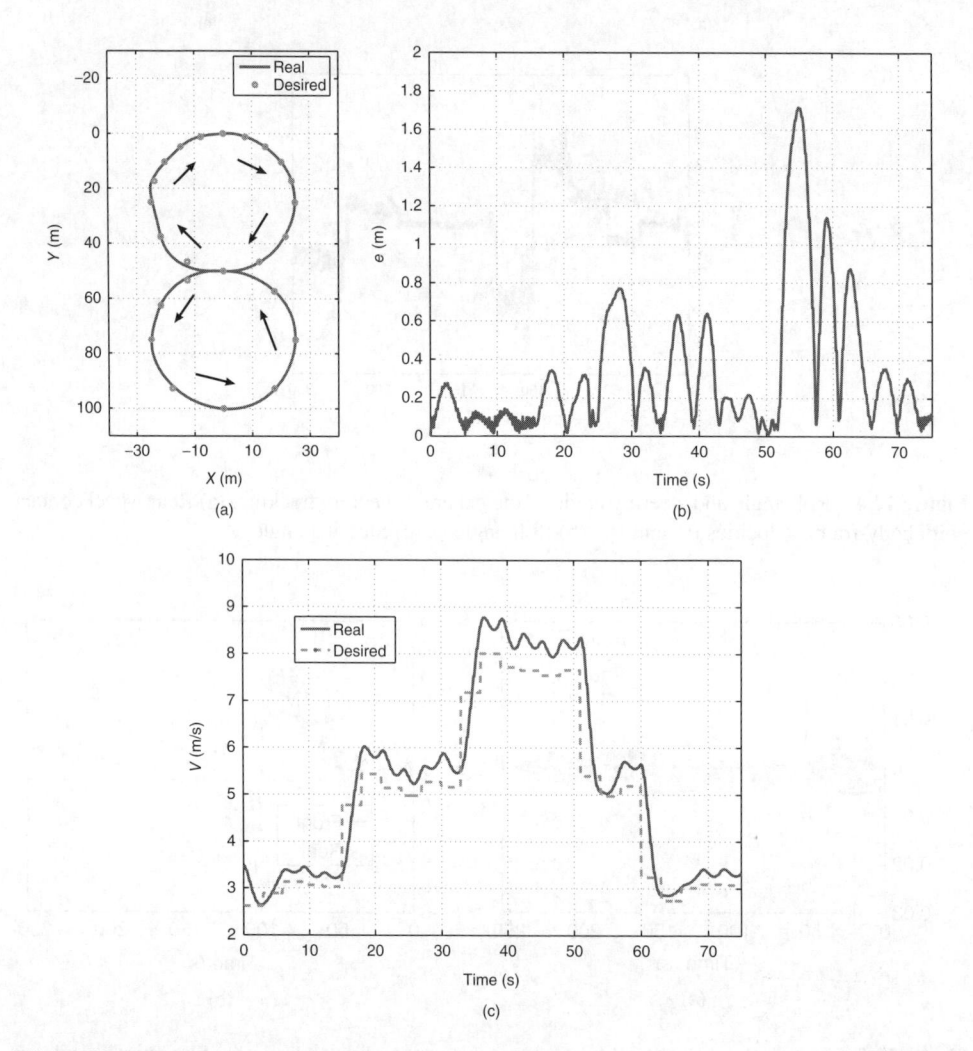

Figure 12.6 An "8"-shape trajectory tracking. (a) Trajectory positions. (b) Tracking position error. (c) Rear wheel contact point velocity magnitude

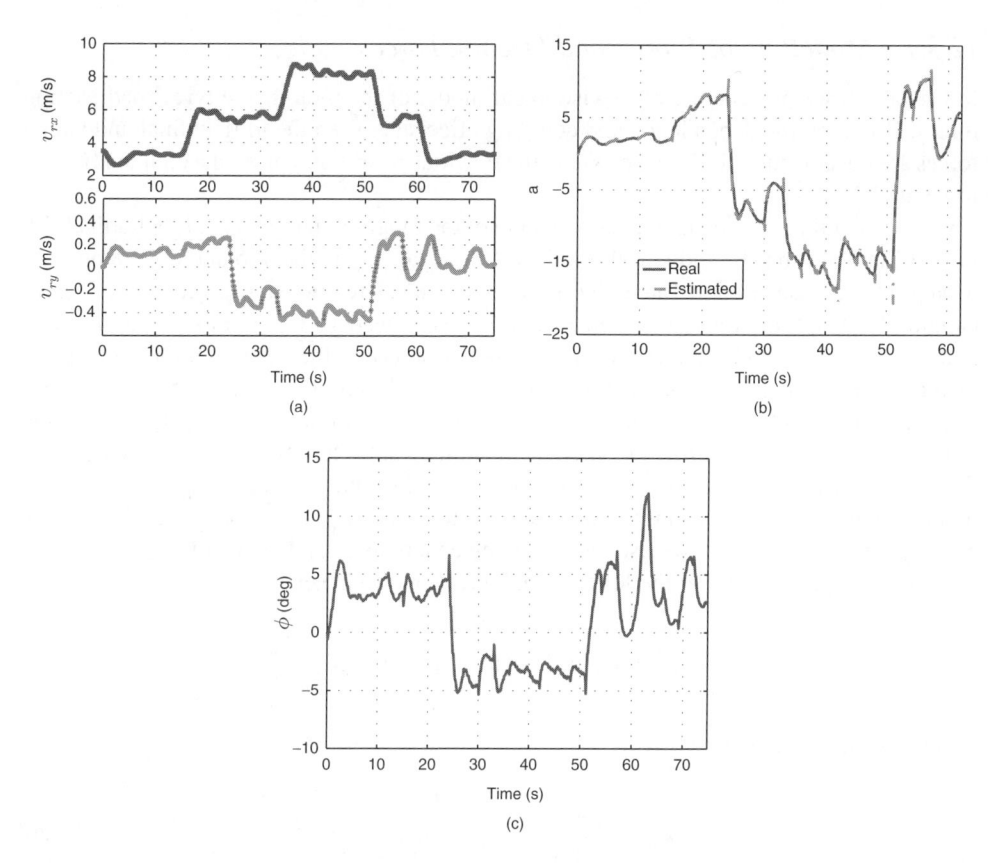

Figure 12.7 Roll angle and steering angle of the "8"-shape trajectory tracking. (a) Rear wheel contact point body-frame velocities v_{rx} and v_{ry}. (b) Roll angle φ. (c) Steering angle ϕ

Figure 12.6c. Comparing with the previous example, the tracking errors of the "8"-shape trajectory are much larger; see Figure 12.6b. This is mainly due to the quick change of the desired velocity profile.

Figure 12.7 shows the body-frame velocity, roll angle and steering angle for the "8"-shape trajectory. We clearly see the change of the lateral velocity during each circle of the trajectory. The lateral velocity magnitude is large due to the smaller turning radius. The maximum roll angle is around 15° and that is much larger than that of the previous example. The steering angle is large as well, to make the motorcycle turn in a tighter circle. The oscillations in both the roll angle (Figure 12.7b) and the steering angle (Figure 12.7c) are probably due to the variations in the desired velocity.

12.3 Path-Following Control System Design

In this section, we extend the modelling approach of Chapter 2 by coupling the longitudinal and the lateral friction forces. We then introduce a velocity-field manoeuvre regulation control in which the goal of the control system design is to follow the trajectory path, while the desired velocity is self-tuned online.

12.3.1 Modelling of Tyre–Road Friction Forces

In Chapter 2, we presented a piecewise linear model of the motorcycle tyre–road friction forces. However, the dependency and coupling effects between the longitudinal and lateral forces are not considered. Here we extend the previous results and present a coupled friction force model.

We consider the pseudo-static friction model of the longitudinal force F_x, longitudinal slip ratio λ_s, lateral force F_y and sideslip ratio λ_γ ($\lambda_\gamma = \tan \gamma$, γ is the slip angle). We propose to approximate the friction forces by a piecewise linear relationship given by (2.29). To capture the coupling effects between F_x and F_y, we consider that the model parameters k and x_m along the x and y directions are dependent on each other. For example, the values of the longitudinal stiffness k_x (k value in (2.29) for F_x) and the maximum slip ratio λ_{sm} (x_m value in (2.29) for F_x) are functions of tyre slip angle ratio λ_γ. Similarly, tyre cornering stiffness k_y (k value in (2.29) for F_y) and the maximum sideslip ratio $\lambda_{\gamma m}$ (x_m value in (2.29) for F_y) also depend on the longitudinal slip ratio λ_s. Denoting k_{0x} and λ_{sm0} (k_{0y} and $\lambda_{\gamma m0}$) as the parameter values of longitudinal (lateral) force F_x (F_y) when coupling effects with F_y (F_x) are not considered, we use the following equations to update parameters k_x and λ_{sm} (parameters k_y and $\lambda_{\gamma m}$). For the longitudinal direction, we have

$$k_x = k_{0x}(a_1 \lambda_\gamma + 1), \quad \lambda_{sm} = \lambda_{sm0} \tag{12.25}$$

and for the lateral direction force,

$$k_y = k_{0y} \frac{a_2 \lambda_s + 1}{a_3 \lambda_s + 1}, \quad \lambda_{\gamma m} = (a_3 \lambda_s + 1)\lambda_{\gamma m0}, \tag{12.26}$$

where a_1, a_2 and a_3 are three parameters in the coupled tyre model. We use (12.25) and (12.26) to capture the coupling effects because such relationships have been observed in experiments.

Figure 12.8 shows the property of the coupled tyre mode when $k_{x0} = 30,000$ N and $k_{y0} = 24,000$ N. In this example, we use $a_1 = -\frac{5}{3}$, $a_2 = -2$ and $a_3 = 10$ for the motorcycle tyres. For slip ratio λ_s, we set $\lambda_{sm0} = 0.15$, $\lambda_{\max} = 0.5$, $\alpha_x = 0.8$, and for sideslip ratio λ_γ, we set $\lambda_{\gamma m0} = 0.11$, $x_{\max} = 1$, $\alpha_x = 0.9$. In the following, we use the coupled coefficients (12.25) and (12.26) in the dynamic model (2.35) to design a path-following controller.

12.3.2 Path-Following Manoeuvring Design

We propose to use velocity field based approach to design the path-following control of the motorcycle system. We assume that the motorcycle motion planning modules such as the one in (Song et al. 2007) generate the desired trajectory $\mathcal{T}: (X_d(\tau), Y_d(\tau))$. Note that the trajectory $\mathcal{T}$ is parameterized by τ, which is not necessarily the same as the time t. Therefore, the desired outcome of the control design is to follow the trajectory path without specifying the velocity trajectory associated with the path. Instead, the desired velocity profile is a part of the control design process using a time-suspension technique.

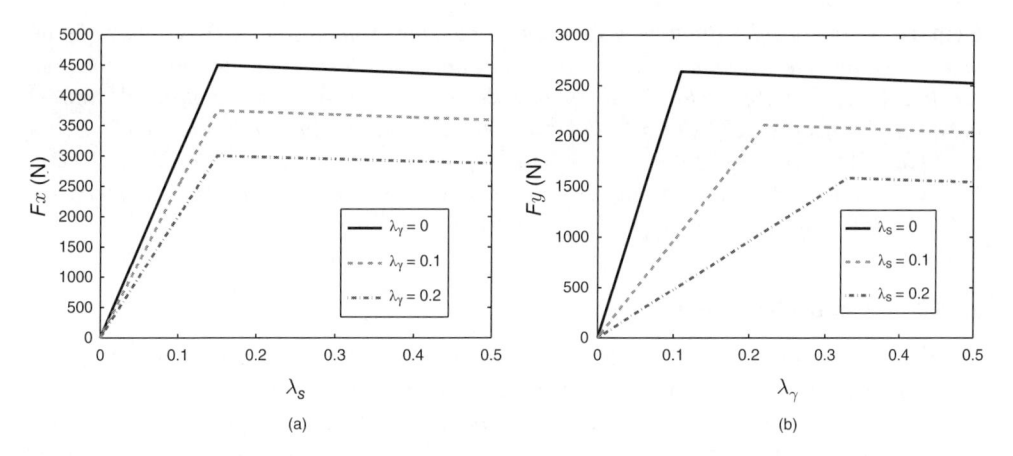

Figure 12.8 Approximate piecewise linear tyre force characteristics. (a) Longitudinal force with various tyre slip angle ratios λ_γ. (b) Lateral force with various tyre slip ratios λ_s. The tyre stiffness parameters are taken from (Sharp et al. 2004)

12.3.2.1 Time Suspension and Velocity Field Design

We use a time suspension technique to design the desired velocity profile. The basic idea of time suspension is to use self-placing technique to adjust the desired rate of the progression of the parameter τ related to $\mathcal{T}$. In other words, we do not need to assign any desired velocity profile in advance, and the motorcycle will instantaneously adjust its velocity according to the changes of the path-following errors. One obvious advantage of using the time suspension technique in our design is to reduce tracking error and thus to improve tracking performance.

We also use a velocity field design concept. The adopted velocity field approach is to define a reference input as a vector of velocities in the moving plane, rather than directly in terms of a reference-parameterized path. The main benefit of using a velocity field design is to further improve tracking performance (Li and Horowitz 2001). To construct the velocity field, we use a potential function-based approach that is similar to those in (Li and Horowitz 2001). We define the following potential function to capture the position errors along the path.

$$U(X, Y) = \frac{1}{2}\beta_1[(1 - \cos(X - X_d)) + (1 - \cos(Y - Y_d))], \quad (12.27)$$

where $\beta_1 > 0$ is a constant gain. At any position, we design the velocity vector by

$$\begin{bmatrix} V_x(\tau) \\ V_y(\tau) \end{bmatrix} = \lambda_1(X, Y) \begin{bmatrix} \frac{dX_d}{d\tau} \\ \frac{dY_d}{d\tau} \end{bmatrix} - \lambda_2(X, Y) \begin{bmatrix} h\sin(X - X_d) \\ h\sin(Y - Y_d) \end{bmatrix}, \quad (12.28)$$

where $\lambda_1(X, Y) = e^{-\beta_2 U(X,Y)}$, $\lambda_2(X, Y) = 2 - e^{-\beta_2 U(X,Y)}$ and $\beta_2 > 0$ is a self-pacing parameter. The time suspension level is defined by the following dynamics of τ:

$$\dot{\tau} = \frac{d\tau}{dt} = \lambda_1(X, Y). \quad (12.29)$$

Remark 3 *We consider the time suspension parameter dynamics (12.29) as a part of augmented motorcycle dynamics (2.35). Note that the τ dynamics is related to the potential function $U(X, Y)$ and therefore to the path-following errors. When the motorcycle follows the desired trajectory, $U(X, Y) = 0$ and $\dot{\tau} = 1$. In this case, τ can be considered as the time variable t. When the path-following errors are large, the progression of desired trajectory (i.e., $\dot{\tau}$) is reduced and the controlled trajectory converges to the desired path with increased $\lambda_2(X, Y)$. It is noted that $0 < \lambda_1(X, Y) \leq 1$ and $1 \leq \lambda_2(X, Y) < 2$.*

12.3.2.2 Controller Design

For motorcycle control systems design, we combine the EIC-based control approach discussed in the previous section with the velocity field approach discussed earlier.

The EIC-based trajectory control design consists of two steps; see Figure 12.2. The velocity vector parameterized by τ, rather than desired trajectory path specified in time t, is used as the reference input to the EIC control. We combine the EIC control and the velocity field design as follows. At any position and on any particular τ, we use (12.27) and (12.28) to calculate the current velocity vector. Then we construct a special trajectory for the EIC controller as

$$\begin{bmatrix} X_d(\tau) \\ X_d^1(\tau) \\ X_d^{(2,4)}(\tau) \end{bmatrix} = \begin{bmatrix} X(\tau) \\ V_x(\tau) \\ 0 \end{bmatrix}, \quad \begin{bmatrix} Y_d(\tau) \\ Y_d^1(\tau) \\ Y_d^{(2,4)}(\tau) \end{bmatrix} = \begin{bmatrix} Y(\tau) \\ V_y(\tau) \\ 0 \end{bmatrix},$$

where τ is updating by (12.29). The basic design idea is to let the velocity vector be the only design components in the desired trajectory space. With this treatment, we can fully inherit the EIC controller design and its properties that are stated in the previous section.

12.3.3 Simulation Results

In this section, we demonstrate the control systems design through two examples: one is for an "8"-shape trajectory-following manoeuvre and the other for a more agile manoeuvre. We use the same racing motorcycle and tyre profiles as in the previous section.

The first example shows that the motorcycle runs under a regular "8"-shape path-following manoeuvre along the trajectory with relatively large curvatures. The desired parameterized trajectory (Figure 12.9a) is given by the following equation parameterized by τ

$$\begin{bmatrix} X_d(\tau) \\ Y_d(\tau) \end{bmatrix} = \begin{bmatrix} 25 \sin(0.1\pi\tau) \\ 40 \cos(0.05\pi\tau) \end{bmatrix} \tag{12.30}$$

Figure 12.9 shows the simulation results. As shown in Figure 12.9a, the starting point of the motorcycle is (0, 40). We use self-pacing parameter $\beta_2 = 100$ in (12.28) and parameter $\beta_1 = 0.0025$ in (12.27), and the initial velocity is 0.1 m/s. In the simulation, we add white noise with standard variations 0.02 m/s, 0.005 m/s^2, 0.3° and 0.6° to velocity, acceleration, roll angle and yaw angle measurements, respectively. By comparing with the desired trajectory, the simulation results show that the motorcycle successfully tracks the desired trajectory under the velocity field control. Figures 12.9b and 12.9c clearly show the desired and actual motorcycle roll angle φ and steering angle ϕ, respectively. The roll angle and steering angle

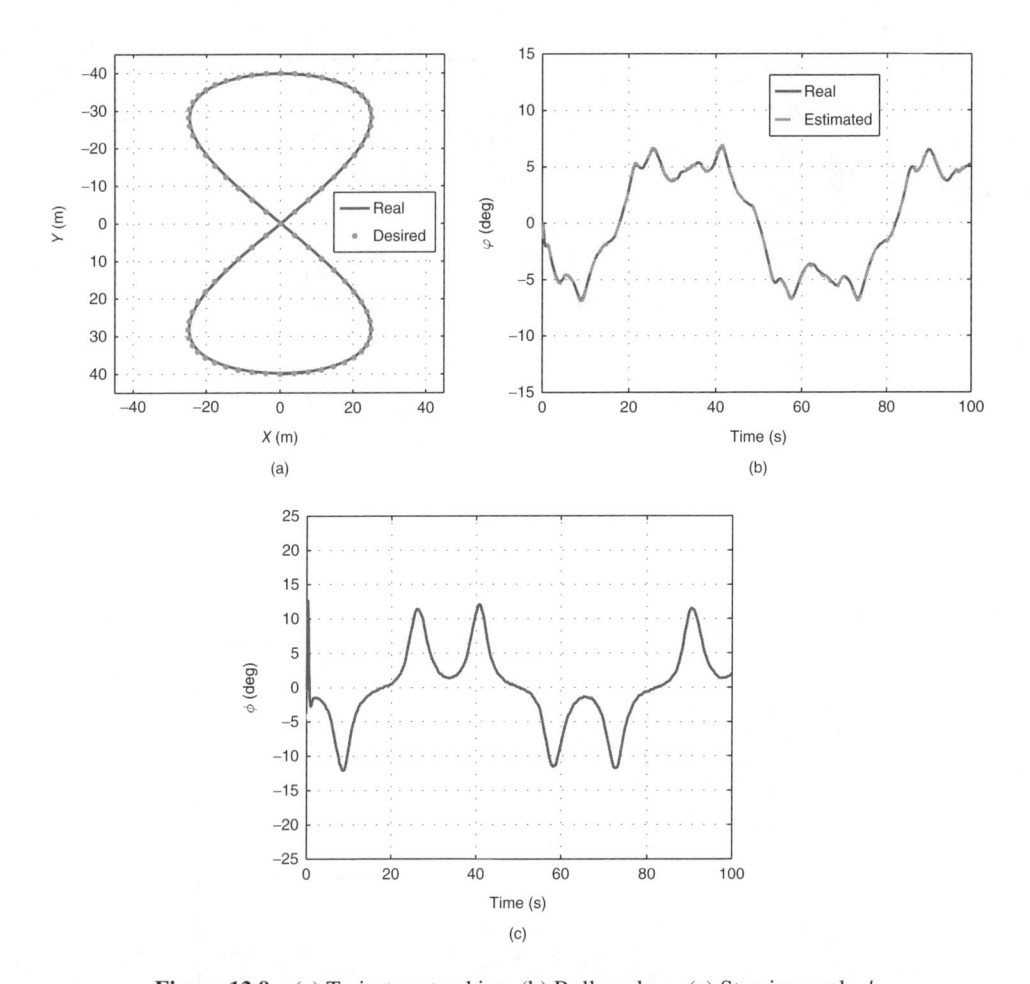

Figure 12.9 (a) Trajectory tracking. (b) Roll angle φ. (c) Steering angle ϕ

are relatively large when turning at small radius curvatures and small along the straight trajectories.

Figure 12.10 shows the positions, the longitudinal velocity and the lateral velocity of point C_2. From Figure 12.10a, the motorcycle takes about 65 s to go through one entire circle. It is quite clear that the motorcycle tunes its own velocity automatically using the self-placing technique. When tuning at a small radius, the tracking errors become large. The motorcycle control system then reduces the rate of the progression in time, namely, its longitudinal velocity, to reduce the errors. Meanwhile, due to the sharp direction change, the lateral velocity is relatively large; see Figure 12.10b.

The tyre slip ratios and angles during the manoeuvre are shown in Figure 12.11. For the front wheel, we see a maximum 15° sideslip angle while for the rear wheel, the slip angle reaches almost 6°. Figure 12.12a shows the tracking error performances of the motorcycle under different values of self-pacing parameter β_2. From this figure, we can see that when self-pacing is increased (i.e, increasing β_2), the tracking errors become smaller, that is, the

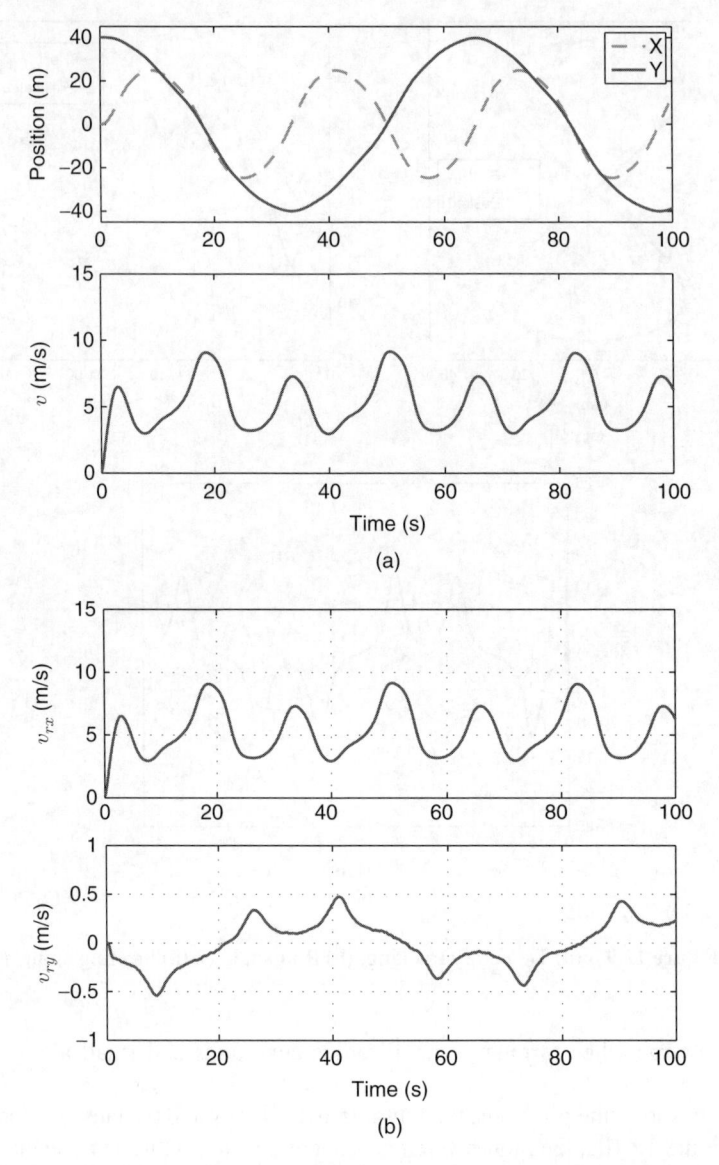

Figure 12.10 A typical motorcycle path-following manoeuvre. (a) Motorcycle position and velocity (b) Longitudinal velocity v_{rx} and lateral velocity v_{ry}

better path-following performance. When $\beta_2 = 0$, the maximal error is always smaller than 0.3 m. Of course, the better performance is traded-off by the smaller motorcycle velocity. This can be observed by the progression factor $\dot{\tau}$ as shown in Figure 12.12b. From Figure 12.12b, we see that increasing β_2 reduces the value of $\dot{\tau}$ in general, which implies that the time has been expanded more. We can clearly see when $\beta_2 = 0$, $\dot{\tau} = 1$ and then the progression always remains at one, which implies that no time suspension exists. In this

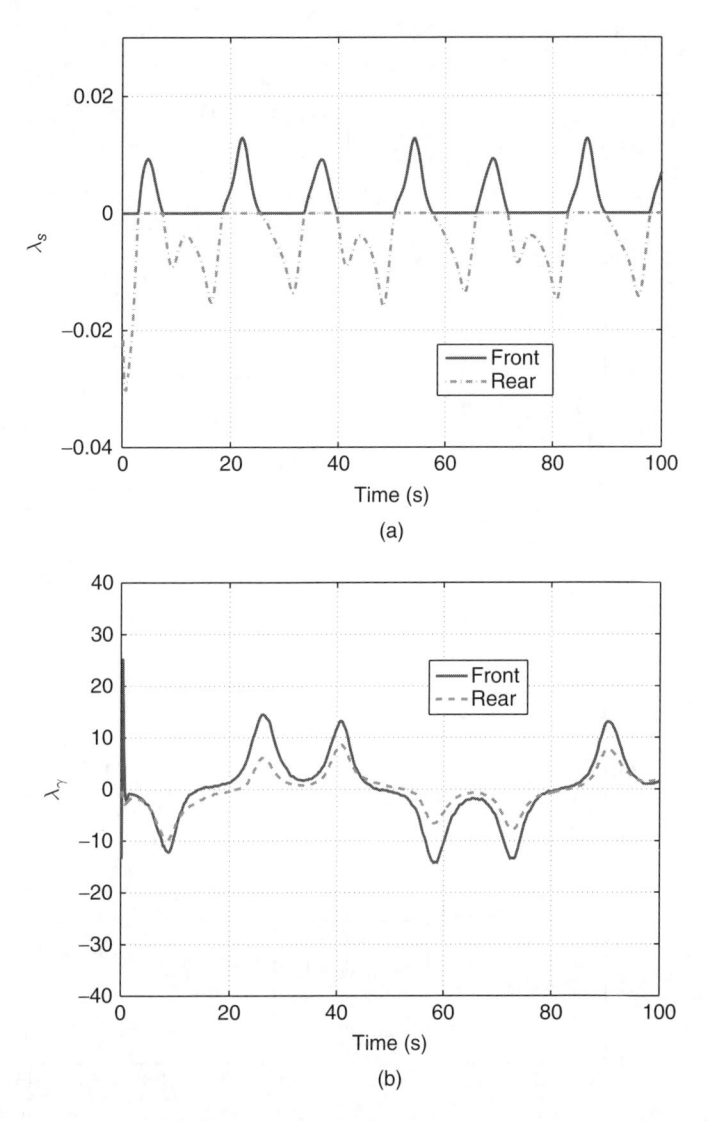

Figure 12.11 Slip ratios and angles at the front and rear wheels during the "8"-shape trajectory tracking. (a) Slip ratio λ_s. (b) Slip angles λ_γ

case, the path-following system is the same as time-based trajectory tracking, as shown in the previous section. Note that the oscillation of both path-following errors and progression $\dot{\tau}$ are due to the repeated motion trajectory.

In the second example, we show that the motorcycle runs with a more agile "8"-shape path-following manoeuvre. In this manoeuvre, the motorcycle will turn sharply at much smaller radii. The desired parameterized trajectory, as shown in Figure 12.13a, is defined as

$$\begin{bmatrix} X_d(\tau) \\ Y_d(\tau) \end{bmatrix} = \begin{bmatrix} 7.5\sin(0.5\pi\tau) \\ 15\cos(0.25\pi\tau) \end{bmatrix}. \tag{12.31}$$

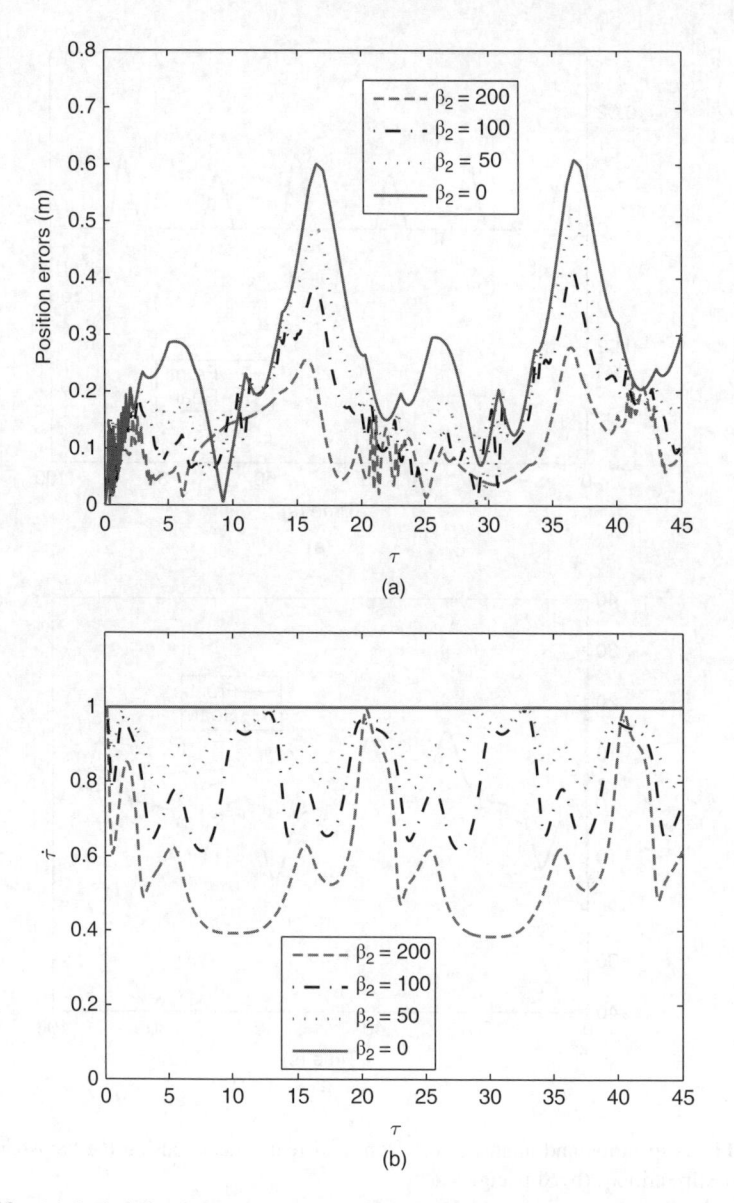

Figure 12.12 (a) Path-following errors under various values of self-pacing parameter β_2. (b) Progression (i.e. $\dot{\tau}$) under various values of self-pacing parameter β_2

The start point of motorcycle is (0, 15). We choose self-pacing parameter $\beta_2 = 80$ and parameter $\beta_1 = 0.00825$. The initial velocity and noise characteristics are the same as those in the previous example.

From Figure 12.13a, we see that even in this extremely tight trajectory case, the motorcycle can still follow the desired trajectory under the velocity field control. From Figures 12.13b and 12.13c, it is clear that the steering angle and roll angle are both larger than those of

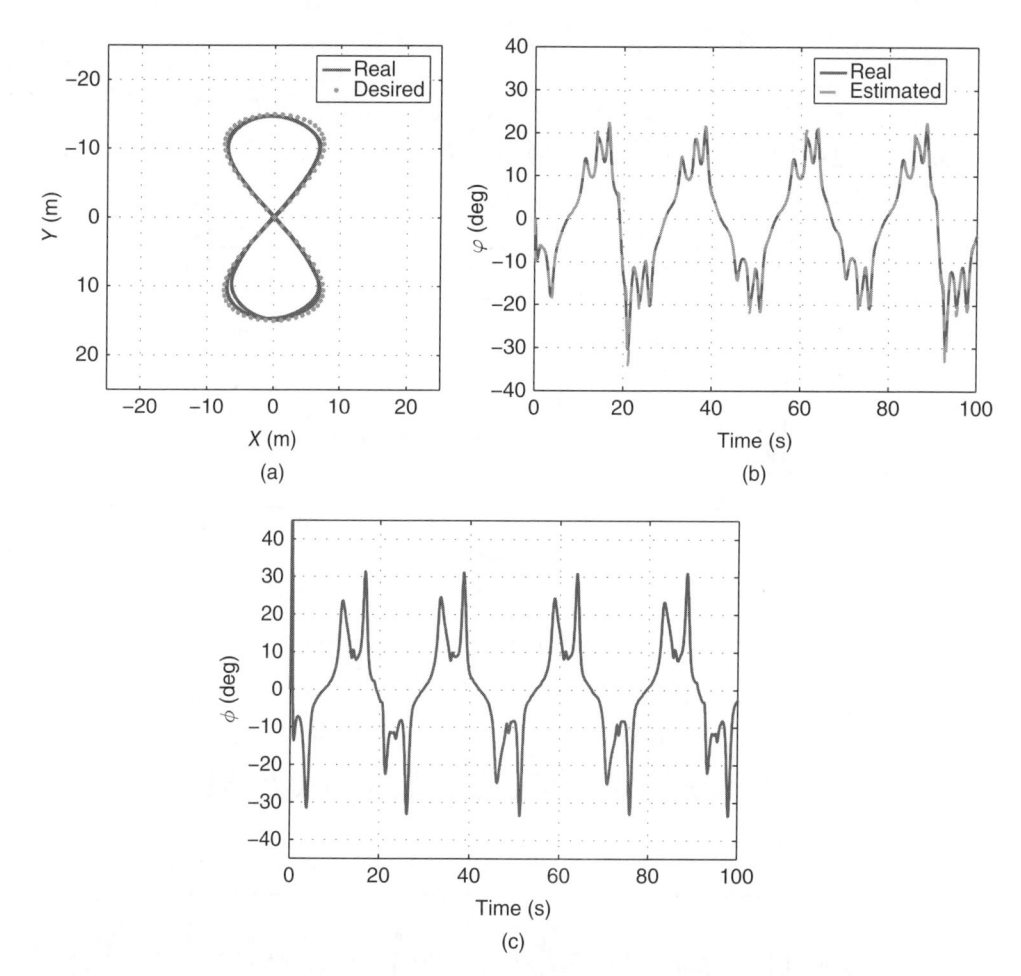

Figure 12.13 Motorcycle agile motion. (a) Path-following performance. (b) Roll angle φ. (c) Steering angle ϕ

the previous example due to the much smaller radius curvatures. Figures 12.14a and 12.14b show the motorcycle position and velocity information for this manoeuvre. We see a large lateral velocity v_{ry}. Figures 12.15a and 12.15b show the tyre slip ratios and slip angles, respectively. It is noted from Figure 12.15a that the required longitudinal slip ratio of the rear tyre nearly reaches 0.15, which is almost the maximal stable slip ratio of the tyre model. From Figure 12.15b, we also see the large slip angles in this agile manoeuvre.

We clearly see the large sideslip angles shown in Figure 12.16a. Particularly for the front wheel, we have seen a 15° sideslip angle. For the rear wheel, the sideslip angle reaches almost 6°, which is around the saturation point of the tyre characteristics. In other words, the motorcycle rear wheel is starting to slide on the ground. If the sideslip angle increases further, the stability of the motorcycle will change significantly. The longitudinal slips are relatively small since the longitudinal acceleration of the motorcycle is not large, and the racing

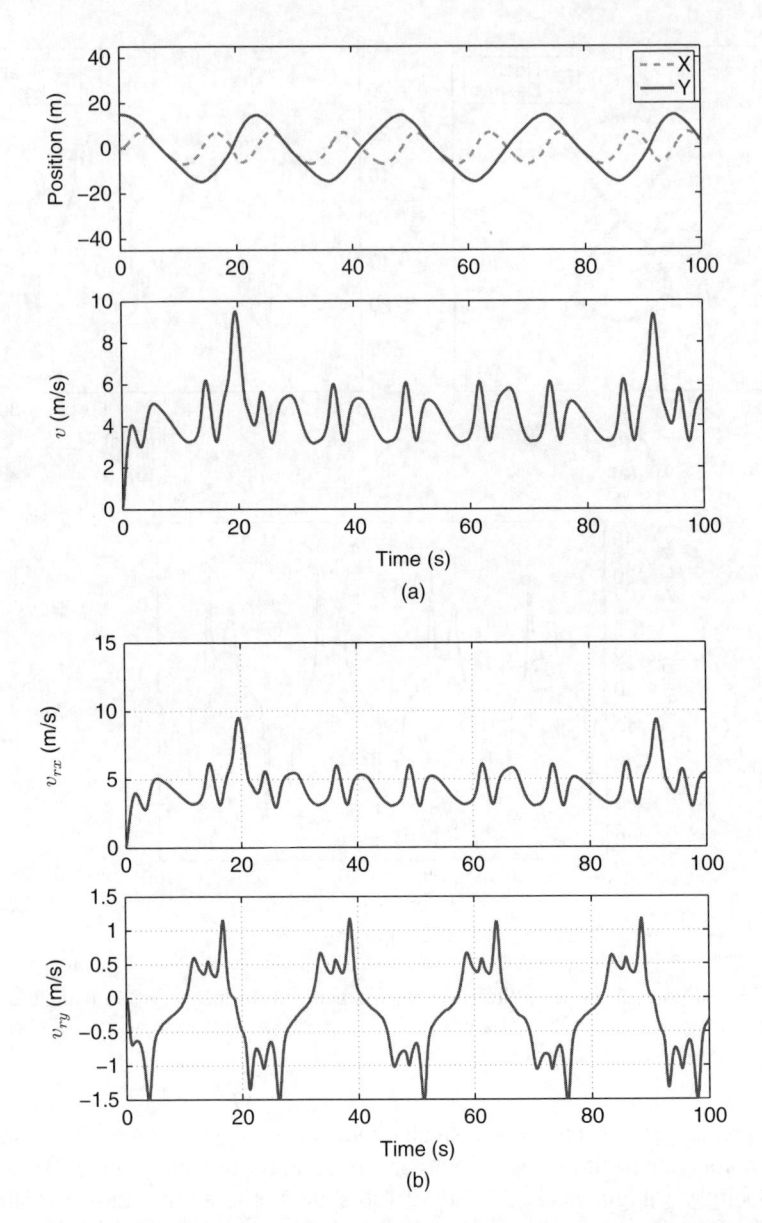

Figure 12.14 Motorcycle agile manoeuvre. (a) Position and velocity (b) Longitudinal velocity and lateral velocity

motorcycle tyre is stiff. This simulation example demonstrates that the proposed dynamic model and control systems capture the realistic aggressive motorcycle manoeuvres.

Compared with the time-based trajectory tracking control design in the previous chapter, the simulation results in this section show that the velocity field based path-following design achieves smoother velocity profiles and much smaller tracking errors.

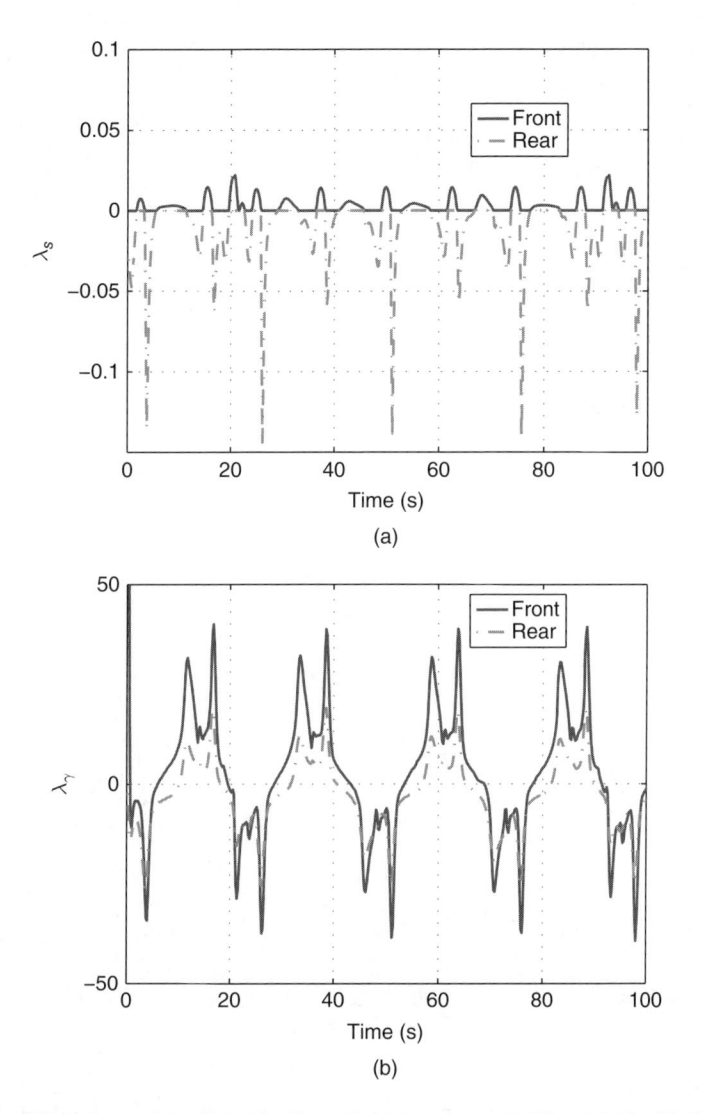

Figure 12.15 Slip ratios and angles under the agile manoeuvre. (a) Slip ratio λ_s. (b) Slip angles λ_γ

12.4 Conclusion

The focus of this chapter was on the control systems design of autonomous motorcycles for agile manoeuvres. Both trajectory tracking and path-following control were presented. The nonlinear trajectory tracking control design took advantage of the external/internal convertible (EIC) dynamical structure of the motorcycle dynamics, and was extended with three control inputs. Such an extension allowed flexibility in control systems design and therefore simplified the complexity of the final calculation. We demonstrated the trajectory tracking control systems design through two simulation examples using a racing motorcycle prototype. To further improve the tracking performance, we presented a velocity field based path-following control design. We first extended the motorcycle dynamics by considering

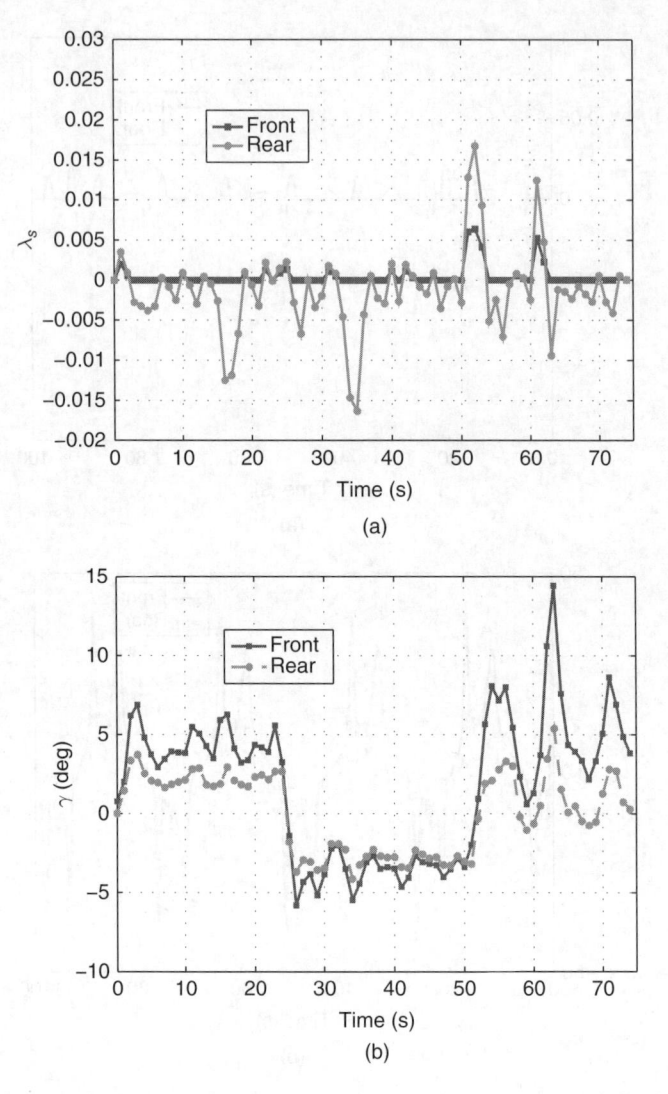

Figure 12.16 Longitudinal slips and slip angles at the front and rear wheels of the "8"-shape trajectory tracking. (a) Slip ratio λ_{fs} and λ_{rs}. (b) Slip angles λ_f and γ_r

the coupled longitudinal/lateral tyre friction forces. A velocity field design was presented to provide a desired velocity profile through a time suspension technique. We then combined the velocity field design with the previously developed EIC motorcycle controller. The control system automatically tuned the velocity profile based on the tracking errors and trajectory properties. The simulation results of a typical manoeuvre and an agile manoeuvre demonstrated that the velocity field based path-following control design reduced the tracking errors.

There are several ongoing research directions. We are currently implementing the proposed control systems on a Rutgers autonomous motorcycle platform. We will report the implementation results in future. We also plan to study how professional racing drivers control motorcycles for agile manoeuvres.

Acknowledgements

Yizhai Zhang thanks Dr N Getz at Inversion Inc. for his helpful suggestions and support. The authors are grateful to Prof. S Jayasuriya of Drexel University, Dr EH Tseng and Dr J Lu at Ford Research and Innovation Center for their helpful discussions and suggestions.

Appendix A: Calculation of the Lie Derivatives

The calculation of $\bar{L}_{\mathbf{N}_{\text{ext}}} u_{rx}^{\text{ext}}$ and $\bar{L}_{\mathbf{N}_{\text{ext}}} u_{ry}^{\text{ext}}$ is obtained by taking the Lie derivative along the nominal external vector field (12.15) and the control input (12.17). The calculation are as follows:

$$\bar{L}_{\mathbf{N}_{\text{ext}}} u_{rx}^{\text{ext}} = \begin{bmatrix} -s_\psi & c_\psi \end{bmatrix} \dot{\psi}(-\mathbf{U} + \mathbf{u}^{\text{ext}}) + \begin{bmatrix} c_\psi & s_\psi \end{bmatrix}$$
$$\left(- \begin{bmatrix} -2u_{rx}^{\text{ext}}s_\psi - 2u_{ry}^{\text{ext}}c_\psi - 3\dot{v}_{rx}\dot{\psi}c_\psi + 3\dot{v}_{ry}\dot{\psi}s_\psi + \ddot{\psi}(v_{rx}s_\psi + v_{ry}c_\psi) \\ 2u_{rx}^{\text{ext}}c_\psi - 2u_{ry}^{\text{ext}}s_\psi - 3\dot{v}_{rx}\dot{\psi}s_\psi - 3\dot{v}_{ry}\dot{\psi}c_\psi - \ddot{\psi}(v_{rx}c_\psi - v_{ry}s_\psi) \end{bmatrix} \dot{\psi} \right.$$
$$\left. + \begin{bmatrix} \bar{L}_{\mathbf{N}_{\text{ext}}} u_X^{\text{ext}} \\ \bar{L}_{\mathbf{N}_{\text{ext}}} u_Y^{\text{ext}} \end{bmatrix} \right)$$
$$= \dot{v}_{rx}\dot{\psi}^2 + (2u_{ry}^{\text{ext}} - u_X^{\text{ext}}s_\psi + u_Y^{\text{ext}}c_\psi)\dot{\psi} + \bar{L}_{\mathbf{N}_{\text{ext}}} u_X^{\text{ext}} c_\psi$$
$$+ \bar{L}_{\mathbf{N}_{\text{ext}}} u_Y^{\text{ext}} s_\psi, \qquad (12.32)$$

$$\bar{L}_{\mathbf{N}_{\text{ext}}} u_{ry}^{\text{ext}} = \dot{v}_{ry}\dot{\psi}^2 - (2u_{rx}^{\text{ext}} + u_X^{\text{ext}}c_\psi + u_Y^{\text{ext}}s_\psi)\dot{\psi} - \bar{L}_{\mathbf{N}_{\text{ext}}} u_X^{\text{ext}} s_\psi$$
$$+ \bar{L}_{\mathbf{N}_{\text{ext}}} u_Y^{\text{ext}} c_\psi. \qquad (12.33)$$

In these equations, we have

$$\begin{bmatrix} \bar{L}_{\mathbf{N}_{\text{ext}}} u_X^{\text{ext}} \\ \bar{L}_{\mathbf{N}_{\text{ext}}} u_Y^{\text{ext}} \end{bmatrix} = \begin{bmatrix} X_d^{(4)}(t) \\ Y_d^{(4)}(t) \end{bmatrix} - b_3 \begin{bmatrix} u_X^{\text{ext}} - X_d^{(3)}(t) \\ u_Y^{\text{ext}} - Y_d^{(3)}(t) \end{bmatrix} - \sum_{i=1}^{2} b_i \begin{bmatrix} X^{(i)} - X_d^{(i)}(t) \\ Y^{(i)} - Y_d^{(i)}(t) \end{bmatrix}.$$

Similarly, we calculate $\bar{L}_{\mathbf{N}_{\text{ext}}}^2 \varphi_e$ by directly taking a directional derivative of $\bar{L}_{\mathbf{N}_{\text{ext}}} \varphi_e$ along the vector field $\mathbf{N}_{\text{ext}}$. From (12.23), we have

$$\bar{L}_{\mathbf{N}_{\text{ext}}}^2 \varphi_e = (h\dot{\psi}c_{\varphi_e} + g\sec^2\varphi_e)^{-1} \left[\frac{gbl_t c_\xi}{h} \left(\frac{\dot{\psi} u_{rx}^{\text{ext}}}{v_{rx}^2} - \frac{2\dot{v}_{rx}^2 \dot{\psi}}{v_{rx}^3} \right) + \dot{\psi} u_{rx}^{\text{ext}} + \right.$$
$$\left. \bar{L}_{\mathbf{N}_{\text{ext}}} u_{ry}^{\text{ext}} + (h\dot{\psi}s_{\varphi_e} - 2g\sec^2\varphi_e \tan\varphi_e)(\bar{L}_{\mathbf{N}_{\text{ext}}} \varphi_e)^2 \right]. \qquad (12.34)$$

References

Aguiar A and Hespanda J 2007 Trajectory-tracking and path-following of underactuated autonomous vehicles with parametric modeling uncertainty. *IEEE Trans. Automat. Contr.* **52**(8), 1362–1379.

Aguiar A, Hespanda J and Kokotović P 2005 Path-following for nonminimum phase systems removes performance limitations. *IEEE Trans. Automat. Contr.* **50**(2), 234–239.

Al-Hiddabi S and McClamroch N 2002 Tracking and maneuver regulation control for nonlinear nonminimum phase systems: Application to flight control. *IEEE Trans. Contr. Syst. Technol.* **10**(6), 780–792.

Åström K, Klein R and Lennartsson A 2005 Bicycle dynamics and control. *IEEE Control Syst. Mag.* **25**(4), 26–47.

Beznos A, Formal'sky A, Gurfinkel E, Jicharev D, Lensky A, Savitsky K and Tchesalin L 1998 Control of autonomous motion of two-wheel bicycle with gyroscopic stabilisation *Proc. IEEE Int. Conf. Robot. Autom.*, pp. 2670–2675, Leuven, Belgium.

Bullo F and Lewis A 2004 *Geometric Control of Mechanical Systems: Modeling, Analysis, and Design for Simple Mechanical Control Systems*. Springer, New York, NY.

Corno M, Savaresi SM, Tanelli M and Fabbri L 2008 On optimal motorcycle braking. *Contr. Eng. Pract.* **16**, 644–657.

Gerdes JC and Rossetter EJ 2001 A unified approach to driver assistance systems based on artificial potential fields. *ASME J. Dyn. Syst., Meas., Control* **123**(3), 431–438.

Getz N 1995 Dynamic inversion of nonlinear maps with applications to nonlinear control and robotics PhD thesis Dept Electr. Eng. and Comp. Sci., Univ. Calif. Berkeley, CA.

Grizzle J, Di Benedetto M and Lamnabhi-Lagarrigue F 1994 Necessary conditions for asymptotic tracking in nonlinear systems. *IEEE Trans. Automat. Contr.* **39**(9), 1782–1794.

Hauser J and Hindman R 1995 Maneuver regulation from trajectory tracking: Feedback linearizable systems *Proc. IFAC Symp. Nonlinear Contr. Syst. Design*, pp. 638–643, Tahoe City, CA.

Isidori A 1995 *Nonlinear Control Systems* third edn. Springer-Verlag, London, UK.

Lee S and Ham W 2002 Self-stabilizing strategy in tracking control of unmanned electric bicycle with mass balance *Proc. IEEE/RSJ Int. Conf. Intell. Robot. Syst.*, pp. 2200–2205, Lausanne, Switzerland.

Li PY and Horowitz R 2001 Passive velocity field control (PVFC): Part II –Application to contour following. *IEEE Trans. Automat. Contr.* **46**(9), 1360–1371.

Sastry S 1999 *Nonlinear Systems: Analysis, Stability, and Control*. Springer, New York, NY.

Sharp RS, Evangelou S and Limebeer DJN 2004 Advances in the modelling of motorcycle dynamics. *Multibody Syst. Dyn.* **12**, 251–283.

Skjetne R, Fossen T and Kokotović P 2004 Robust output maneuvering for a class of nonlinear systems. *Automatica* **40**, 373–383.

Song D, Lee HL, Yi J and Levandowski A 2007 Vision-based motion planning for an autonomous motorcycle on ill-structured roads. *Auton. Robots* **23**(3), 197–212.

Tanaka Y and Murakami T 2004 Self-sustaining bicycle robot with steering controller *Proc. 2004 IEEE Adv. Motion Contr. Conf.*, pp. 193–197, Kawasaki, Japan.

Tanaka Y and Murakami T 2009 A study on straight-line tracking and posture control in electric bicycle. *IEEE Trans. Ind. Electron.* **56**(1), 159–168.

Yi J, Song D, Levandowski A and Jayasuriya S 2006 Trajectory tracking and balance stabilization control of autonomous motorcycles *Proc. IEEE Int. Conf. Robot. Autom.*, pp. 2583–2589, Orlando, FL.

Yi J, Zhang Y and Song D 2009 Autonomous motorcycles for agile maneuvers: Part II: Control systems design *Proc. IEEE Conf. Decision Control*, pp. 4619–4624, Shanghai, China.

Zhang Y and Yi J 2010 Dynamic modeling and balance control of human/bicycle systems *Proc. IEEE/ASME Int. Conf. Adv. Intell. Mechatronics*, pp. 1385–1390, Montreal, Canada.

13

Estimation Problems in Two-Wheeled Vehicles

Ivo Boniolo[a], Giulio Panzani[b], Diego Delvecchio[b], Matteo Corno[b], Mara Tanelli[b], Cristiano Spelta[a], and Sergio M. Savaresi[b]
[a]Dipartimento di Ingegneria, Università degli Studi di Bergamo, Italy
[b]Dipartimento di Elettronica, Informazione e Bioingegneria, Politecnico di Milano, Italy

13.1 Introduction

Electronic control systems are present on all commercial cars: anti-lock braking systems (ABS) and traction control (TC) are employed to improve performance and safety in acceleration and braking manoeuvres (Borrelli et al. 2006; Savaresi and Tanelli 2010), while electronic stability control (ESC) systems actively modify the vehicle dynamics in order to restore vehicle stability in the face of dangerous manoeuvres (Abe et al. 2001; Canale et al. 2009, 2007).

The development of active control systems for two-wheeled vehicles, however, has started with a significant time delay. This is due to economic, cultural and technical factors. In particular, the Powered-Two-Wheelers (PTW) market is limited and the amount of investment on R&D is limited. Moreover, many riders believe that they do not need any help riding their bikes and that electronic control systems alter the riding experience. Finally, and most interesting from an engineering point of view, dealing with motorcycle dynamics is more complex than it is for four-wheeled vehicles (Cossalter et al. 2004; Limebeer et al. 2001). Only in recent times have motorcycle manufacturers started working on production versions of traction control systems, ABS and slow-adaptive control of steering dampers (Kazuhiko et al. 2010; Savaresi et al. 2010a; Wakabayashi and Sakai 2004).

Modelling, Simulation and Control of Two-Wheeled Vehicles, First Edition.
Edited by Mara Tanelli, Matteo Corno and Sergio M. Savaresi.
© 2014 John Wiley & Sons, Ltd. Published 2014 by John Wiley & Sons, Ltd.

In the scientific literature, some preliminary results have been obtained that address the control of two-wheeled vehicles. The problem of controlling the traction of motorcycles is addressed in (Cardinale et al. 2009; Tanelli et al. 2009c). The design of braking control systems is addressed in (Tanelli et al. 2009a; Corno et al. 2009). Moreover, some interesting results have also been obtained with semi-active control strategies: in (De Filippi et al. 2011b; De Filippi and Savaresi 2011) an industrially amenable control strategy to damp the weave and wobble modes by acting on a semi-active steering damper was proposed. In (Evangelou et al. 2010), burst oscillations are suppressed with a mechanical steering compensator. In (Evangelou 2010), cornering weave oscillations are reduced by controlling the geometry of the rear suspension. In (Yi et al. 2009) the control of an autonomous motorcycle using the steering torque and the wheel angular velocity as control parameters is presented. Recently, the first results in active stability control system were presented (De Filippi et al. 2010; De Filippi et al. 2011a).

To implement such control systems, it is crucial to have a reliable set of measurements or estimations of the variables involved in the motion of interest. As discussed in the previous chapters, when moving on curves there is a crucial variable which determines a motorcycle behaviour: the *roll angle*. This angle, also known as the *tilt* angle, is the inclination of the vehicle with respect to the vertical direction and it represents the amount of inclination that the vehicle needs in order to ensure the force balance on the curve and hence to reach a steady-state cornering condition.

Besides the roll angle, the longitudinal speed is fundamental for describing traction and braking conditions, when significant accelerations or decelerations make the wheels slip. As discussed in Chapters 3 and 8, when the wheel does slip, it means that there is a difference between the wheels and vehicle speed, so that both these variables are needed to design slip control systems. While wheel speed can be derived from a direct processing of the wheel encoder signals, vehicle speed cannot be directly measured (optical sensors can be used for this purpose, but can be employed only for test and prototyping purposes).

In view of this discussion, this chapter is devoted to illustrating an approach to the estimation of the roll angle that is based on a reduced sensor configuration, and an approach for deriving a reliable estimate of the vehicle speed both in traction and in braking conditions. Further, in order to provide a wider view on estimation problems in two-wheeled vehicles, the estimation of the suspension stroke is also addressed, as it allows a significant simplification of the vehicle setup for suspension control systems. In fact, by estimating this variable one can eliminate the stroke sensor, which is both costly and delicate, thus hardly usable on commercial vehicles.

13.2 Roll Angle Estimation

As the previous chapters of this book have revealed, a reliable and real time estimation of the vehicle roll angle, that is the inclination of the vehicle with respect to the vertical direction, is the key to implementing effective and reliable control systems to enhance both safety and performance on board two-wheeled vehicles. This variable greatly influences the tyre–road contact forces, which are the means by which traction and braking forces are transmitted to the ground. As it is not possible to obtain a direct measure of the lean angle on commercial vehicles (a direct measurement can be achieved only by means of optical sensors that are only suitable for racing bikes, see (Norgia et al. 2009), it is important to devise effective estimation methods.

Thus, to move a step further in active control systems design for two-wheeled vehicle, the enabling technology comes from an effective estimation method that is able to provide the roll angle value in a reliable way and in real time. Moreover, to suit industrial cost constraints, such a method should rely on a low-cost sensor configuration.

Besides its usefulness in control system design, a reliable online measure of the roll angle can also be employed in the racing context to assess tyre performance. In fact (Cossalter 2002; Cossalter et al. 2002; Sharp et al. 2004), the roll angle has a major impact in determining the tyre–road contact forces that ensure the stability of a motorcycle on a curve.

Note that, in principle, one may think of estimating the roll angle by simply integrating the output of a gyroscope that measures the roll velocity. This choice, even though viable in principle, has many drawbacks, the most significant of which is that numerical integration is particularly sensitive to measurement errors, which cause a drift in the integrated signal. Hence, more refined methods must be devised to obtain a reliable roll angle estimate under all driving conditions.

For this, the reader can refer to (Boniolo and Savaresi 2010), where the problem of roll angle estimation is treated in detail and in different contexts. Further (Boniolo et al. 2009b) presented a first solution to estimate the lean angle, based on the wheel speed signal and four gyroscopes, while in (Boniolo et al. 2009a) an analysis of the most suitable signals to be employed for this purpose was carried out within a neural network framework. Besides these results, (Tseng et al. 2007) employs kinematics-based observers to estimate the roll and pitch angles using an inertial measurement unit, while (Gasbarro et al. 2004) proposes an approach for estimating the whole vehicle trajectory. Both these attempts need expensive sensor systems and, most importantly, cannot provide a roll angle measure in real time. Some other approaches may be found in the patent literature. For example, in (Gustaffson et al. 2002; Hauser et al. 1995; Schiffmann 2003; Schubert 2005) several methods are described, the common purpose of which is to devise robust estimation approaches with low cost (and small size) equipment.

In this section, the problem of the estimation of the roll angle of a two-wheeled vehicle with a reduced and low-cost set of sensors is addressed see also (Boniolo et al. 2012). This specific approach has been chosen as it has the merit of offering an amenable solution for practical applications on commercial vehicles. To present the method, an incremental approach is followed: first, a solution requiring only a two-axis accelerometer is presented, which offers limited accuracy but can be employed with control systems that need threshold-based information on the lean angle value (*e.g.*, a classification such as small, medium, large). Then a refined solution is devised, which employs two additional gyroscopes and an ad hoc algorithm to realize the data fusion of the different sensors. The main feature of this more advanced approach is that the speed measurement is not needed. This has several advantages: first, the vehicle speed cannot be directly measured, but it must be estimated via indirect measurements when the vehicle is subject to accelerations/decelerations (in this case, the wheel speed is no longer a good estimate of the vehicle speed, due to the presence of wheel slip), and vehicle speed estimation for two-wheeled vehicles is a difficult problem, see (Savaresi and Tanelli 2010) and the dedicated section of this chapter. Further, the wheel speed signal is not always accessible to the lean angle estimation system, especially if this is performed by an additional plug-in system not deployed with the bike, but added, for example, as an after-market kit. These systems cannot, in fact, access the vehicle bus where the internal signals are transmitted. Finally, the proposed system can be used also in combination with other sensors that enable more sophisticated estimation methods as a redundant scheme that allows the detection of possible malfunctioning of the additional sensors and provides an additional estimation to be used in case of faults.

The section is organized as follows. Section 13.2.1 describes the reference frames and the needed notation, while Section 13.2.2 illustrates the vehicle set-up used in the tests. Section 13.2.3 discusses the accelerometer-based estimation method, and Section 13.2.4 shows how the use of a refined estimation scheme comprising two additional gyroscopes can yield improved performance.

13.2.1 Vehicle Attitude and Reference Frames

As is well known, the attitude of a rigid body can be represented in different ways, see (Shuster 1993), where a survey on the attitude representation is reported. In this application, the motorcycle attitude is described based on Euler angles. Thus, to define the vehicle attitude, the reference frames must be first introduced (Figure 13.1) (Cossalter 2002). Specifically, the roll axis is considered, without loss of generality, as the longitudinal axis of the vehicle. The pitch axis is the axis of rotation due to the lowering of the motorcycle steering head when it is turned, while the vertical rotations of the vehicle act on the yaw axis. Conventionally, the roll angle is denoted by φ, the pitch angle by ϑ and the yaw angle by ψ.

Now that the rotation axes have been defined, the body reference frame (xyz) of the vehicle is chosen as a dextral, time-varying coordinate system positioned at the centre of gravity (COG) of the vehicle. The body reference frame has its x axis being the roll axis of the

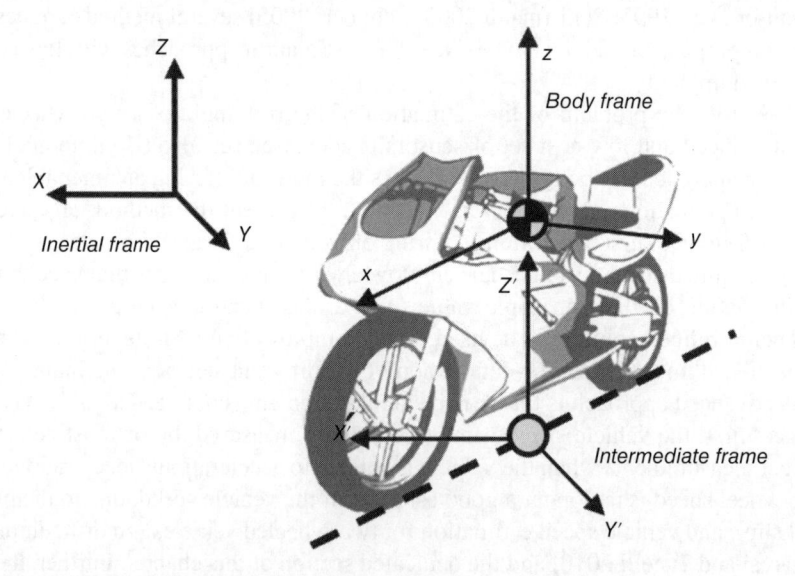

Figure 13.1 Definition of the measurement axes, adapted from (Boniolo et al. 2012). Reproduced with permission from IEEE

vehicle and the z axis being the yaw one. The orientation of the body reference frame can thus be described with respect to the inertial reference frame (XYZ), that is a dextral, fixed and time-invariant coordinate system. In this case, the attitude angles of the vehicle are defined as *absolute* Euler attitude angles φ, ϑ and ψ (or simply Euler attitude angles). The rotation between the two frames (XYZ) and (xyz) is described by the following rotation matrix

$$R_{ZXY}(\varphi, \vartheta, \psi) = R_Y(\vartheta)R_X(\varphi)R_Z(\psi) =$$
$$\begin{bmatrix} c_\vartheta c_\psi - s_\varphi s_\vartheta s_\psi & c_\vartheta s_\psi + s_\varphi s_\vartheta c_\psi & -c_\varphi s_\vartheta \\ -c_\varphi s_\psi & c_\varphi c_\psi & s_\varphi \\ s_\vartheta c_\psi + s_\varphi c_\vartheta s_\psi & s_\vartheta s_\psi - s_\varphi c_\vartheta c_\psi & c_\varphi c_\vartheta \end{bmatrix}, \quad (13.1)$$

where $R_X(\varphi)$, $R_Y(\vartheta)$ and $R_Z(\psi)$ represent the elementary rotations around the X axis of an angle φ, around the Y axis of an angle ϑ and around the Z axis of an angle ψ, and s_ε and c_ε stand for $\sin_\varepsilon$ and $\cos_\varepsilon$, respectively. Further, the models of the angular rates measurements expressed in the body frame are given by (Boniolo and Savaresi 2010, Shuster 1993)

$$\tilde{\omega}(t) = \omega(t) + \Delta_\omega(t) + \eta_\omega(t), \quad (13.2)$$

where

$$\omega(t) = \begin{bmatrix} \omega_x \\ \omega_y \\ \omega_z \end{bmatrix} = \begin{bmatrix} c_\vartheta \dot{\varphi} - s_\vartheta c_\varphi \dot{\psi} \\ \dot{\vartheta} + s_\varphi \dot{\psi} \\ s_\vartheta \dot{\varphi} + c_\varphi c_\vartheta \dot{\psi} \end{bmatrix}, \quad (13.3)$$

while the accelerations can be written as

$$\tilde{a}(t) = a(t) + \Delta_a(t) + \eta_a(t), \quad (13.4)$$

where

$$a(t) = \begin{bmatrix} a_x \\ a_y \\ a_z \end{bmatrix} = \begin{bmatrix} -c_\varphi s_\vartheta(\dot{V}_z + g) + c_\vartheta \dot{V}_x + s_\varphi s_\vartheta(\dot{\psi} + V_x + \dot{V}_y) \\ s_\varphi(\dot{V}_z + g) + c_\varphi(\dot{\psi} + V_x + \dot{V}_y) \\ c_\varphi c_\vartheta(\dot{V}_z + g) + s_\vartheta \dot{V}_x - s_\varphi c_\vartheta(\dot{\psi} + V_x + \dot{V}_y) \end{bmatrix} \quad (13.5)$$

In (13.5), $\Delta_\omega(t)$ is such that $\dot{\Delta}_\omega(t) = \eta_{\Delta_\omega}(t)$ and represents the model of the offset of the gyroscopes, $\Delta_a(t)$ is such that $\dot{\Delta}_a(t) = \eta_{\Delta_a}(t)$ and is the model of the offset of the accelerometers, and $\eta_\omega(t)$, $\eta_{\Delta_\omega}(t)$, $\eta_a(t)$ and η_{Δ_a} are zero-mean Gaussian white noises with variance σ_ω, σ_{Δ_ω}, σ_a and σ_{Δ_a}, respectively. Further, $\eta_\omega(t)$ and $\eta_{\Delta_\omega}(t)$ are assumed to be independent, and the same holds for $\eta_a(t)$ and η_{Δ_a}. Finally, V_x, V_y and V_z are the longitudinal, lateral and vertical velocities of the vehicle, while g is the gravitational acceleration.

It is worth mentioning that the vehicle attitude can be also defined with reference to the road reference frame, which is a dextral, fixed coordinate system, the vertical axis of which is always perpendicular to the road plane, and the yaw rotation with respect to the inertial reference frame is null. In this case, the attitude angles are referred to as *road* attitude angles and in what follows they will be indicated with the subscript r. A difference between the Euler and road attitude angles arises in the presence of slopes and banks on the road, that act

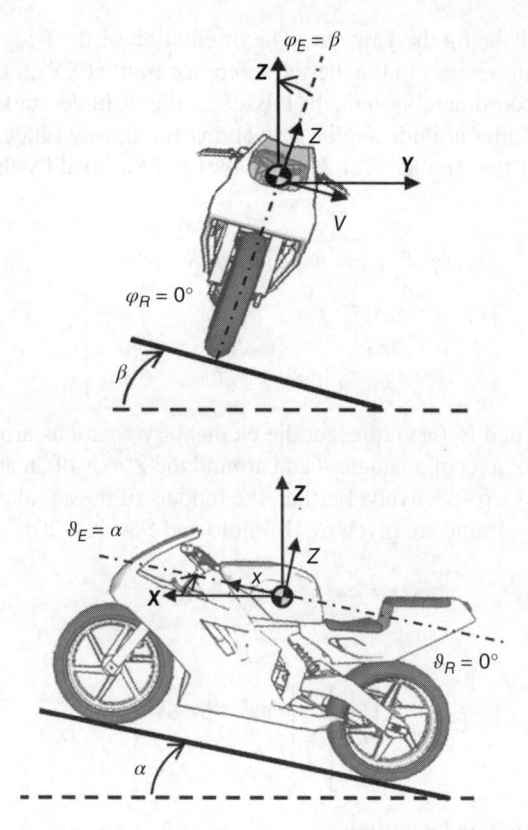

Figure 13.2 Schematic view of the attitude angles, adapted from (Boniolo et al. 2012). Reproduced with permission from IEEE

on the inertial measurements as pitch and roll dynamics, respectively (Figure 13.2). Denoting the road slope by α and the bank angle by β, the inertial and road attitude angles are related by

$$\varphi = \varphi_r + \beta, \quad \vartheta = \vartheta_r + \alpha, \quad \psi = \psi_r. \tag{13.6}$$

From a practical viewpoint, the principal drawback of the inertial measurements provided by a strap-down Inertial Measurement Unit (IMU) is that they provide information on the Euler attitude angles, but it is not possible to gather separate knowledge of the road inclination and of the road attitude angles. Thus, all the methods based on inertial measurements are affected by errors introduced by the road inclination.

13.2.2 Experimental Set-up

The results to be presented were obtained from data measured on an instrumented sports motorcycle, namely an Aprilia Tuono1000 Factory (Figure 13.3).

On the test vehicle, the following set of sensors was available: an IMU composed of three single-axis silicon sensing micro electro-mechanical systems (MEMS) gyroscopes

(a) (b) (c) (d)

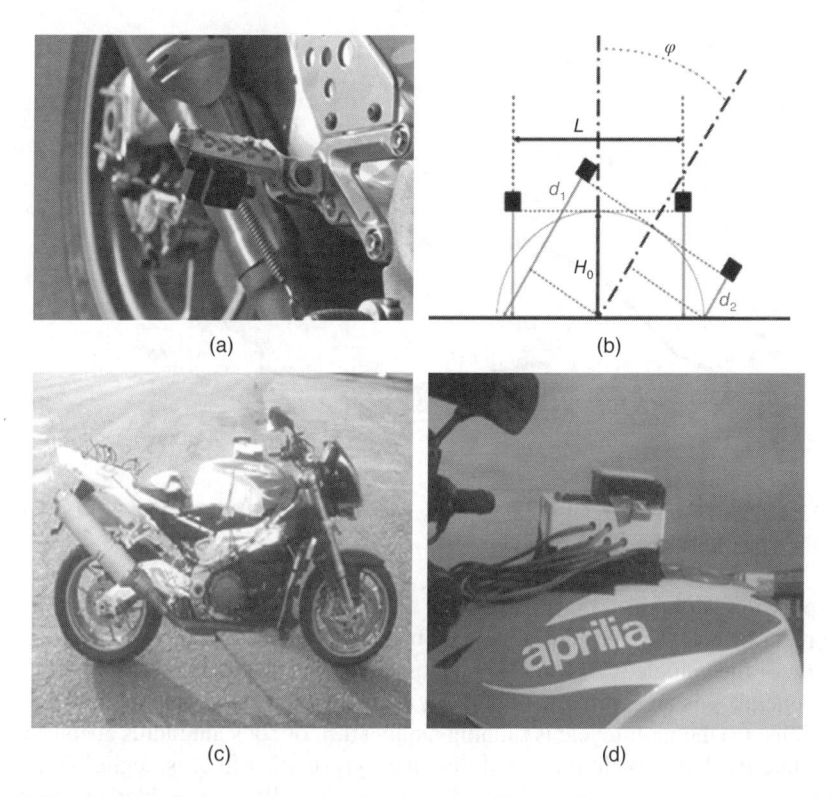

Figure 13.3 Close-up of the electro-optical triangulator (a) and schematic view of the triangulation principle used to measure the road roll angle with electro-optical sensors (b); picture of the test vehicle (c) and close-up of the inertial measuring unit (d), adapted from (Boniolo et al. 2012). Reproduced with permission from IEEE

(CRS-07), a three-axis ST-Microelectronics MEMS accelerometer (LIS3L02AS4) and two Hall-effect wheel encoders with 48 teeth to measure the front and rear wheel rotational speed. For comparison purposes, the real value of the lean angle was measured using electro-optical techniques as (Figure 13.3) $\varphi_r = \arctan(d_1 - d_2/L)$, where d_1 and d_2 are the distances measured on the left side and on the right side of the vehicle, respectively, and L is the mounting distance between them; see (Norgia et al. 2009) for more details.

The signals were logged with a sampling frequency of 100 Hz and filtered with a second-order low-pass filter having a cut-off frequency of 7 Hz. The experimental tests were carried out on the Enzo Ferrari circuit at Imola, Italy (Figure 13.4).

13.2.3 Accelerometer-Based Roll Angle Estimation

To start the presentation of the proposed estimation approach, let us recall that in steady-state turning conditions the roll angle φ can be expressed as (Cossalter 2002)

$$\varphi = \arctan\left(\frac{\dot{\psi} V_x}{g}\right). \tag{13.7}$$

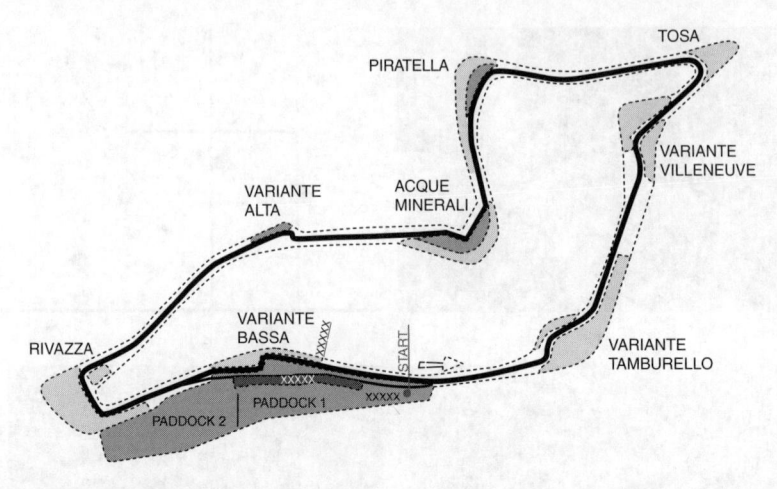

Figure 13.4 Enzo Ferrari circuit in Imola, Italy, adapted from (Boniolo et al. 2012). Reproduced with permission from IEEE

This expression is obtained considering that a motorcycle in turning conditions is subject to a moment balance at the point of contact between the tyre and the road that defines the lean angle of the vehicle (Cossalter 2002; Cossalter et al. 1999; Tanelli et al. 2009a). Note that, in general, φ is an approximation of φ_r, as the two are equal only under the following assumptions: (a) the motorcycle is running along a turn of constant radius at constant speed (steady-state working conditions), and thus the gyroscopic effect is negligible (Cossalter 2002); (b) the track is plane; (c) the tyre thickness is null; (d) the rider's COG is in the principal symmetry plane of the motorcycle.

Further, note that, in steady-state cornering, the lateral and vertical accelerations given in (13.5) reduce to

$$a_y \simeq s_\varphi g + c_\varphi \dot{\psi} V_x \tag{13.8}$$

$$a_z \simeq c_\varphi g - s_\varphi \dot{\psi} V_x. \tag{13.9}$$

In turn, this implies

$$a_y^2 + a_z^2 \simeq g^2 + \dot{\psi}^2 V_x^2, \tag{13.10}$$

which yields

$$|\dot{\hat{\psi}}| = \sqrt{\frac{(\tilde{a}_y^2 + \tilde{a}_z^2) - g^2}{V_x^2}} \arctan\left(\frac{\dot{\psi} V_x}{g}\right). \tag{13.11}$$

Substituting (13.11) into (13.7), the absolute value of the roll angle φ can be obtained as

$$|\hat{\varphi}| = \arctan\left(\frac{(\tilde{a}_y^2 + \tilde{a}_z^2) - g^2}{g}\right). \tag{13.12}$$

The main advantage of the estimate given in (13.12) is that it does not require the speed measurement, and it is obtained using only two single-axis accelerometers.

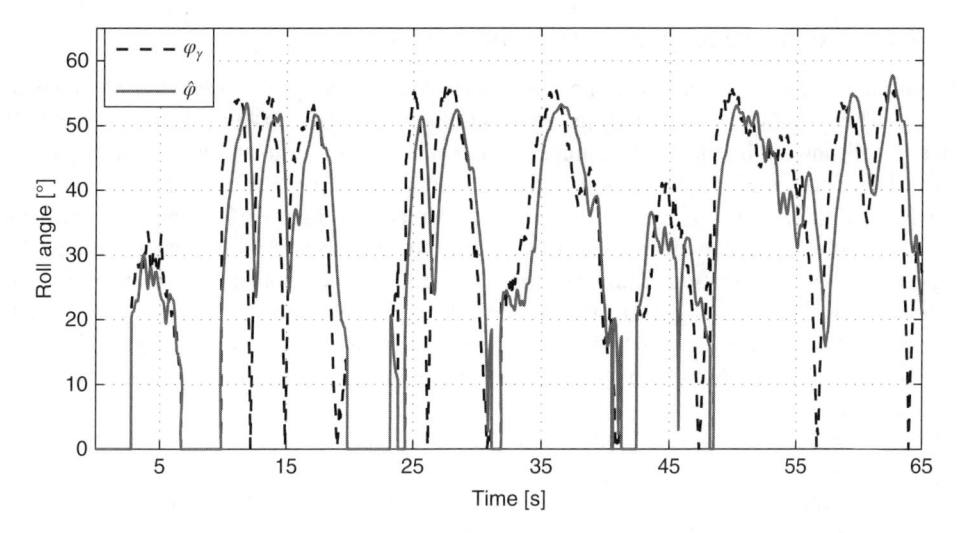

Figure 13.5 Detail of the estimation results obtained with the estimate in (13.12) in the experimental tests on the Imola circuit, adapted from (Boniolo et al. 2012). Reproduced with permission from IEEE

Figure 13.5 reports a detail of the estimation results obtained by means of Equation (13.12) in an experimental test on the Imola circuit. These tests disclosed the two main limitations related to the estimate based on (13.12):

- The proposed algorithm is based on the assumption of steady-state conditions, and as such the accuracy of the roll angle estimate lowers during transients.
- Typically, the acceleration signals measured on a motorcycle are affected by significant noise due to the chassis vibrations. This makes it necessary to employ a low-pass filter before using the signals for the estimation. As a consequence, the roll angle can be reconstructed with the desired accuracy only for values of φ_r larger than 20° and for dynamic variations only up to 0.5 Hz.

Under these constraints, the error-to-signal ratio* of the estimate obtained in the experimental tests is 9% (consider that the ESR of the estimation method proposed in (Boniolo et al. 2009b), which used four gyros and the speed signal, was 8%). This performance can be regarded as appropriate for use with traction and braking control systems, which in general need a threshold-based indication of the roll angle value (an estimation error up to 5°, compatible with the obtained performance, can be easily tolerated, as in general it translates into a 1.5–2% variation of the wheel slip, that is within the natural oscillations exhibited by the measured signals; see (Tanelli et al. 2009a)). Note, further, that traction and braking controllers also need an estimate of the magnitude of the roll angle, based on which the tyre–road forces vary, while the sign is not mandatory. The same holds true for roll-over detection systems. However, it is interesting to investigate possible improvements to the estimation obtained with (13.12) that may overcome the aforementioned limitations.

---

* The error-to-signal ratio is defined as the ratio between the mean square error of the estimation error $\varphi_r - \hat{\varphi}$ and the mean square error of the reference signal, in this case φ_r. This cost function gives an indication of the effect of the estimation error on the overall signal reconstruction performance.

13.2.4 Use of the frequency separation principle

To provide a more accurate roll angle estimate, one can adopt the approach depicted in Figure 13.6, which was first analysed in (Boniolo et al. 2009a, 2009b). It is based on the idea of splitting the input inertial signals and the vehicle speed into an high frequency (subscript HF) and a low frequency (subscript LF) component, using a linear filter. This filter is based on the *frequency separation principle* depicted in Figure 13.6: the filter extracts the low frequency information about the variable to be estimated from one signal and the high-frequency information from the other. These signal components are then processed independently, after having been split by the frequency separation block and then, at each sampling instant, the LF estimate $\hat{\varphi}_{LF}$ and the HF estimate $\hat{\varphi}_{HF}$ are added to build the final roll angle estimate $\hat{\varphi}$. In this way, one expects to obtain a more accurate estimation over all the frequency domain.

It should be noticed that, as underlined also in (Boniolo and Savaresi 2010), to extract high and low frequency components without losing information, the two filters ought to be *complementary* in the frequency domain sense: $HP_{filter}(j\omega) + LP_{filter}(j\omega) = 1$, $\forall \omega$. This ensures that no loss of information occurs at any frequency.

As can be seen in Figure 13.7, the estimate obtained from (13.12) can be used to reconstruct the absolute value of the roll angle $\hat{\varphi}_{LF}$, while its sign can be retrieved using the measurement ω_z obtained from a vertical gyroscope (note that if the sign of the roll angle is not relevant for the considered application, this gyroscope is not needed). The HF component $\hat{\varphi}_{HF}$ is then obtained by integrating the HF component of the body-fixed roll rate ω_x (note that the HF components does not contain the bias term and thus does not pose a drift problem during integration). The frequency separation block is made of a standard first-order high pass (HP) filter with cut-off frequency f_{sp} and of a summing junction that splits the signal into the LF and HF components. Only the value of the parameter f_{sp} must be tuned, and its value has been determined to minimize the ESR of the final estimation error, yielding an optimal value of $f_{sp} = 0.2$ Hz.

To evaluate the performance of the overall approach, Figure 13.8 shows the roll angle estimate: the ESR of the estimation amounts to 7.5%, thus yelding approximately a 20% improvement while maintaining a low-cost sensor configuration.

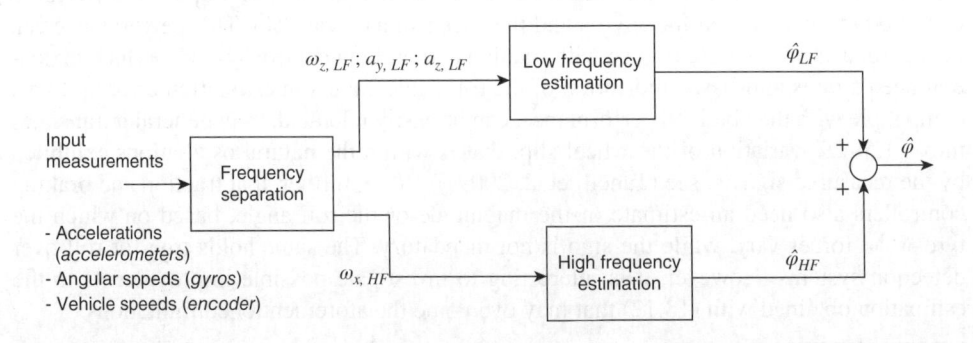

Figure 13.6 High-level architectural view of the proposed estimation algorithm, adapted from (Boniolo et al. 2012). Reproduced with permission from IEEE

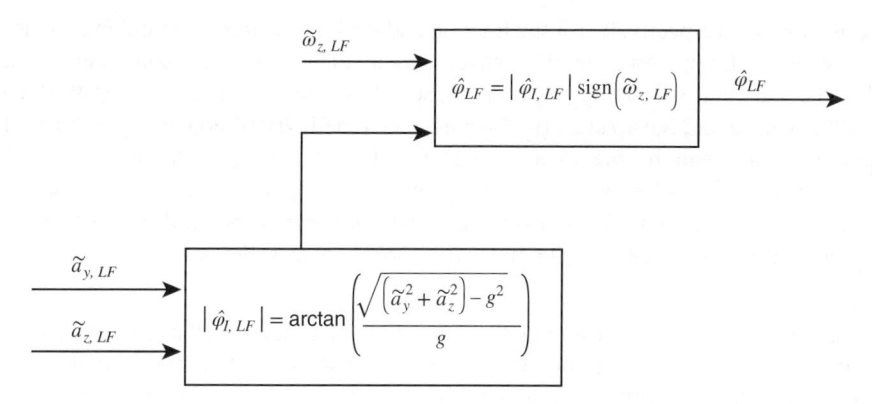

Figure 13.7 Estimation of the LF component of the roll angle from lateral and vertical acceleration measurements, adapted from (Boniolo et al. 2012). Reproduced with permission from IEEE

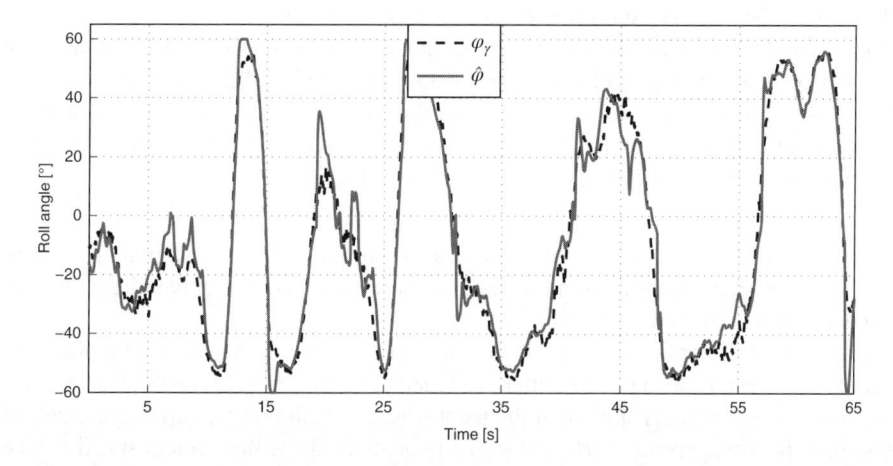

Figure 13.8 Detail of the results obtained with the method shown in Figure 13.6 in the experimental tests on the Imola circuit, adapted from (Boniolo et al. 2012). Reproduced with permission from IEEE

13.3 Vehicle Speed Estimation

If we consider the safety-oriented control of longitudinal vehicle dynamics – braking and traction control – the application of the *wheel slip control* paradigm is widespread (Corno et al. 2009; Panzani et al. 2013; Savaresi and Tanelli 2010; Tanelli et al. 2009a, 2009b): the torque applied to the wheel is regulated in order to prevent excessive slipping of the wheel(s) (Johansen et al. 2003; Petersen et al. 2003; Savaresi et al. 2007). According to its definition, the wheel slip (usually referred to as λ) is given by

$$\lambda = \frac{\omega R - v}{\omega R} \text{ or } \lambda = \frac{v - \omega R}{v}, \quad (13.13)$$

where R is the wheel radius, ω the wheel angular speed and v the speed of the wheel centre (that is, from another perspective, the vehicle longitudinal speed). The different definitions above are applied, respectively, for the traction and the braking control problem, and lead to a positive wheel slip in both cases. The wheel slip cannot be directly measured; in fact, while for the wheel angular speed ω it has been already shown in (Bascetta et al. 2009; Bélanger et al. 1998; Corno and Savaresi 2010; Savaresi and Tanelli 2010) how it is possible to have accurate measurements by means of a shaft encoder, an accurate estimate of the vehicle speed v is very difficult to obtain; see (Jiang and Gao 2000; Ryu et al. 2002; Tanelli et al. 2008, 2009b), also on four-wheeled vehicles. There are several possibilities to directly and precisely measure this quantity. The most widespread are as follows:

- Using optical sensors: they estimate the vehicle speed (longitudinal and lateral), applying a correlation analysis between two subsequent photographs (Ator 1966). Although they provide an accurate speed measurement, these systems are expensive. Moreover, they are quite sensitive to weather, surface and light conditions.
- Using a GPS system (Ryu et al. 2002): in this case several drawbacks, such as the loss of accuracy in speed estimation, for example in urban paths or in tunnels, or the high cost of high accuracy/sampling GPS devices can be pointed out.
- Model-based velocity estimation: the existing methods are based on filtering techniques and sensor fusion (Jiang and Gao 2000; Kobayashi et al. 1995; Semmler et al. 2002). They have been shown to be rather successful in four-wheeled vehicles. The difficulties of obtaining a control-oriented model for motorcycles has not, up to this point, allowed the extensions of these methods to two-wheeled vehicles.

The aim of this section is to present innovative solutions for the estimation of vehicle speed in traction control applications–based on the wheel slip control paradigm–by means of standard vehicle low-cost sensors.

In traction control applications, the front wheel speed $v_f = \omega_f R_f$ has been typically used to estimate the vehicle speed v. As a matter of fact, as all motorcycles are rear wheel driven, during traction no torque is applied at the front wheels and the wheel slip is thus negligible. So the forward wheel speed $\omega_f R_f$ provides a good estimate of the vehicle speed v. Hence, applying the wheel slip definition (13.13), the rear wheel slip can be expressed as

$$\lambda_r = \frac{\omega_r R_r - \omega_r R_r}{\omega_r R_r}. \tag{13.14}$$

Equation (13.14) is usually referred to as *relative slip*, see (Tanelli et al., 2009). In the scientific literature, this solution is proposed to estimate the longitudinal slip during traction. However, in the last few years the standard motorcycle equipment has been experiencing an important enrichment, following the path traced by four-wheeled vehicles. This is due, on the one hand, to the decreasing cost of standard electronic devices, and, on the other hand, to the increased awareness that more effective safety systems require more resources and dedicated equipment. An inertial measurement unit (IMU), which measures acceleration and angular rates, is now typically present on the newest generation of motorcycles. In Section 13.3.1, we will show how the basic estimate $v = \omega_f R_f$ can be considerably improved by exploiting the longitudinal accelerometer signal a_x.

13.3.1 Speed Estimation During Traction Manoeuvres

In this section, the traction control oriented vehicle speed estimation is discussed. To this end, a sensor fusion approach is undertaken: the wheel velocity and longitudinal accelerometer measurements are fused to overcome their respective limitations. Particular attention is devoted to obtaining reliable, easily implementable and cost-effective techniques that rely on standard electronics and sensors. All the proposed methods are introduced and discussed, before being experimentally validated on an instrumented sports motorcycle.

13.3.2 Experimental Setup

The proposed approach well suites a high-end sport motorcycle. The experimental vehicle used to validate the vehicle speed estimation algorithm is equipped with the following sensors:

- Front and rear wheel encoders. The motorcycle is equipped with two Hall-effect encoders with 48 teeth; the angular wheel velocity is estimated using the $1/\Delta T$ method, reported for example in (Corno and Savaresi 2010). It is worth noticing that the resulting speed estimate is affected by considerable disturbances at high frequency. As shown in (Panzani et al. 2012) it is not uncommon to measure a periodic high frequency noise related to the wheel rolling frequency and its harmonics, caused by an eccentricity of the Hall sensor. As a result, this algorithm provides an accurate estimation of the wheel velocity at low frequency.
- A single-axis MEMS longitudinal accelerometer. Typically, acceleration measurements are affected by low frequency noise and drift. The most important source of this noise is the effect of gravity on the acceleration measurement due to a non-perfect horizontal alignment of the measurement axis.
- An optical sensor that provides an accurate measurement of the vehicle velocity (with an error lower than 0.1 km/h); this device is, however, very expensive and not available for production vehicles; in this context, it has been employed as a *reference* to validate the proposed method.

13.3.3 Vehicle Speed Estimation via Kalman Filtering and Frequency Split

Wheel encoders and longitudinal acceleration measurements show somewhat complementary characteristics from a frequency domain perspective. The wheel velocity measurement is affected by high frequency noise; conversely, the accelerometer is affected by low frequency noise and it carries most of the valuable information in the high frequency range. The low frequency noise actually makes the open loop integration of the acceleration signal impossible (this would cause a rapid drift in the estimated speed). This situation is rather common in kinematic estimation problems (Boniolo and Savaresi 2010a; Euston et al. 2008; Pascoal et al. 2000), and was discussed in Section 13.2 also for the roll angle estimation problem. Similarly to what was done in Section 13.2 to refine the estimate, a *complementary filter* will be used to solve this problem.

Applied to vehicle speed estimation, the estimated speed is the sum of high and low frequency components (respectively v_{hf} and v_{lf}): the high-frequency component is obtained

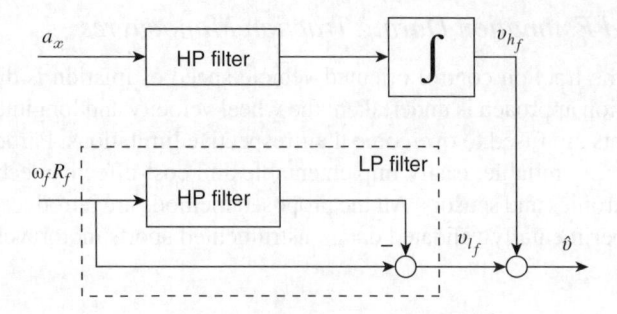

Figure 13.9 Block diagram representation of the frequency separation principle

by the integration of a high-pass filtered longitudinal acceleration a_x. The low-frequency component comes from the low-pass filtered front wheel speed $v_f = \omega_f R_f$.

According to the proposed approach (Section 13.2.4), the designer has two degrees of freedom: the order of the high-pass filter, and its cut-off frequency. Most of the times, a first-order high-pass filter is used and the cut-off frequency is experimentally tuned, giving

$$HP_{filter}(s) = \frac{s}{s+\tau} \qquad LP_{filter}(s) = 1 - HP_{filter} = \frac{\tau}{s+\tau}, \qquad (13.15)$$

with τ being the location of the pole that has to be tuned. This approach is very intuitive, easy and efficient to implement, but there is no guarantee of optimality, and tuning is essentially a trial and error procedure.

In the following, we aim to prove that the *frequency separation principle* can be recast into the wider class of Kalman filter estimators. Such an interpretation is advantageous for several reasons.

- A more rigorous interpretation is given to the *frequency separation principle*, for which the original idea stemmed from heuristic and intuitive reasoning.
- The Kalman filter approach sets the order of the filter to be used in the frequency separation scheme, thus fixing one of the degrees of freedom in the design of the vehicle speed estimator.
- The choice of the filter cut-off frequency is no longer dependent on the experimental tuning but depends, at least in principle, on the characteristic of the noise that is accounted for by the Kalman filter approach.

To properly set the estimation problem within the Kalman filtering theory, a dynamic model of the system is needed. To this end, the following equations are introduced

$$\dot{v} = b + a_x + \eta_1$$
$$\dot{b} = \eta_2$$
$$v_f = v + \varepsilon_1. \qquad (13.16)$$

The first equation represents the relationship between vehicle speed (v) and acceleration (a_x) (here considered as an input): in an ideal set, the vehicle speed derivative would be equal to the measured acceleration. A constant term b and a white noise disturbance η_1 are added to account for unmodelled dynamic and noise effects (such as mounting offset, pitch dynamics and sensor drift). The offset dynamic is described as a Brownian motion driven by η_2. The output of the system is the front wheel speed, affected by a Gaussian measurement noise modelled with the term ε_1. The following noise characteristics are assumed:

$$\eta_1 \sim \mathcal{N}(0, q_1), \quad \eta_2 \sim \mathcal{N}(0, q_2), \quad \varepsilon_1 \sim \mathcal{N}(0, r_1)$$

The dynamic system can be written in the compact matrix form

$$\dot{x} = A\,x + B\,u + w$$
$$y = C\,x + D\,u + \varepsilon, \tag{13.17}$$

where

$$x = \begin{bmatrix} v & o \end{bmatrix}^T \quad u = a_x \quad y = v_f$$

$$A = \begin{bmatrix} 0 & -1 \\ 0 & 0 \end{bmatrix} \quad B = \begin{bmatrix} 1 & 0 \end{bmatrix}^T \quad C = \begin{bmatrix} 1 & 0 \end{bmatrix} \quad D = [0]$$

$$w \sim \mathcal{N}(0, Q) \quad \varepsilon \sim \mathcal{N}(0, R) \quad Q = \begin{bmatrix} q_1 & 0 \\ 0 & q_2 \end{bmatrix} \quad R = [r_1].$$

The steady-state Kalman filter is

$$\dot{\hat{x}} = A\hat{x} + B\,u + K(y - C\,\hat{x}), \tag{13.18}$$

where K is the Kalman gain, that is computed according to

$$K = P\,C^T\,R^{-1}. \tag{13.19}$$

Since the steady-state filter is here considered, P is the symmetric positive semi-definite covariance matrix that satisfies the associated algebraic Riccati equation (ARE), which can be solved symbolically. Let us define

$$P = \begin{bmatrix} \alpha & \beta \\ \beta & \gamma \end{bmatrix}, P > 0,$$

where α, β and γ are unknown constants. The ARE equation results in

$$0 = -\frac{1}{r_1} \begin{bmatrix} \alpha^2 & \alpha\beta \\ \alpha\beta & \beta^2 \end{bmatrix} + \begin{bmatrix} -2\beta & -\gamma \\ -\gamma & 0 \end{bmatrix} + \begin{bmatrix} q_1 & 0 \\ 0 & q_2 \end{bmatrix}, \tag{13.20}$$

the solution of which yields

$$\alpha = \sqrt{q_1\,r_1 + 2r_1\sqrt{r_1\,q_2}}$$
$$\beta = -\sqrt{r_1\,q_2} \tag{13.21}$$
$$\gamma = -\frac{\alpha\beta}{r_1}.$$

The final expression of the Kalman filter will be

$$\dot{\hat{x}} = (A - KC)\hat{x} + B\,u + K\,y.$$

The filter outputs are the estimate of the vehicle speed and the sensor time-varying offset, while its two inputs are the longitudinal acceleration (through vector B) and the front wheel velocity (through vector K), i.e.,

$$K = \frac{1}{r_1}\begin{bmatrix}\alpha\\\beta\end{bmatrix},$$

where α and β depend only on the white noise variance. The complete filter can be written as

$$\dot{\hat{x}} = (A - KC)\hat{x} + B\,a_x + Kv_f = \tilde{A}\hat{x} + B\,a_x + Kv_f =$$

$$= \begin{bmatrix}-\dfrac{\alpha}{r_1} & -1\\ -\dfrac{\beta}{r_1} & 0\end{bmatrix}\hat{x} + \begin{bmatrix}1\\0\end{bmatrix}a_x + \begin{bmatrix}\dfrac{\alpha}{r_1}\\ \dfrac{\beta}{r_1}\end{bmatrix}. \tag{13.22}$$

The state-space expression of the Kalman filter (13.22) can be easily recast into an I/O representation, where the two inputs are the front wheel speed and the vehicle acceleration, and the single output of interest is the vehicle speed. In particular, the relationship between the longitudinal acceleration and the vehicle speed estimate is given by

$$\hat{V}_{ax}(s) = \frac{s}{s^2 + \dfrac{\alpha}{r_1}s - \dfrac{\beta}{r_1}}\,A_x(s), \tag{13.23}$$

While that between front wheel speed and vehicle speed estimate is of the form

$$\hat{V}_{vf}(s) = \frac{\dfrac{\alpha}{r_1}s - \dfrac{\beta}{r_1}}{s^2 + \dfrac{\alpha}{r_1}s - \dfrac{\beta}{r_1}}\,V_f(s). \tag{13.24}$$

Analysing expressions (13.23) and (13.24), it can be seen that the overall Kalman filter estimate can be represented with the block diagram shown in Figure 13.9 with

$$HP_{filter}(s) = s\hat{V}_{ax}(s) = \frac{s^2}{s^2 + \dfrac{\alpha}{r_1}s - \dfrac{\beta}{r_1}},\quad LP_{filter}(s) = \hat{V}_{vf}(s) = \frac{\dfrac{\alpha}{r_1}s - \dfrac{\beta}{r_1}}{s^2 + \dfrac{\alpha}{r_1}s - \dfrac{\beta}{r_1}}. \tag{13.25}$$

From the above expressions a complementary second-order filter is easily recognized (notice that $HP_{filter}(j\omega) + LP_{filter}(j\omega) = 1, \forall \omega$). It has thus been shown that the *frequency separation principle* – and its complementary filter application – can be interpreted as a Kalman filter estimation technique. It is interesting to notice that for the complementary filter to be optimal for system (13.16), it has to be implemented as a second-order filter. Furthermore, expression (13.21) provides the optimal tuning for the parameters of the low- and high-pass filters.

13.3.4 Experimental Validation

To test the effectiveness of the proposed estimators, data from an instrumented sports motorcycle has been used. Figure 13.10 compares the estimated velocity with a more traditional low-pass filter (i.e., a low-pass filter using only the front wheel velocity tuned to have the same noise rejection properties), during a sudden motorcycle acceleration (from approximately 78 km/h). This manoeuvre helps to appreciate the effect of the high-frequency correction brought by the acceleration signal: it helps smoothing the front wheel measurement, removing the effect of the noise without introducing phase lag. Conversely, the solution based only on the front wheel introduces a considerable amount of phase lag. The advantages of the proposed data fusion in terms of noise filtering are better appreciated in Figure 13.11, where the Kalman filter estimate and the measured velocity during an acceleration at high speed are depicted, together with the squared estimation error. As can be seen, the proposed sensor fusion method considerably reduces the estimation error without introducing phase lag: almost a 10-fold reduction of the squared estimation error is obtained.

It is also interesting to consider the difference between the optimal second-order complementary filter and the first-order one (Figure 13.12). Since the same cut-off frequency has been used for both filters, both first- and second-order estimates show a similar noise suppression level, but the first-order estimate exhibits a low-frequency estimation error due to the non-correct filtering of the low-frequency bias affecting the accelerometer.

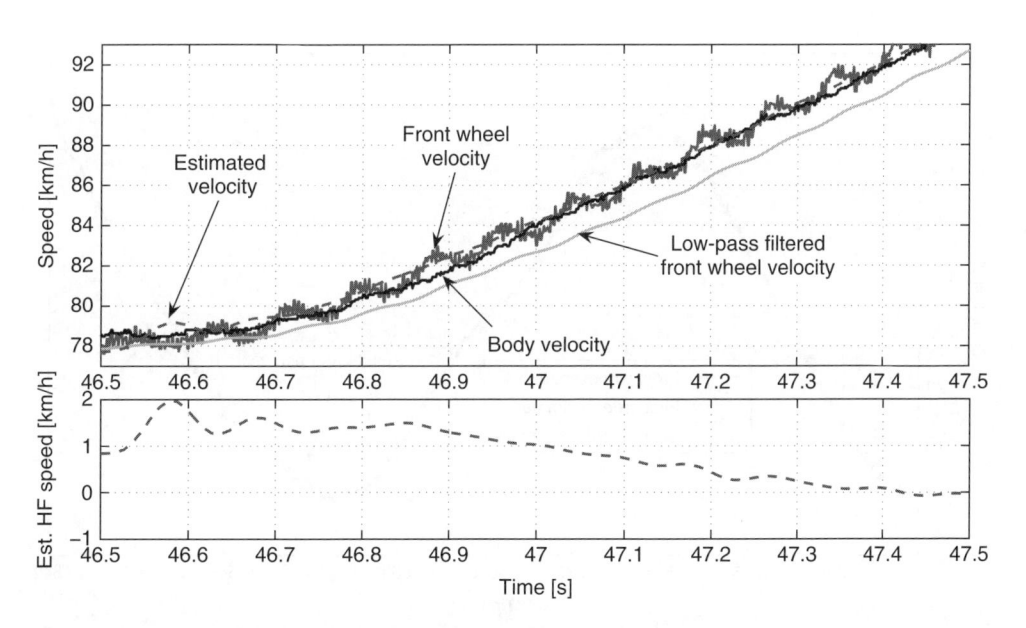

Figure 13.10 Track test comparison between different speed signals (top plot): measured vehicle speed, front wheel speed, low-pass filtered and estimated vehicle speed; high-frequency estimate component constructed from the longitudinal acceleration (bottom plot)

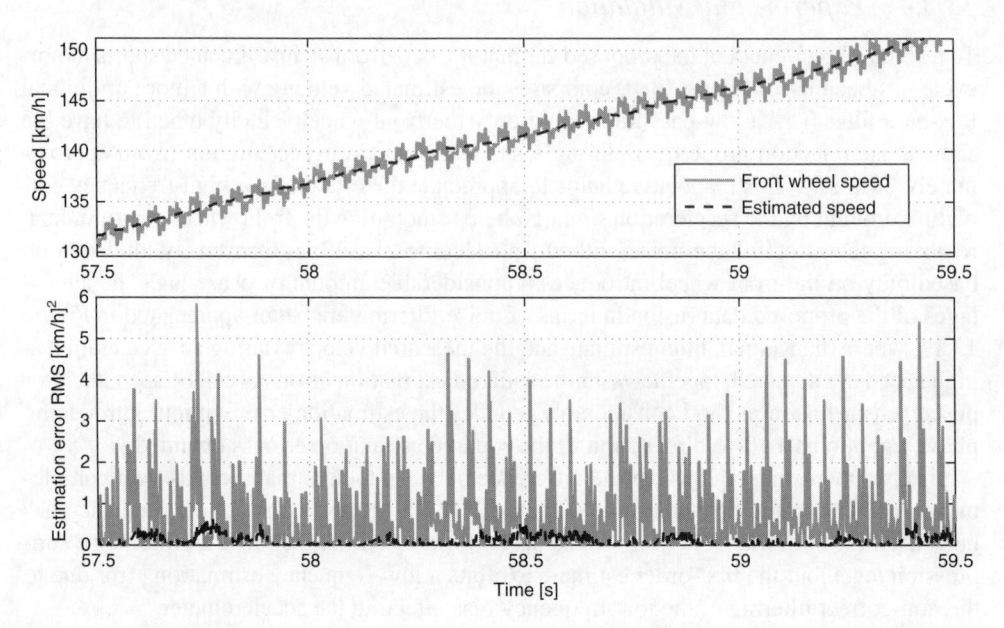

Figure 13.11 Estimated and front wheel speed (top plot) and measured vehicle speed squared estimation error (bottom plot)

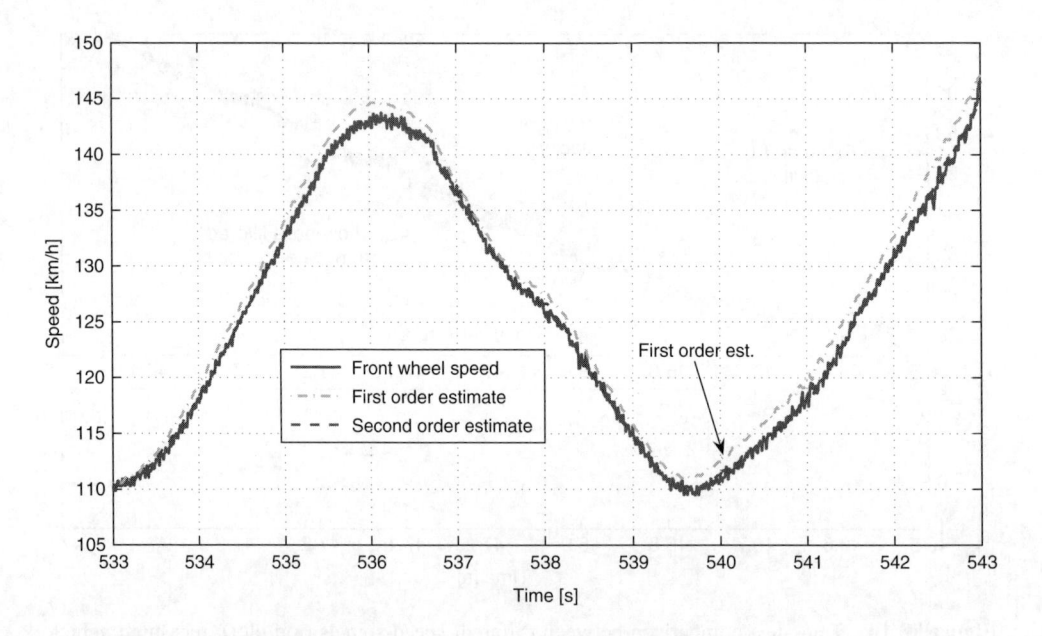

Figure 13.12 Comparison between the first- and second-order vehicle speed estimators

13.4 Suspension Stroke Estimation

In motorcycle applications, the suspension stroke measure represents key information for control systems such as the traction control and semi-active suspension control systems. The knowledge of the stroke suspension is the basis for many strategies of semi-active suspension control existing in the literature (Hrovat 1997; Poussot-Vassal et al. 2012; Savaresi et al. 2010b). Furthermore, the measure of the front suspension elongation helps to detect the *wheelie* phenomenon (Panzani et al. 2013) and thus it is helpful in traction control systems.

One of the most common solution for the stroke measurement is the use of a linear potentiometer placed in parallel to the suspension itself. This kind of solution is able to provide an exact measure of the stroke, but it suffers from a low reliability due to its weakness in intensive usage and to the adverse road and weather conditions. For this reason, the search for alternative solutions has been of interest for both academic and industrial researchers. The research focused mainly on solutions oriented to the stroke speed estimation based on the use of accelerometers and Kalman filtering (Delvecchio et al. 2011; Koch et al. 2010). On the one hand, the use of accelerometers leads to a reliable solution for industrial applications, while on the other hand it suffers from the impossibility of estimating the absolute value of the stroke (unless a specific zeroing system is available). Note that this kind of information would be of interest in the case of end-stop management (Spelta et al. 2011).

This section presents a system to estimate the absolute stroke of a suspension, based on the information of the pressure and temperature of the gas included in the damper compensation chamber. In fact, it can be shown that a suspension elongation is strictly related to the gas physical conditions of the compensation chamber. The estimation procedure, based on the adiabatic transformation assumption, is able to guarantee a level of accuracy adequate for semi-active suspension control. Some experimental results are also presented to demonstrate the effectiveness of the proposed solution.

13.4.1 Problem Statement and Estimation Law

The suspension system on a motorcycle generally includes a hydraulic shock absorber as represented in Figure 13.13. The shock absorber includes two parts: a tube filled with a fluid (usually an oil) and a piston attached to a rod moving up and down in the fluid. The tube is linked to the wheel, and the piston rod is linked to the vehicle chassis. The damping effect is achieved by the viscosity of the fluid flowing through the valves of the piston rod. The variation of the suspension elongation is related with the variation of the volume of the piston rod within the tube. As the fluid is incompressible, this volume variation must be compensated by an air-spring, usually called compensation chamber. Therefore the gas volume, and thus the gas pressure and temperature, may vary due to the piston rod movements. The relationship between suspension elongation and the gas volume in the compensation chamber can be exploited for the estimation of the stroke suspension (this idea was first introduced by Savaresi et al. 2009).

To define the mathematical relationship between the stroke suspension and the internal gas status, consider the following assumptions:

- The gas in the compensation chamber can be viewed as a perfect gas. Note that this kind of assumption is common to describe the air behaviour.

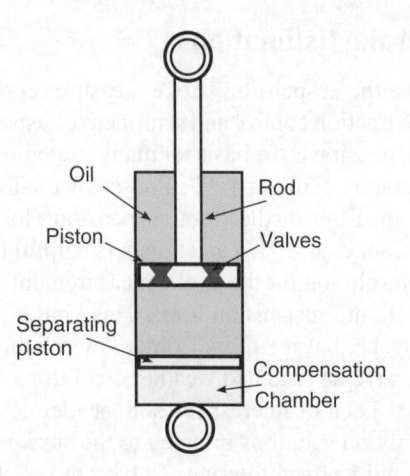

Figure 13.13 Description of a hydraulic damper with pressure/temperature sensor embedded in the compensation chamber

- The air chamber is adiabatically isolated from the surrounding environment. This kind of assumption sounds reasonable as the suspension dynamics (thus the gas dynamics) are much faster than the heat exchange dynamics between the chamber and the environment.
- Although the separating piston can be subjected to some non-uniform deformations, the volume of the air chamber can be assumed as cylindrical.

Under the aforementioned assumptions, the equations describing the relationship between the suspension stroke and the gas pressure are

$$pV = kT \tag{13.26}$$

$$pV^\gamma = p_0 V_0 \tag{13.27}$$

$$V = r^2 \pi (h - az), \tag{13.28}$$

where p, V and T are the pressure, volume and temperature of the gas respectively, and k is a constant gain, depending on the gas mass. The constant γ is the adiabatic constant of the gas (for air, $\gamma = 1.4$), while p_0 and V_0 are the gas pressure and the gas volume, respectively, in a specific nominal condition. Further, r is the diameter of the damper, z is the suspension compression, h is the air chamber height, while the compression of the suspension is null (fully extended shock absorber) and α is the ratio between the stroke compression and the resulting air chamber compression due to the compensation of the piston rod volume. Equations (13.26)–(13.28) can be solved as a function of z, namely

$$z = (h - \alpha z_0)\left(\frac{P_0}{P}\frac{T}{T_0}\right)^{1/\gamma}, \tag{13.29}$$

where z_0 represents the value of the stroke in relation to a known condition (V_0, p_0). Equation (13.29) may represent the stroke estimation based on the pressure and temperature of the compensation gas spring. By inspecting (13.29), the following remarks can be made:

- T and P are the measures coming from the sensors installed on the compensation chamber.
- T_0, p_0 are the gas pressure and the gas temperature measured for a known suspension stroke z_0. In practice, during an end-of-line procedure, the suspension elongation can be mechanically set at the nominal condition z_0, then the gas pressure p_0 and the gas temperature T_0 can be measured and saved. Usually, a simple nominal condition is given by the fully extended damper, namely $z_0 = 0$.
- h and α are internal mechanical parameters of the damper. Note that, since the estimation given by (13.29) is linear with respect to h and α, these can be identified with a regression of the experimental data.

13.4.2 Experimental Results

The experimental tests to assess the performance of the proposed approach were carried out on a hypersport motorcycle. Specifically, the considered experimental set-up is the following (Figure 13.14):

- A combined temperature–pressure sensor was installed in proximity to the compensation chamber of one leg of the front fork (Figure 13.14). The sensor is distributed by Bosch GmbH (Germany) for automotive applications (part number: AZ0261230). Its range is 0–3 bar.
- A linear potentiometer was installed in parallel to the front suspension, in order to have the correct information of the suspension stroke. This sensor is a 150 mm SLS 130 potentiometer by Penny and Giles Ltd (UK).

Figure 13.14 Installation of the pressure/temperature sensor on the front shock absorber in a motorcycle

- The signals of the gas temperature, the gas pressure and of the potentiometer are acquired by a data acquisition system with a 12-bit resolution and 1 kHz sampling frequency by e-Shock SRL (Italy)

The vehicle with the sensor and the data acquisition system was tested during standard driving conditions. An example of the data acquisition is shown in Figure 13.15. Note that, as expected, the temperature variation shows a dynamic much slower than the dynamics of the stroke sensor. The test was repeated several times: part of the dataset was used for the parameter identification of (13.29); part of the dataset was exploited for the validation of the estimation rule. The data was evaluated by the numerical evaluation and optimization of the following cost function

$$J_S = \frac{\sum_{i=0}^{n} (z(i) - \hat{z}(i))^2}{\sum_{i=0}^{n} (z(i))^2},$$

where z and $\hat{z}$ are the measured and estimated stroke, respectively. Note that index J_s stands for the error-to-signal ratio of the stroke elongation estimation. For evaluation purposes, the following cost function is also introduced:

$$J_V = \frac{\sum_{i=0}^{n} (\dot{z}(i) - \hat{\dot{z}}(i))^2}{\sum_{i=0}^{n} (z(i))^2},$$

where $\dot{z}$ and $\hat{\dot{z}}$ are the measured and estimated stroke speed, respectively. Note that, similarly to index J_V, index J_s represents the error-to-signal ratio of the stroke elongation estimation.

Figure 13.16 shows the experimental comparison between the acquired data (from the potentiometer) and the estimated data given by (13.29) fed by the combined pressure–temperature sensor. To improve the validation of the estimation rule, the stroke speed information was also reported. Note that this kind of measure is given by the numerical derivation

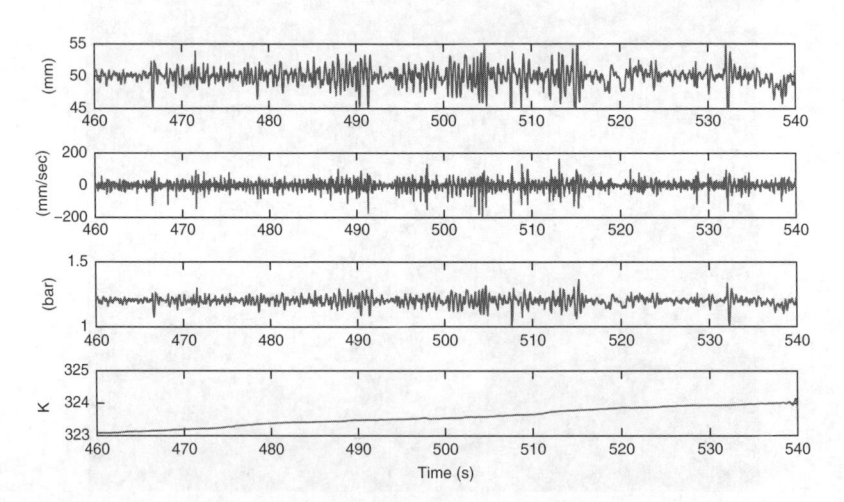

Figure 13.15 Data Acquisition. From top to bottom: stroke displacement; stroke speed; gas pressure; gas temperature

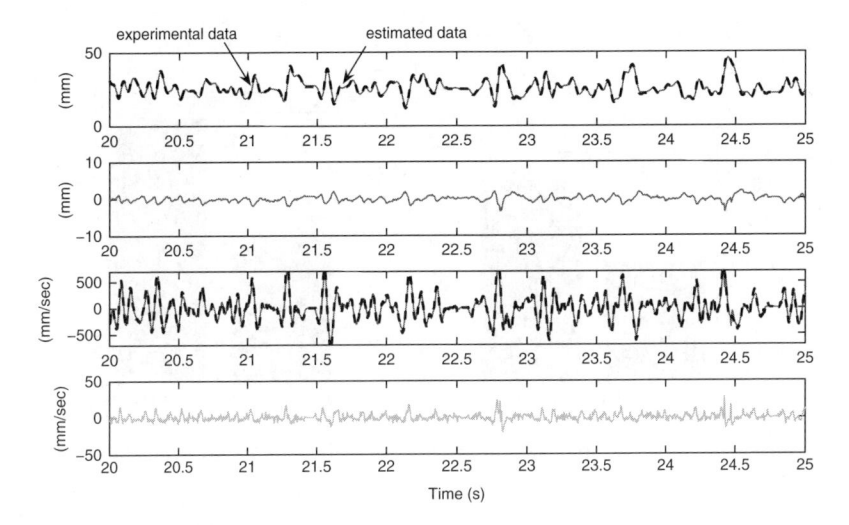

Figure 13.16 Experimental results. Comparison between experimental data and estimated data (dotted line). From top to bottom: stroke elongation; estimation error of the stroke elongation; stroke speed; estimation error of the stroke speed

of the stroke elongation measure, compared with the numerical derivation of the stroke elongation estimation. By inspecting Figure 13.16, some considerations can be made:

- The estimation of the stroke of the shock absorber can be considered as accurate. Both the stroke elongation and the stroke speed are well described by the estimated signals.
- The error of the stroke elongation estimation is within 5 mm; the performance index is $J_s = 3.45\%$. The error of the stroke speed elongation is within 20 mm/sec.
- The accuracy of the estimation suggests the correctness of the adiabatic assumption.

Remark 1 (On the temperature sensor) *As already mentioned, the temperature signal features a dynamic much slower than the pressure signal dynamic (Figure 13.15). For a safe motorcycle usage, the range of the environmental temperature might be in the range $0-40°C$ equal to a range of about $T = 273 - 313K$. By considering a nominal temperature of $20°C$ ($T_0 = 293K$) the ratio T/T_0 can be considered as unitary, thus the temperature signal is negligible. Under this assumption the estimated stroke elongation is given by*

$$z = (h - \alpha z_0)\left(\frac{P_0}{P}\frac{T}{T_0}\right)^{1/\gamma} \approx z = (h - \alpha z_0)\left(\frac{P_0}{P}\right)^{1/\gamma}. \qquad (13.30)$$

To assess the validity of the simplified solution given by (13.30), some on-road tests were carried out. These were performed in several environmental conditions. The estimation performances in terms of indexes J_s and J_v are shown in Figure 13.17, which compares the achievable results with the estimation rule (13.29), and the results achieved by the estimation rule (13.30). As expected, in the case of lack of the temperature information, the estimation performance degrades. This is a clear trade-off between the complexity of the system and

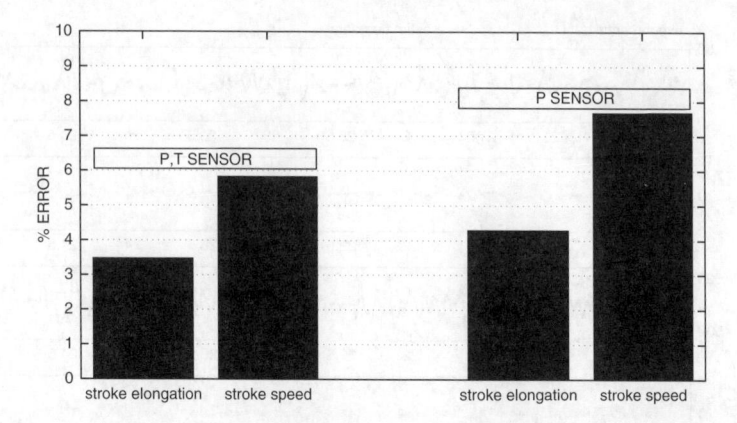

Figure 13.17 Estimation performance in the case of combined pressure-temperature sensor (left) and in the case of pressure sensor only

the achievable results. However, the overall performances suggest that the simple use of a pressure sensor can be an interesting solution for industrial applications.

13.5 Conclusions

This chapter presented three estimation problems that are relevant for control systems design of two-wheeled vehicles. Specifically, an effective approach for the estimation of the roll angle of the vehicle starting from a very limited set of sensors was presented, and its validity assessed on experimental data measured on the Imola circuit. Further, the longitudinal speed estimation was considered, and the problem was solved for traction manoeuvres, again proving the effectiveness of the method on a test vehicle. Finally, the problem of estimating the suspension stroke was addressed, and an innovative solution based on measurements of the pressure and temperature of the gas in the damper compensation chamber was discussed and validated on an instrumented motorcycle.

References

Abe M, Kano Y, Suzuki K, Shibahata Y and Furukawa Y 2001 Side-slip control to stabilize vehicle lateral motion by direct yaw moment. *JSAE Review* **22**(4), 413–419.
Ator J 1966 Image velocity sensing by optical correlation. *Applied Optics* **5**(8), 1325–1331.
Bascetta L, Magnani G and Rocco P 2009 Velocity estimation: Assessing the performance of non-model-based techniques. *Control Systems Technology, IEEE Transactions on* **17**(2), 424–433.
Bélanger P, Dobrovolny P, Helmy A and Zhang X 1998 Estimation of angular velocity and acceleration from shaft-encoder measurements. *The International Journal of Robotics Research* **17**(11), 1225–1233.
Boniolo I and Savaresi S 2010 *Estimate of the Lean Angle of Motorcycles*. VDM Verlag.
Boniolo I, Panzani G, Savaresi S, Scamozzi A and Testa L 2009a On the roll angle estimate via inertial sensors: Analysis of the principal measurement axes *ASME Dynamic Systems and Control Conference, DSCC 2009*.
Boniolo I, Savaresi S and Tanelli M 2009b Roll angle estimation in two-wheeled vehicles. *IET Control Theory & Applications* **3**(1), 20–32.
Boniolo I, Savaresi SM and Tanelli M 2012 Lean angle estimation in two-wheeled vehicles with a reduced sensor configuration *2012 IEEE International Symposium on Circuits and Systems (ISCAS)*, pp. 2573–2576.

Borrelli F, Bemporad A, Fodor M and Hrovat D 2006 An MPC/hybrid system approach to traction control. *IEEE Transactions on Control Systems Technology* **14**(3), 541–552.

Canale M, Fagiano L, Ferrara A and Vecchio C 2009 Comparing Internal Model Control and Sliding-Mode Approaches for Vehicle Yaw Control. *IEEE Transactions on Intelligent Transportation Systems* **10**(1), 31–41.

Canale M, Fagiano L, Milanese M and Borodani P 2007 Robust vehicle yaw control using active differential and IMC techniques. *Control Engineering Practice* **15**(8), 923–941.

Cardinale P, D'Angelo C and Conti M 2009 Traction control systems for motorcycles. *EURASIP Journal of Embedded Systems* **2009**(3), 393–438.

Corno M and Savaresi S 2010 Experimental identification of engine-to-slip dynamics for traction control applications in a sport motorbike. *European Journal of Control* **16**(1), 88–108.

Corno M, Panzani G and Savaresi S 2013 Traction control-oriented state estimation for motorcycles. *IEEETCST* p. To appear.

Corno M, Savaresi S and Balas G 2009 On linear-parameter-varying (lpv) slip-controller design for two-wheeled vehicles. *International Journal of Robust and Nonlinear Control* **19**(12), 1313–1336.

Cossalter V 2002 *Motorcycle Dynamics*. Race Dynamics, Milwaukee, USA.

Cossalter V, Doria A and Lot R 1999 Steady Turning of Two-Wheeled Vehicles. *Vehicle System Dynamics* **31**, 157–181.

Cossalter V, Lot R and Maggio F 2002 The influence of tire properties on the stability of a motorcycle in straight running and curves *SAE Automotive Dynamics & Stability Conference and Exhibition (ADSC)*, Detroit, Michigan, USA, May 7–9.

Cossalter V, Lot R and Maggio F 2004 On the Stability of Motorcycle during Braking *SAE Small Engine Technology Conference & Exhibition*, Graz, Austria, September 2004. SAE Paper number: 2004-32-0018 / 20044305.

De Filippi P and Savaresi S 2011 A mixed frequency/time-domain method to evaluate the performance of semi-active steering damper control strategies during challenging maneuvers *Proceedings of the 11th ASME Dynamic Systems and Control Conference (DSCC)*.

De Filippi P, Tanelli M, Corno M and Savaresi S 2010 Towards electronic stability control for two-wheeled vehicles: a preliminary study *Proceedings of the 10th ASME Dynamic Systems and Control Conference (DSCC)*, Boston, Massachusetts, USA, pp. 133–140.

De Filippi P, Tanelli M, Corno M and Savaresi S 2011a Enhancing active safety of two-wheeled vehicles via electronic stability control *Proceedings of the 18th IFAC World Congress on Automatic Control*, Milan, Italy.

De Filippi P, Tanelli M, Corno M, Savaresi S and Fabbri L 2011b Semi-active steering damper control in two-wheeled vehicles. *IEEE Transactions on Control Systems Technology* **19**(5), 1003–1020.

Delvecchio D, Spelta C and Savaresi S 2011 Estimation of the tire vertical deflection in a motorcycle suspension via Kalman-filtering techniques *Proceedings of the 2011 IEEE Multi-conference on Systems and Control*, Denver, Colorado.

Euston M, Coote P, Mahony R, Kim J and Hamel T 2008 A complementary filter for attitude estimation of a fixed-wing uav *Intelligent Robots and Systems, 2008. IROS 2008. IEEE/RSJ International Conference on*, pp. 340–345 IEEE.

Evangelou S 2010 Control of motorcycles by variable geometry rear suspension *IEEE International Conference on Control Applications (CCA)*, pp. 148–154.

Evangelou S, Limebeer D and Tomas-Rodriguez M 2010 Suppression of burst oscillations in racing motorcycles *49th IEEE Conference on Decision and Control (CDC)*, pp. 5578–5585.

Gasbarro L, Beghi A, Frezza R, Nori F and Spagnol C 2004 Motorcycle trajectory reconstruction by integration of vision and MEMS accelerometers *Proceedings of the 43th Conference on Decision and Control, Nassau, Bahamas*, pp. 779–783.

Gustaffson F, Drevo M and Forssell U 2002 Methods for estimating the roll angle and pitch angle of a two-wheeled vehicle, system and computer program to perform the methods International Patent WO 02/01151.

Hauser B, Ohm F and Roll G 1995 Motorcycle ABS using horizontal and vertical acceleration sensors US Patent 5,445,44.

Hrovat D 1997 Survey of advanced suspension developments and related optimal control application. *Automatica* **33**(10), 1781–1817.

Jiang F and Gao Z 2000 An adaptive nonlinear filter approach to the vehicle velocity estimation for abs *Control Applications, 2000. Proceedings of the 2000 IEEE International Conference on*, pp. 490–495 IEEE.

Johansen T, Petersen I, Kalkkuhl J and Ludemann J 2003 Gain-scheduled wheel slip control in automotive brake systems. *Control Systems Technology, IEEE Transactions on* **11**(6), 799–811.

Kazuhiko T, Yutaka N, Takehiko N, Kazuya T and Makoto T 2010 Control technology of brake-by-wire system for super-sport motorcycles *SAE 2010 World Congress and Exhibition*, Detroit, Michigan, USA, April 13–15.

Kobayashi K, Cheok K and Watanabe K 1995 Estimation of absolute vehicle speed using fuzzy logic rule-based Kalman filter *American Control Conference, 1995. Proceedings of the*, vol. 5, pp. 3086–3090 IEEE.

Koch G, Kloiber T, Pellegrini E and Lohmann B 2010 A nonlinear estimator concept for active suspension control *Proceedings of the 2010 American Control Conference*, Baltimore, USA.

Limebeer DJN, Sharp RS and Evangelou S 2001 The stability of motorcycles under acceleration and braking. *Proc. I. Mech. E., Part C, Journal of Mechanical Engineering Science* **215**, 1095–1109.

Norgia M, Boniolo I, Tanelli M, Savaresi S and Svelto C 2009 Optical sensors for real-time measurement of motorcycle tilt angle. *IEEE Transactions on Instrumentation and Measurement* **58**(5), 1640–1649.

Panzani G, Corno M and Savaresi S 2012 On the periodic noise affecting wheel speed measurement *16th IFAC Symposium on System Identification, 2012*, pp. 1695–1700.

Panzani G, Formentin S and Savaresi S Wheelie detection for single-track vehicles. Proceedings of the 2013 European Control Conference, Zurich, Switzerland, July 17–19, 2013, pp. 956–961.

Pascoal A, Kaminer I and Oliveira P 2000 Navigation system design using time-varying complementary filters. *Aerospace and Electronic Systems, IEEE Transactions on* **36**(4), 1099–1114.

Petersen I, Johansen T, Kalkkuhl J and Ludemann J 2003 Wheel slip control using gain-scheduled lq-lpv/lmi analysis and experimental results *Proceedings of IEE European Control Conference, Cambridge, UK, September 1*, vol. 4.

Poussot-Vassal C, Spelta C, Sename O, Savaresi S and Dugard L 2012 Survey and performance evaluation on some automotive semi-active suspension control methods: A comparative study on a single-corner model. *Annual Reviews in Control* **36**, 148–160.

Ryu J, Rossetter E and Gerdes J 2002 Vehicle sideslip and roll parameter estimation using gps *Proceedings of the International Symposium on Advanced Vehicle Control (AVEC), Hiroshima, Japan*, pp. 373–380.

Savaresi S and Tanelli M 2010 *Active Braking Control Systems Design for Vehicles*. Springer Verlag.

Savaresi S, Corno M, Formentin S and Fabbri L 2010a System and method for controlling traction in a two-wheeled vehicle US Patent 20100312449.

Savaresi S, De Filippi P, Tanelli M, Corno M and Fabbri L 2009 Method for estimating the suspension stroke of a vehicle and apparatus implementing the same US Patent Application 20090024270.

Savaresi S, Poussot-Vassal C, Spelta C, Sename O and Dugard L 2010b *Semi-Active Suspension Control Design for Vehicles*. Butterworth-Heinemann: Oxford.

Savaresi S, Tanelli M and Cantoni C 2007 Mixed slip-deceleration control in automotive braking systems. *Journal of Dynamic Systems, Measurement, and Control* **129**, 20.

Schiffmann J 2003 *Vehicle roll angle estimation and method* European Patent EP1,346,883.

Schubert P 2005 Vehicle rollover sensing using angular accelerometer US Patent 1,139,83.

Semmler S, Fischer D, Isermann R, Schwarz R and Rieth P 2002 Estimation of Vehicle Velocity using Brake-by-Wire Actuators *Proceedings of the 15th IFAC World Congress, Barcelona, Spain*.

Sharp RS, Evangelou S and Limebeer DJN 2004 Advances in the modelling of motorcycle dynamics. *Multibody System Dynamics* **12**, 251–283.

Shuster M 1993 A survey of attitude representations. *The Journal of Astronomical Sciences* **41**(4), 439–517.

Spelta C, Previdi F, Savaresi S, Bolzern P, Cutini M, Bisaglia C and Bertinotti S 2011 Performance analysis of semi-active suspension with control of variable damping and stiffness. *Vehicle System Dynamics* **42**(1–2), 237–256.

Tanelli M, Corno M, Boniolo I and Savaresi S 2009a Active braking control of two-wheeled vehicles on curves. *International Journal of Vehicle Autonomous Systems* **7**(3), 243–269.

Tanelli M, Piroddi L and Savaresi S 2009b Real-time identification of tire-road friction conditions. *Control Theory & Applications, IET* **3**(7), 891–906.

Tanelli M, Prandini M, Codecà F, Moia A and Savaresi S 2008 Analysing the interaction between braking control and speed estimation: the case of two-wheeled vehicles *Proceedings of the 47th IEEE Conference on Decision and Control, Cancun, Mexico*, pp. 5372–5377.

Tanelli M, Vecchio C, Corno M, Ferrara A and Savaresi S 2009c Traction control for ride-by-wire sport motorcycles: a second order sliding mode approach. *IEEE Transactions on Industrial Electronics* **56**(9), 3347–3356.

Tseng H, Xu L and Hrovat D 2007 Estimation of land vehicle roll and pitch angles. *Vehicle System Dynamics* **45**(5), 433–443.

Wakabayashi T and Sakai K 2004 Development of electronically controlled hydraulic rotary steering damper for motorcycles *Proceedings of the 5th International Motorcycle Safety Conference*.

Yi J, Zhang Y and Song D 2009 Autonomous motorcycles for agile maneuvers, Part ii: Control systems design *48th IEEE Conference on Decision and Control and 28th Chinese Control Conference, Shanghai, China*.

Index

Acceleration
　lateral, 90
　longitudinal, 90, 183, 201
Aerodynamic
　forces, 224
Aerodynamics
　drag coefficient, 50
　drag force, 94

Braking
　force, 51, 89

Constraint
　holonomic, 103
　holonomic, non-ideal, 103
　non-holonomic, 125
　unilateral, 145
Control
　admissible, 124
　bang-bang, 127
　constrained, 124
　dynamic inversion, 300
　feedforward action, 211
　Ground-Hook, 253, 280
　Linear-Quadratic-Regulator
　　(LQR), 107
　of autonomous motorcycle, 293
　optimal, 106, 124, 166
　path following, 294
　Proportional-Integral-Derivative (PID),
　　164
　Rotational Ground-Hook, 254

Rotational Sky-Hook, 253
Sky-Hook, 252, 279
underactuation, 293

Driver
　assistance systems, 192
　grip force, 164
　seat contact force, 164
Driving torque
　see also, Traction
　　torque, 203

Electronic throttle, 60
　control, 74
　position, 62, 201
Engine, 88
　ignition spark, 62
　sound, 188
　spark advance, 62, 201
　　control, 74
　torque, 62, 86, 201

Frame compliance, 37

Generalized velocity, 47

Handlebar, 4
Hardware-in-the-loop, 185
Human Machine Interface (HMI), 192

Inertial Measurement Unit (IMU),
　61, 324

Modelling, Simulation and Control of Two-Wheeled Vehicles, First Edition.
Edited by Mara Tanelli, Matteo Corno and Sergio M. Savaresi.
© 2014 John Wiley & Sons, Ltd. Published 2014 by John Wiley & Sons, Ltd.

Lean angle
    see, Roll
        angle, 227

Maneuver
    µ-jump, 260
    "8"-shape trajectory, 304
    autonomous, 43
    coasting down, 61
    constraint, 86, 123
    control, 84
    definition, 123
    feasible, 124
    high-side, 41, 264
    kick-back, 40
    kickback, 262
    leaning kickback S, 262
    low-side, 260
    optimal, 85, 120
    panic brake, 258
    preview, 193
    regulation, 107
    safest, 120
    slalom, 190
    stoppie, 5, 90, 171
    time, 126
    wheelie, 5, 90, 171
Mass
    quarter-car, 202
    unsprung, 4
Model identification
    experiment
        frequency-sweep input, 62, 281
        pseudo-random excitation, 275
        sinusoidal input, 273
        step input, 62
    experimental, 59
    Input/Output, 61
    Kalman filter, 332
    parameter estimation, 248
    spectrogram, 62
Motorcycle
    see Two-wheeled vehicle, 3
    autonomous, 43

Pitch
    angle, 3
Powered two-wheeler, 199
    see Two-wheeled vehicle, 59

Rider
    ability, 124
    biomechanical model, 155
    control action, 184
    frequency response, 168
    hip shake, 176
    human body model, 163
    impedance, 39
    limit, 128
    lower body, 39
    modal properties, 39
    muscular tension, 176
    passive mobility, 37
    perception, 192
    physical strenght, 176
    preference, 120
    sensation, 191
    skill, 120
    upper body, 39
    virtual, 85, 102, 120
Road
    plane, 4
Roll
    angle, 3, 6, 48, 125, 299
        estimation, 320
        observer, 321
    dynamics, 49
    rate, 125

Simulator, 183
    audiovisual cues, 186
    Bikesim, 244
    mock-up, 183
    riding, 183
    SIMPACK, 161
    tuning, 188
    VehicleSim, 225
    virtual scenario, 188
    visual scenario, 188

Index

Speed
  longitudinal, 62
Steer
  inertia, 248
Steering
  angle, 6, 48, 244, 305
  axis, 4
  compensator, 221
    inerter, 229
  control, 243
  damper
    passive, 245
    semi-active, 245
  damping coefficient, 244
  dynamic model, 245
  instability, 243
  mechanism, 47
  system, 130
  torque, 184, 244
  virtual, 49
Suspension, 13
  control, 271
    comfort-oriented, 277
    handling-oriented, 277
  damper, 13
    electro-hydraulic, 271
    electro-rheological, 273
    magneto-rheological, 273
    passive, 272
  damping force, 273
  equivalent damping, 16
  equivalent stiffness, 16
  front, 13
    telescopic fork, 14
  preload, 16
  rear, 13
    cantilever mono-shock, 15
    swingarm, 15
  semi-active, 271
  spring, 13
  stroke, 281
  stroke measure estimation, 337

Traction
  control, 59, 199
  control system, 203
  force, 51, 115
Trajectory
  control, 297
  curvature, 6, 86, 167
  curvature profile, 91
  minimum time, 86
  optimization, 104
  tracking, 107
  turning radius
    see, curvature, 6
Two wheeled vehicle
  dynamics
    in-plane, 62
Two-wheeled vehicle
  model
    linearized, 124
  degree of freedom (DOF), 4
  dynamics, 3, 123
    bounce mode, 18
    burst oscillation, 222
    capsize mode, 33, 222
    chatter, 27
    countersteering, 29
    engine to slip, 59
    engine-to-slip, 24
    hop mode, 18
    in-plane, 18
    out-of-plane, 18
    oversteering, 31
    pitch mode, 18
    roll equilibrium, 29
    spark-to-slip, 69
    understeering, 31
    weave mode, 33, 176, 221, 243
    wobble mode, 33, 176, 221, 243
  kinematics, 3
    attitude, 323
    caster angle, 5, 47
    normal trail, 5, 248
    wheelbase, 5
  model
    black-box, 64
    full-vehicle, 21
    half-vehicle, 19
    Lagrangian, 48
    mathematical, 44, 224

Two-wheeled vehicle (*continued*)
    multibody, 19, 161
    non-holonomic, 115
    point mass, 90
    rigid bodies, 35
    single corner, 202
    sliding plane motorcycle, 102
  powertrain, 22
  transmission, 75
Two-wheeleed vehicle
  kinematics
    normal trail, 128
Tyre, 6
  absolute slip, 72
  Burckhardt force model, 10
  camber angle, 7
  combined slip, 11
  deflection, 288
  effective rolling radius, 8
  lateral slip, 7, 51, 87
  lateral velocity, 8
  longitudinal slip, 7, 51, 62, 87, 302
  longitudinal velocity, 8
  Pacejka force model, 9, 87, 185
  piecewise linear model, 52
  radial deflection, 7
  relative slip, 72
  relaxation dynamics, 13, 185, 225
  rolling radius, 61
  rolling resistance moment, 7
  Side slip angle, 302
  sideslip angle, 11, 51, 87
  spin rate, 7, 8
  stiffness, 248
  yawing moment, 7
Tyre-road
  contact force, 7
    lateral, 7, 53, 86
    longitudinal, 7, 53
    vertical, 7, 53, 203
  frictional force
    see also, contact force, 52
  kinematics, 51
  longitudinal friction coefficient, 203
  overturning moment, 7
  surface, 78

Velocity
  lateral, 125
  longitudinal, 203
    estimation, 331
  optimal profile, 90
    computation, 99
  see also, speed, 90

Wheel
  encoder, 60
  front, 4
  rear, 3
  slip control, 205
  slip reference generation, 208

Yaw
  angle, 3
  inertia, 248
  rate, 125